Thinking Chemistry

Michael Lewis
Guy Waller

Radley College
Radley
Oxfordshire

Oxford University Press

Oxford University Press, Walton Street,
Oxford OX2 6DP.

Oxford University Press, Walton Street, Oxford OX2 6DP
Oxford London
New York Toronto Melbourne Auckland
Kuala Lumpur Singapore Hong Kong Tokyo
Delhi Bombay Calcutta Madras Karachi
Nairobi Dar es Salaam Cape Town

and associated companies in
Beiruit Berlin Ibadan Mexico City Nicosia

Oxford is a trade mark of Oxford Univerity Press

ISBN 0 19 914074 X

First published 1980
Reprinted 1981, 1982, 1983, 1984, 1985
GCSE Edition, 1986

The design on the cover shows a
crystal of sodium chloride (common
salt) dissolving in water. The orange
spheres are sodium ions, the green
ones are chloride ions.

Printed and bound in Great Britain by
Butler & Tanner Ltd,
Frome and London

Introduction

Thinking Chemistry is a textbook for GCSE students. It sets out to present the syllabus material in a way that is clear, logical and visually interesting. Like modern syllabuses it emphasizes both the concepts and applications of chemistry. At each stage it encourages the student to think and understand, rather than simply memorize facts. Instructions for experimental work are not included: the book can be used alongside any GCSE practical course.

Thinking Chemistry is divided into five sections. The first four deal with the key concepts: structure, periodicity and reactivity, and chemical change and its characteristics. Each concept is developed through analysis of experimental facts. Facts and theory are kept carefully distinct, and presented in a way that reflects the scientific approach, where observation comes first, then inference.

The fifth section deals with the chemistry of the elements, and emphasizes its connection with the concepts developed in earlier sections. Each element or group of elements is treated in the same order: physical structure and properties, occurrence and extraction, compound formation, solubility, acid/base behaviour, and redox behaviour. Small 'reminder' diagrams in the margins reinforce the link with earlier sections.

Thinking Chemistry is highly illustrated with diagrams and photographs. The diagrams include conventional diagrams of experimental apparatus, and more unusual diagrams that show the behaviour of atoms, molecules and ions during physical and chemical change. The photographs have been chosen to relate the content both to the school laboratory and to the outside world.

Each chapter of **Thinking Chemistry** ends with a summary and a set of questions.

Acknowledgements

The publishers wish to thank the following for permission to reproduce photographs:

Air Products, p. 358; Associated Press, p. 228;
Barnaby's Picture Library, pp. 204, 226, 281, 306, 312 right;
Boosey & Hawkes, p. 308 left;
Paul Brierley, p. 18 centre; British Airways, pp. 309 top, 385;
British Alcan Lynemouth Ltd., p. 226; British Leyland, p. 174;
British Petroleum, p. 269; British Shipping/F. A. Hughes & Co., p. 312 left;
British Steel, pp. 306, 307; Camera Press, pp. 93 bottom, 155, 272, 300 left, 321 top;
Cavendish Laboratories p. 53; CEGB SE Region, p. 227;
Bruce Coleman, pp. 4 both, 18 left, 236, 337 left; Courtaulds Ltd., p. 383;
Cumberland Graphics Ltd., p. 321 bottom;
John Day and Peter Dart, Rothamsted Experimental Station, p. 360 left;
FAO, UN, p. 222; Griffin & George Ltd., p. 8 right;
Hercal Ltd., p. 309 bottom right;
ICI, pp. 300, 344 right, 360 right, 362, 364, 375 left, 379;
ICI Plant Protection Division, p. 399;
Institute of Geological Sciences, pp. 2 bottom left, 294, 337 right;
Keystone Press Agency, pp. 308 right, 346, 363;
Laboratory for Solid State Physics, Swiss Federal Institute of Technology, Zürich, p. 1; London Brick Company, p. 121; London Underground Ltd., p. 228;
Moet et Chandon/A. C. Cooper, p. 325; Needham Chalks, p. 139 right;
Northamptonshire Newspapers Ltd., p. 226;
Oxford University Press, p. 332, © Oxford University Press 1980, from *Crosstalk 2* by M. M. Webster published by Oxford University Press;
Permutit, p. 300 right; Picturepoint, pp. 2 top, 32 both; Popperfoto, pp. 11, 225, 322;
Press Office of the White House, Washington, p. 289;
Rothamsted Experimental Station, p. 360; Shell (UK) Ltd., pp. 338, 343;
Syndication International/Dick Williams, p. 309 bottom left;
United Kingdom Atomic Energy Authority, p. 291;
Van den Berghs Ltd., p. 349 left; Vencel Resil Ltd., p. 340 right;
P. Waegli, Swiss Federal Institute of Technology, Zürich, p. 171;
R. Wessicken, Swiss Federal Institute of Technology, Zürich, p. 311;
R. Wessicken and H. G. Wiedemann, Swiss Federal Institute of Technology, Zürich, p. 92; Diana Wyllie, p. 323.

Cartoons on pages 112 and 114 by Rory Stormouth-Darling
Additional studio and location photography by Chris Honeywell

Contents

Opposite: a crystal of tellurium oxide, TeO_2, photographed by a scanning electron microscope. The crystal has been magnified × 340

Section one

The properties and structure of matter

1 Studying chemistry

1.1 Matter

Matter: what is it?

Look around you. You can probably see things made of metal, wood, glass and plastic. All these substances are made of **matter**. You can tell the amount of matter in any substance just by weighing it. Anything which is made of matter has a mass, which can be measured by weighing. Heavy things have more matter in them than light things.

Solids and liquids are easy to weigh, and it is also possible to weigh gases. This shows that solids, liquids and gases are all different forms of matter.

Matter is made of particles

Beaches are made of tiny grains of sand. Each grain of sand is made of millions of even smaller particles.

The dictionary says that a **particle** is the smallest possible amount of a system. We think that all matter is made up from particles. A single particle is so small that we cannot see it. We can only see collections of particles. The point of the sharpest needle is thought to be made up of over 100000000 particles of metal!

A sandy beach

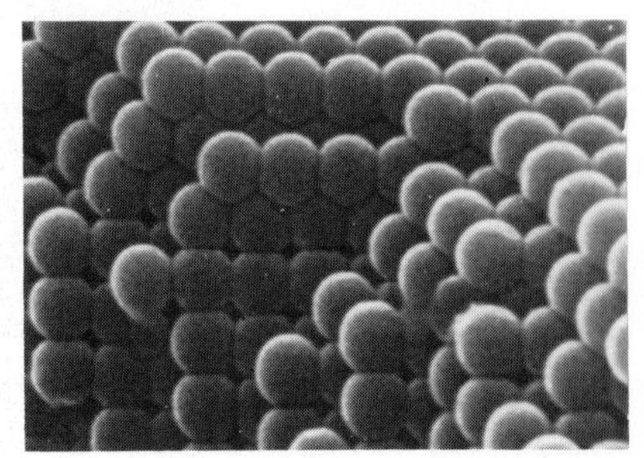

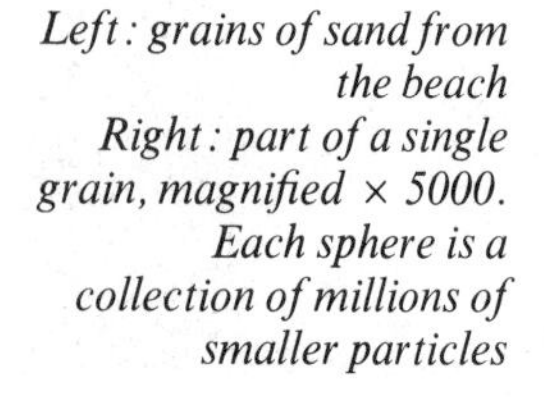

Left: grains of sand from the beach
Right: part of a single grain, magnified × 5000. Each sphere is a collection of millions of smaller particles

1.2 Facts and theories

The difference between fact and theory

Human beings find out things by using their senses. These senses are touch, hearing, sight, smell, and taste. When we describe things that we have sensed in any of these ways, we are stating **facts**. For example:

solids are hard
grass is green.

We can weigh solids, liquids, and gases. So it is a fact that solids, liquids, and gases are all forms of matter. But is it a fact that matter is made of particles? We cannot touch or see each particle. Yet all scientists believe that matter is made of particles! This is not a fact, but a **theory**. Theories are ideas which are used to explain facts.

A non-scientific example will show you the difference more clearly. A farmer shuts his cows in a field. On coming to collect them at milking time, he finds them on the road. The gate of the field is open. The facts are simple: he shut them in but they got out. A number of possible theories could be put forward to explain these facts: a passing walker may have left the gate open; the latch on the gate may not have caught properly; a clever cow may know how to open the gate. Each of these theories explains the facts, but we cannot prove any of them without more information.

The same is true of chemical facts and theories. We believe that matter is made of particles because it helps us to explain the facts about matter. In the rest of this book, we shall study the facts about matter and try to explain them using the particle theory.

Systems and particles

Systems are real things that we can observe.
The solar system is made up of the sun and the planets. Each part of the solar system affects the behaviour of the rest of it. The class you are in at school could also be called a system. If you all work hard, then the teachers think the class is a good one. If you are all noisy the class gets a bad name.

In a chemical sense, the word 'system' is used to describe a particular substance or reaction that is being studied. A typical system is a test-tube of solid being heated, or two substances reacting together in a beaker.

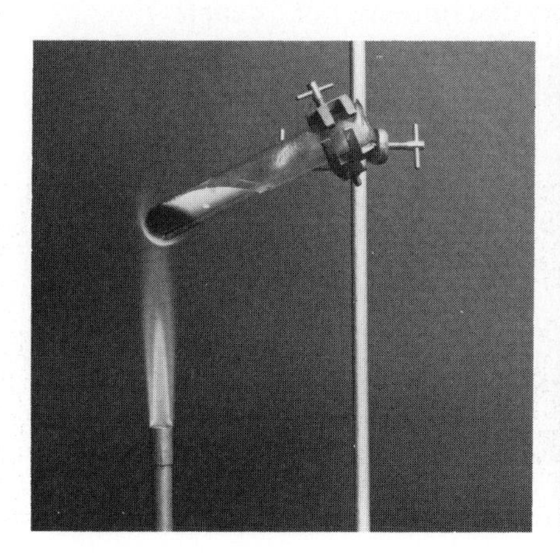

Two familiar systems

By using all our senses, we can obtain a number of facts about the properties of any system. For example:

the system changed from solid to liquid (melted)
the system turned from a green colour to a black colour
the system gave off a gas which had a choking smell.

The particle theory

Consider another simple non-scientific example:

System	*Particle*
A swarm of bees	A single bee

Fact	*Theory*
A swarm of bees can be a dangerous system.	A bee stings when disturbed.

To explain why a swarm is dangerous, we think about the individual bees. In the same way, to explain the facts about chemical systems we can think about the individual particles. We use the particle theory, which states:
The properties of a system are the result of how its particles behave.

The behaviour of particles is often best shown in a series of cartoon-style pictures. We shall use these pictures in many of the later chapters, instead of giving lengthy descriptions. For example:

1 Two particles approach.	2 They collide.	3 They bounce off.	4 They move away.
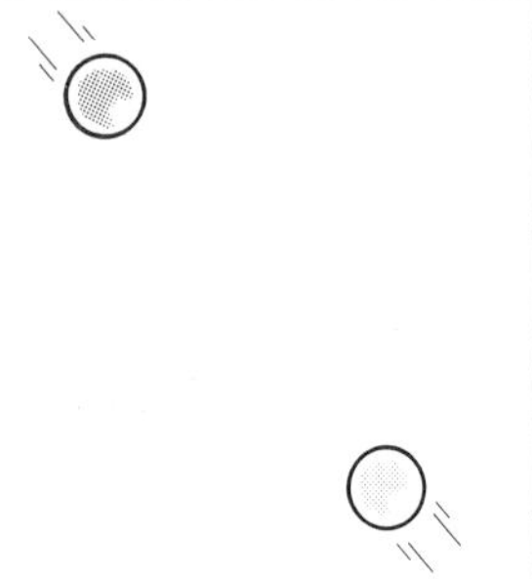		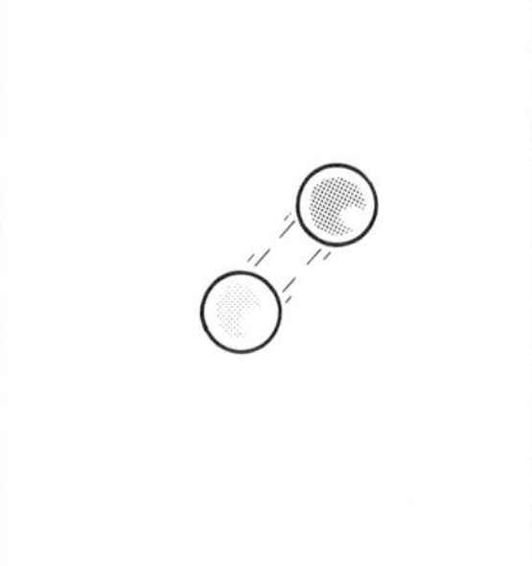	

1.3 Studying chemistry

Chemistry is an experimental science. The two main aims of chemists are:

(i) to observe and measure the properties of systems under changing conditions;
(ii) to explain the observed properties using the particle theory.

The first aim is the easier one. It simply involves finding out the facts about system properties. The second aim is more difficult. It involves suggesting theories that explain and fit the facts.

Summary

All things are made of matter.

Matter is made of particles.

The observed properties of a system are facts: is it large or small? hot or cold? liquid or solid?

Theories are used to explain facts.

Theories are explanations for the behaviour of particles.

2 The physical properties of matter

This chapter is about some of the physical properties of substances. It will give you an idea of the way some substances change when they are heated or mixed together. In Chapter 3 we will explain these changes by looking at the behaviour of the particles.

2.1 States of matter

Solid, liquid, and gas

Table 2.1 Some common substances

Solid	*Liquid*	*Gas*
Iron	Water	Air
Chalk	Petrol	Steam
Ice	Alcohol	Natural gas
Coal	Cooking oil	Carbon dioxide

Table 2.1 shows some of the most common everyday substances. All these substances are matter (they can all be weighed), but at room temperature they exist in different **states of matter**. Solid, liquid and gas are the three different states of matter and each has its particular properties. These properties are shown in Table 2.2.

Table 2.2 Properties of the states of matter

*State**	*Properties*	*Examples*
Solid (s)	Definite mass Definite volume Definite shape	A 50p coin stays the same shape in your pocket or purse.
Liquid (l)	Definite mass Definite volume Changing shape	A pint of milk can take the shape of a bottle or carton or jug.
Gas (g)	Definite mass Changing volume Changing shape	Puncture a balloon: all the air escapes. Steam from a kettle goes all over the place.

* In chemistry, the state of a substance is shown by the initial letter of the state, written in brackets after the name of the substance.

The word **vapour** is often used instead of the word gas. Strictly, a vapour is a gas which can be compressed into a liquid without cooling. In this book both words may be used when talking about the gas state.

Solids melt

If a solid is heated, it often melts. Ice and candlewax melt if warmed; even steel melts when heated enough. One state can therefore change into another state if the temperature is changed.

When a substance melts, it changes state from solid to liquid. This change always happens at one particular temperature for a pure substance. Pure water, for example, always changes from solid to liquid at 0°C.

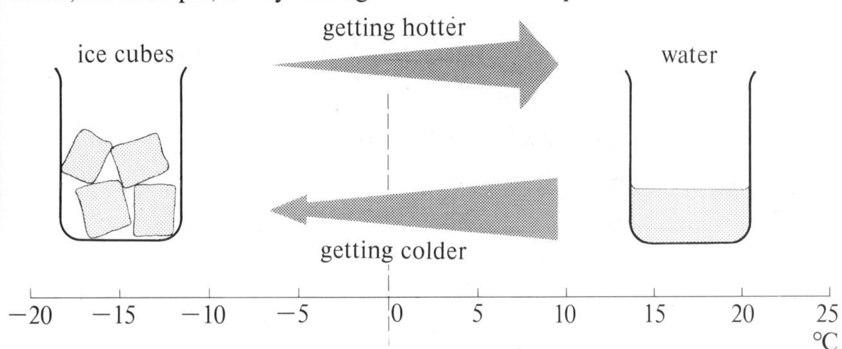

The temperature at which a pure solid turns to the liquid state is called the **melting-point** (m.p.). The temperature at which a pure liquid turns to the solid state is called the **freezing-point** (f.p.). The melting-point and freezing-point of any given substance are both the same temperature.

A few solids do not melt when they are heated but instead turn directly into gas. A change of state from solid to gas is called **sublimation**. Like melting, this change also happens at a particular temperature for a pure solid. A substance which changes from solid to gas **sublimes**. Carbon dioxide is one example of a substance that sublimes.

Liquids evaporate and boil

Unlike solids, which melt at a single temperature, liquids change to gases over a range of temperatures. Splashes of water evaporate at room temperatures. After rain, puddles dry up. Where has the water gone? A simple experiment can show that the water has gone into the air as a gas.

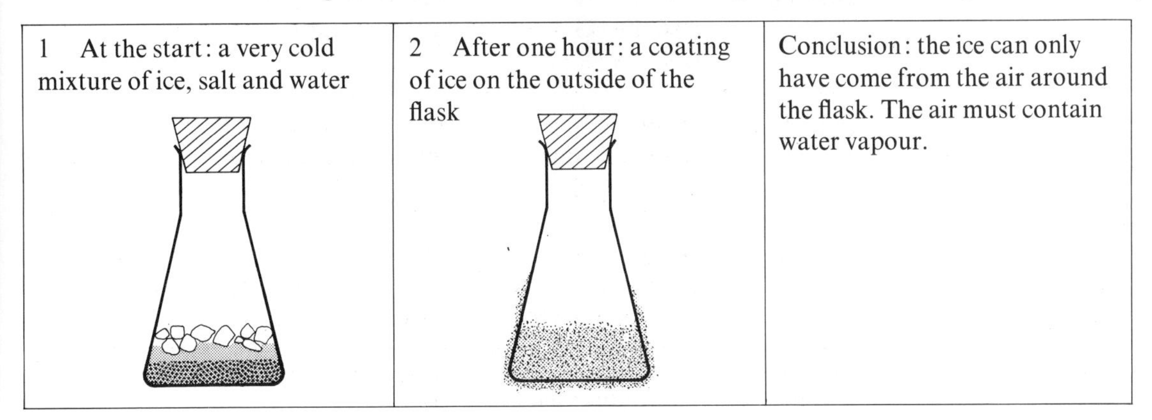

1 At the start: a very cold mixture of ice, salt and water
2 After one hour: a coating of ice on the outside of the flask
Conclusion: the ice can only have come from the air around the flask. The air must contain water vapour.

When liquids change into their gases over a range of temperatures, the process is called **evaporation**. The hotter the liquid, the faster it evaporates. Finally it boils. Gas bubbles appear in the liquid itself. The temperature of the liquid stops going up while it boils. Increased heating makes the liquid boil faster but does not make it any hotter. It is at its **boiling-point**. The boiling-point (b.p.) of a liquid is the lowest temperature at which bubbles of its gas can exist in the liquid.

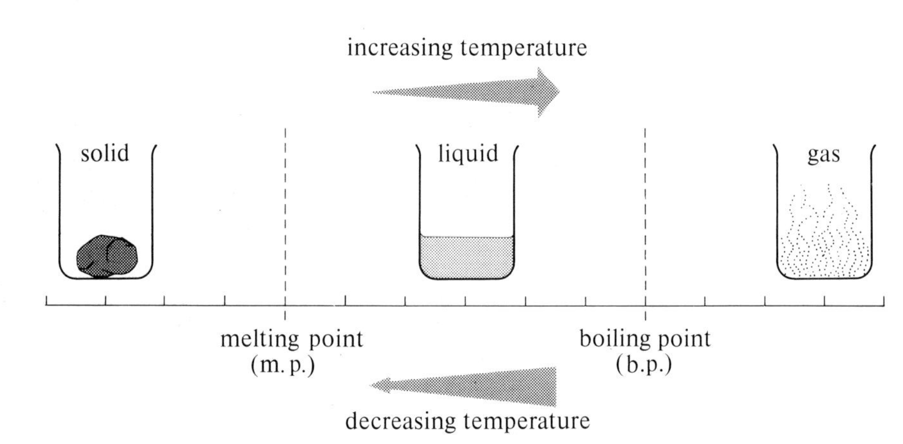

Boiling-points can change if the surrounding pressure changes. Normally boiling-points are measured at the pressure of the atmosphere at sea level. This pressure is called **standard pressure** or **atmospheric pressure**. The boiling point of water is 100°C, at standard pressure; but up mountains the pressure drops below standard, so water boils at a lower temperature. Mountaineers on Everest have to use a pressure cooker to boil an egg! Water can be made to boil even at room temperature. This is done by lowering the pressure above it with a vacuum pump.

Left: 'cold' boiling water. The tube leads to a vacuum pump
Right: by boiling water in a pressure cooker you can make it 'extra hot'. Its boiling-point increases with pressure

Fixed points A pure substance melts and boils at definite temperatures. These temperatures are called the **fixed points** for that substance. Table 2.3 shows the fixed points of some common substances. They are all measured at standard pressure.

Table 2.3

Substance	*m.p./°C*	*b.p./°C*
Oxygen	−219	−183
Nitrogen	−210	−196
Alcohol (ethanol)	−114·5	78
Water	0	100
Sulphur	119	444
Salt (sodium chloride)	801	1465
Copper	1083	2600
Carbon dioxide		sublimes at −78°C
Carbon		sublimes at 4200°C

2.2 Purity

Fixed points and purity A system is **pure** when there is only one type of substance present in it. All pure substances have their own fixed points—m.p. and b.p. values. These values do not change as long as the substance remains pure and the pressure is kept constant. For example, at standard pressure:

pure water always melts at 0°C and boils at 100°C;
pure salt always melts at 801°C and boils at 1465°C.

Lists of fixed points have been made for most substances. So if you know the fixed points, you can find out what a substance is by searching the lists. For example, an unknown solid that melted at 119°C and boiled at 444°C could be recognized as sulphur by looking through lists of fixed points in a book of data.

Impurities Sea water is impure water. If you boil away the water, a solid powder will be left behind. Sea water freezes at a temperature well below 0°C and boils at a temperature above 100°C. The impurities have altered the m.p. and b.p. values of the water. Fixed points can therefore be used to check whether water is pure or not. The fixed points of all substances can be used to check for purity in the same way.

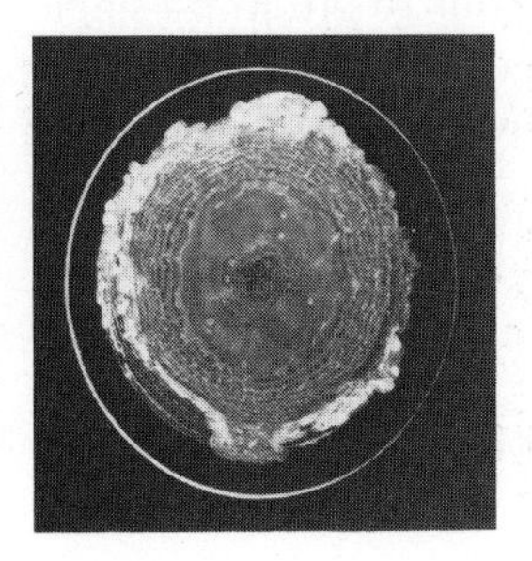

Solids left when sea water evaporates. It boils above 100°C

2.3 Mixtures

Solutions and suspensions

If two substances mix to give a mixture in one state only (see (a) and (c) below), the mixture is called a **solution**. We say there is only one **phase** present. Now look at (d) below. The two substances being mixed are both in the same *state*, but the resulting mixture has two *phases*. If a mixture has more than one phase, it is called a **suspension**. Both (b) and (d) are suspensions. Mixture (b) is a suspension of a solid in a liquid; (d) is a suspension of a liquid in a liquid.

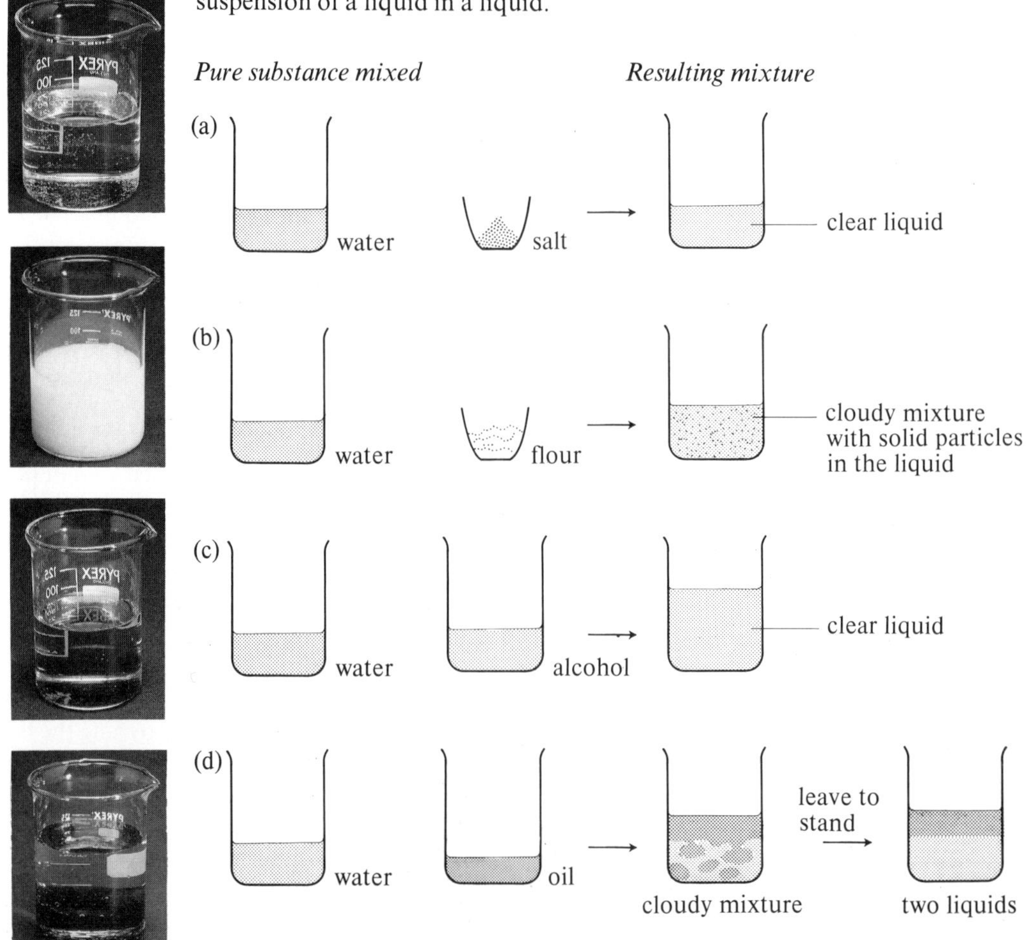

A special name is given to two liquids that will not mix to produce a single phase. They are said to be **immiscible** with one another. The suspension they produce on being shaken together is called an **emulsion**. The two liquids in (d) would give an emulsion if shaken together.

To summarise:

a solution is a mixture of matter in one state only—it is clear;
a suspension is a mixture of matter in more than one state or phase—it is cloudy.

Table 2.4 Some common mixtures: (g) stands for gas, (l) for liquid, (s) for solid.

Solutions		*Suspensions*		*Emulsions*	
Mixture of	*Product*	*Mixture of*	*Product*	*Mixture of*	*Product*
Nitrogen (g) Oxygen (g)	Air	Water (l) Air (g)	Cloud	Coloured oils (l) Water (l)	Paint
Petrol (l) Oil (l)	2-stroke fuel	Dust (s) Water (l) Air (g)	Smog	Fatty oils (l) Water (l)	Milk
Carbon dioxide (g) Water (l)	Soda-water	Earth (s) Water (l)	Muddy water	Oil (l) Vinegar (l)	Salad dressing

The substances making up a solution are often called **solute** and **solvent**. The solute is the part of the solution that is dissolved, while the solvent is the part that does the dissolving. The solutions from the list above are made up like this:

Solution	*Solute*	*Solvent*
Air	Oxygen 20 %	Nitrogen 80 %
2-stroke fuel	Oil 5 %	Petrol 95 %
Soda-water	Carbon dioxide 0·2 %	Water 99·8 %

There is both a solution and a suspension in this scene. Can you identify them?

2.4 Separating mixtures

Separating suspensions

A two-state mixture, like gravel chippings in water, can easily be separated by filtering. The gravel pieces are too big to go through the hole in the filter funnel:

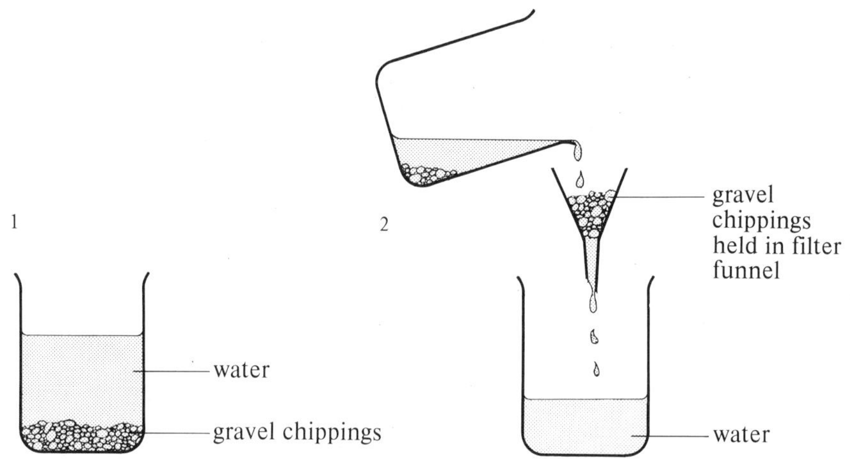

For a fine suspension, like clay or mud in water, the same idea can be used but this time a filter paper is put in the funnel:

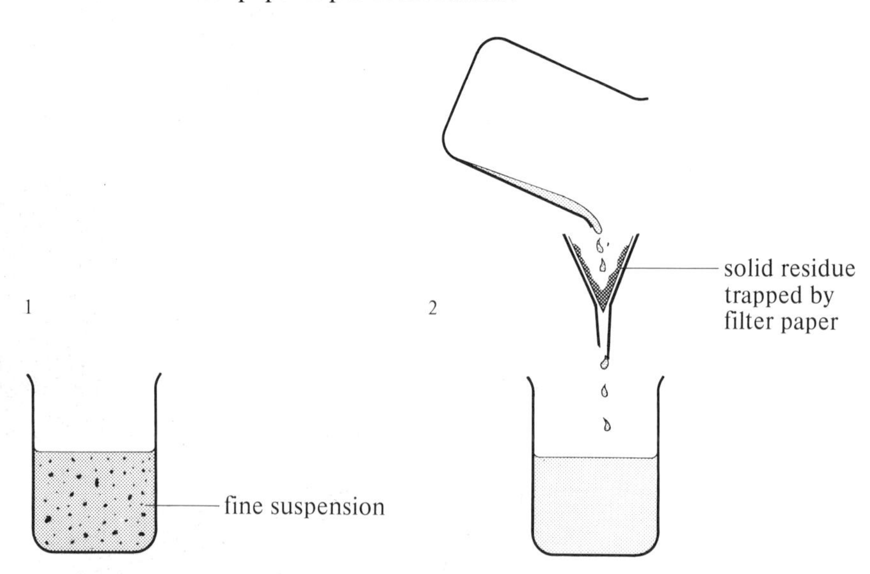

The solid left behind on the filter paper is called the **residue**. The liquid that passes through the filter paper is called the **filtrate**.

An emulsion—or suspension of two liquids—can be separated in the same sort of way, if one of the liquids is water. This time, the filter is made of special water-proof paper that allows only liquids other than water to pass through it. This type of filter paper is called **phase separation paper**.

Separating solutions

Suspensions are easy to separate because there are two phases present and one phase can be trapped in some way. Solutions are less easy to separate because they exist in only one phase. To separate them, we must turn them into mixtures of two phases, somehow. The only way is to make one part of the solution change state, but not the rest.

Simple distillation

Take the sea-water mixture, which is mostly salt dissolved in water. From Table 2.3 and the diagram below, you can see that the fixed points of salt and water are far apart.

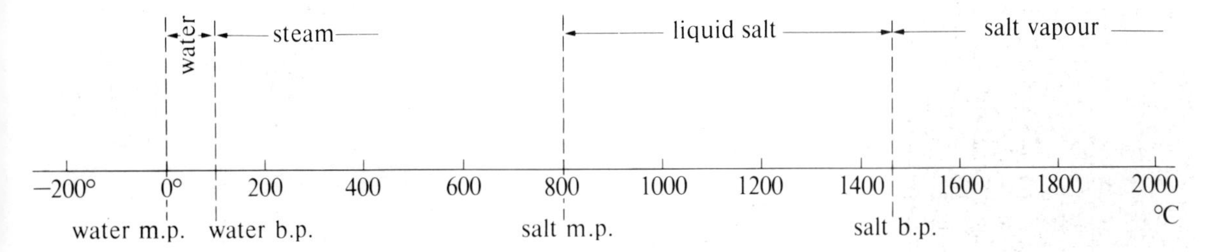

If a mixture of salt and water is heated in a flask, these things happen.

1 The temperature of the solution rises to about 100°C.
2 At 100°C the water boils, but the salt stays in the flask.
3 At this temperature water turns into steam and leaves the flask.
4 The steam is collected and cooled down to give pure liquid water.
5 When all the water has left the flask, solid salt remains.

This process is called **simple distillation**. The apparatus used for it is shown below. The steam cools and condenses to water in the condenser.

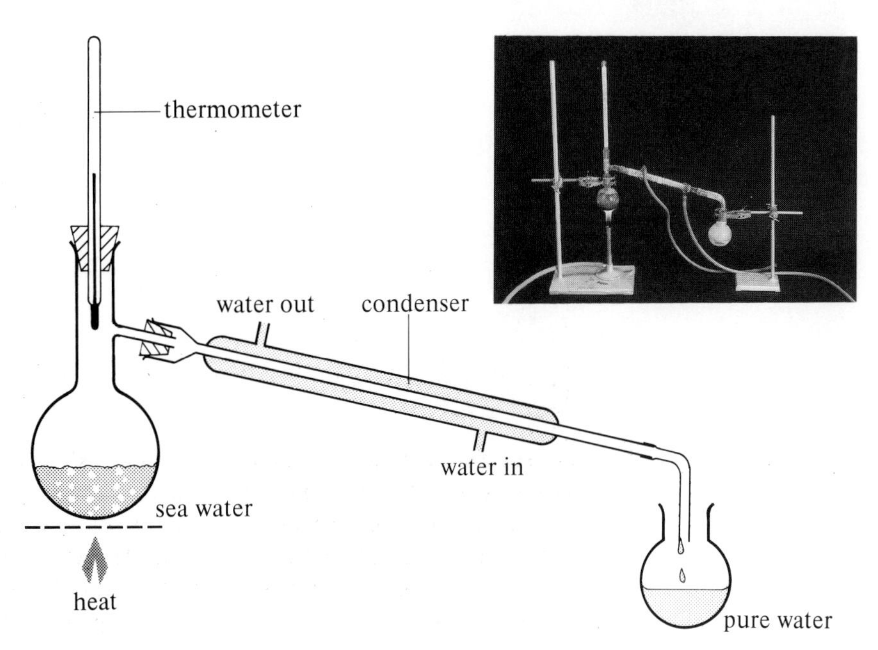

Fig. 1
Simple distillation

Fractional distillation

An alcohol and water solution is more difficult to separate. This is because their boiling-points are close together.

alcohol b.p. water b.p.

0 100 200 300 400 500 600 700 800 900
°C

If we heat an alcohol and water solution, both parts of the mixture will evaporate at the same time. The alcohol will evaporate faster but there will be a lot of water evaporating as well. A special sort of distillation is used to solve this problem. It is called **fractional distillation**.

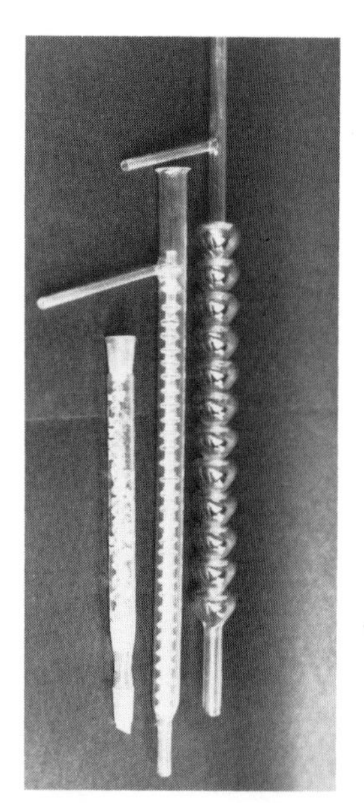

Fig. 2
Fractional distillation. Three different types of column are shown above

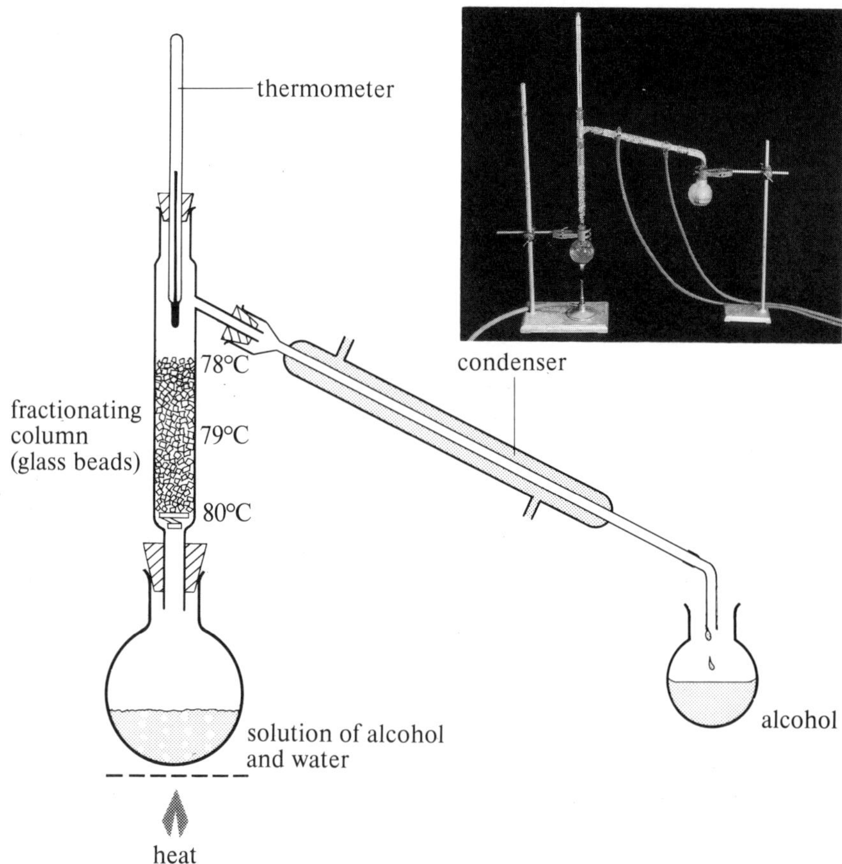

These are the steps in a fractional distillation:

1. The mixture in the flask boils.
2. A mixture of both alcohol and water vapour heats the column.
3. The alcohol goes all the way up the column as a gas, because the column is at, or above, its boiling-point.
4. The water vapour turns back to liquid on the glass beads of the column, because the temperature of the beads is below 100°C.
5. Eventually all the alcohol is gone, and only water remains in the flask.

Fractional distillation is used to separate any solutions made from substances whose boiling-points are close together. For example, liquid air is separated by fractional distillation into oxygen and nitrogen; crude oil is separated into its various fractions such as petrol, paraffin, and lubricating oil. Both these processes are carried out on a very large scale in industry. We shall meet both again in later chapters.

Separating a solid mixture You can separate a solid mixture, like salt and sugar, by the **method of solvents**.

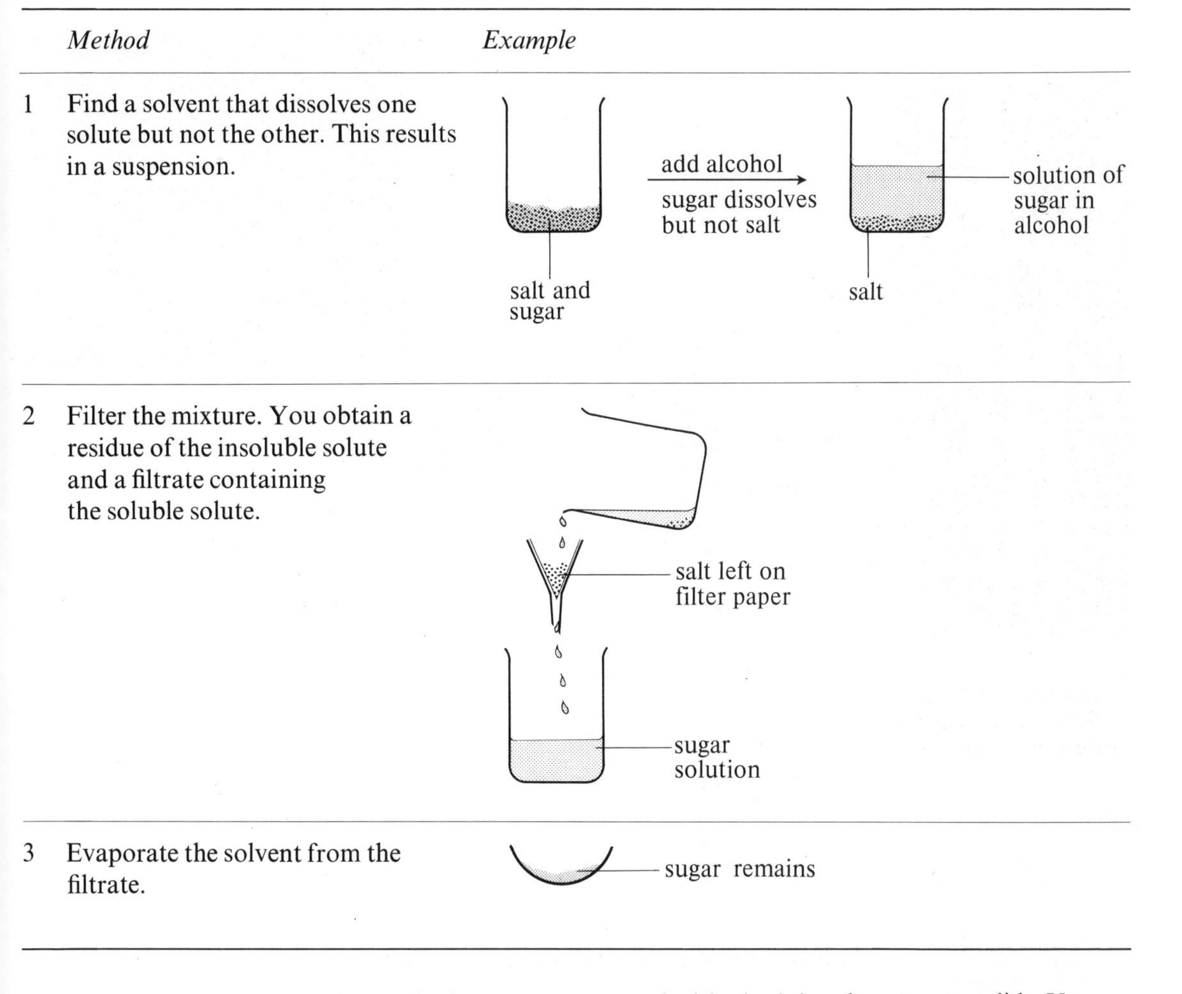

Method
Example
1 Find a solvent that dissolves one solute but not the other. This results in a suspension.
add alcohol
sugar dissolves but not salt
solution of sugar in alcohol
salt and sugar
salt
2 Filter the mixture. You obtain a residue of the insoluble solute and a filtrate containing the soluble solute.
salt left on filter paper
sugar solution
3 Evaporate the solvent from the filtrate.
sugar remains

In this method, you are concerned with obtaining the separate solids. You are not concerned with keeping the solvent, so it can be evaporated off.

Water can sometimes be used to separate a solid mixture (for example salt and sand) because many substances do not dissolve in water. Some other useful solvents, besides water and alcohol, are ether, acetone, toluene, and trichloromethane. Trichloromethane is the solvent used in 'dry-cleaning'. You will learn more about these solvents in later chapters.

Separating more than one solute from solution

The method of solvents can be used to separate two solutes from a solution. You simply evaporate off the solvent from the solution. This leaves a mixture of two solids. Then use the method of solvents shown on page 15. However, it is not always necessary to evaporate off the solvent, as is shown in the two separation methods that follow: solvent extraction and chromatography.

Solvent extraction

If a second solvent can be found that is immiscible with the original solvent, a process usually called **solvent extraction** might work. This is how to carry out the process:

Method — *Example*

1 Take the original solution that contains the two solutes.

A brown solution of salt and iodine in water

2 Add a second solvent that
—dissolves one solute but not the other;
—is immiscible with the original solvent.

Add tetrachloromethane, to give two liquid phases.

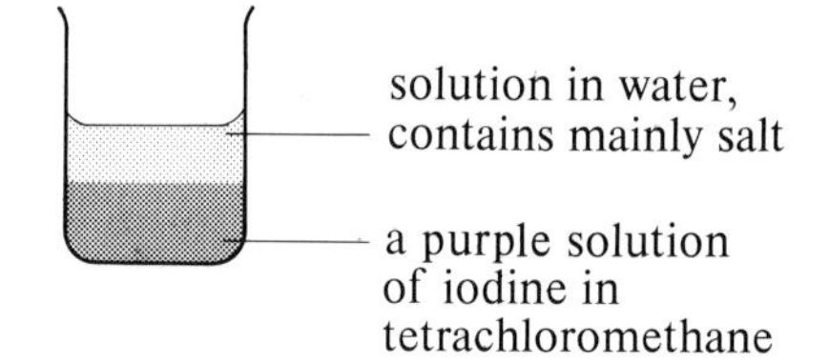

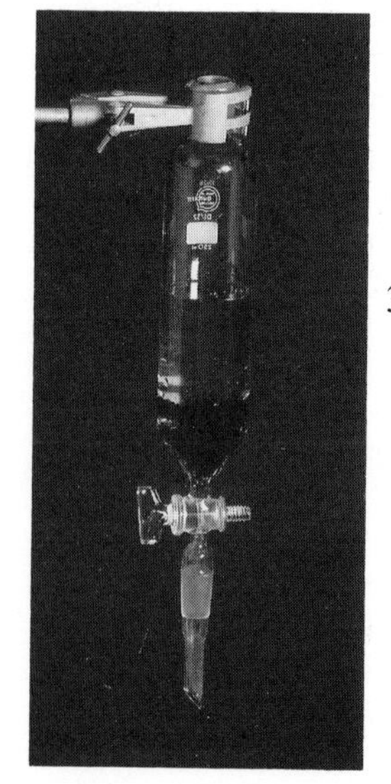

3 Separate the two immisicible solutions from one another in a separating funnel. Open the tap and drain off the bottom layer; close the tap when the bottom layer has gone.

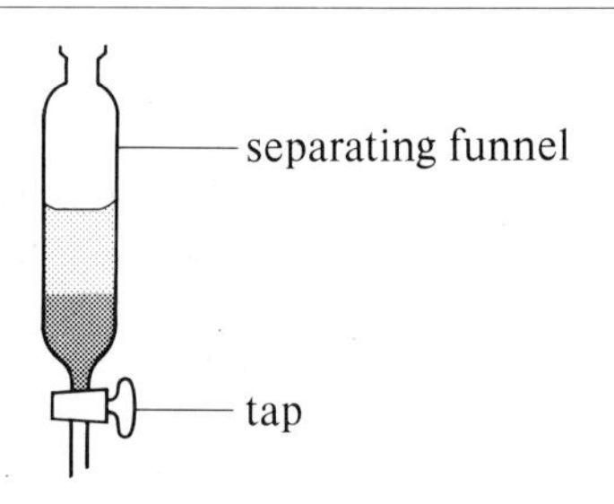

4 Evaporate off each of the solvents to obtain the two solutes.

Chromatography You will have noticed how drops of ink spread out on blotting paper. You may also have noticed that at the edges of the spreading ink spots, slightly different colours appear. The dyes in the ink are being separated. This is an example of **chromatography**.

Chromatography is used to separate solutes when there is only a small volume of solution available. There are several different ways to carry it out. One method, called **paper chromatography**, is described below.

1 Absorb a small drop of the solution on to some porous material, like filter paper. The solvent spreads out, leaving the solutes behind. In the photograph below, the solutes are the dark spot near the bottom of the paper.

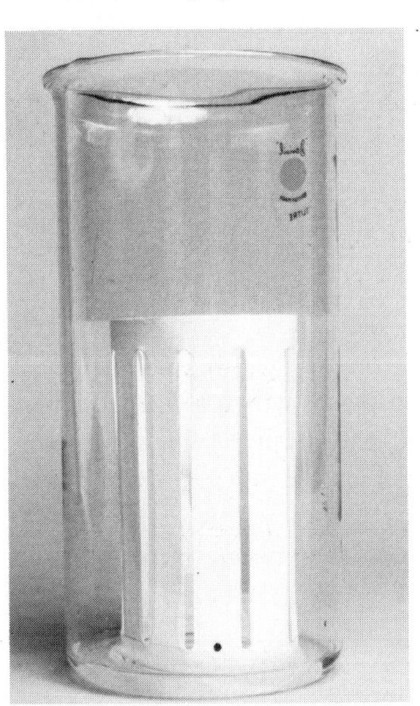

2 Allow another solvent to flow up the paper. This solvent gradually redissolves the solutes back into solution. They begin to spread out and move up with the solvent.

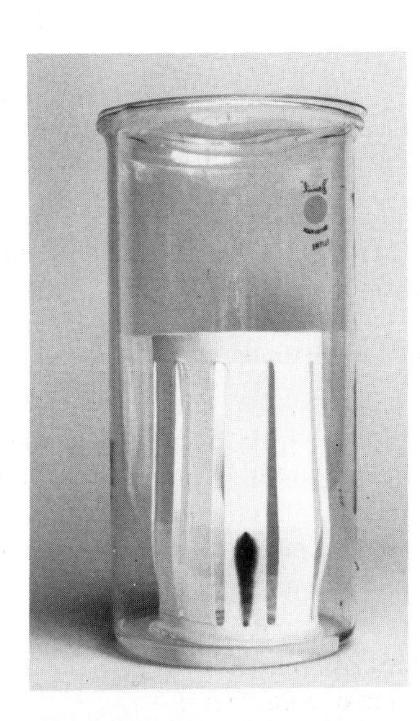

3 The most soluble solutes dissolve easily and are carried quickly up the paper. The least soluble solutes get left behind. When the solvent evaporates, the solutes are left on the paper in different places.

Chromatography can be used when you have only a very small volume of solution. It is also used when a mixture contains so many solutes that a single pair of solvents would not produce good separation. It is most widely used for analysing mixtures in biology. To **analyse** a mixture means to find out what substances it contains. In analysis, the substances are separated so that they can be recognised, but are not usually extracted in large quantities.

Some examples are:

the separation of pigments from plants
the separation of antibiotic drugs from their growing media
the separation of amino acids from proteins
the separation of metals from minerals.

Usually, drops of several different solutions are put on the paper, along a line. Some of these are unknown solutions to be analysed. The others contain known substances, that we *guess* are in the unknown solutions. Paper chromatography is detective work!

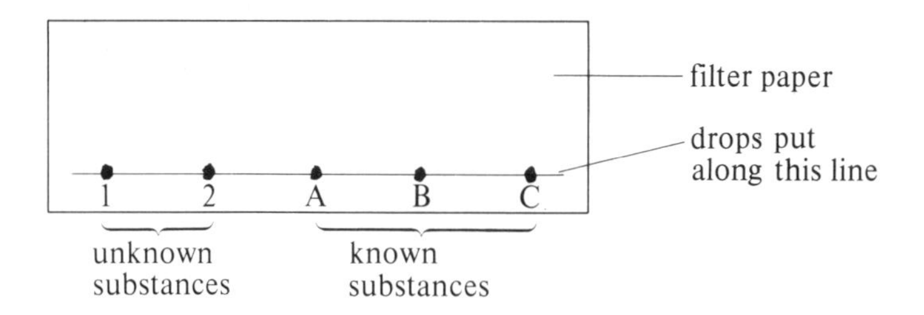

After chromatography the paper might look like this:

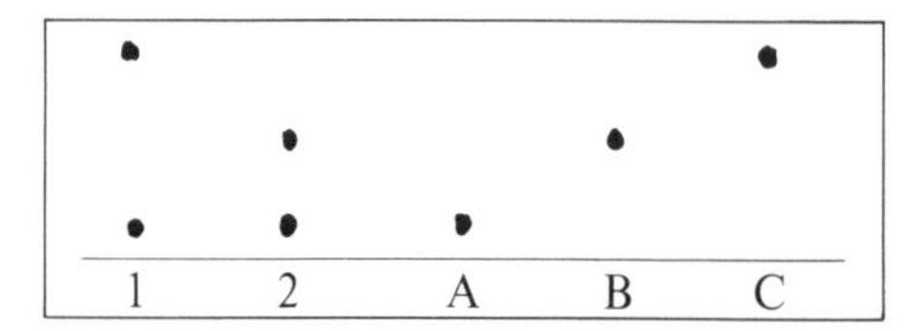

These results show that unknown 1 is a mixture of A and C, and unknown 2 is a mixture of A and B.

2.5 Crystals

Solids are often found in small, regular-shaped pieces which reflect the light. Solids of this type are said to be **crystalline** and the small regular pieces are called **crystals**. Diamonds, grains of salt, and sugar are all crystals.

Dull, powdery solids that do not have regular-shaped pieces are said to be **amorphous**. This means literally 'having no definite structure'. Rust and flour are amorphous solids.

Three crystalline substances. Left to right: snow, salt, diamond

Solubility in water

Many crystalline solids dissolve in water. It is generally found that the mass of crystals that dissolves in a fixed amount of water increases if the temperature of the system is raised. Table 2.5 shows the mass of potassium nitrate crystals that dissolves in 100 grams of water at different temperatures.

Table 2.5

Temperature/°C	*Mass dissolved/g*
20	30
30	44
40	60
50	80
60	104

Results like these are usually shown on a graph. The curve which is produced is called the **solubility curve** for the particular solid. The solubility curves for potassium nitrate, hydrated copper (II) sulphate, common salt, and potassium sulphate are plotted below.

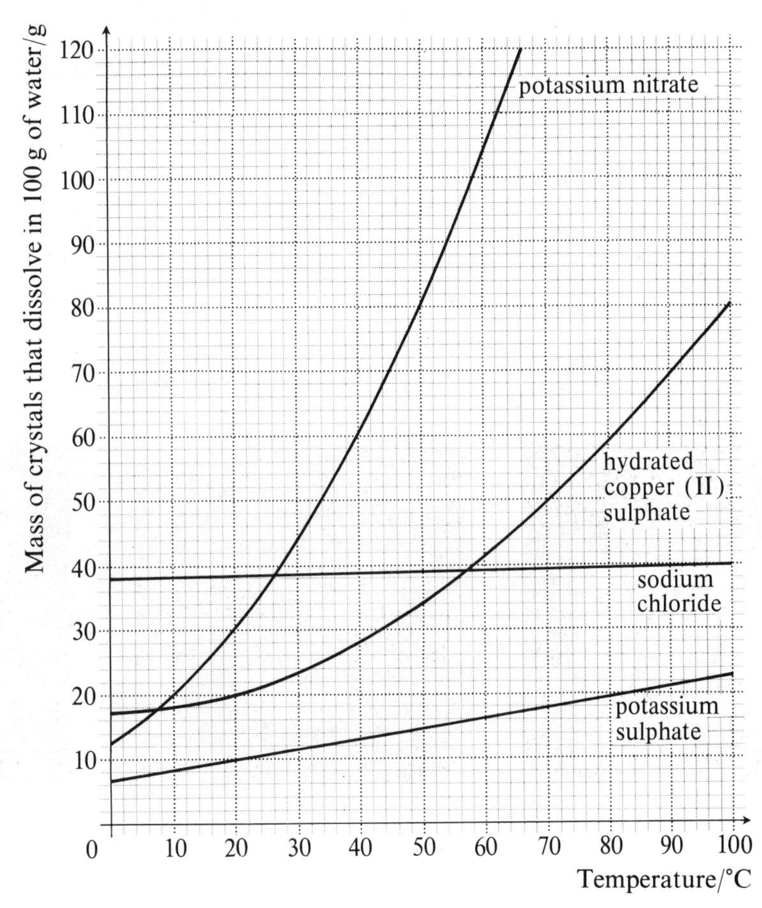

Fig. 3
Solubility curves

The maximum number of grams of any solid that dissolves in 100 grams of solvent at a given temperature is called the **solubility** of the solid at that temperature. So the solubility of potassium sulphate in water at 20°C is 10 g. Notice that the solubility of potassium nitrate increases sharply with increasing temperature, whereas almost as much salt will dissolve in water at 0°C as at 100°C.

A solution that contains as much dissolved solute as possible is said to be **saturated**.

Crystallisation

At 60°C, a saturated solution of potassium nitrate contains 104 grams of crystals dissolved in 100 grams of water. What happens if this saturated solution is cooled to 20°C?

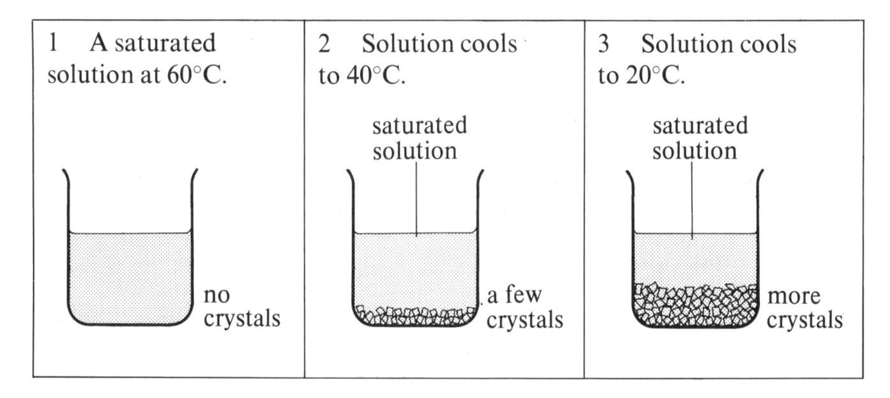

Less solid dissolves at 40°C than 60°C. Less solid dissolves at 20°C than at 40°C. As the liquid cools, the amount of solid it can hold in solution gets smaller and smaller. The rest of the solid gradually appears as crystals, which grow on the bottom and sides of the beaker. This process is called **crystallisation**. The slower the cooling, the larger the crystals that grow.

Note that all three solutions shown in the diagram above are saturated, but at *different temperatures*, so they will contain different amounts of solid dissolved in them.

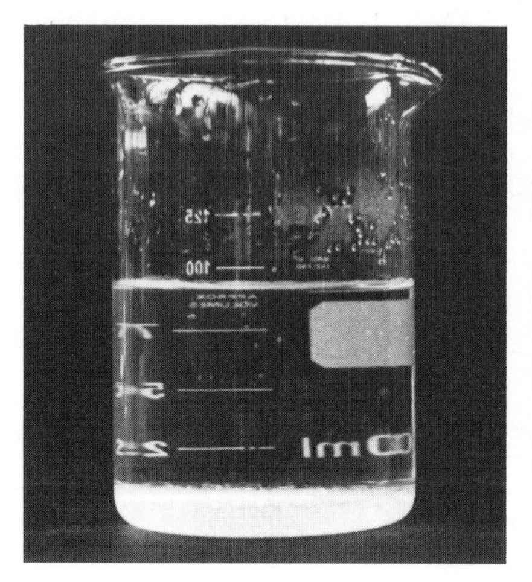

A saturated solution of potassium nitrate at 40°C (left) and at 20°C (right)

Summary

All matter can exist in the three states, solid, liquid, and gas. The states can be changed from one to another by changing the temperature (or sometimes the pressure) of the system.

The temperature at which a pure solid melts or a pure liquid boils is called a fixed point; these are the melting-point (m.p.) and boiling-point (b.p.).

No two pure substances have the same fixed points. Therefore fixed points are used to identify substances and to check their purity.

Substances can be mixed together. Mixtures are either clear and in one phase (solutions) or cloudy and in more than one phase (suspensions).

Liquids that will not mix to give one phase are immiscible.

When you shake immiscible liquids together you get an emulsion.

The substances in a suspension are easy to separate by filtering, because they are in two different phases.

The substances in a solution are less easy to separate, because there is only one phase present. To separate them, one substance must be made to change state.

Some ways to separate the substances in a solution are:

simple distillation
fractional distillation
method of solvents
solvent extraction
chromatography.

Crystals are solid particles with regular shapes. They grow when a solution is cooled or evaporated.

The solubility of a substance at any temperature is the amount in grams that will dissolve in 100 grams of solvent at that temperature. Solubility usually increases with temperature.

Questions

1 Draw a table with headings like this:

System	*Type of mixture*	*States present*

In the first column write these substances: foam; whisky; smoke; sea water; cloud; soda water.
In the second column write either *Solution* or *Suspension*. In the third column write down the two states present.

2 Here is a list of mixtures:
(a) salt in water
(b) water in salt
(c) mud in water
(d) propanone in water (they are miscible)
(e) salt and sand
(f) tetrachloromethane in water (they are immiscible)
What is the best way to separate each mixture to obtain the last part pure?
In each case explain why your chosen method works.

3 Carbon dioxide sublimes at −78°C. It is called **dry ice**, and is widely used for keeping ice-cream cold.
(a) Why do you think it is called dry ice?
(b) Why is it used instead of ordinary ice, to keep ice-cream cold? Try to give two reasons.

4 What does the word *pure* mean? Do you think you can have a pure mixture?

5 What effect does adding an impurity have on the m.p. of a substance? If salt is added to ice, what happens to the melting point of ice?
When there is a bad frost, council lorries often spread salt on frozen roads. Why do you think they do this?

6 (a) What is the difference between the freezing point and melting point of a pure substance?
(b) Give two uses for fixed points in chemistry.

7 The composition of air is roughly:

Substance	%	*b.p./°C*
Nitrogen	78	−192
Oxygen	20·9	−183
Water	varies	100
Carbon dioxide	~0·03 (varies)	−78 (sublimes)

(a) How could you remove water and carbon dioxide from the air?
Once these have been removed, the air is cooled right down until it becomes a liquid solution of oxygen in nitrogen. It is then fractionally distilled.
(b) Why is fractional distillation used instead of simple distillation?
(c) Will oxygen or nitrogen boil first?
(d) Which gas will come off the top of the fractionating column?
(e) Can you think of a use for pure oxygen?
(f) Can you think of a use for pure nitrogen?

8 Use the solubility curve of potassium nitrate on page 19 to answer these questions:
(a) Find the solubility of potassium nitrate at: (i) 25°C (ii) 45°C.
(b) What mass of crystals will form if a saturated solution in 100 g of water is cooled: (i) from 45°C to 25°C? (ii) from 50°C to 20°C?
(c) What mass of potassium nitrate will dissolve in 1 kg (1000 g) of water at 50°C?
(d) What mass of water is needed to dissolve 100 g of potassium nitrate at 45°C?

9 The diagram shows the result of a chromatography experiment. 1 and 2 are mixtures. A, B, C, D and E are pure substances. Which pure substance is:
(a) not in either mixture?
(b) in both mixtures?

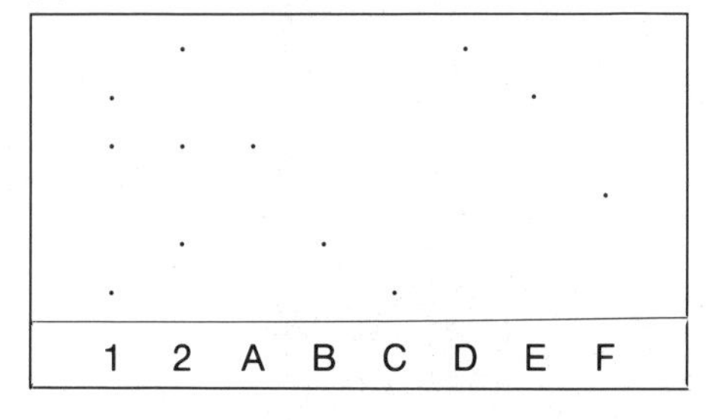

3 The physical structure of matter

In Chapter 2, we looked at the physical properties of some systems. In this chapter, we will try to explain these properties by thinking about the behaviour of the particles.
Remember the two aims of the chemist:

(i) to know the facts about chemical systems;
(ii) to find a theory that explains these facts.

The sections of this chapter follow the order and content of those in Chapter 2. You might find it helpful at times to look back at Chapter 2. In any case you should revise its summary on page 21.

3.1 States of matter

Solid, liquid, and gas

We saw in Chapter 2 that 'pure' means that only one substance is present. A pure substance can exist as a solid, liquid, or gas, depending on the temperature and pressure. If the substance is pure, the same particles must be present in each state. So why do the three states look and act differently? It must be because the particles are arranged differently in each state. Let us look at the different states again.

Facts	*Theory*
A solid has a definite shape. It is hard. 	The particles are held together in a definite, fixed pattern; they are unable to move past one another. There must be strong forces of attraction between the particles. 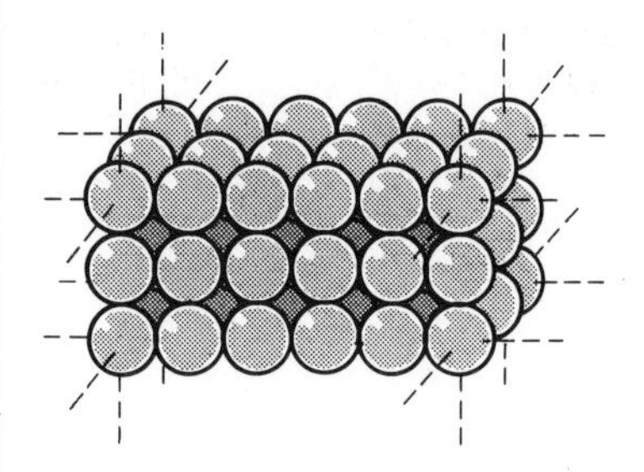 An ordered arrangement of particles in three dimensions is called a **lattice**.

Facts	*Theory*
A liquid has no fixed shape. It has a fixed volume. It pours easily.	The particles are still quite close together, but they can move past one another. The forces of attraction must be weaker. The particles exist in 'clumps' which can slide past each other. There is no fixed lattice. There are gaps between the clumps. 'clumps' of particles
A gas has no fixed shape or volume. It can easily be compressed. It is much lighter than the same volume of liquid or solid. 	The particles are far apart. They can be pushed closer together. They must be moving fast because when they hit the walls of a container, they exert a pressure. 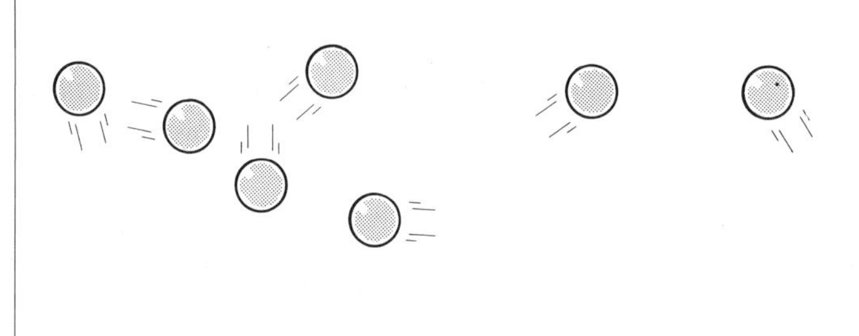
Solids expand when they are heated: 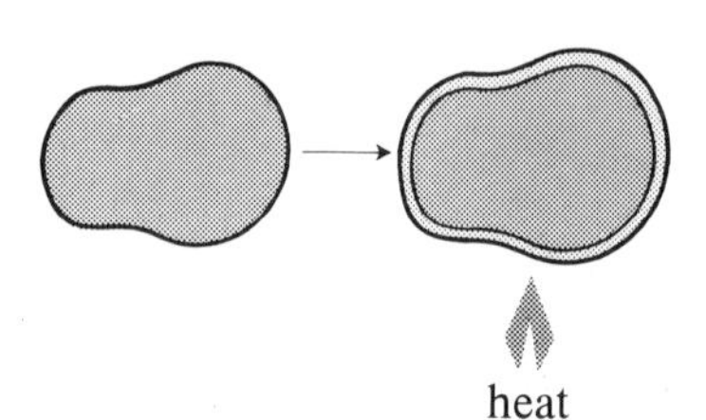 The hot solid does not weigh any more than when it was cold. It simply occupies more space.	The particles move further apart: 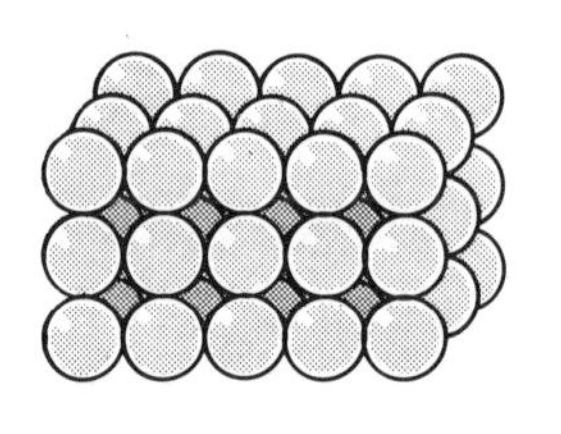 heat energy 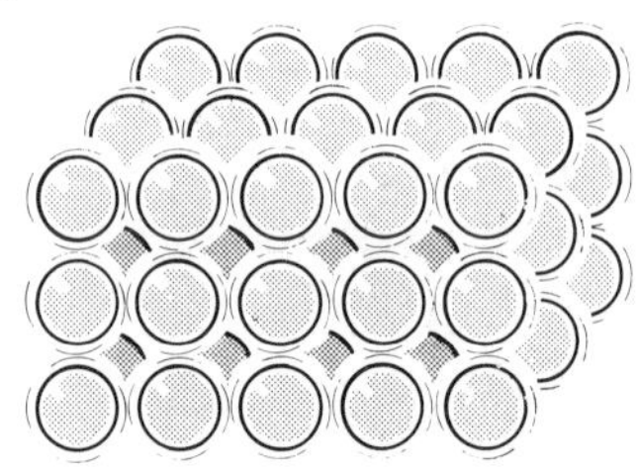 particles held in fairly rigid lattice, vibrating a bit particles vibrate more so they jostle one another more and take up more space The number of particles has not changed. The particles have moved further apart and are moving about more.

Temperature and state changes

Heat is a form of energy. When a solid is heated, the particles take in this energy. It makes them move—they vibrate in all directions, hitting each other. This energy of movement is called the **kinetic energy** of the particles. Not all the particles take the same amount of energy. Some get a little, some get more. Some particles lose energy by hitting others. Others gain more energy by being hit.

The temperature of a system is a measure of the average energy of the particles in the system.

In a cold system, the average energy of the particles is lower than in a hot system. What happens when a system changes state?

Facts | *Theory*

Melting:
When a solid is being heated, the temperature suddenly stops rising. Instead of getting hotter the solid melts.

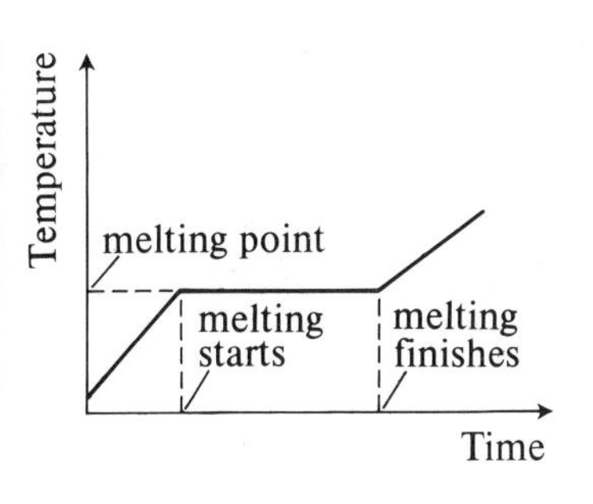

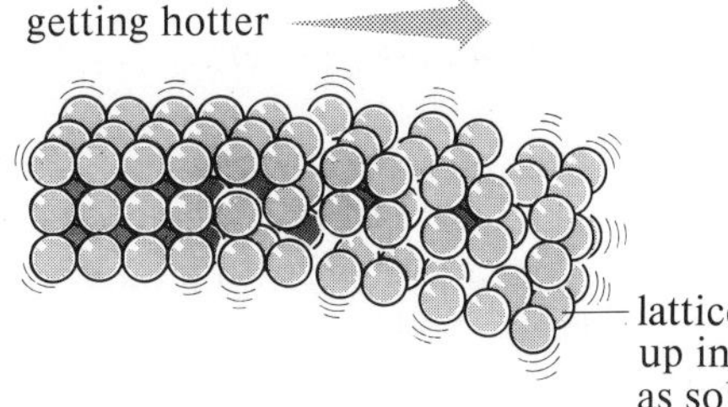

As the temperature goes up, the kinetic energy of the particles increases. Eventually they vibrate enough to leave their positions in the lattice.
The lattice breaks into smaller clumps of particles which can move past each other.
It needs energy to break up the lattice into clumps. That is why the temperature does not rise further until the melting has finished.

Evaporating:
A liquid open to the air slowly evaporates.
At the same time the rest of the liquid gets colder.
The higher the temperature, the faster the liquid evaporates.

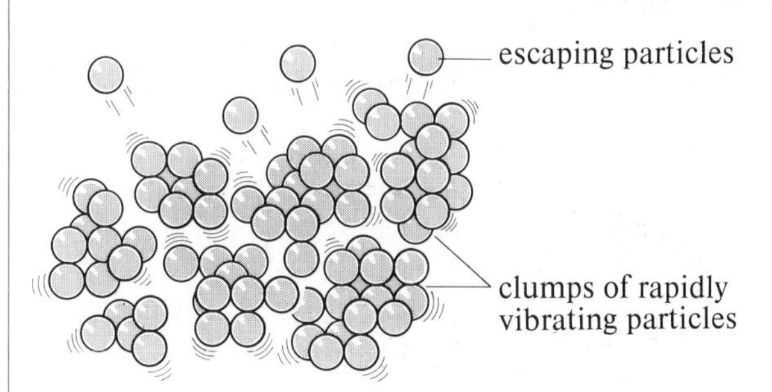

Some particles with a lot of energy can break off from a clump. They can even escape completely into the vapour state.
This means that the average energy of the rest will be less, so the temperature of the system falls.
At a higher temperature, there are more high energy particles likely to escape.

Facts	*Theory*
Boiling: The same pattern of temperature rise happens on heating a liquid as on heating a solid. Bubbles can be seen to form in the liquid at its boiling point.	many particles escaping from the surface large clumps of particles moving in opposite directions leave almost empty spaces - bubbles - containing only a few fast-moving particles
Temperature boiling-point boiling starts boiling finishes Time	As the temperature increases, the average energy of the particles increases. More and more particles have enough energy to escape. When a particle with the average energy can escape, the liquid is said to boil.

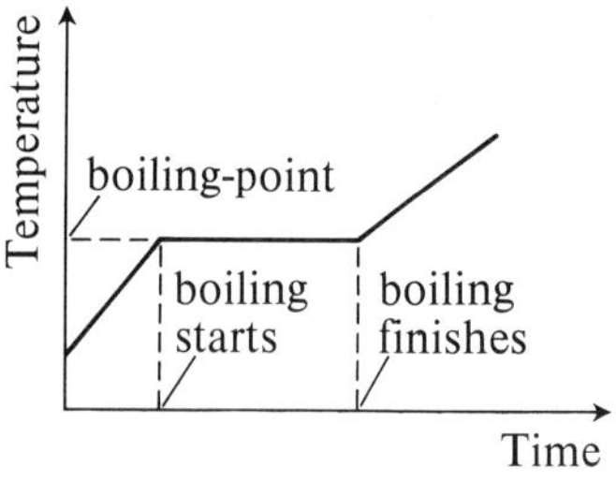

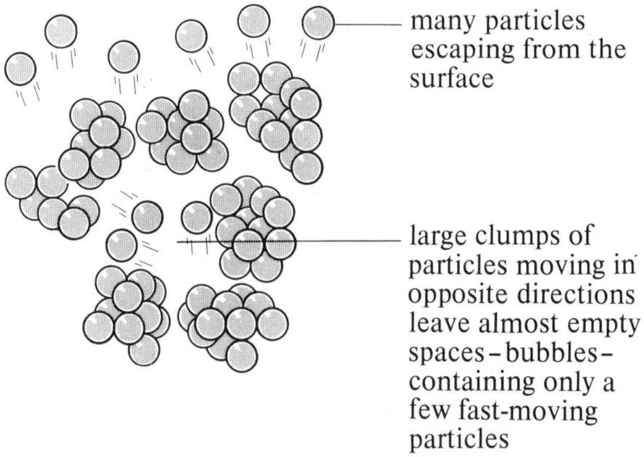

3.2 Purity

Fixed points Why do substances have constant melting-points and boiling-points?

Facts	*Theory*
Pure substances have constant m.p. and b.p. values, called fixed points.	A pure substance contains only one type of particle. These are packed together in a particular way. The energy needed to break up the lattice depends only on the type of particle and the way it is packed.
Each substance has its own unique fixed points.	Other substances, packed in other ways and made of other particles, need different amounts of energy to break out of their lattices.
Any impurity will change these fixed points.	Impurity particles disturb the arrangement of the lattice and so the amount of energy needed to break it up becomes different.

3.3 Mixtures

Suspensions A suspension is a cloudy mixture. The cloudiness is caused by pieces of one substance floating in another. These pieces are big enough to be seen, so they must contain vast numbers of particles. In fact there are so many particles grouped together in each piece that the pieces keep their own properties. Separating a suspension involves trapping the pieces of one sort of particle while letting the others pass.

Solutions When you look at a solution you can see it is clear. There are no pieces big enough to be seen. The particles of each sort are evenly mixed together.

Fact	*Theory*
A solution is a mixture of matter in one state only.	Single particles of one sort are surrounded by those of a different sort. Both particles are usually of roughly the same size. There is an even spread of one type of particle through the other. Separating solutions is not as easy as separating suspensions, because the mixing process is much more complete in a solution. It is impossible to trap just one kind of particle from a solution, using a filter funnel.

3.4 Crystals

Why do crystals have their special properties?

Facts	*Theory*
Crystals have regular shapes.	Particles forming the crystal are packed in a very exact and ordered pattern. When this pattern is repeated many millions of times, the crystal is produced.
Crystals dissolve in solvents.	solvent particles Solvent particles surround the lattice and attract the crystal particles away from the lattice. This finally results in an even mixture of solvent and crystal particles.
Crystals begin to form in a solution as the solvent evaporates—crystallisation.	As more and more solvent particles leave the surface, the crystal particles are forced closer and closer together. Small pieces of lattice begin to form.

Summary

The behaviour of particles is summarised in the table below.

Particle behaviour	*Example*
Particles attract one another.	solid lattice Solids are hard and often found as regular-shaped crystals.
Particles move, vibrate, and spin. They possess kinetic energy; heating the system leads to an increase in their kinetic energy.	Heat energy supplied to a lattice makes it break up. The lattice breaks up as particles escape from the system by increased moving, vibrating, and spinning.
Particles collide with one another.	When a saturated solution is cooled or evaporated, crystals form. This is because the particles collide, then stay together.

Questions

1 (a) Draw simple diagrams to show the differences in the arrangements of particles in the solid, liquid, and gas states.
(b) In which state are the particles moving most?
(c) In which state are the particles closest together?

2 Each of the following sentences contain a mistake. Each can be corrected by taking out the word in italics and putting in another word. Think about the sentences, and then rewrite them putting in the correct word.
(a) Solid *particles* are harder than liquid ones.
(b) Suspensions are *clear* mixtures.
(c) The residue contains the *soluble* part of a mixture.
(d) The filtrate is a *suspension*.
(e) When a liquid evaporates, *all* its particles have enough energy to escape.
(f) The pressure exerted by a gas is due to collisions between the *gas* and the walls of the container.
(g) When a solid is heated the particles *expand* more.

3 When a solid melts, the liquid produced is usually about 5–10% larger in volume. For example when cubes of ice melt, the water in the container occupies about the same space as the ice did.
When a liquid boils, the gas produced is 100 000–500 000% larger in volume at standard pressure. For example, when a kettle boils, the steam from a little of the water can fill a whole room.
Give reasons for these two very different volume changes by explaining what happens to the particles during melting and boiling.

4 Explain in terms of the behaviour of particles:
(a) Why liquids do not have a definite shape like solids.
(b) Why steam can burn you very badly.
(c) Gases always mix completely when they are put in the same container.

5 The diagrams show hot and cold solids, in different positions.

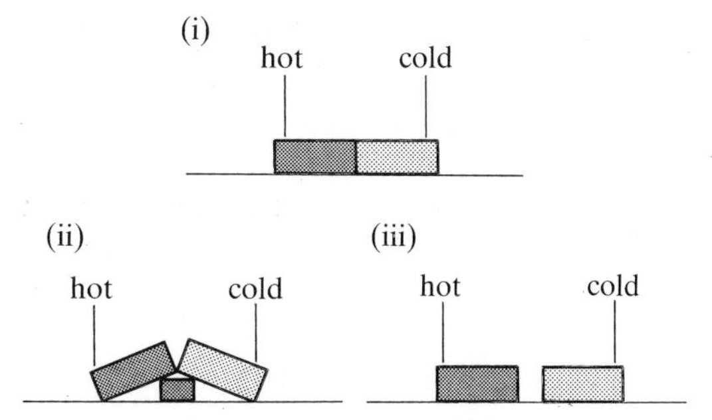

(a) In which case will the cold solid heat up fastest?
(b) What is the difference between the particles in the hot solid and those in the cold solid?

6 Define the following words:
fixed point; melting point; evaporation; subliming; solution; solvents; miscible.

7 Explain the following facts by describing what happens to the particles involved.
(a) On a high mountain, where the pressure is less, pure water boils at a lower temperature than it does at sea-level.
(b) When a bicycle pump is used, the valve and the pump get hot.

8 Pure lead melts at 327°C, and pure tin melts at 232°C. Solder, which is a mixture of tin and lead melts at 183°C.
(a) What effect has the impurity, tin, had on the melting point of lead?
(b) What effect has the impurity, lead, had on the melting point of tin?
Assume that in a metallic lattice the atoms are round and packed together as tightly as possible.
(c) Draw a diagram of part of a tin lattice and draw another diagram, this time of a lead lattice, using another colour and making the atoms a different size.
(e) Now draw part of a lattice of solder. It does not matter how many of each atom you draw, but they must be different sizes.
(f) Are the atoms in the solder lattice as tightly packed together as they are in the lattices of the pure metals?
(g) Try to explain why solder has a lower melting point than either of the pure metals.

4 Chemical change: elements and compounds

In the last two chapters we looked at many physical changes. Now we are going to meet chemical changes for the first time.

4.1 Solids which decompose instead of melting

Decomposition: a type of chemical change

If ice or candlewax is heated, it changes into the liquid state at its melting-point. If sugar or dynamite is heated, very different results are obtained. The sugar does melt but quickly starts to turn yellow and then brown; finally a black, charred solid is left. Dynamite explodes.

Here are some more examples of what heat can do to solids:

Table 4.1 Heat on some solids

Solid	*Observations on heating*
Blue, hydrated copper sulphate	A white solid is produced; steam is given off.
Green copper carbonate	A black solid is produced; a heavy gas which turns lime water milky is given off.
Blue, hydrated copper nitrate	A black solid is produced; dense fumes of a brown, choking gas, water vapour, and oxygen are given off.
Orange ammonium dichromate	* A feathery green solid is thrown out of the reaction-tube; flames can be seen and heat is given out from the system.

* This change can be violent. It should only be done with very small amounts of ammonium dichromate.

All these changes have certain things in common:

They are all caused by heating the system.
The starting substance is no longer there, after heating.
Two or more new substances have been produced.

These properties are typical of a **chemical change**. There are many sorts of chemical change. This sort is called **decomposition**.

Decomposition is a chemical change in which one substance is broken down into two or more simpler substances.

Reactants and products

The original substances at the start of any chemical change are called **reactants**. The new substances present at the end are called **products**. The chemical change itself is called a **reaction**.

Reactions are written down like this:

reactants ⟶ products

The arrow means **react to give**. The reactions in Table 4.1 would be written like this:

Reactants	*react to give*	*Products*
Copper sulphate (blue)	⟶	White solid, and water
Copper carbonate (green)	⟶	Black solid, and gas
Copper nitrate (blue)	⟶	Black solid, brown gas, oxygen, and water
Ammonium dichromate (orange)	⟶	Green solid, and gas

Properties of chemical change

Although decomposition is not the only type of chemical change, *all* chemical changes have the same three important properties in common:

(i) **A chemical change produces new substances.**
Reactants ⟶ products

(ii) **There is an energy exchange between the reacting system and its surroundings, during the reaction.**
All the decompositions above needed heat energy from a bunsen burner to start them off. One of them, the last one, gave out heat energy once it started. Some chemical changes take in energy, others giving out energy:

Energy exchange	*Energy exchange*
Surroundings ⟶ system	*System ⟶ surroundings*

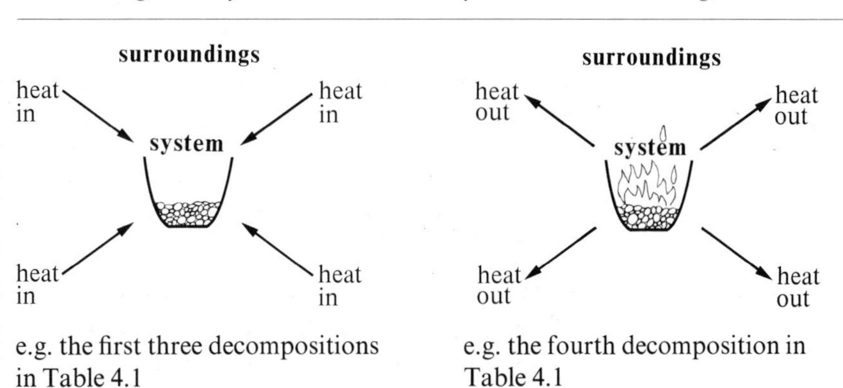

e.g. the first three decompositions in Table 4.1	e.g. the fourth decomposition in Table 4.1
Chemical reactions in which energy is taken in from the surroundings are called **endothermic** reactions.	Chemical reactions in which energy is given out to the surroundings are called **exothermic** reactions.

Above, vegetables cooking and coal burning. Which reaction is endothermic?

(iii) **In every reaction, the mass of a product is always a fixed proportion of the mass of reactant you start with.**

This third important property of chemical reactions can best be shown by looking closely at the first decomposition described in Table 4.1. Here is a simple experiment that you could do in the laboratory:

Different amounts of hydrated copper sulphate are weighed out and then heated. When all the water vapour is driven off, the remaining products are weighed. Some typical results are shown below as pie charts:

Table 4.2 The decomposition of copper sulphate

Mass of reactant	*reacts to give*	*Mass of solid product*	*Mass of steam*
blue  5 grams	→	white 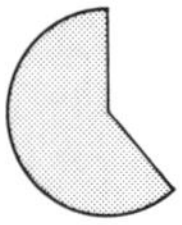 3.2 grams	vapour  1.8 grams
blue  10 grams	→	white 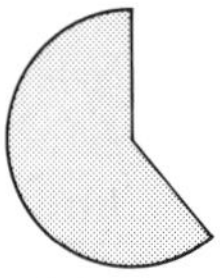 6.4 grams	vapour  3.6 grams
blue 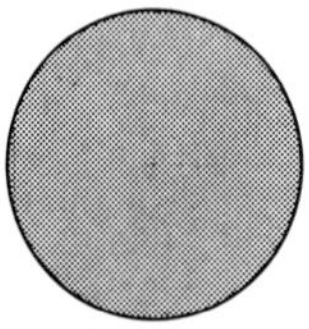 15 grams	→	white 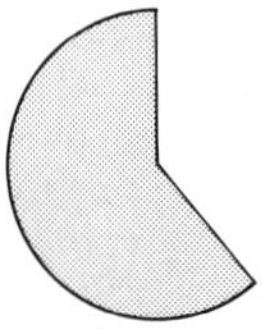 9.6 grams	vapour  5.4 grams

In each case, the amount of water vapour produced is a fixed proportion (or percentage) of the amount of blue solid you start with. It is always 36%:

$$\frac{1{\cdot}8}{5} \times 100 = 36\%; \quad \frac{3{\cdot}6}{10} \times 100 = 36\%; \quad \frac{5{\cdot}4}{15} \times 100 = 36\%$$

So the mass of white solid is always 64% of the mass of blue solid you start with. This pattern of fixed proportions applies to all chemical changes.

Elements

There are certain substances which decompose when electrical energy passes through them. These substances do not conduct electricity when they are in the solid state but do when in the liquid state.

One example is lead bromide. At room temperature it is a white solid, that does not conduct electricity. Suppose that solid lead bromide is put into the first tube in the apparatus below. The steel and carbon rods can carry the current to and from the lead bromide.

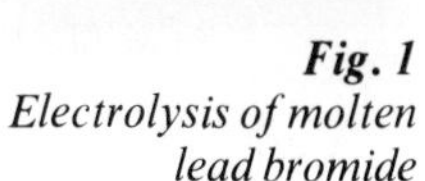

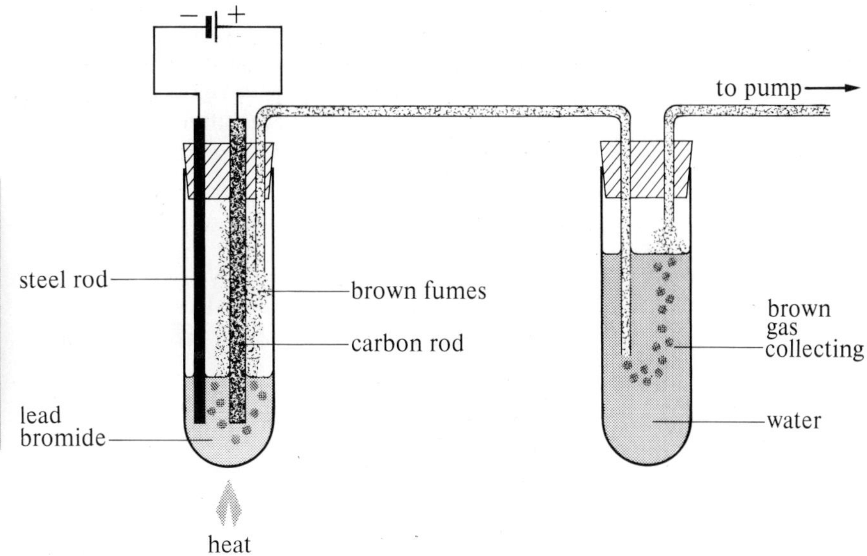

Fig. 1
Electrolysis of molten lead bromide

If the first tube is heated, *without* the battery being connected, the lead bromide melts. Nothing else happens. But once the battery is connected, small bubbles start to appear around the carbon rod. Slowly a brown gas is produced which is sucked into the second tube. This brown gas is bromine. After about ten minutes a small silvery lump can be seen at the bottom of the first tube. This lump is molten lead. So for this reaction:

Reactant	*Products*
Liquid lead bromide	(i) A metallic solid (ii) A brown gas

The system has taken in electrical energy from the surroundings—an endothermic reaction—and new substances have been formed. Lead bromide has been decomposed.

The two products of the decomposition, lead and bromine, cannot themselves be decomposed any further. In no way can either of them be broken down into simpler substances. They are examples of **elements**.

An element is a substance that cannot be decomposed by heat or electrical energy into simpler substances.

There are just over one hundred elements. You probably already know:

oxygen	carbon	iron	gold
hydrogen	sulphur	silver	copper

A bigger list is given on page 37.

4.2 Some mixtures have unexpected properties

Combination: another type of chemical change

If we heat mixtures of salt and sand or salt and water, no new substances are produced. Changes of state may occur, but we should expect these.

There are other mixtures, however, that behave quite differently:

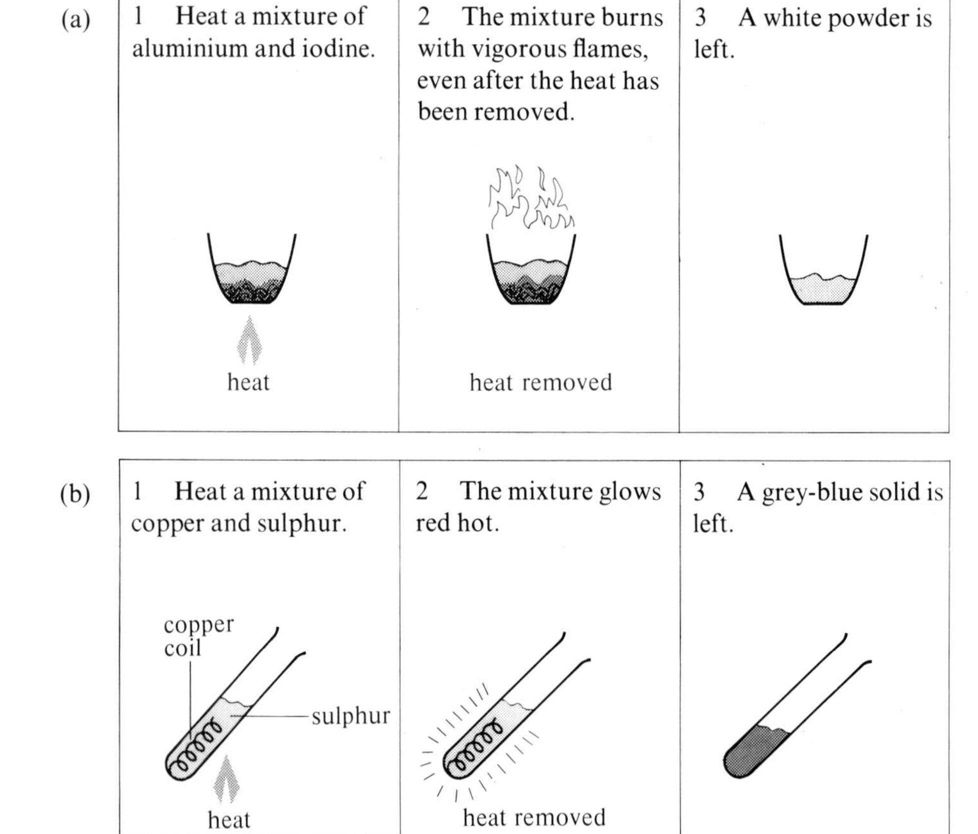

(a)
1 Heat a mixture of aluminium and iodine.
heat
2 The mixture burns with vigorous flames, even after the heat has been removed.
heat removed
3 A white powder is left.
(b)
1 Heat a mixture of copper and sulphur.
copper coil
sulphur
heat
2 The mixture glows red hot.
heat removed
3 A grey-blue solid is left.

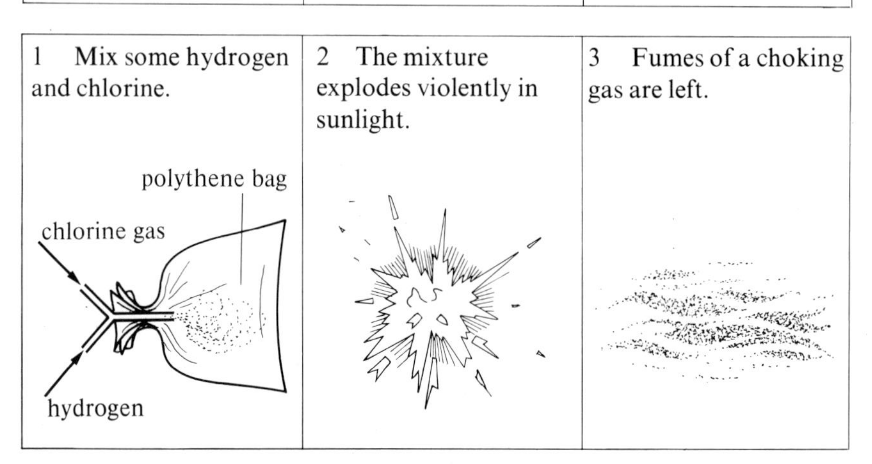

(c)*
1 Mix some hydrogen and chlorine.
polythene bag
chlorine gas
hydrogen
2 The mixture explodes violently in sunlight.
3 Fumes of a choking gas are left.

* This explosive experiment is very dangerous.

In all three of these examples, we can recognise the important properties of chemical change—

(i) Formation of a new substance: in (a) the white powder, in (b) the grey-blue solid, and in (c) the choking gas.

(ii) Energy exchange: all three processes were exothermic once started; they heated up their surroundings.

(iii) Reactants and products in fixed proportions by mass: careful weighing of the reactants and products shows this to be true. For mixture (b), we could do the following experiment. Typical results are shown in Table 4.3.

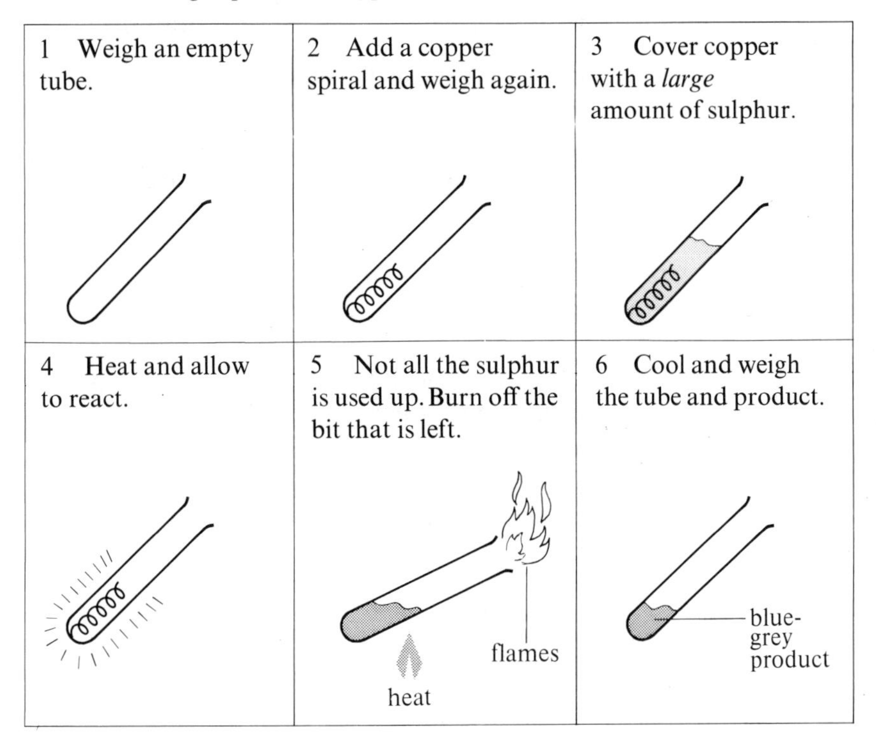

Table 4.3 The reaction of copper with sulphur

Mass of copper /g	*Mass of sulphur /g*	*react to give*	*Mass of blue-grey product/g*
1	0·25		1·25
2	0·5		2·5
3	0·75		3·75

From Table 4.3 you can see that 2 grams of copper react with exactly 0·5 grams of sulphur. So if 2 grams of copper are mixed with 1 gram of sulphur, there will be 0·5 grams of sulphur left in excess.

Notice again the fixed proportions in the reaction. As the mass of copper doubles so do the other masses. As it trebles so do they.

In each reaction shown on page 34, two reactants give a single product. This type of reaction is called **combination**.
Combination is a chemical change in which two substances react to produce a single new substance.

Synthesis The word **synthesis** is also used when talking about combination reactions.
These two sentences, which mean the same thing, show the use of the word:

'Hydrogen and oxygen react together to make water.'

'Hydrogen and oxygen **combine** to **synthesise** water.'

So **combination** describes what the reactants (hydrogen and oxygen) do, while **synthesis** describes the making of the product (water).

4.3 The names of elements and compounds

There are over a hundred known elements. Below is a list of the more common ones. It is divided into two columns. On the left are elements which are metals, and on the right are those which are not metals.

Table 4.4 The more common elements

Metal	*Symbol*	*Non-metal*	*Symbol*
Aluminium	Al	Argon	Ar
Barium	Ba	Bromine	Br
Calcium	Ca	Carbon	C
Chromium	Cr	Chlorine	Cl
Copper	Cu	Fluorine	F
Gold	Au	Helium	He
Iron	Fe	Hydrogen	H
Lead	Pb	Iodine	I
Lithium	Li	Neon	Ne
Magnesium	Mg	Nitrogen	N
Manganese	Mn	Oxygen	O
Mercury	Hg	Phosphorus	P
Nickel	Ni	Silicon	Si
Platinum	Pt	Sulphur	S
Potassium	K		
Silver	Ag		
Sodium	Na		
Tin	Sn		
Titanium	Ti		
Zinc	Zn		

Notice that after the name of each element there is a symbol. These symbols are short ways of writing the names. Each symbol always consists of a capital letter and sometimes a small letter. For example carbon is C, chlorine is Cl. Most of the symbols are the initial letters of the names of the element. Some are based on the Latin names: sodium is Na from Latin *natrium*; potassium is K from Latin *kalium*.

The Periodic Table of elements

In Table 4.4, the elements are listed in alphabetical order. But the usual way of showing them is very different. On the next page is the usual list of elements. It is called the **Periodic Table of elements**. It may seem an odd way to list them, but there are many good reasons for doing it like this. One good reason is that it will make chemistry easier for you to understand.

The vertical columns of elements in the table are called **groups**. Each group has a number, which is always written in Roman numerals. The first column, Li to Fr, is the Group I elements. The second column, Be to Ra, is the Group II elements, and so on.

The horizontal rows of elements in the Periodic Table are called **periods**. They are numbered with ordinary numerals. Period 2 is the elements Li to Ne.

Each element in the table also has its own number. Calcium, for example, is number 20.

The stepped line divides the metals from the non-metals. The metals are on the left.

Table 4.5 The Periodic Table of elements

	Group I	II											Group III	IV	V	VI	VII	VIII
1				1 H hydrogen														2 He helium
2	3 Li lithium	4 Be beryllium											5 B boron	6 C carbon	7 N nitrogen	8 O oxygen	9 F fluorine	10 Ne neon
3	11 Na sodium	12 Mg magnesium											13 Al aluminium	14 Si silicon	15 P phosphorus	16 S sulphur	17 Cl chlorine	18 Ar argon
4	19 K potassium	20 Ca calcium	21 Sc scandium	22 Ti titanium	23 V vanadium	24 Cr chromium	25 Mn manganese	26 Fe iron	27 Co cobalt	28 Ni nickel	29 Cu copper	30 Zn zinc	31 Ga gallium	32 Ge germanium	33 As arsenic	34 Se selenium	35 Br bromine	36 Kr krypton
5	37 Rb rubidium	38 Sr strontium	39 Y yttrium	40 Zr zirconium	41 Nb niobium	42 Mo molybdenum	43 Tc technetium	44 Ru ruthenium	45 Rh rhodium	46 Pd palladium	47 Ag silver	48 Cd cadmium	49 In indium	50 Sn tin	51 Sb antimony	52 Te tellurium	53 I iodine	54 Xe xenon
6	55 Cs caesium	56 Ba barium	57 La lanthanum	72 Hf hafnium	73 Ta tantalum	74 W tungsten	75 Re rhenium	76 Os osmium	77 Ir iridium	78 Pt platinum	79 Au gold	80 Hg mercury	81 Tl thallium	82 Pb lead	83 Bi bismuth	84 Po polonium	85 At astatine	86 Rn radon
7	87 Fr francium	88 Ra radium	89 Ac actinium															

58 Ce cerium	59 Pr prae-sodymium	60 Nd neodymium	61 Pm promethium	62 Sm samarium	63 Eu europium	64 Gd gadolinium	65 Tb terbium	66 Dy dysprosium	67 Ho holmium	68 Er erbium	69 Tm thulium	70 Yb ytterbium	71 Lu lutetium
90 Th thorium	91 Pa protactinium	92 U uranium	93 Np neptunium	94 Pu plutonium	95 Am americium	96 Cm curium	97 Bk berkelium	98 Cf californium	99 Es einsteinium	100 Fm fermium	101 Md men-delevium	102 No nobelium	103 Lw lawrencium

Compounds of two elements

The name of a compound is a combination of the names of the elements in it—just as the compound itself is a chemical combination of the elements:

a compound of copper and sulphur is called copper sulphide;
a compound of sodium and chlorine is called sodium chloride.

Each compound has a **formula** containing the symbols of the elements in it. If the numbers of particles of each element in the compound are the same, then the formula is just the atomic symbols:

CuS is the formula for copper sulphide; there are equal numbers of copper and sulphur particles in copper sulphide.
$NaCl$ is the formula for sodium chloride; there are equal numbers of sodium and chlorine particles in sodium chloride.

If the numbers of particles of each element are not the same, the formula shows this by subscript numbers written after the symbols:

H_2O means twice as many hydrogen particles as oxygen ones;
Al_2O_3 means two aluminium particles for every three oxygen particles.

Naming them

Naming compounds with only two elements in them is quite easy. Here are the rules:

1. Which of the two elements is nearer the left of the Periodic Table? Write down its name.
2. Write half the name of the second element in the compound.
3. Add the ending *-ide*.

Check that the names given for these compounds are correct:

Na_2S	Sodium sulph*ide*
$MgBr_2$	Magnesium brom*ide*
$FeCl_3$	Iron chlor*ide*
Al_2O_3	Aluminium ox*ide*

Compounds of three elements

When a compound has three elements in it one of them is nearly always oxygen. Oxygen combines so often with other elements in this way that a special method of naming the compounds is used. Here are the steps:

1. Which element is nearer the left in the Periodic Table? Write down its name.
2. Write half the name of the second element (not oxygen).
3. Do not use any part of oxygen's name.
4. Add the ending *-ate*.

Some examples are:

KNO_3	Potassium nitr*ate*	
$MgSO_4$	Magnesium sulph*ate*	
$NaClO_4$	Sodium chlor*ate*	
$CaCO_3$	Calcium carbon*ate*	(Note: carb*onate* instead of carb*ate*)

There is an important exception to these rules. Many compounds are formed from a metal, hydrogen, and oxygen. Instead of calling these metal hydrates, they are called **metal hydroxides**:

NaOH sodium hydroxide
KOH potassium hydroxide

If we examined a long list of *-ate* compounds, we should find that nearly all of them contain:

(i) a metal, named first because metals are on the left of the Periodic Table;
(ii) a non-metal—only half its name is given;
(iii) oxygen, therefore the name ends in *-ate*.

But a really close look at the list of the *-ate* compounds shows up another most important point. For the *-ate* compounds of a particular non-metal, the ratio of non-metal particles to oxygen particles in the compounds always seems to be the same:

Table 4.6 The *-ate* compounds

Non-metal in the compound, along with oxygen	*Ratio of the numbers of particles*		*Name*
Hydrogen	$-OH$	One oxygen particle to every one hydrogen particle	hydroxide
Carbon	$-CO_3$	Three oxygen particles to every one carbon particle	carbonate
Nitrogen	$-NO_3$	Three oxygen particles to every one nitrogen particle	nitrate
Sulphur	$-SO_4$	Four oxygen particles to every one sulphur particle	sulphate

Summary

All chemical changes have three characteristics:
a new substance is formed;
an energy exchange occurs between the system and surroundings;
reactants and products are in fixed proportions to one another by mass.

Decomposition is a chemical change, in which a substance is broken down into two or more simpler substances, by the action of heat or electricity.

An element is a substance that cannot be decomposed.

Combination is a chemical change, in which two substances react to produce a single new substance.

When elements combine, they produce compounds.

You should also know the rules for naming compounds.

Questions

1 Some solids melt when they are heated, others decompose when they are heated.
(a) Write down four everyday substances that melt when they are heated.
(b) Write down four everyday substances which decompose when heated.

2 Write down three ways in which a mixture of iron and sulphur differs from the compound iron sulphide.

3 The diagrams below describe a simple experiment that you may have seen.

1

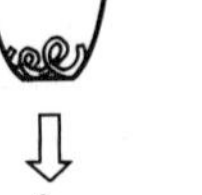

A piece of magnesium ribbon is weighed.

2

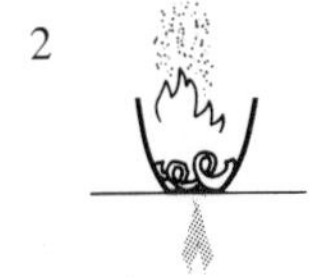

heat

The magnesium is heated until it burns.

3

A white solid is left.

4

The white solid is weighed.

Magnesium is an element.
(a) What does the word *element* mean?
(b) Is the white solid a mixture, an element, or a compound?
(c) Is this reaction a physical change or a chemical change?
(d) Give as many reasons as you can for your answer in (c) above.
(e) Will the white solid weigh more, less, or the same as the magnesium ribbon? Explain your answer.

4 The *decomposition* of T.N.T. (an explosive) is *exothermic*. When T.N.T. explodes, the solid decomposes very very quickly. The whole reaction is over in less than one thousandth of a second. All the products are gases. When 5 g of T.N.T. explodes, 10 000 cm^3 of gas is produced.
(a) Rewrite the first sentence in your own words, to show that you understand what the words in italics mean.
(b) Give three reasons for believing that the explosion of T.N.T. is a chemical reaction.

5 Explain the meaning of the following words: reaction, endothermic, element, compound.

6 Try to think of some everyday chemical changes. They may take place at home, in factories, or in the streets.
(a) Write down three changes that are exothermic.
(b) Write down one change that needs energy and so is endothermic.
(c) Write down three changes that are combinations where an element combines with a compound.
(d) Write down one exothermic combination between two elements.

7 (a) Draw an outline of the first four periods of the Periodic Table.
(b) Divide up the table into its groups and periods. Number the groups with Roman numbers, and the periods with normal numbers.
(c) Draw the stepped line that divides the metal elements from the non-metal ones.
(d) Which sort of elements are there more of, metal or non-metal?
(e) Now learn the names and symbols of the first twenty elements in the table.

8 Try to write the names of the substances given here, using Table 4.4 on page 36, and the rules for naming on page 39.

(a) Ca (b) CO (c) $ZnCl_2$ (d) NaCl
(e) $CaCO_3$ (f) CuS (g) KOH (h) Mg
(i) PBr_3 (j) $ZnSO_4$ (k) PbO (l) Na_2SO_4

9 Some very common substances have everyday names, which do not tell you what elements are in them. Using the rules on page 39, try to work out the chemical names for:
(a) water, H_2O; (b) methane, CH_4;
(c) ammonia, NH_3; (d) salt, NaCl;
(e) lime, CaO; (f) limestone, $CaCO_3$.

10 Look at and read Table 4.3 on page 35.
(a) What does the diagram numbered 4 tell you about this reaction?
(b) What substance is being made in the reaction at the mouth of the tube in the diagram numbered 5?
(c) What is the blue-grey product shown in the diagram numbered 6 called?
(d) Three sets of figures showing the reacting masses are given. Calculate the % of copper in each sample of the blue-grey product.
(e) What do you notice about your result?

5 Elements, compounds, and atoms

5.1 The particles of elements and compounds

Remember the two aims of a chemist, described in Chapter 1:

(i) to know the facts about chemical systems;
(ii) to find a theory that explains the facts.

Chapter 4 described some chemical changes—these were facts—and now we are going to look at theories to explain them. So the order of this chapter follows the order of Chapter 4.

Decomposition In Chapter 3, we put forward the theory that a single, pure substance contained only the particles of that substance. Since the decomposition of a compound usually results in the formation of a new solid substance and a gaseous substance, this must mean that new particles are being produced.

Facts	*Theory*	
When copper carbonate—a green solid—is heated, it decomposes. A black solid and a dense colourless gas are produced.	1 Section of a solid lattice: 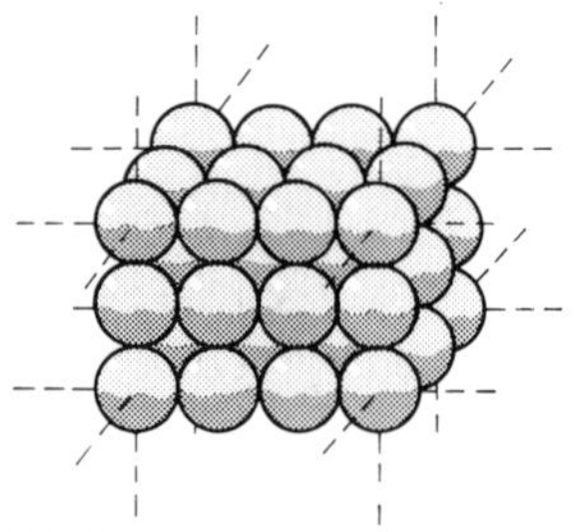	2 Heat causes the particles to vibrate more. 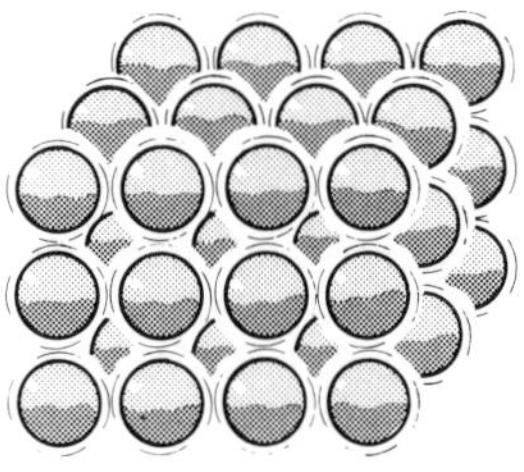
	3 The particles themselves start to break up. 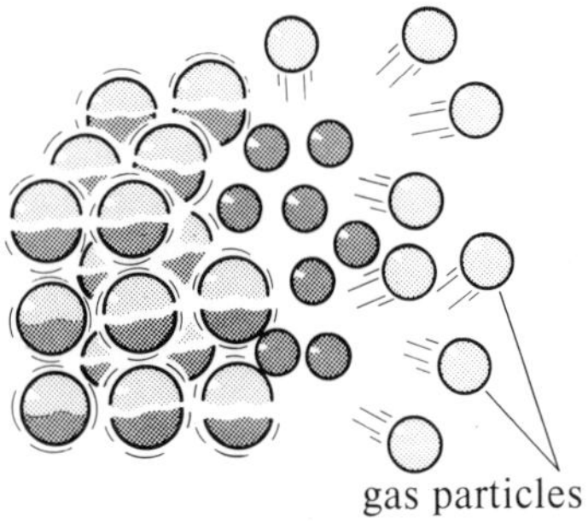	4 A new solid lattice and a gas are formed. 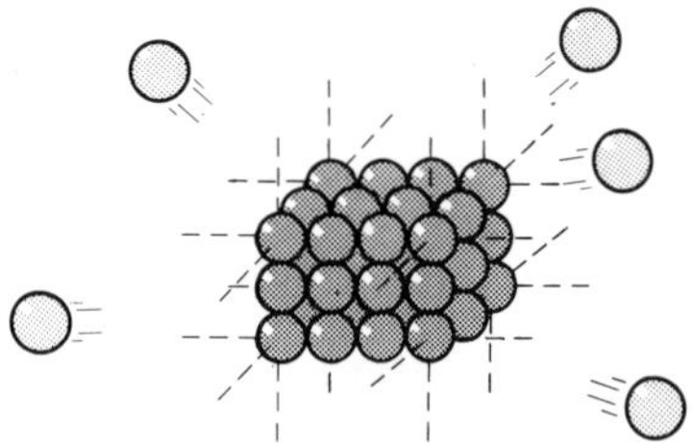

It is worth comparing the behaviour of particles in a solid that is melting with that of particles in a solid that is decomposing. Compare the diagrams on page 42 with the diagram for melting on page 25. Can you see the difference? When a solid melts, it is the forces of attraction *between* the particles that are being overcome. When a solid decomposes, it is the forces of attraction *inside* each particle that are being overcome. New particles are produced as a result of the old ones breaking up.

Elements and atoms

Facts	*Theory*
An element is a substance that cannot be broken down into anything simpler.	The particles of an element cannot be broken down into smaller pieces. Particles of elements must therefore be the building blocks for all the particles of compounds. The smallest particle of an element that can exist on its own is called an **atom**.
Elements combine to produce compounds. The chemical combination of elements is called synthesis.	The atoms of elements join together to produce new particles—the particles of a compound.

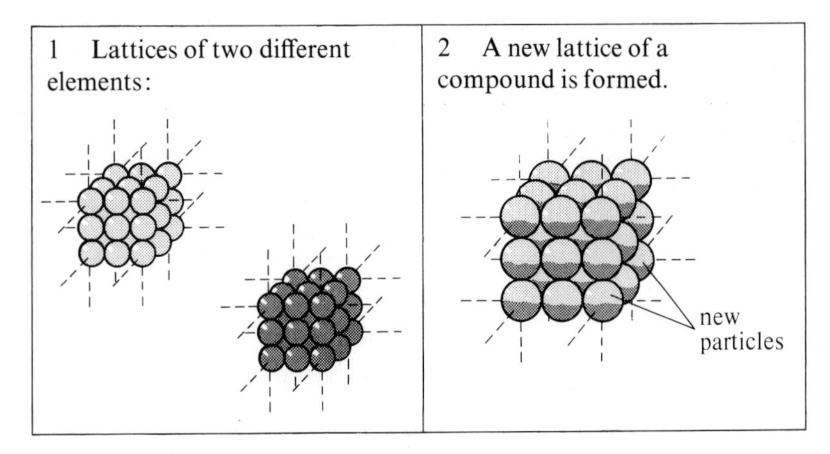

1 Lattices of two different elements:
2 A new lattice of a compound is formed.
new particles

5.2 The formulas of compounds

The ratio of atoms in each particle of the compound

In Chapter 4, page 39, we saw that formulas of compounds often have small subscript numbers, for example $ZnCl_2$. The formula $ZnCl_2$ tells you there are twice as many chlorine atoms as zinc atoms in the compound zinc chloride. It also tells you there are twice as many chlorine atoms as zinc atoms in *each particle* of the compound.

The formula of a compound tells you the ratio of the numbers of each type of atom in its particles.

Here are some more examples:

Name	*Formula*	*Ratio of atoms present*
Sodium chloride	NaCl	One atom of sodium for every one atom of chlorine
Iron chloride	$FeCl_3$	One atom of iron for every three atoms of chlorine
Calcium sulphate	$CaSO_4$	One atom of calcium for every one atom of sulphur and four atoms of oxygen

Naming different compounds of the same element

Unfortunately, the naming of compounds is complicated by the fact that the same two elements often combine in more than one way. This produces *different* compounds from the same elements:

Elements	*can combine to give*	*Compounds*	
Iron and chlorine Fe Cl	⟶	$FeCl_2$ or $FeCl_3$	iron chloride
Calcium, sulphur and oxygen Ca S O	⟶	$CaSO_3$ or $CaSO_4$	calcium sulphate

In each pair above, both compounds have been given the same name, although they are clearly different. So they should not both have the same name. The old-fashioned solution to the problem was to change the name in some way. This was done by adding a prefix or by changing some of the letters.

Prefixes (still often used in naming carbon compounds):

Prefix	*Meaning*	*Example*
mon-	one	CO, carbon monoxide
di-	two	CO_2, carbon dioxide
tri-	three	NCl_3, nitrogen trichloride
tetra-	four	CCl_4, carbon tetrachloride
penta-	five	PCl_5, phosphorus pentachloride

Changing some of the letters:

$CaSO_3$	calcium sulphite
$CaSO_4$	calcium sulphate
$NaNO_2$	sodium nitrite
$NaNO_3$	sodium nitrate

You can see that the compound with less oxygen ends in *-ite* and the one with more oxygen ends in *-ate*. And look at these more complicated examples, that used the Latin names:

$FeCl_2$	ferrous chloride
$FeCl_3$	ferric chloride
$CuCl$	cuprous chloride
$CuCl_2$	cupric chloride

To avoid all these different ways of naming compounds, the modern method is to use a set of numbers called **oxidation numbers**. The rules for working these out may appear to be very strange at first! They are given here so that all the rules for naming compounds are close together in the book; but you will not have to use them much until you reach Chapter 18.

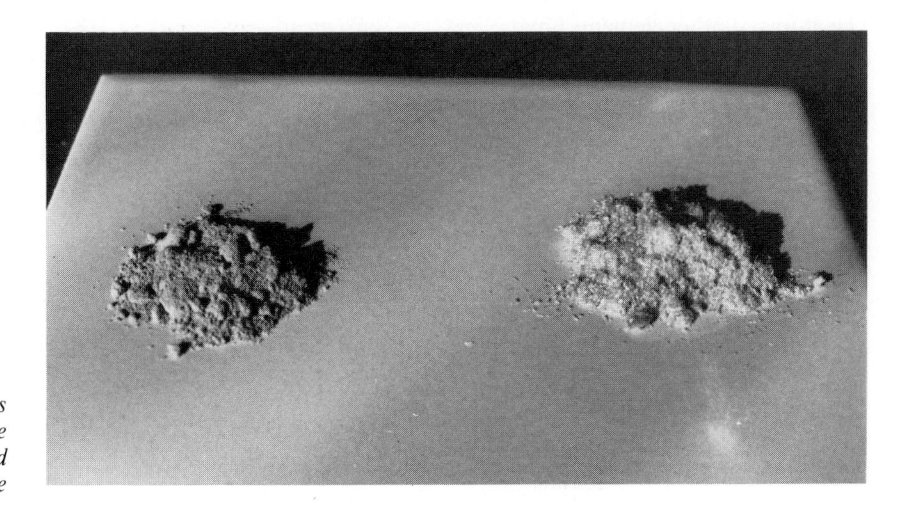

Two different compounds made from the same elements, copper and chlorine

Table 5.1 Working out oxidation numbers

	Rules	*Example*
1	Each atom of an element in the formula of a substance has an oxidation number.	Consider iron chloride, $FeCl_2$, which has one atom of iron for every two atoms of chlorine.
2	The oxidation numbers are always written in Roman numerals. They range from −IV to +VII. ← negative | positive → −IV −III −II −I 0 +I +II +III +IV +V +VI +VII	In $FeCl_2$, the oxidation number of the iron atom is +II; the oxidation number of each chlorine atom is −I.
3	The total of all the oxidation numbers in the formula must be zero.	$FeCl_2$ +II 2×(−I) Total = 0
4	The atoms of some elements have the same oxidation number in all compounds.	*Element* — *Oxidation no. in compounds* Hydrogen — +I Group I metals* — +I Group II metals — +II Aluminium — +III Group VII non-metals (in compounds without oxygen) — −I Oxygen (except in peroxides) — −II
5	The atoms of other elements have different oxidation numbers in different compounds.	*Element* — *Oxidation no. in compounds* Iron — +II or +III Copper — +I or +II Lead — +II or +IV Nitrogen — −III, +III or +V Sulphur — −II, +IV or +VI

* For the groups, see the Periodic Table on page 38.

We can use these oxidation number rules to work out the oxidation numbers of all the elements in any formula.

Using oxidation numbers in names

The elements given beside rule 5 can have different oxidation numbers in their compounds. These are the elements that can combine with different amounts of another element. They are the ones that cause naming problems.

Nowadays, for compounds of these elements, the oxidation number is made part of the name. For example:

	Old name	*Name with oxidation number*
CuCl	Cuprous chloride	Copper(I) chloride
$CuCl_2$	Cupric chloride	Copper(II) chloride

The names with oxidation numbers are read as 'copper one chloride' and 'copper two chloride', respectively.

No oxidation number is written after chloride, because the oxidation number of chlorine is always −I in chlorides.

The examples in Table 5.2 will help to make this method of naming compounds clear.

Table 5.2 Naming compounds using oxidation numbers

Formula	*How to work out the oxidation numbers*	*Name*
$FeCl_2$	1 Chlorine is in Group VII ∴ Cl = −I. 2 There are two atoms of chlorine 2×(−I). 3 Total oxidation number must be zero. 4 ∴ Iron must be +II (+II+2×(−I) = 0).	Iron(II) chloride
$FeCl_3$	1 Chlorine is in Group VII ∴ Cl = −I. 2 There are three atoms of chlorine 3×(−I). 3 Total oxidation number must be zero. 4 ∴ Iron must be +III (+III+3×(−I) = 0).	Iron(III) chloride
$CaSO_3$	1 Calcium is in Group II ∴ Ca = +II. 2 Oxygen has oxidation number −II. 3 There are three oxygen atoms 3×(−II). 4 Total oxidation number must be zero. 5 ∴ Sulphur must be +IV (+II+IV+3×(−II)=0).	Calcium sulphate(IV)
$CaSO_4$	1 Calcium is in Group II ∴ Ca = +II. 2 Oxygen has oxidation number −II. 3 There are four oxygen atoms 4×(−II). 4 Total oxidation number must be zero. 5 ∴ Sulphur must be +VI (+II+VI+4×(−II)=0).	Calcium sulphate(VI)

The oxidation numbers are shown for iron and sulphur above, because they can vary. They are not shown for calcium and chlorine, because these do not vary.

Just as the name of a compound can be worked out from its formula, it is also possible to work out the formula from the name. This is sometimes necessary.

Table 5.3 Find the formula from the name

Name	*Method of working out formula*	*Formula*
Potassium manganate(VII) (Its old name is potassium permanganate)	1 Potassium is in Group I ∴ K = +I. 2 The number VII follows manganate ∴ Mn = +VII. 3 Ending *-ate* means oxygen is present. 4 Oxygen is always −II. 5 Total oxidation number is zero. 6 K +I; Mn +VII; O 4×(−II); Total = 0	$KMnO_4$
Sodium carbonate(IV)	1 Sodium is in Group I ∴ Na = +I. 2 The number IV follows carbonate ∴ C = +IV. 3 Ending *-ate* means oxygen is present. 4 Oxygen is always −II. 5 Total oxidation number is zero. 6 Na 2×(+I); C +IV; O 3×(−II); Total = 0	Na_2CO_3

5.3 Chemical change: writing equations

Just as the formula of a compound tells you about the atoms in its particles, an equation can tell you how the particles change during a chemical change.

Reactants and products

Fact	*Theory*
When calcium carbonate is heated, it decomposes to give calcium oxide and carbon dioxide.	Particles in calcium carbonate split up to give new particles.

Calcium carbonate ($CaCO_3$) *Calcium oxide* (CaO) *Carbon dioxide* (CO_2)

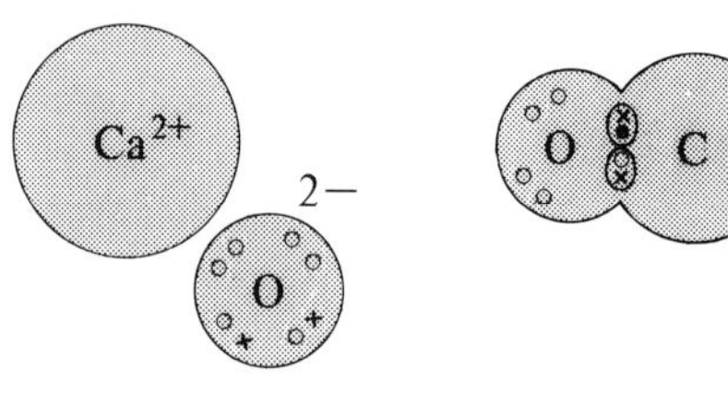

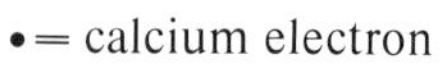

• = calcium electron
○ = oxygen electron
× = carbon electron

During decomposition, each carbonate particle in calcium carbonate splits in the same way:

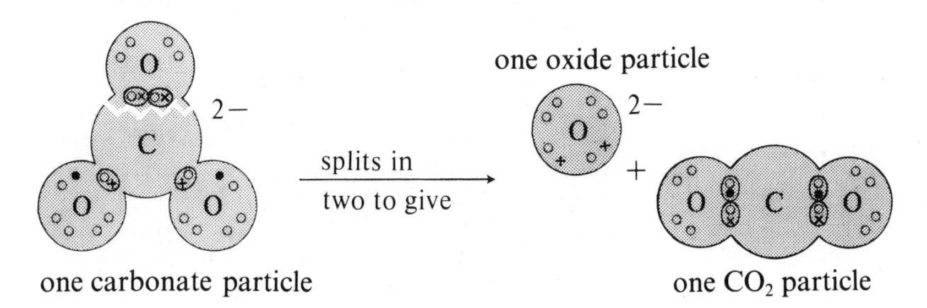

Chemical equations Chemical equations are used as the easiest way of showing the reactants and products and their ratios.

Using the reaction above as our example, here are the steps in writing an equation.

(i) The formulas of the reactants are written on the left.
(ii) The formulas of the products are written on the right.
(iii) An arrow, ⟶, is used instead of the 'equals' sign, =.
(iv) Like equations in maths, there must be the same number of each type of particle on the left as on the right. For example:

	Left		*Right*		
	$CaCO_3$		CaO		$+CO_2$
Number of particles:	one Ca one C three O	⟶	one Ca one O	+	one C two O

Adding gives:	one Ca one C three O	⟶C	one Ca one three O

A chemical equation in which the amount of each element on the left-hand side does not equal the amount on the right-hand side is said to be **unbalanced**. Writing and balancing equations needs plenty of practice! Some rules to help you are given on the next page. Study these carefully, then go through all the examples that follow.

Table 5.4 How to write chemical equations

Rules to follow	*Example of a reaction*
1 Find out the formulas of all the reactants and products in the reaction. Sometimes their names help. Sometimes 'dot-cross' diagrams may be needed to work out electronic structures.	Copper(II) oxide reacts with carbon when heated strongly. Pure copper and carbon dioxide are produced. *Reactants* *Products* CuO; C Cu; CO_2
2 Write the reactant formulas on the left and product formulas on the right. Connect them by an arrow ⟶	$CuO + C \longrightarrow Cu + CO_2$
3 Balance the equation: (i) Choose the first element in the unbalanced equation.	$CuO + C \longrightarrow Cu + CO_2$ (i) Cu
(ii) Count the atoms of it on the left and on the right. Are they equal?	(ii) One on the left and one on the right.
(iii) If not, scale up one side or the other to make them equal.	(iii) They are equal ∴ no scaling needed.
(iv) Now do this for the next element in the equation ...	(iv) O: one to the left but two on the right. We need 2CuO on the left to make the amounts equal. $2CuO + C \rightarrow Cu + CO_2$
(v) and for the next one ... and so on until all amounts balance. Check again that it *all* balances.	(v) C: one on the left and one on the right. But now Cu is unbalanced—there are two on the left but one on the right. We need 2Cu on the right to make the amounts equal.
4 Make sure that you have not changed any of the formulas.	$2CuO + C \rightarrow 2Cu + CO_2$ The equation is balanced.

Example 1. When ammonia is passed down a hot steel tube, it decomposes to its elements nitrogen and hydrogen.

Step 1 The formula of ammonia is NH_3; products are N_2 and H_2.

Step 2 Equation: $NH_3 \rightarrow N_2 + H_2$
This is unbalanced.

Step 3 Balance for nitrogen by doubling ammonia:

$$2NH_3 \rightarrow N_2 + H_2$$

Balance for hydrogen: there are six on the left so there should be six on the right: $2NH_3 + N_2 + 3H_2$

Step 4 This equation is now balanced, and the formulas are correct.

Example 2. Magnesium burns brightly in oxygen gas to give the white solid magnesium oxide.

Step 1 Reactants are Mg and O_2; product is MgO.

Step 2 Equation: $Mg + O_2 \rightarrow MgO$
This is unbalanced.

Step 3 Check magnesium—it is balanced.
Balance for oxygen by doubling MgO:

$$Mg + O_2 \rightarrow 2MgO$$

The equation is now unbalanced for magnesium! Rebalance by doubling Mg on the left: $2Mg + O_2 \rightarrow 2MgO$

Step 4 This equation is now balanced, and the formulas are correct.

Summary

Energy can cause particles themselves to break up, as well as causing them to move apart.

The simplest possible particles are atoms.

A pure element is made of only one kind of atom.

Atoms can join together to form more complicated particles, the particles of a compound.

Oxidation numbers are used in the modern method of naming compounds.

Some elements have the same oxidation number in all their compounds, but others have different oxidation numbers in different compounds.

Oxidation numbers can be used to distinguish between two different compounds made from the same elements.

The changes in a chemical reaction can be shown in a chemical equation. Like maths equations, the numbers of particles on each side of the equation must balance.

Questions

1 The sentences below have pairs of words or phrases separated by a sloping line, e.g. hot/cold. One of the pair will make the sentence correct, the other will make it wrong. Choose the correct word(s) for each sentence, and then write down the full corrected sentence.
(a) When a cold solid is heated, the particles in it (expand/vibrate) more.
(b) When a solid melts, the particles in it (melt/move apart).
(c) A cold solid has (cold/slowly moving) particles.
(d) When a solid decomposes, the particles (decompose/break up).
(e) Two pure substances mix to form a (compound/mixture).
(f) (A pure/An impure) substance is made up of (one/more than one) sort of particle.
(g) A solution is a (uniform/non-uniform) mixture.
(h) A solution contains particles in (one/two) phase(s).
(i) A suspension contains particles in (one/two) phase(s).
(j) Two pure substances combine to form (a compound/an element).
(k) If heated strongly enough, all elements will (decompose/melt).
(l) The particles in a gas are moving (faster/slower) and are (closer/further apart) than the particles in a liquid.

2 Below are some simple factual statements. For each one, try to write a simple theory to explain the fact.
(a) A solid melts when it is heated.
(b) A solid decomposes when it is heated.
(c) Two substances combine chemically when they come into contact.

3 Compare and contrast the particles in:
(a) a mixture of solid elements
(b) a pure liquid element
(c) a gaseous compound
(d) a miscible mixture

4 Turn to Table 5.1 on page 46, which tells you how to work out oxidation numbers. Using the rules given there, work out the oxidation numbers of the underlined atoms in the formulas below.

$\underline{Cu}Cl_2$ $H_2\underline{S}$ $\underline{C}O_2$ $\underline{Hg}O$ $H_2\underline{S}O_4$
$\underline{Mn}O_2$ $H\underline{N}O_3$ $\underline{N}H_3$ $\underline{C}O$ $\underline{Cr}Cl_2$

5 Learn the list of oxidation numbers given beside rule 4 in Table 5.1 on page 46. Now try to work out the oxidation numbers of the underlined atoms without the help of the book.

6 Below are the formulas and old-fashioned names of some substances. Using the rules and examples on pages 46 and 47 to help you, work out the proper modern names for the following:
(a) ferrous oxide, FeO
(b) cupric chloride, $CuCl_3$
(c) ferric hydroxide, $Fe(OH)_3$
(d) sodium nitrite, $NaNO_2$
(e) potassium sulphite, K_2SO_3

7 Write balanced equations for these reactions:
(a) Calcium metal + oxygen gas → calcium oxide
(b) Sodium metal + oxygen gas → sodium oxide
(c) Copper metal + chlorine gas → copper(II) chloride
(d) Lead(II) oxide + hydrogen gas → lead + water
(e) Copper(II) carbonate → copper(II) oxide + carbon dioxide

8 Write balanced equations for these reactions:
(a) Calcium hydroxide → calcium oxide + water
(b) Aluminium metal + iodine solid → aluminium iodide
(c) Iron metal + oxygen gas → iron(III) oxide
(d) Phosphorus solid + oxygen gas → phosphorus(V) oxide
(e) Sodium hydrogen carbonate → sodium carbonate + carbon dioxide + water
(f) Sulphur(IV) oxide + oxygen gas → sulphur(VI) oxide

6 The structure of the atom

Our solar system is made up of the sun and the planets. The sun is at the centre of the solar system because it has a larger mass than all the planets put together.

Chemists believe that the structure of the atom is rather like this. At the centre of the atom is the nucleus. This has a much larger mass than all the rest of the atom. Just as most of the solar system is empty space with only a few planets here and there, so most of the atom is empty space, we think, with only a few particles here and there:

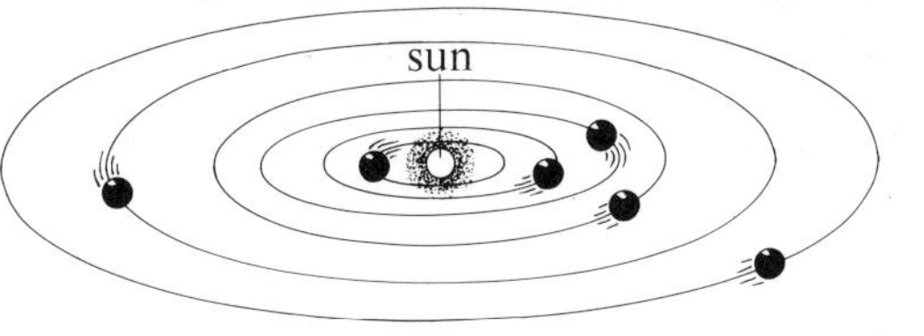

the solar system

nucleus

the atom

6.1 Rutherford's scattering experiments

It was some famous experiments by Rutherford (later Lord Rutherford) that gave chemists this idea of the atom. Here are some diagrams showing what he did.

He knew that the radiation from a radioactive substance would make a mark on photographic paper. (Marie Curie discovered this.) He decided to put a piece of gold foil in the way of the beam of radiation.

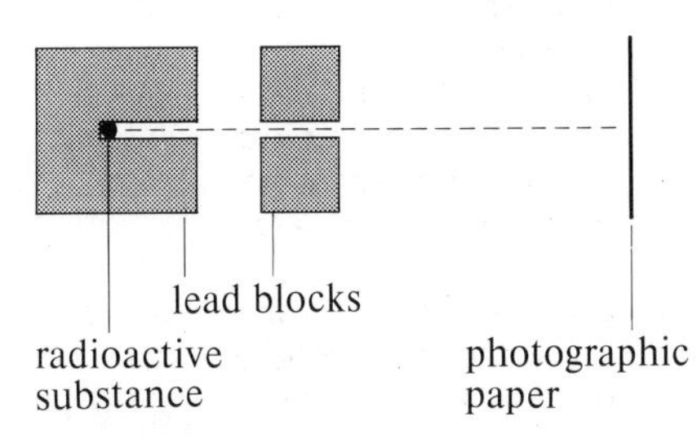

He found that the pattern on the photographic paper remained almost exactly the same even though the gold was in the way.

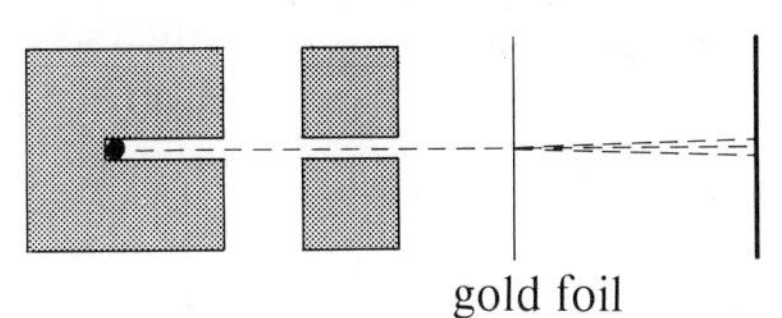

Lord Rutherford, photographed at the Cavendish laboratory, Cambridge, in 1935

Rutherford concluded that even substances as heavy as gold were mostly empty space. The radiation just passed straight through them as though they were not there. Later he invented some better radiation detectors and another experiment was done.

These very accurate detectors showed that a small part of the radiation was being bounced off in other directions.

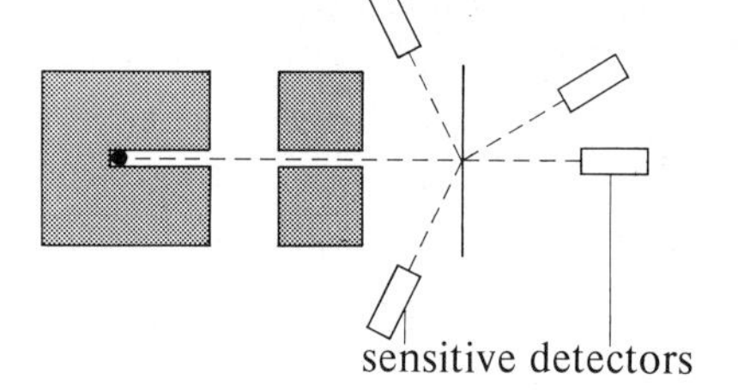

Rutherford explained this by saying that although most of the gold was empty space, it contained some very small massive particles. If the radiation made a direct hit on one of these particles it bounced off, like a bullet ricochetting off a rock.

The size of the nucleus

By working out the fraction of radiation that did bounce off instead of going straight through the gold foil, Rutherford calculated the size of the nucleus at the centre of an atom. It is so small that we cannot imagine it, but this scaled-up drawing (Fig. 1) gives some idea:

Fig. 1

Think about somewhere that is 1 kilometre away. If you held the scaled-up nucleus in your hand, all the other particles of the atom would move around you within a sphere of radius 1 km!

Forces of attraction in atoms

In the solar system the planets are held round the sun by the force of gravity. In an atom the forces holding it together are called **electrostatic forces.**

These electrostatic forces are also found in everyday things. Try rubbing a plastic ruler or comb against the sleeve of your pullover. Now hold it just over a tiny piece of paper. The plastic should be able to pick up the paper. The force of attraction between the plastic and the bit of paper is an electrostatic force, the same kind that is found in atoms. Rubbing the plastic against

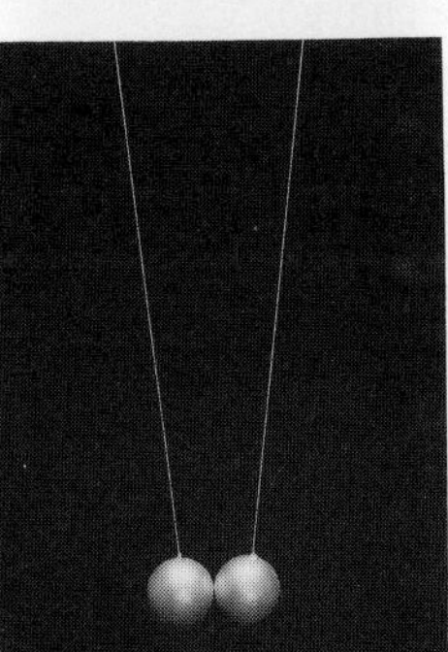

your sleeve produces an electrical property called **charge**. Only when things are charged will there be electrostatic forces.

Experiments show that there are two kinds of charge. They are called **positive** and **negative** charge. The forces between positively and negatively charged things are shown in the diagrams below. These show a pair of small, very light polystyrene balls hanging on nylon threads. The balls are charged in different ways.

No charge: no attraction

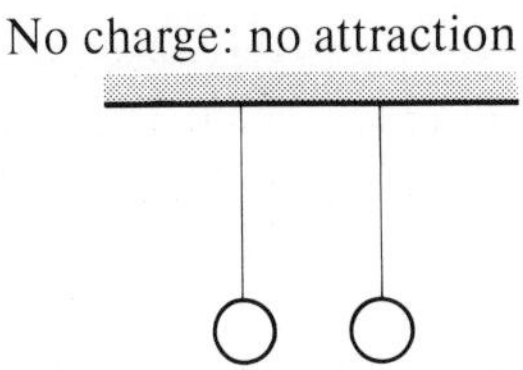

One charged: some attraction

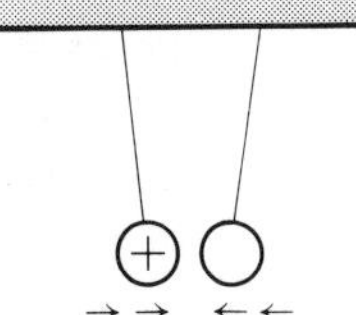

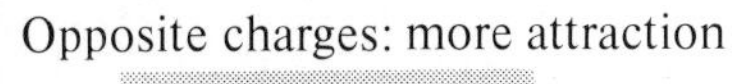

Opposite charges: more attraction

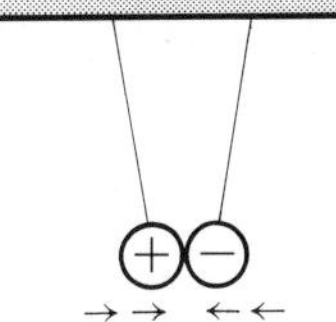

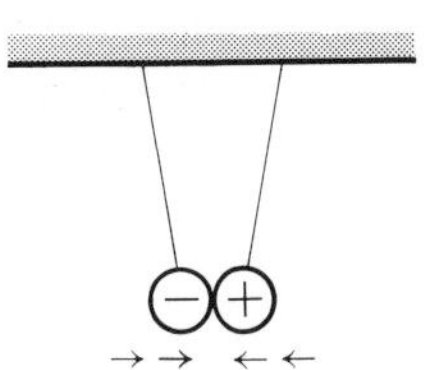

Same charges: repulsion. They push apart.

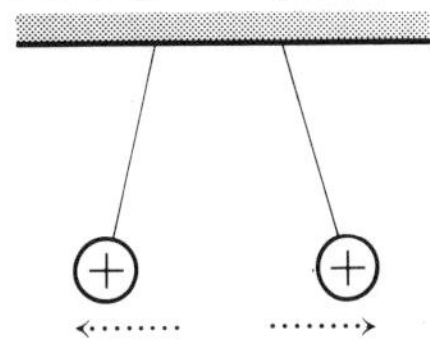

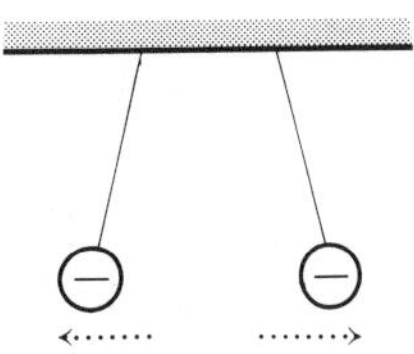

These results can be summed up like this:

two things charged in the same way repel each other;
two things charged in opposite ways attract each other.

This is often shortened to:
Like charges repel, unlike charges attract.

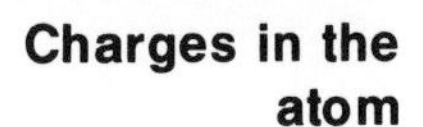

Charges in the atom

Rutherford was also able to show that the centre of the atom—the nucleus—was positively charged. So the structure of the atom is:

positively charged nucleus

surrounding negative parts

The positive charge on the nucleus attracts and holds in the surrounding negative parts of the atom because they have opposite charge. It can be shown that the whole atom is **neutral**. This means that the amount of positive charge in the nucleus is exactly cancelled out by the amount of negative charge around it.

6.2 Protons, electrons, and atomic number

The nucleus of an atom contains positively-charged particles. These particles are called **protons**. Each proton has one unit of positive charge.
Around the nucleus of the atoms are negatively-charged particles. These are called **electrons**. Each electron has one unit of negative charge.
In a neutral atom there is always the same number of protons and electrons:

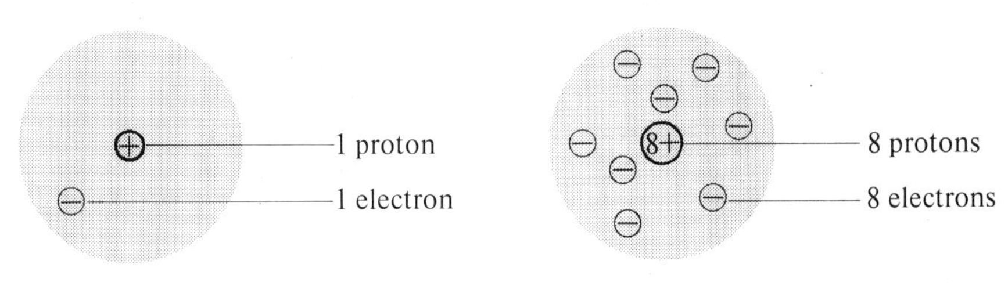

a hydrogen atom

an oxygen atom

All the atoms of one element are identical. They all have the same number of protons. All carbon atoms have 6 protons, for example. Atoms of different elements have different numbers of protons: carbon atoms have 6 protons but oxygen atoms have 8 protons.

The number of protons in an atom is therefore very important. It tells us which element the atom belongs to. Because this number is so important it is given a special name. It is called the **atomic number** and has the symbol Z.
The atomic number of an element is the number of protons in an atom of the element.

The elements are listed in the Periodic Table in order of their atomic numbers:

Table 6.1

Element	*Number in Periodic Table*	*Atomic number*	*Number of protons*	*Number of electrons*
Hydrogen	1	1	1	1
Helium	2	2	2	2
Carbon	6	6	6	6
Oxygen	8	8	8	8
Chlorine	17	17	17	17
Lead	82	82	82	82

The atomic number is sometimes written as a small subscript number in front of the symbol of the element, for example:

$_{1}H$ $_{2}He$ $_{6}C$ $_{8}O$ $_{17}Cl$ $_{82}Pb$

6.3 The mass of atoms

In 1918, a scientist called Aston invented a machine that was able to compare the masses of atoms. The machine is called a **mass spectrometer**. He made an important discovery about the atoms of a pure element. They did not all weigh the same! Not for the first time, scientists were faced with two ideas that did not agree with one another:

(i) All the atoms of a pure element are neutral and have equal numbers of protons and electrons.
(ii) A sample of a pure element contains atoms of different masses.

How could this puzzle be solved? The answer suggested was: *there is more in an atom than just protons and electrons; there are also neutral particles.* If there were neutral particles present as well, their numbers would not need to be fixed. A pure element could then have atoms with the same numbers of protons and electrons, but differing numbers of these neutral particles.

The neutral particles were called **neutrons**. Clear proof of their existence was not discovered until 1932.

Table 6.2 Summary of the structure of the atom

Particles in an atom	*Charge*	*Mass*	*Position in atom*	*Model of the atom*
Proton	+	1	Nucleus	cloud of very light electrons
Neutron	0	1	Nucleus	tiny but heavy nucleus of protons and neutrons
Electron	—	Very light compared with protons and neutrons	Always moving very fast around the nucleus	

Atomic number, mass number and isotopes

On page 56, we saw that Z—the atomic number—is the number of protons in the nucleus of an atom. A second number, from which the number of neutrons can be found, is needed to describe an atom fully. This is the **mass number**, *A*.

The mass number is the total number of particles in the nucleus (that is, the total number of neutrons and protons).

But it is the amount of charge in an atom, and not the mass, that decides which element it belongs to. So it is possible to have two atoms of the same element with different mass numbers,

Atom	*Protons*	*Neutrons*	*Electrons*	*Atomic number* (Z)	*Mass number* (A)
A	17	18	17	17	35
B	17	20	17	17	37

Both A and B above are atoms of chlorine—they both have atomic numbers of 17. Their mass numbers differ because they have different numbers of neutrons. Atoms A and B are examples of **isotopes.**

Isotopes are atoms with the same atomic number but different mass number.

Hydrogen has three isotopes:

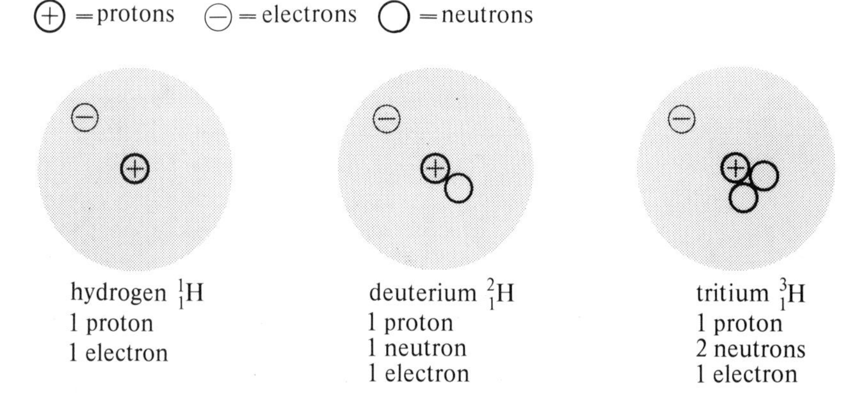

When they are needed, mass numbers are written just above and before the symbol of an atom:

$^{35}_{17}Cl$ $^{37}_{17}Cl$ $^{\text{Mass number } (A)}_{\text{Atomic number } (Z)}$ Element's symbol

Summary

An atom consists of a small, heavy nucleus of protons and neutrons, and a surrounding region of space containing fast moving electrons.

The forces of attraction that hold the particles in an atom together are electrostatic.

Protons have a charge of $+1$ and a mass of 1.
Electrons have a charge of -1 and almost no mass.
Neutrons have no charge, and a mass of 1.

The atomic number, Z, is the number of protons in an atom.
The mass number, A, is the number of protons + neutrons in an atom.

All the atoms of an element have the same atomic number, but they can have different mass numbers because the number of neutrons can vary.

Isotopes are atoms of the same element that have different numbers of neutrons.

Questions

1 The nucleus of an atom is much smaller than the atom itself. It is about 100 000 times smaller. Imagine an atom scaled up so that its circumference is the size of a dart-board 1 m (100 cm) across. The nucleus is at the centre, like the bull's-eye. How big is it? Give the answer in cm. Would there be much chance of hitting the 'nucleus' with a dart?

2 The mass of the hydrogen atom is about $1{\cdot}601 \times 10^{-24}$ g. The mass of the proton is about $1{\cdot}600 \times 10^{-24}$ g. What % of the total mass of the atom is the proton?

3 Static electricity and electrostatics affect all sorts of everyday situations. For example, your pullover crackles when you pull it over your head; records attract dust which causes crackles when they are played; after walking on nylon carpets you can get an electric shock when you touch something metal like a hand-rail. Write down at least three more examples.

4 For each of the following pairs of objects, write down whether there will be attraction, repulsion, or no force between them.

(a) ⊕ ⊕ (b) ⊕ ⊖ (c) ⊕ ○
(d) ○ ⊖ (e) ⊖ ⊖ (f) ⊖ ⊕
(g) ○ ⊕ (h) ○ ○ (i) ⊖ ○

5 Here is a diagram representing the helium atom, whose atomic number is 2. It shows only the charged particles. Copy the diagram. On your copy mark in all the forces of attraction and repulsion in the atom. Will they all be the same strength? Explain.

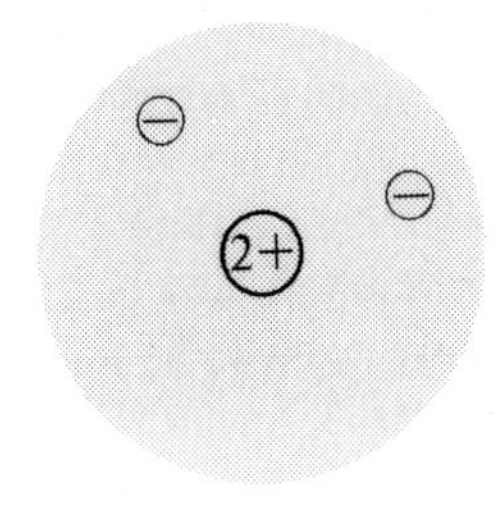

6 (a) Draw diagrams representing the three isotopes of hydrogen.
(b) How do they differ?
(c) What do the isotopes have in common?

7 (a) Explain the difference between mass number and atomic number.
(b) When you have written the symbol for an element, where do you write the mass and atomic numbers? Give an example.
(c) Write down the following atoms in a column.

$^{1}_{1}H$ $^{7}_{3}Li$ $^{14}_{7}N$ $^{80}_{35}Br$ $^{197}_{79}Au$

Besides each atom in separate columns, write down the number of protons, the number of protons + neutrons, and the number of electrons.

8 (a) Naturally occurring chlorine consists of two stable isotopes, 75% $^{35}_{17}Cl$ and 25% $^{37}_{17}Cl$. Calculate the mass of the average chlorine atom on the relative atomic mass scale.

9 Naturally occurring copper consists of two stable isotopes, $^{63}_{29}Cu$ and $^{65}_{29}Cu$. The relative atomic mass of copper is 63·5. Calculate the % of each of these isotopes in natural copper.

10 Naturally occurring boron consists of the isotopes $^{10}_{5}B$ and $^{11}_{5}B$. The relative atomic mass of boron is 10·8.
(a) Which isotope is most abundant.
(b) Calculate the relative number of atoms of each isotope in the natural element.

11 Magnesium has three stable isotopes: $^{24}_{12}Mg$, $^{25}_{12}Mg$, and $^{26}_{12}Mg$.
(a) If there is 79% of $^{24}_{12}Mg$, and equal amounts of the other two isotopes, calculate the % of each of the other two.
(b) Calculate the relative atomic mass of magnesium.

7 The periodic properties of the elements

7.1 The structure of the Periodic Table

You will remember from Chapter 4, page 37, that all the elements are listed in a special table. This is called the Periodic Table. Here is an outline of it to remind you.

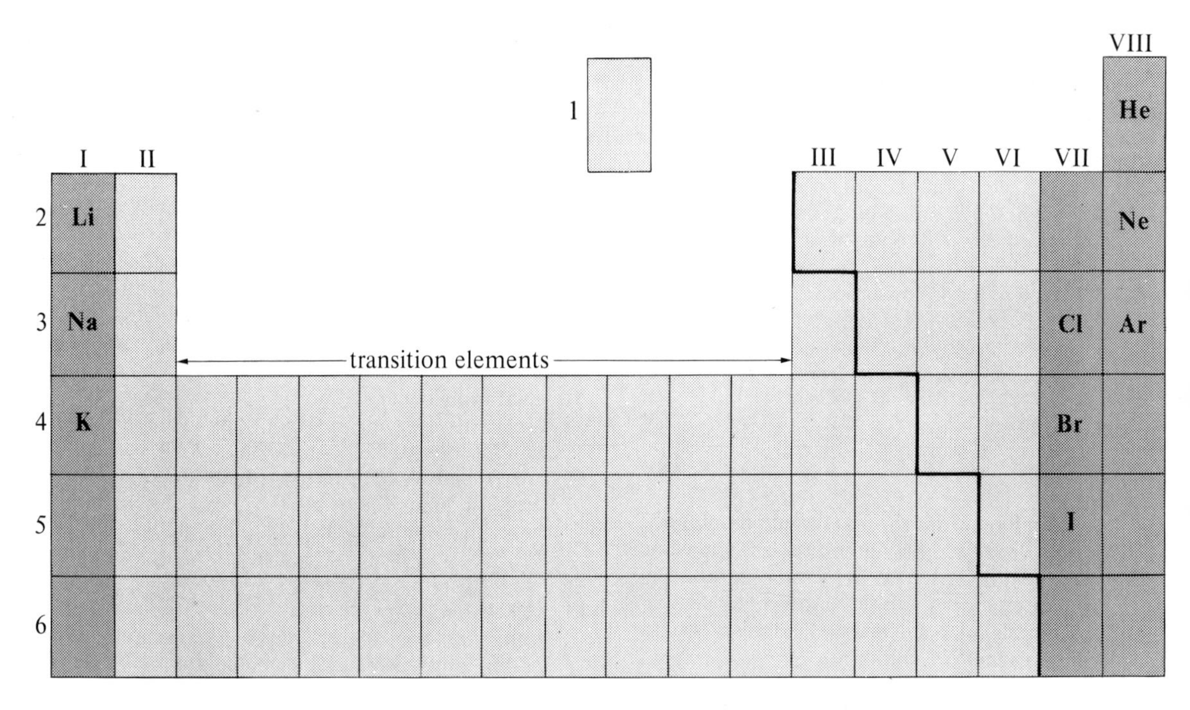

You may remember that the long vertical columns of elements in the table are called **groups**. Each group has a number, which is always written in Roman numerals. The horizontal rows of elements in the Periodic Table are called **periods**. They are numbered with ordinary numerals.

The elements between Groups II and III are not usually put into groups. Instead they are called the **transition elements**.

In this chapter we are going to look at the properties of the elements in the three shaded groups above:

Group I which is sometimes called the alkali metal group;
Group VII which is sometimes called the halogen group;
Group VIII which is sometimes called the noble gas group. It used to be called the inert gas group. (It is often called Group O, instead of Group VIII.)

Sodium, like all Group I metals, is stored under oil to prevent it from reacting with the air

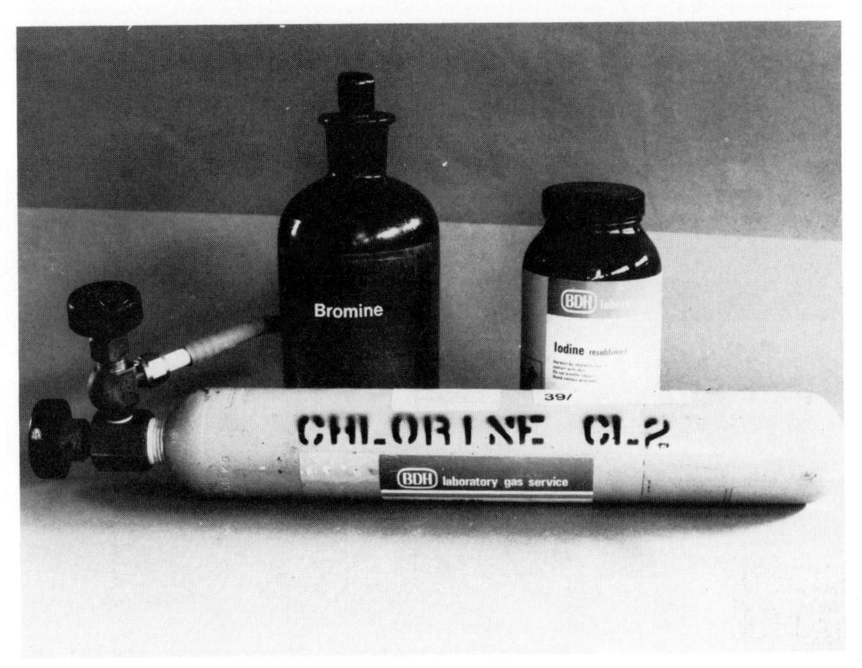

Three of the halogens, the reactive non-metals in Group VII

7.2 Group I: lithium, sodium and potassium

Physical properties

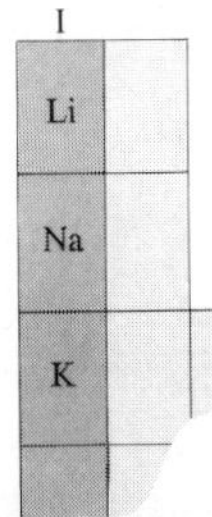

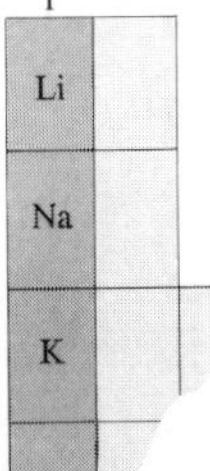

Lithium, sodium, and potassium are solids that are stored in jars of oil. When their surfaces are freshly cut (they must always be cut under the surface of the oil), the solids appear silvery and shiny. They are all soft and melt easily. They are all conductors of electricity. They are all metals.

Table 7.1 compares their physical properties with those of steel, a more common metal. Notice how alike the physical properties of the Group I metals are. They are quite different from those of steel.

Table 7.1

Element	*Symbol*	*m.p./°C*	*b.p./°C*	*Hardness* /Moh*	*Density** /g cm^{-3}*
Lithium	Li	180	1336	0·6	0·53
Sodium	Na	98	883	0·4	0·97
Potassium	K	64	759	0·5	0·86
Compare with steel		1400	3100	6·0	7·7

* Hardness is measured on a scale called the Moh scale. It ranges from 0 for things like gases to 10 for the very hardest substance, diamond.
** The density is the mass of a 1-cm cube of the element. 1 cm^3 of water has a mass of 1 g, so all of these elements are less dense than water.

Reactivity†

(a) With oxygen and chlorine

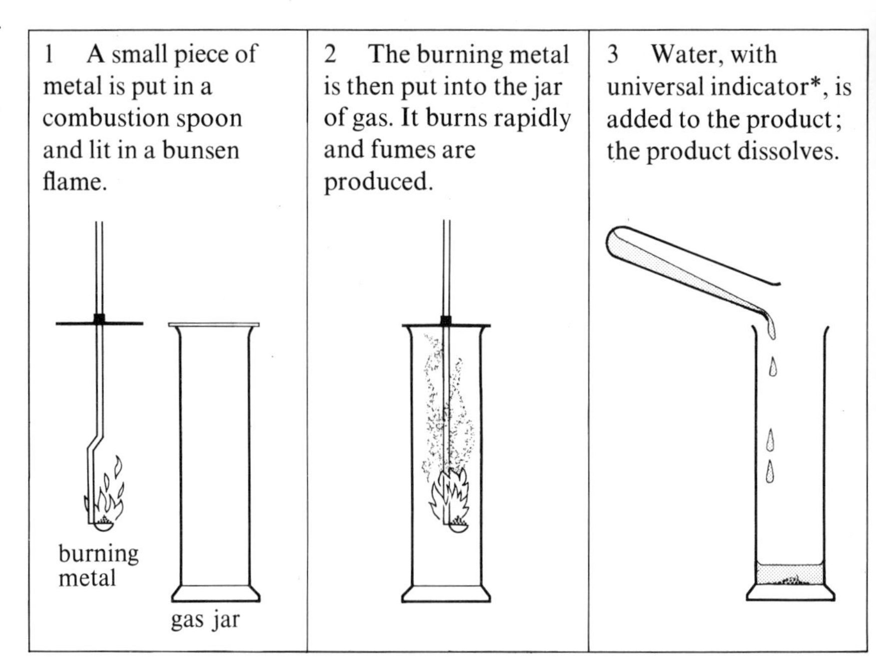

1 A small piece of metal is put in a combustion spoon and lit in a bunsen flame.
burning metal
gas jar
2 The burning metal is then put into the jar of gas. It burns rapidly and fumes are produced.
3 Water, with universal indicator*, is added to the product; the product dissolves.

* Universal indicator is a green liquid. When added to something acid (like lemon juice) it goes red. When added to something alkaline (like garden lime) it goes blue.

When chlorine or oxygen is used, similar results are obtained for all three elements. These are shown in Table 7.2. The elements all burn vigorously in the gas to produce metal chlorides or metal oxides.

Table 7.2

Element	*Reaction with oxygen*	*Reaction with chlorine*
Lithium	Burns brightly. White solid produced which dissolves in water, turning the indicator blue.	Burns steadily. Smoky white solid produced which dissolves in water, turning the indicator green.
Sodium	Burns brightly with yellow flame. White solid produced which dissolves in water, turning the indicator blue.	Burns brightly. Smoky white solid produced which dissolves in water, turning the indicator green.
Potassium	Burns brightly with lilac flame. White solid produced which dissolves in water, turning the indicator blue.	Burns very brightly. Smoky white solid produced which dissolves in water, turning the indicator green.

† The reactions on these two pages can be dangerous, and should be done only by your teacher.

(b) With water

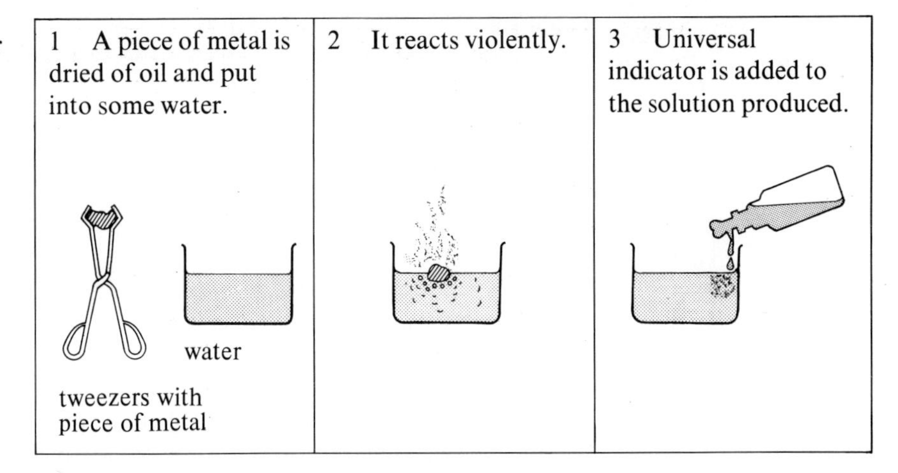

1 A piece of metal is dried of oil and put into some water.
water
tweezers with piece of metal
2 It reacts violently.
3 Universal indicator is added to the solution produced.

Similar results are obtained for each element, as shown below.

Table 7.3

Metal	*Observations when put in water*
Lithium	Metal floats. Bubbles of gas are produced around the metal, which slowly reacts. The indicator goes blue.
Sodium	Metal floats, but shoots across the surface of the water. It reacts rapidly, and bubbles of gas can be seen. The indicator goes blue.
Potassium	The floating metal reacts so violently that it catches fire and burns with a lilac flame. The indicator goes blue.

Each product makes the indicator go blue. This means that it is alkaline. That is why these elements are called **alkali metals**.

Conclusions These reactions of lithium, sodium, and potassium show that they are all metals with similar properties. *They are all in the same group: they have similar properties*. The properties change slightly down the group. Lithium is less reactive than sodium which is less reactive than potassium.

7.3 Group VII: chlorine, bromine and iodine

Physical properties These three elements are all coloured. One is a gas, one a fuming liquid and one a solid. They do not conduct electricity: they are insulators. They are all non-metals. When in the gas phase, all three have a rather similar pungent and irritating smell. This smell is the one you notice at swimming pools, where chlorine is used to kill germs in the warm water. They are all poisonous. Their physical properties are compared in Table 7.4.

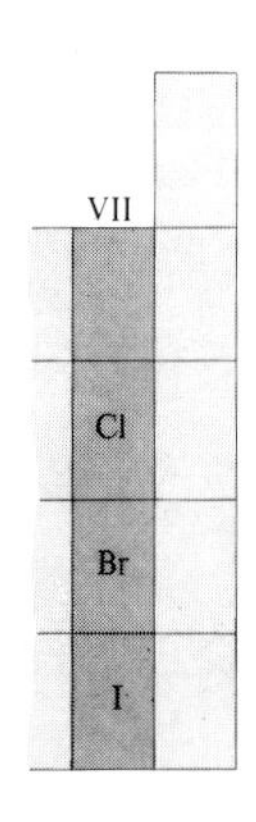

Table 7.4

Element	*Symbol*	*m.p./°C*	*b.p./°C*	*Appearance*	*Density /g cm^{-3}*
Chlorine	Cl	−101	−35	Green gas	0·003
Bromine	Br	−7	59	Red-brown liquid	3·14
Iodine	I	114	184	Grey-black solid	4·94

Reactivity

(a) With litmus or indicator paper

If a piece of damp, blue litmus paper is held in the gas phase of any of these elements, the same observations are made: the paper starts to turn pinkish-red but then rapidly goes white—the colour of the litmus is bleached.

(b) With aluminium

Element *Apparatus used for the reaction**

Chlorine

aluminium foil
white fumes
chlorine gas
heat

Bromine

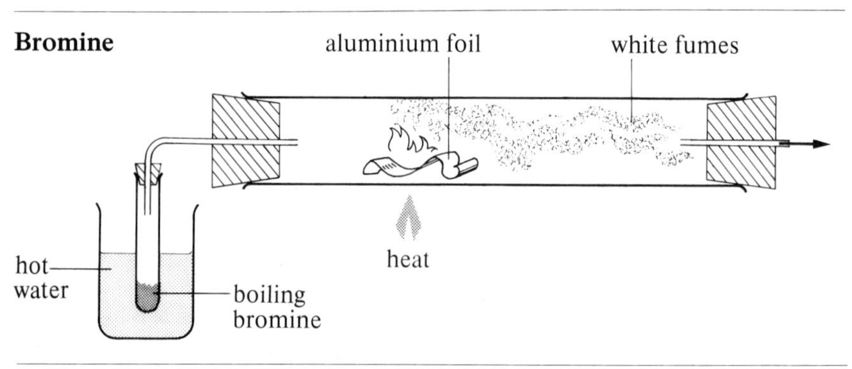

Iodine

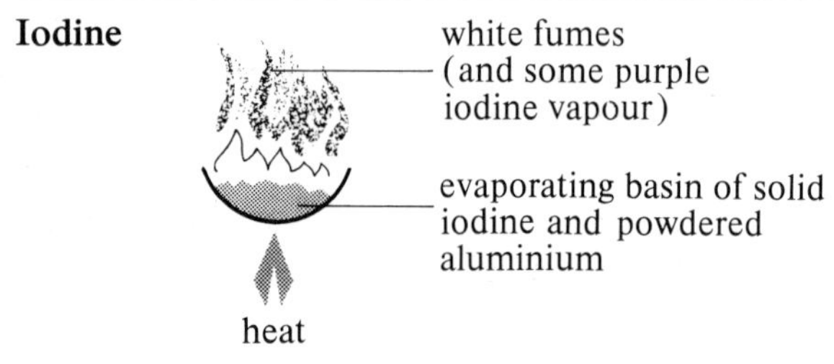

* Note: all these experiments *must* be done in a fume cupboard.

In all three of these experiments:

(i) no reaction was seen until the aluminium was heated;
(ii) once reaction had started, the aluminium went on glowing after the bunsen was turned off;
(iii) a whitish, powdery solid was produced.

The products from these exothermic reactions are aluminium chloride, aluminium bromide and aluminium iodide.

(c) With phosphorus Reactions with phosphorus are carried out as shown below. Only your teacher may do these experiments as they can be dangerous.

Chlorine gas	*Bromine liquid* (in water)	*Iodine solid*
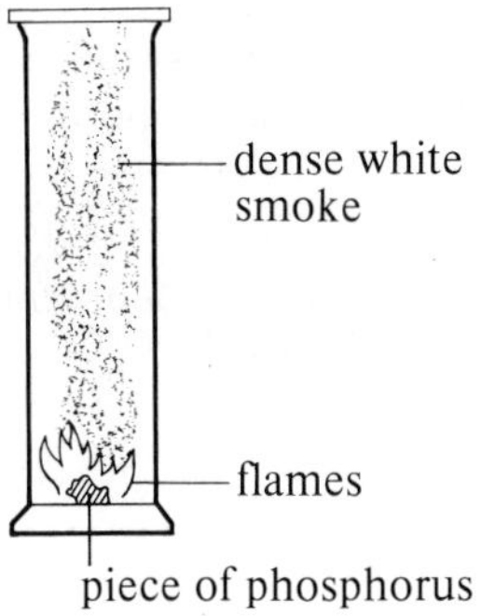 dense white smoke; flames; piece of phosphorus	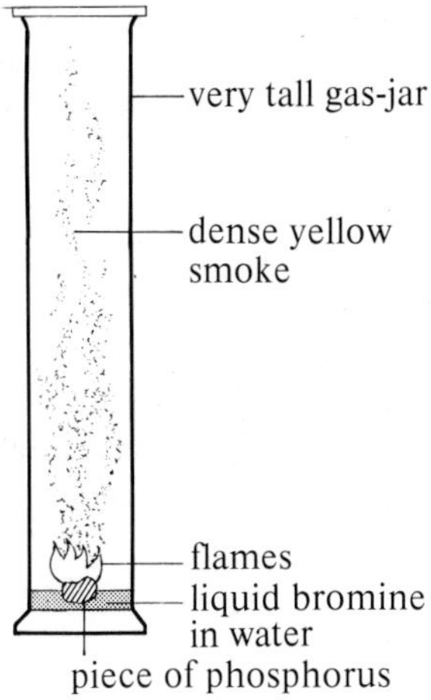 very tall gas-jar; dense yellow smoke; flames; liquid bromine in water; piece of phosphorus	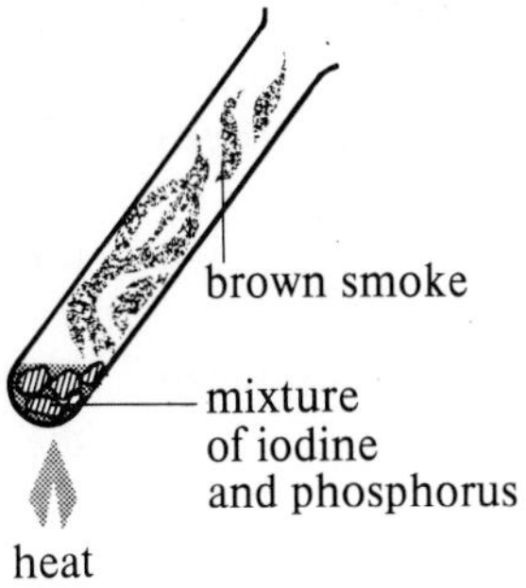 brown smoke; mixture of iodine and phosphorus; heat
A small piece of phosphorus is dropped into a gas-jar of chlorine.	A small piece of phosphorus is dropped into a tall gas-jar with a little bromine and water.	A small piece of phosphorus is mixed with iodine crystals in a test-tube.
It bursts into flames without heating.	It bursts into flames without heating.	No reaction is seen until the mixture is warmed.
White smoke is produced.	Yellow smoke is produced.	Brown smoke is produced.
Product: phosphorus chloride.	Product: phosphorus bromide.	Product: phosphorus iodide.

Conclusions These reactions of chlorine, bromine, and iodine show that they all have much the same properties. *They are all in the same group: they have similar properties.*

7.4 Group VIII: helium, neon and argon

Physical properties

These elements are all gases found in small quantities in the air around us. They have no smell or colour. They are all non-metals. Their physical properties are compared in Table 7.5 below.

Table 7.5

Element	*Symbol*	*m.p./°C*	*b.p./°C*	*State at room temperature*	*Density compared to air**
Helium	He	−272	−269	Gas	0·14
Neon	Ne	−249	−246	Gas	0·69
Argon	Ar	−189	−186	Gas	1·38

* Throughout the book the densities of all gases are compared to air (with the exception of chlorine in Table 7.4). If the number is less than 1 the gas is lighter than air; if more than 1 the gas is heavier than air.

Lack of reactivity

No compounds of these three elements have ever been found. They are completely unreactive. It was only in 1965 that an English chemist called Bartlett managed to make the last two noble gases, xenon and krypton, react for the very first time.

Conclusions

These facts show that helium, neon and argon have properties that are much the same. *They are in the same group: they have similar properties.*

Summary

All the elements can be arranged in a Periodic Table.

The columns of the table are called groups. The rows are called periods.

Elements in the same group have similar properties.

Metallic elements are found on the left of the Periodic Table, and non-metals on the right. The dividing line is the 'steps' in the diagram on page 60.

Lithium, sodium and potassium are all part of Group I, which is often called the alkali metal group.

Chlorine, bromine and iodine are all part of Group VII, which is often called the halogen group.

Helium, neon and argon are all part of Group VIII, which is often called the Noble gas group. It is also sometimes called Group O.

Questions

1 Neatly draw an outline of the Periodic Table. Mark accurately on your outline:
(a) the dividing line between metals and non-metals
(b) Group I
(c) Group VII
(d) Group VIII
(e) Period 2
(f) transition elements

2 This paragraph is about Group I elements. Rewrite it, choosing the correct word from each pair in brackets, so that it makes sense.
The Group I elements are all (metals/non-metals) and (exclude/include) the element hydrogen. They are all (hard/soft) substances which will (float/sink) when put into water. They are all very (reactive/unreactive) and have to be kept in a jar of (oil/water). They become (more/less) reactive going down the group.

3 This paragraph is about Group VII elements. Rewrite it, choosing the correct word from each pair in brackets, so that it makes sense.
The Group VII elements are called the (noble gases/halogens). They are all (metallic/non-metallic) elements and so are (conductors/insulators) of electricity. Going down the group their melting points (increase/decrease) and they get (lighter/darker) in colour. (Like/Unlike) the Group I (metals/non-metals) they are very (reactive/unreactive). They react (exothermically/endothermically) with aluminium and phosphorus.

4 (a) Describe what experiments you might carry out to compare the reactivity of some metallic elements.
(b) What do you understand by *chemical reactivity?*

5 This paragraph is about Group VIII elements. Rewrite it, putting in the correct word from each pair in brackets, so that it makes sense.
The Group VIII elements are all (gases/solids) with very (high/low) fixed points. They are all very (reactive/unreactive) and so there are (few/many) compounds formed by these elements. The elements in this group all have (similar/different) properties. The group is sometimes called Group (I/O).

6 Study the table on page 61, and then complete the sentences below, by filling in the blanks with suitable words.
Rubidium is a .... element which .... electricity. It is .... potassium in Group ... of the Periodic Table. When it reacts with water containing indicator, the indicator will turn .... and .... of gas will be seen. The reaction will be very ....

7 Study the table on page 64, and then complete the sentences below with suitable words.
Fluorine is a non-metallic .... at the .... of the .... group. Its colour is .... and at room temperature, it exists as a ....
Astatine at the bottom of the halogen .... will be .... in colour, and will exist as a .... at room temperature. When damp litmus paper is held in the vapour phase of the element it will turn .... then go ....

8 Sodium is in Group I and chlorine is in group VII. The formula of sodium chloride is NaCl. Remembering what you have learnt about the properties of elements in a group, write down the formulas of the compounds:
potassium bromide; lithium fluoride; sodium iodide; rubidium chloride.

9 You have read about three groups of elements.
(a) Compare how chemical reactivity changes as you go down each group. Try to explain the differences in group behaviour.
(b) Describe how chemical reactivity changes as you go across a period.

8 Atomic structure and the Periodic Table

8.1 Forces between atoms

Think again about the aims of a study of chemistry:

(i) to know the facts about chemical systems;
(ii) to find a theory that explains the facts.

In this chapter we will suggest answers to the questions:

Why are the elements of Group VIII so unreactive?
Why are the Group I and Group VII elements so violently reactive?
Why do the elements of a group have similar properties?

The chemical bond

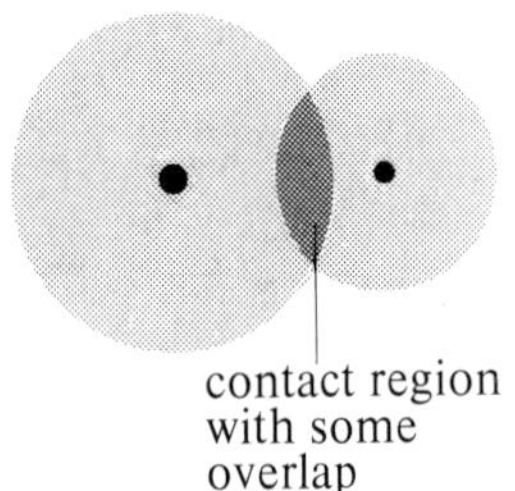

We have already looked at the forces of attraction that hold a single atom together. But we have not yet asked: *What holds two atoms together?*

If two atoms come close enough to each other, their outsides overlap. This overlap is possible because most of the outer part of the atom is empty space. You will remember this from Chapter 6. The overlap is shown in the small diagram on the left.

The overlap results in a whole series of electrostatic forces which are shown in the larger diagram below. The diagram shows the two atoms at an instant when an electron from each of them is in the region of overlap.

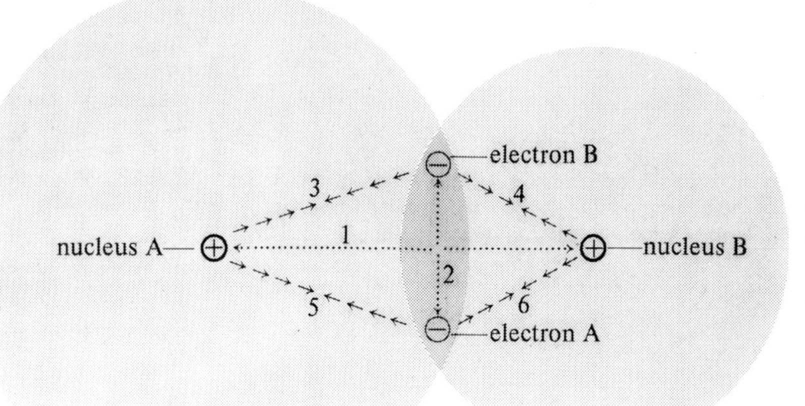

Fig. 1
The forces between two neighbouring atoms

You can see that there are forces of repulsion:

between nucleus A and nucleus B, marked 1,
between electron A and electron B, marked 2;

and also forces of attraction:

between nucleus A and electron B, marked 3,
between nucleus B and electron B, marked 4,
between nucleus A and electron A, marked 5,
between nucleus B and electron A, marked 6.

Of course the electrons A and B are moving all the time, so will not always be in the overlap area as we have shown. This does not matter. The same sort of forces will be present—look at this smaller diagram:

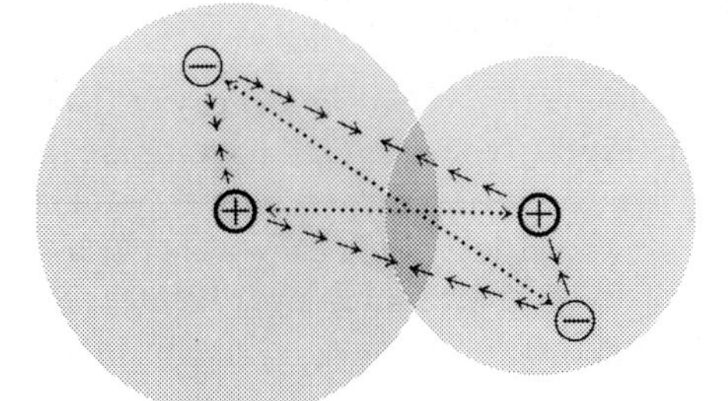

Sometimes the repulsive forces between two atoms are stronger than the attractive forces, so the atoms move apart again. Sometimes the attractive forces are stronger than the repulsive forces and the atoms stay together; we say a **chemical bond** has been formed.
A chemical bond is any force of attraction that holds two atoms together. It is always electrostatic in nature.

8.2 The atoms of the Group VIII elements

Lack of reactivity

For the chemist, the strangest fact from the last chapter is not the reactivity of the elements in Groups I and VII. Much odder is the fact that the elements of Group VIII do not react. Practically all elements form a whole range of compounds; but this group, the noble gases, form hardly any compounds. We shall start with this key fact.

Fact	*Theory*
Hydrogen gas explodes with chlorine gas to produce hydrogen chloride.	Atoms of hydrogen are able to join with those of chlorine to produce new particles—the particles of the compound hydrogen chloride.
Helium gas will not combine chemically with any element.	Atoms of helium are unable to join with any other atoms.

Helium compared with hydrogen

These results suggest that hydrogen atoms can form chemical bonds. When hydrogen atoms overlap with other atoms the attractive forces are greater than the repulsive forces, so the atoms are held together. Helium atoms do not form bonds. When helium atoms overlap with other atoms the repulsive forces are greater than the attractive forces, so the atoms move apart again. Hydrogen and helium atoms are therefore different. What makes them different? Compare their structures:

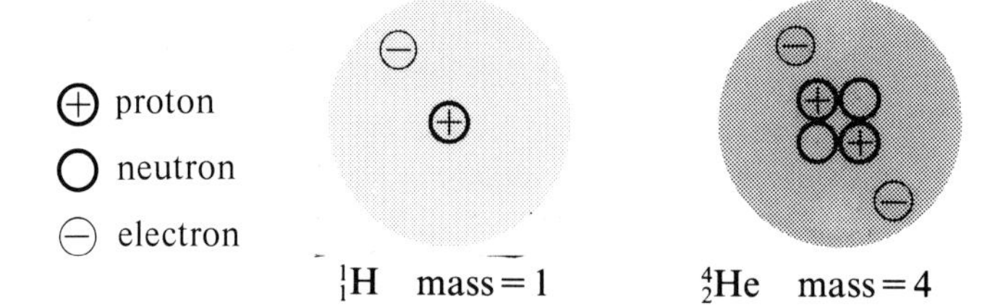

Notice that a helium atom is *four times* heavier than a hydrogen atom. But when a mass spectrometer is used to compare the masses of particles present in the two gases, there are some unexpected results. The particles in helium gas are found to be only *twice* as heavy as the particles in hydrogen gas. This must mean that hydrogen gas consists of pairs of atoms bonded together.

Particle in hydrogen gas	*Mass*	*Particle in helium gas*	*Mass*
	1+1 = 2		4

Notice that both particles now have *a pair of electrons.*

We do not find two helium atoms joined together, or more than two hydrogen atoms joined together. This leads to the conclusion: *a particle with a single pair of electrons has some special property that makes it unlikely to bond to other particles.*

Neon compared with lithium

So it is possible that helium atoms do not bond with other atoms because they have already got a single pair of electrons each. We might expect a similar lack of reactivity in atoms with two pairs of electrons. The element beryllium ($Z = 4$) has four electrons in each atom, but there are many known compounds of this element. It is not unreactive like helium. So there is nothing special about two pairs of electrons.

The next element which is unreactive like helium, is the gaseous element neon ($Z = 10$). Neon has ten electrons. There are no known compounds of neon. The table at the top of the next page compares neon ($Z = 10$) with lithium ($Z = 3$).

Facts	*Theory*		
	Behaviour of atoms	*Number of electrons*	*Conclusion*
Lithium reacts with oxygen, chlorine, and water.	Lithium atoms can form bonds with other atoms.	For Li, $Z = 3$ $\therefore$ 3 electrons $2+1 = 3$ First pair + one more	The extra electron is the one that is used for bonding.
Neon does not react at all.	Neon atoms can not form bonds with other atoms.	For Ne, $Z = 10$ $\therefore$ 10 electrons $2+8 = 10$ First pair + eight more	The extra 8 electrons, called an **octet**, have the same properties as the first pair:

We can conclude that: *a particle with one pair and one octet of electrons has some special property that makes it unlikely to bond to other particles.*

Argon compared with sodium

The pattern that we have seen with hydrogen and helium, and then with lithium and neon, is repeated with the elements sodium and argon, as shown in the table below.

Facts	*Theory*		
	Behaviour of atoms	*Number of electrons*	*Conclusion*
Sodium reacts violently with oxygen, chlorine, and water.	Sodium atoms can form bonds with other atoms.	For Na, $Z = 11$ $\therefore$ 11 electrons $2+8+1 = 11$ First pair + first octet + one more	The extra electron is the one that is used for bonding.
Argon does not react at all.	Argon atoms cannot form bonds with other atoms.	For Ar, $Z = 18$ $\therefore$ 18 electrons $2+8+8 = 18$ First pair + first octet + second octet	The second octet has the same properties as the first pair and the first octet.

These results lead us to the conclusion: *a particle with a single pair and two octets of electrons has some special property that makes it unlikely to bond to other particles.*

These conclusions led scientists directly to the idea that electrons are fitted into 'shells' within each atom. This very important theory is outlined in the next section.

8.3 Periodicity

In many sports, definite numbers are needed to make up a team. In tennis singles, each side has one player; a full football team is eleven people; a full cricket team is also eleven people. If your team is short of players you have to get someone else to join in. If your team is already complete there is no room for anyone else.

In the same sort of way the electrons in an atom are arranged in definite groups. These groups are called **shells**. This idea is called the **shell theory**.

The shell theory

The ideas in the shell theory are these:

1 Electrons are arranged in shells around the nucleus.
2 Each shell has a limit to the number of electrons that it can hold.
3 An atom with full shells cannot fit any more electrons into these shells. The atom therefore does not bond to other atoms.

Using this theory, the structures of the Group VIII elements are examined in Table 8.1.

Table 8.1 Using the shell theory

Element	*Behaviour of atoms*	*Conclusion*
Helium $^{4}_{2}He$	Helium atoms do not bond.	Helium atoms have full shells of electrons.
	Helium atoms have 2 electrons.	The first shell can hold two electrons.
Neon $^{20}_{10}Ne$	Neon atoms do not bond.	Neon atoms have full shells of electrons.
	Neon atoms have 10 electrons. $2+8 = 10$	The second shell can hold eight electrons.
Argon $^{40}_{18}Ar$	Argon atoms do not bond.	Argon atoms have full shells of electrons.
	Argon atoms have 18 electrons. $2+8+8 = 18$	The third shell can hold eight electrons.

Using these conclusions, an argon atom will look like this:

Argon: ${}^{40}_{18}Ar$

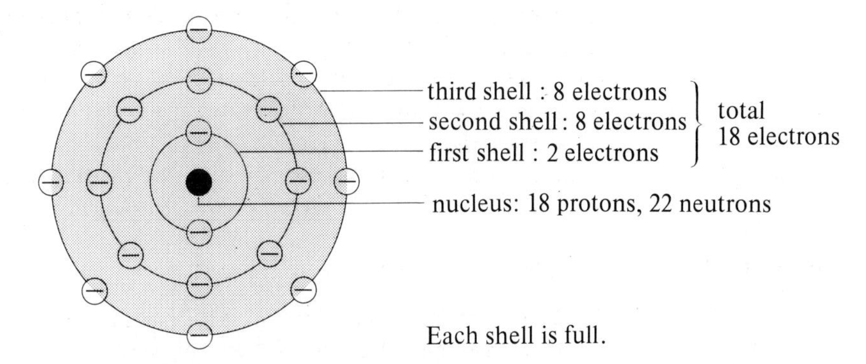

Note that the shells are numbered from the inside. The first shell is the one nearest the nucleus. The outer shell in the case of argon is the third shell.

Not all atoms have full outer shells like argon. We can draw a similar diagram for any atom. Take magnesium, for example:

Magnesium ${}^{24}_{12}Mg$ 12 electrons

A magnesium atom's 12 electrons must be fitted around the nucleus. The first two go into the first shell, which is then full. The next eight go into the second shell, which is then full. That leaves $12-(8-2) = 2$ electrons. These go into the third shell which is therefore not full. And because magnesium atoms do not have full outer shells, they are able to bond to other atoms.

Magnesium ${}^{24}_{12}Mg$

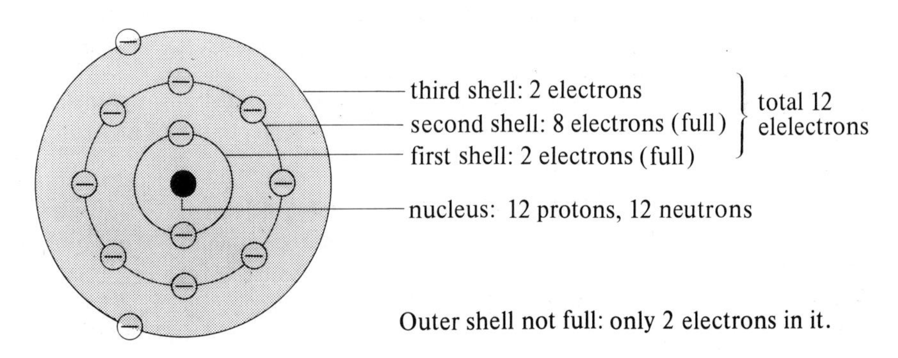

Instead of drawing a diagram like the one above, we often write the numbers of electrons in each shell after the element's symbol:

Ar 2,8,8 Mg 2,8,2

Atoms of elements in the same group

Using the shell theory, the arrangement of electrons in all atoms can be worked out. Here they are for the first twenty elements in the Periodic Table:

Group I	Group II	Transition block	Group III	Group IV	Group V	Group VI	Group VII	Group VIII
*Hydrogen **1** $^{1}_{1}H$								Helium **2** $^{4}_{2}He$
Lithium 2,**1** $^{7}_{3}Li$	Beryllium 2,**2** $^{9}_{4}Be$		Boron 2,**3** $^{11}_{5}B$	Carbon 2,**4** $^{12}_{6}C$	Nitrogen 2,**5** $^{14}_{7}N$	Oxygen 2,**6** $^{16}_{8}O$	Fluorine 2,**7** $^{19}_{9}F$	Neon 2,**8** $^{20}_{10}Ne$
Sodium 2,8,**1** $^{23}_{11}Na$	Magnesium 2,8,**2** $^{24}_{12}Mg$		Aluminium 2,8,**3** $^{27}_{13}Al$	Silicon 2,8,**4** $^{28}_{14}Si$	Phosphorus 2,8,**5** $^{31}_{15}P$	Sulphur 2,8,**6** $^{32}_{16}S$	Chlorine 2,8,**7** $^{37}_{17}Cl$	Argon 2,8,**8** $^{40}_{18}Ar$
Potassium 2,8,8,**1** $^{39}_{19}K$	Calcium 2,8,8,**2** $^{40}_{20}Ca$							

* Hydrogen has been shown here in Group I, for convenience. However you will learn later that hydrogen does not really belong to any group.

There are two important conclusions to be drawn from this table:

(i) **Atoms in the same group have the same number of outer-shell electrons.**
(ii) **The number of outer-shell electrons of an atom is equal to its group number.**

We have already seen that the elements in a group have similar properties. As the elements also have the same number of outer shell electrons, it may be these electrons that are responsible for the similar properties. We shall be returning to this idea quite often, later on.

8.4 Diagrams of atoms

To show the full structure

We now understand the full structure of an atom to be:

(i) positive nucleus, containing positive protons and neutral neutrons
(ii) negative electrons, spreading out in shells from the nucleus; maximum of 2 in first shell, 8 in second shell, 8 in third shell.

The first, second and third shells are often called the K, L and M shells, respectively. Diagrams of the full structure often show the electrons in these shells grouped in pairs. For example:

	Aluminium	
$^{27}_{13}Al$	Al	2,8,3

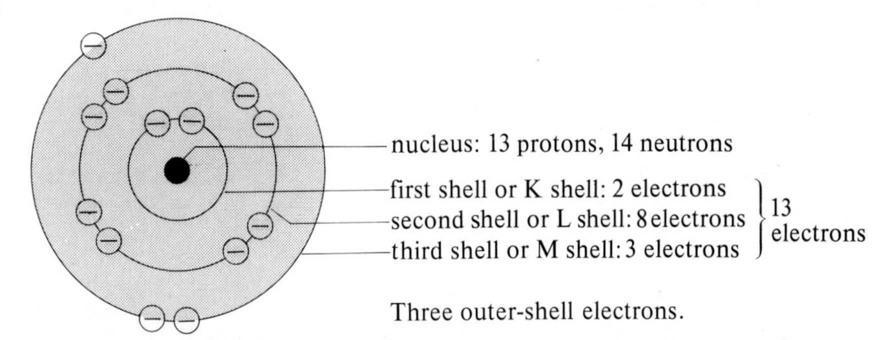

Three outer-shell electrons.

To show outer-shell structure The most important property of an atom is often the number of outer-shell electrons it has. To emphasize this, the symbol for the element can be used to stand for the nucleus and inner-shell electrons; the outer-shell electrons are then shown as dots, very small circles or crosses spaced out around the symbol:

```
 •        •         ××
Mg      • N :     ×Cl×
 •        •         ××
```

Summary

Atoms attract one another.

The attraction is an electrostatic one.

The force of attraction between atoms is called a bond.

Electrons are grouped within an atom in regions of space called shells.

Each shell has a maximum number of electrons that it can hold.
The first shell, or K shell, can hold 2 electrons.
The second shell, or L shell, can hold 8 electrons.
The third shell, or M shell, can hold 8 electrons.

A particle whose outer shell is full is unlikely to bond to any other atoms.

Atoms of elements in the same group of the Periodic Table have the same number of outer-shell electrons.

The group number of an element in the Periodic Table is the same as the number of outer-shell electrons in its atom.

Questions

1 (a) Draw three diagrams of two atoms overlapping like these:

(a)

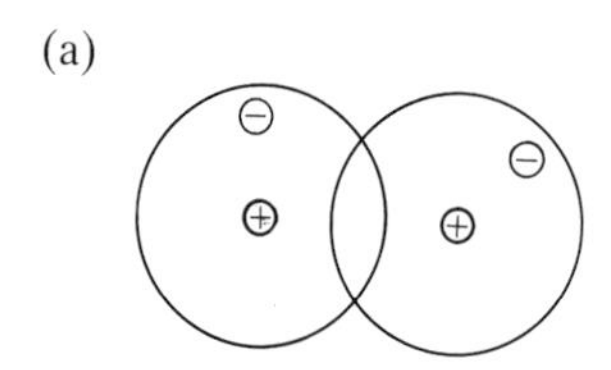

(b)

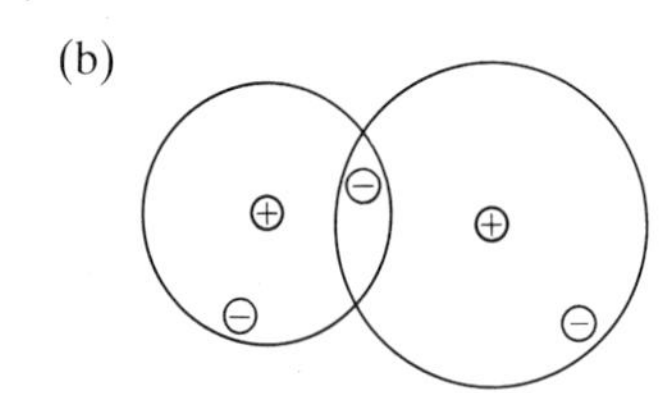

(c)

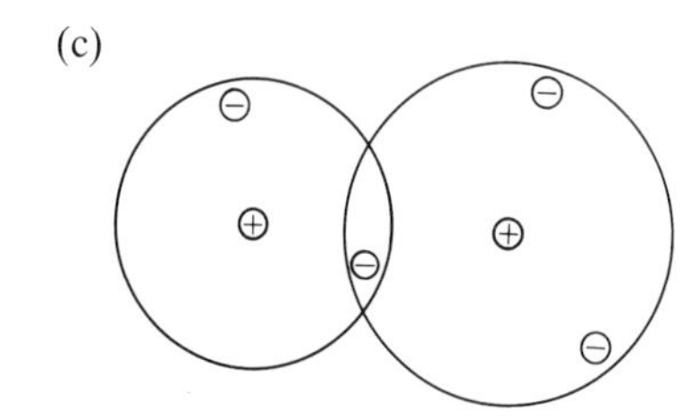

(b) Now neatly mark in all the forces of attraction using this kind of arrow →→←←, and all the forces of repulsion using this kind of arrow <······>.

2 A hydrogen atom consists of a proton and an electron. A helium atom consists of two protons, two neutrons, and two electrons. Hydrogen atoms are often joined in pairs.
(a) What similarities are there between a hydrogen atom pair and a helium atom?
(b) In what ways do they differ?

3 Remember that according to the shell theory, the maximum number of electrons in the shells are: first shell, 2; second shell, 8; third shell, 8.
(a) Draw the electronic structures of the Group I elements $_{3}Li$, $_{11}Na$, and $_{19}K$. What do they all have in common? How do they differ from each other?
(b) Draw the electronic structures of the Group VII elements $_{9}F$, and $_{17}Cl$. What do they have in common? How do they differ from each other?
(c) Draw the electronic structures of the Group VIII elements $_{2}He$, $_{10}Ne$, and $_{18}Ar$. What do they all have in common? How do they differ from each other?

4 The shell theory makes three main statements. Write them down in your own words.

5 In the Periodic Table, how does:
(a) the inner-shell structure vary down a group?
(b) the outer-shell structure vary down a group?
(c) the inner-shell structure vary across a period?
(d) the outer-shell structure vary across a period?

6 (a) Do the atoms of metal elements have nearly full or nearly empty outer-shells?
(b) Do the atoms of non-metal elements have nearly full or nearly empty outer-shells?
(c) As you go down a group, do you think the atoms of that group get bigger or smaller? Why?
(d) As you go down a group, do you think the force of attraction between the nucleus and an electron in the outer shell gets bigger or smaller? Why?

7 (a) How does the outer-shell structure of an atom relate to the number of the group that its element is found in?
(b) How does the number of shells in the structure of an atom relate to the period that its element is found in?

8 Explain the meaning of the following words.
(a) shell
(b) chemical bond
(c) electrostatic force
(d) an octet of electrons

9 (a) What does the word periodic mean?
(b) Try to explain why the table of elements is called a *Periodic* Table.
(c) What facts does the shell theory and the position of elements in the Periodic Table try to explain?

9 The properties of some common solids

Why are diamonds so hard and shiny?
Why does salt dissolve in the sea?
Why can't sand dissolve in the sea?
Why can copper carry current?
Why doesn't sand on a hot beach melt?
Why do candles melt so easily?
You should be able to answer these questions, and others like them, when you have studied this chapter and the two that follow. In this chapter we will examine some common solids and find out what properties they have in common. We will then sort them into groups.

9.1 The physical properties of some common solids

Table 9.1 Some common solids and their uses.

Solid	*Chemical name*	*Uses*
Aluminium	Aluminium	Saucepans, aircraft
Candlewax	Mixture of hydrocarbons	Wax paper, lighting
Carborundum	Silicon carbide	Grindstones
Copper	Copper	Pipes, wires
Copper sulphate	Copper(II) sulphate	Copper plating, fungicides
Diamond	Carbon	Cutting, grinding, jewels
Lead	Lead	Roofing
Moth-balls	Naphthalene	Moth-proofing
Salt	Sodium chloride	Flavouring and preserving food
Sand	Silicon dioxide	Glass making, sand paper
Sulphur	Sulphur	Making sulphuric acid
Washing soda	Sodium carbonate	Water softening

These are the solids listed in Table 9.1. Can you identify them?

Comparing their physical properties We can compare the physical properties of these solids, to see what properties they have in common, and what groups they can be divided into. In Table 9.2 their hardness and melting-points are compared, as well as the effect on them of mechanical energy—how they behave when struck with a hammer.

Table 9.2

Solid	*Appearance*	*Hardness /Moh*	*Effect of being struck with a hammer*	*m.p./°C*
Aluminium	Silvery, shiny	2–3	Bends easily but does not break	660
Candlewax	Waxy, white	0·2	Crumbles, breaks up, flattens	65
Carborundum	Black-grey	9–10	Unaffected	2600
Copper	Reddish-brown, shiny	2–3	Bends easily but does not break	1083
Copper sulphate	Blue, crystalline	3–4	Shatters into pieces	Decomposes
Diamond	Sparkling, crystalline	10	Unaffected	3500
Lead	Greyish solid	1	Bends very easily but does not break	327
Moth-balls	White, crystalline	0·5	Break up and flatten	80
Salt	White, crystalline	2–3	Shatters into pieces	801
Sand	Brownish-white grains	8–10	Unaffected	1700
Sulphur	Yellow, crystalline	1·5–2·5	Gets crushed into a fine powder	119
Washing soda	White, crystalline	1	Gets crushed into a fine powder	Decomposes

Table 9.3 Action of heat (thermal energy) on the solids

Solid	*Heating*	*Observations*
Aluminium rod	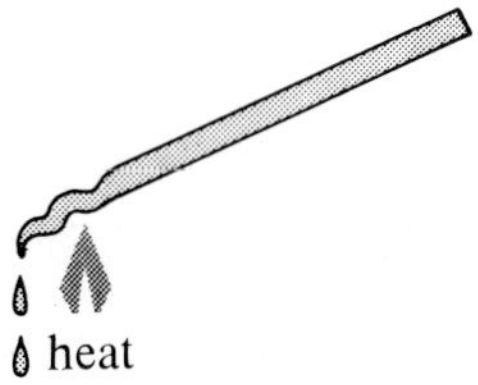 heat	The far end of the rod quickly heats up. With strong heat, the rod starts to melt.
Candle	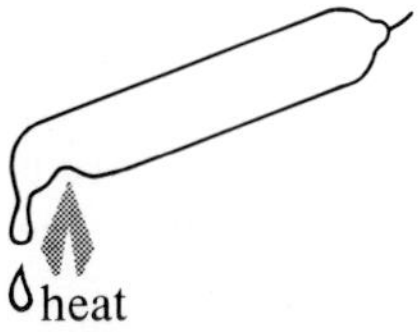 heat	Gentle heating causes the solid to melt. The far end remains cool.
Carborundum (grindstone)	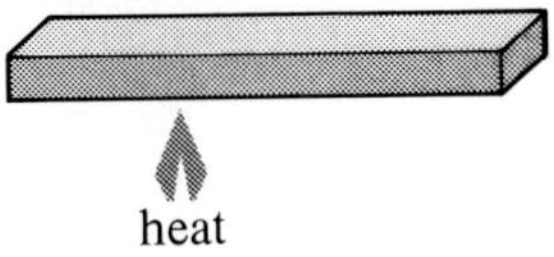 heat	No visible effects at all. The far end of the grindstone remains cool.
Copper rod	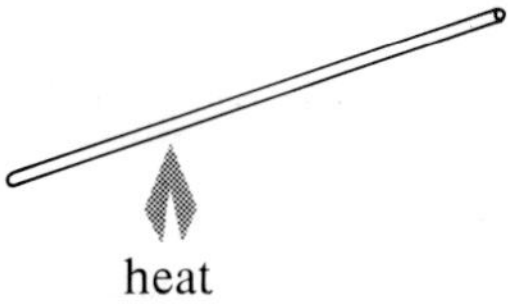 heat	The far end of the copper rod quickly heats up. It darkens and goes black.
Copper sulphate crystal	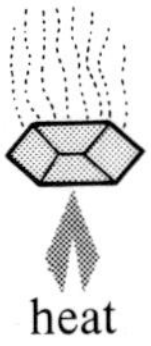 heat	The crystal crackles and steam is produced. The crystal turns white, powdery and falls apart.
Diamonds	 heat	No visible effects at all.
Lead rod	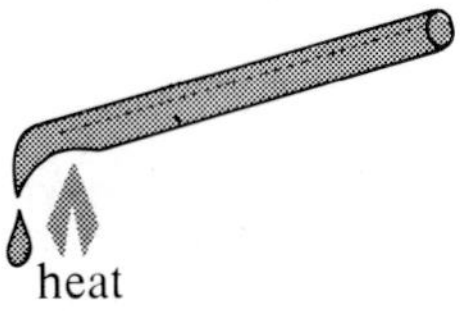 heat	The far end of the rod heats up quickly. The rod is easily melted.
Moth-balls	 heat	The solid melts easily to give a clear liquid. Stronger heating causes this liquid to boil.

Solid	*Heating*	*Observations*
Large salt crystal	heat	Strong heating makes the crystal crackle and start to melt.
Sand	heat	No visible effects at all.
Sulphur rod	heat	The sulphur rod melts easily into a dark brown liquid, then catches fire. The far end remains cool.
Washing soda	heat	The crystal crackles and steam is produced. The crystal starts to turn powdery and then falls apart.

Conclusions These facts help us to divide the solids into different groups:

(i) Solids which bend but do not break—they are said to be **malleable**—and which are shiny and conduct heat well along their lengths.
Aluminium, copper, and lead belong to this group.

(ii) Solids which are very hard and have very high m.p.'s.
Diamond, carborundum, and sand belong to this group.

(iii) Solids which are soft and have low m.p.'s.
Candlewax and naphthalene are in this group.

(iv) Solids which decompose on heating.
Copper sulphate and washing soda belong to this group.

It is rather difficult to decide into what group salt and sulphur should go. We have only looked at a few properties so far, however, and we shall find more information about these solids later.

9.2 The solubility of the solids in some solvents

Before looking at the solubility of the solids listed in Table 9.1, we must discover a little more about solvents themselves.

Types of solvent In Chapter 6 we met the idea of electrostatic charge and the forces it produces. Do you remember that a piece of plastic, like a ruler, can be charged up by rubbing it on your sleeve? It will then attract things like dust, hairs and small bits of paper. Have you ever tried to make a balloon stick to the ceiling by the same method? Rub a blown-up balloon against your clothes so that its surface becomes charged. Climb on a chair and push it against the ceiling. When you let go it will stay there.

It is not only solids that are attracted by electrostatic charge. Certain liquids can also be attracted. If a charged object is put next to a fine jet of solvent flowing out of a burette, the solvent may be attracted towards the charged object:

Solvents that are attracted

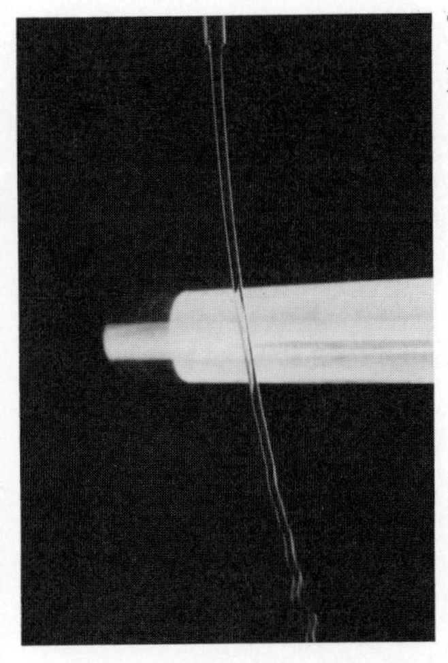

These are called **polar solvents**.

Examples:
water
alcohol (ethanol)
acetone (propanone)

Solvents that are not attracted

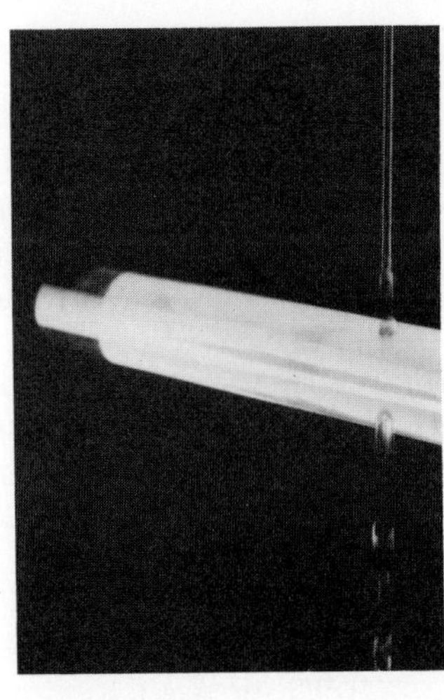

These are called **non-polar solvents**.

Examples:
petrol
xylene
tetrachloromethane

Solubility of the solids

Now suppose we try dissolving the solids in our list in two different solvents, water and xylene. Water is a polar solvent, and xylene a non-polar solvent. We get the results shown below.

Table 9.4 Solubility of the solids

Solid	*Solubility in polar solvent (water)*	*Solubility in non-polar solvent (xylene)*
Aluminium	Insoluble	Insoluble
Candlewax	Insoluble	Soluble
Carborundum	Insoluble	Insoluble
Copper	Insoluble	Insoluble
Copper sulphate	Soluble	Insoluble
Diamond	Insoluble	Insoluble
Lead	Insoluble	Insoluble
Moth-balls	Insoluble	Soluble
Salt	Soluble	Insoluble
Sand	Insoluble	Insoluble
Sulphur	Insoluble	Soluble
Washing soda	Soluble	Insoluble

Conclusions These results help us divide our solids into three types:

(i) Solids which dissolve in a polar solvent (water).
These are copper sulphate, salt and washing soda.
Solutions in water are often called **aqueous solutions**.
(In Latin, *aqua* is water.)

(ii) Solids which dissolve in a non-polar solvent (xylene).
These are candlewax, moth-balls, and sulphur.

(iii) Solids which do not dissolve in either solvent.
These are aluminium, carborundum, copper, diamond, lead, and sand.

9.3 The action of electricity on the solids

On page 78, we described the effects of *mechanical energy* on the solids. On page 79 we described the effects of *thermal energy* on them. We must now look at the effect of *electrical energy* on them. We will try to pass a current from a battery through each solid. For a solid that dissolves in either solvent, we will also try to pass a current through the solution.

The electrical circuits We can try to pass a current through each solid using circuit 1 below, and through each solution using circuit 2 below. The current will only flow around the circuit if it can pass through the solid or solution. If it can, the light bulb will light up.

Circuit 1

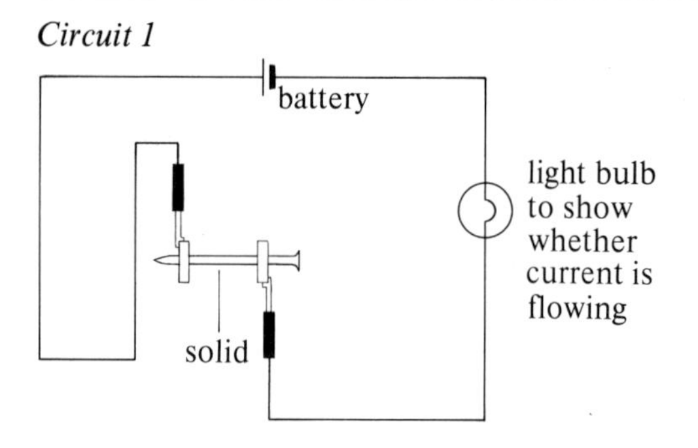

Each solid in turn is connected into the circuit.

Circuit 2

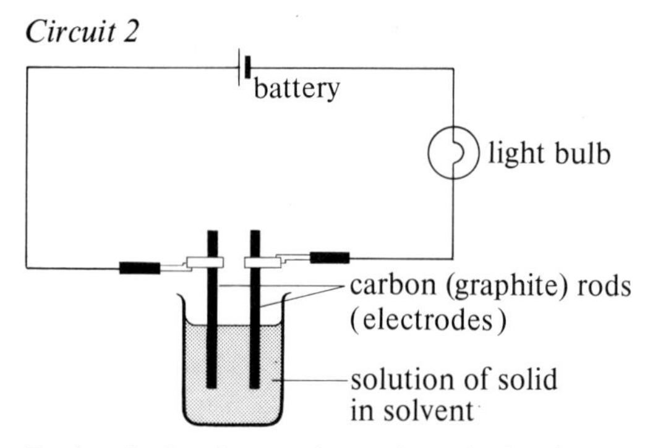

Each solution in turn is put into the beaker.

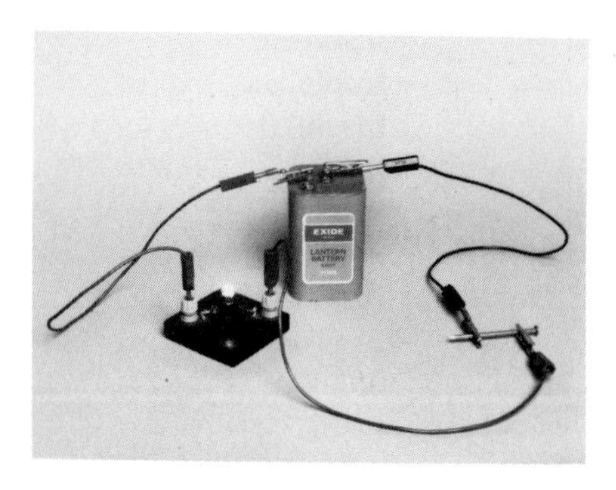

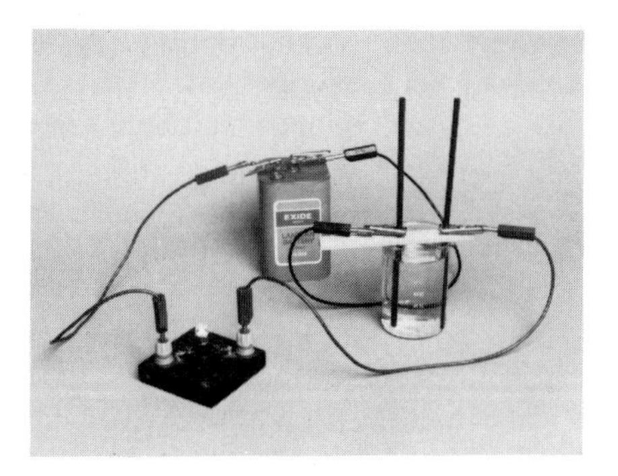

The electrical properties of the solids and their solutions

The results of these experiments are shown in the table below.

Table 9.5 The effect of electricity on the solids and their solutions

Solid	*Using the solid and circuit 1*	*Using the solution and circuit 2*
Aluminium	Current flows.	—
Candlewax	No current flows.	No current flows.
Carborundum	No current flows.	—
Copper	Current flows.	—
Copper sulphate	No current flows.	The aqueous solution allows current to flow. Bubbles appear on one rod.
Diamond	No current flows.	—
Lead	Current flows.	—
Moth-balls	No current flows.	No current flows.
Salt	No current flows.	The aqueous solution allows current to flow. Bubbles appear on rods.
Sand	No current flows.	—
Sulphur	No current flows.	No current flows.
Washing soda	No current flows.	The aqueous solution allows current to flow. Bubbles appear on rods.

Conclusions

These results help us divide the substances in the list into three types:

(i) Solids which allow a flow of electricity.
These are aluminium, copper, and lead.
Note that all three solids are insoluble in both water and xylene.

(ii) Solids whose solutions in water allow a flow of electricity.
These are copper sulphate, salt, and washing soda.
Note that the solids themselves do not let electricity flow.

(iii) Substances which do not allow a flow of electricity, either as solids or in solution (if they dissolve).
These are candlewax, carborundum, diamond, moth-balls, sand, and sulphur.

If a current flows through it, a substance is a **conductor**. So the substances in (i) above are conductors. The substances in (ii) are conductors when in solution; these substances are called **electrolytes**.

If no current flows, the substance is an **insulator**. So the solids in (iii) are insulators.

9.4 Classifying the solids

If you compare the three sets of conclusions in this chapter, you will find that some patterns emerge. For example aluminium, copper, and lead are all ductile, shiny, and conduct heat well; they are all insoluble in water; they all conduct electricity. Using these patterns, we can divide our solids into four classes:

	Examples	*Properties*
Molecular solids	Candlewax Moth-balls (naphthalene) Sulphur	They are not very hard. They do not conduct heat or electricity. They have low m.p. values. They tend to dissolve in non-polar rather than polar solvents.

Macromolecular solids	Carborundum Diamond Sand	They are very hard. They do not conduct heat or electricity. They have very high m.p. values. They do not dissolve in solvents.

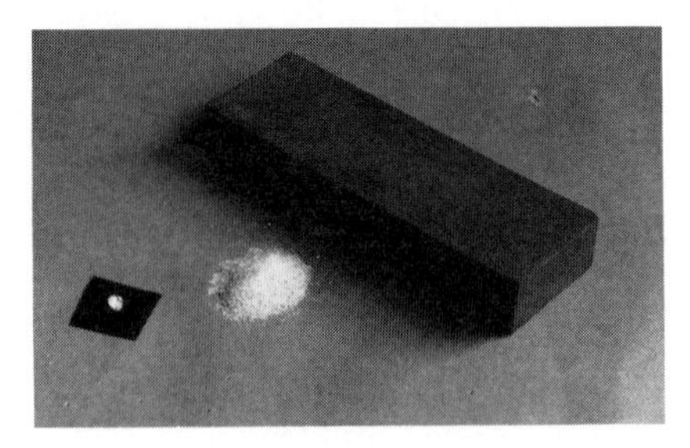

Metallic solids	Aluminium Copper Lead	They are quite hard. They bend but do not break when put under stress. They conduct heat and electricity. They have a range of m.p. values. They do not dissolve in solvents.

Ionic solids

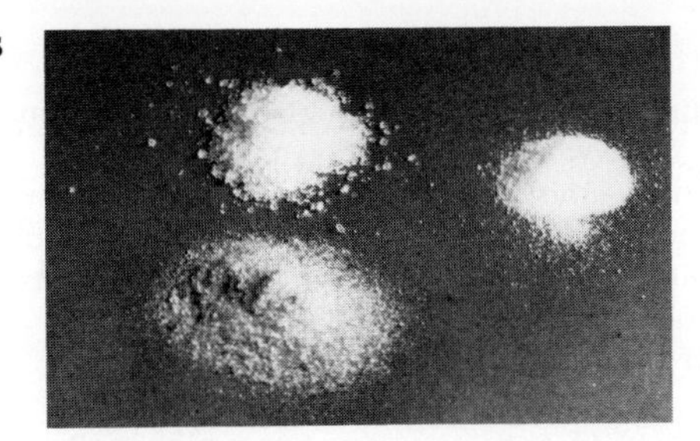

Copper sulphate
Salt (sodium chloride)
Washing soda (sodium carbonate)

They are quite hard but brittle (they shatter when hit).
They do not conduct heat or electricity.
They have fairly high m.p. values, or decompose when heated.
They dissolve in polar solvents giving solutions that conduct electricity—they are electrolytes.

Summary

It is possible to classify solids into different types. This can be done by testing their physical properties and solubilities.

We test their physical properties by looking at the effects on them of mechanical energy, thermal energy and electrical energy.

We test solubilities by trying to dissolve them in polar and non-polar solvents.

We find solids fall into four classes:

(i) molecular solids
(ii) macromolecular solids
(iii) metallic solids
(iv) ionic solids.

You should know the properties of each class of solid.

Questions

1 In Table 9.1 on page 77, the third column lists the uses of the solids. Make another table with three columns headed *Solid, Use* and *Reason for use*. Fill in the first column with the same list of solids as in Table 9.1, and copy down their uses in the *Use* column. In the last column try to explain why the solid is chosen for those uses.

2 A solid may be hard and bendable, it may be hard but easily bent (malleable), or it may be hard and easily shattered (brittle).
(a) Look at the four classes in our final list on pages 84 and 85. Choose one of them to fit each of the three descriptions.
(b) Match each of the following to one of the three descriptions above:
granite glass wire brick brass concrete

3 Here are two lists, one of polar solvents and one of non-polar solvents.

Polar solvents	*Non-polar*
Water	Oil
Alcohol (meths)	Paraffin
Propanone (acetone)	White spirit (turps)

Answer the following questions, from your everyday experience.
(a) Will polar solvents mix and dissolve in each other?
(b) Will non-polar solvents mix and dissolve in each other?
(c) Will polar and non-polar solvents mix and dissolve in each other?

4 The hardness of steel on the Moh scale is about 6.
(a) Which substances in Table 9.2 on page 78 could be used to scratch steel? What are the main uses of these substances?
(b) Which metals in the table could be marked by steel, because it is harder than they are? What are most metal-working tools made of?

5 On page 85 in the summary, solids fall into four classes.
Which type of substance:
(a) conducts in the solid state?
(b) insulates in the solid state, but conducts when melted?
(c) melts easily?
(d) is usually soft?
(e) has a high melting point and dissolves in water?
(f) is bendy?
(g) tends to dissolve in non-polar solvents?
(h) is an electrolyte?

6 (a) Many large shops have a fire-control system in the ceilings. These consist of water jets, with plugs of special plastic sealing their ends. If the temperature in the shop rises above a certain point, these plastic plugs melt, setting off the sprinkler system. Which class of solids does this plastic belong to? Explain.
(b) The stylus tips in record players are often made of artificial sapphire. It is used because it is very hard, and keeps it shape. Which class of solids does artificial sapphire belong to? Explain.
(c) Tubes for glue and toothpaste are made of a special alloy that can bend easily without breaking. Which class of solids does this alloy belong to? Explain

7 Explain why each of the following substances is chosen for the use described.
(a) Glass or china is used to make insulators on electricity pylons.
(b) Wire is used to make paper clips and safety pins.
(c) Plastic is used to make saucepan handles.
(d) Copper is used to make central-heating pipes.
(e) Sodium carbonate is used to make bath-salts.
(f) Sand is used to clean badly encrusted objects by sand-blasting.
(g) Beeswax is dissolved in methylated spirits to make furniture polish.

10 Bonding between atoms

Remember again the two aims of the chemist:

(i) to know the facts about chemical systems;
(ii) to find a theory that explains the facts.

In earlier chapters, we found theories that explained:

why solids are hard, and melt to give liquids (Chapter 3);
why solids sometimes decompose (Chapter 5);
how atoms attract one another (Chapter 8).

Now we have to explain why the four classes of solids from page 84—molecular solids, macromolecular solids, metallic solids, and ionic solids—have such different properties. We can explain some of the differences easily, for example:

Facts	*Theory*
Sulphur: m.p. 119°C Diamond: m.p. 3500°C Solid sulphur melts at a much lower temperature than solid carbon (diamond).	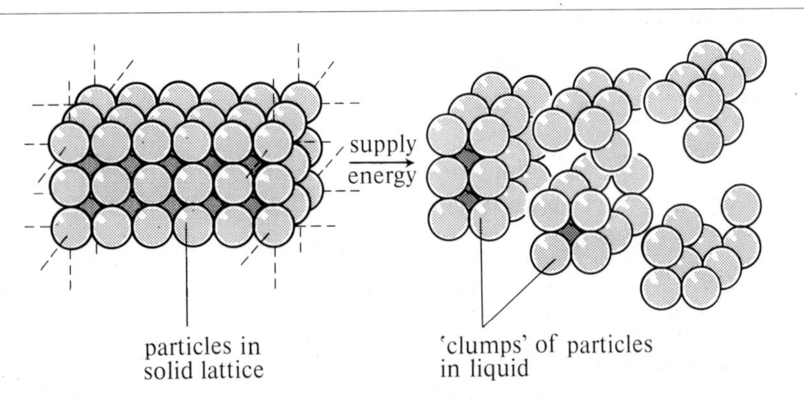 Far less energy is needed to break up a lattice of sulphur particles than the lattice of carbon particles in diamond.

The theory above certainly explains the facts, but it does not go far enough. *Why* are the forces of attraction between particles in a sulphur lattice so much weaker than those between particles in a carbon lattice? That is the sort of question we must answer in this chapter. To do so we will be using the shell theory (page 72). Do you remember it?

10.1 Molecular solids

Iodine: a molecular solid

The first class of solids in our list in Chapter 9 (page 84) was the molecular solids. To try to understand their properties we will take a simple example: the element iodine. You may have already met iodine as a brown liquid to put on cuts. This brown liquid is in fact a solution of iodine in alcohol—the High Street chemist calls it tincture of iodine. But we want to look at the pure element. Since it is an element there is only one type of atom present; that makes our task simpler.

Iodine is a halogen from Group VII. It has all the typical properties of a molecular solid:

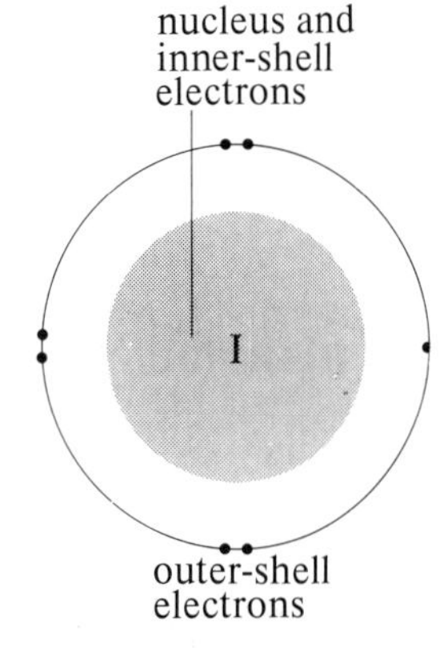

it is soft and easily broken
it has low fixed points (m.p. 114°C and b.p. 185°C)
it is a very poor conductor of heat and electricity
it dissolves well in non-polar solvents such as xylene and tetrachloromethane.

Since iodine is in Group VII, the atom must have seven outer-shell electrons. The outer shell is not full. We would expect one iodine atom to attract another strongly because the outer shell is not full. And yet the properties of iodine suggest that the forces between the particles in its solid lattice are weak:

Fact	*Theory*
Iodine has a low m.p. and is soft and flaky.	The forces of attraction between the particles of the solid are easily overcome.

Both these ideas can be made to agree. Consider the effect of two atoms coming together and attracting one another:

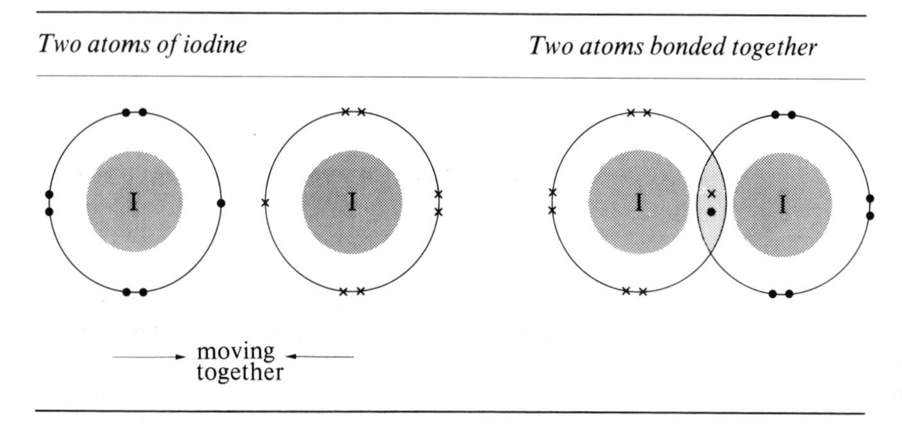

The outer shells overlap and a pair of electrons is shared. By sharing a pair of electrons, each atom now has a full outer shell.

Covalent bonding

As a result of their outer shells overlapping, a pair of electrons is shared by the two atoms. This sharing of a pair of electrons can result in an overall force of attraction between the two atoms. This force of attraction is called a **covalent bond**.

A covalent bond is the force of attraction between two atoms as a result of their sharing a pair of electrons.

So there is a covalent bond between the pair of iodine atoms. Notice that, as the result of sharing a pair of electrons, each atom has a full shell like a Group VIII atom. Each started with seven electrons of its own, and now has a share in an eighth electron. The pair of iodine atoms bonded together is an example of a new type of particle that we have not named before. It is a **molecule**.

Molecules

A molecule is an uncharged particle containing atoms that have full shells of electrons. It is produced when atoms bond covalently together.

Usually, molecules are drawn in a simpler way than the iodine molecule above. Only the outer-shell structure of the bonding atoms is shown:

two iodine atoms

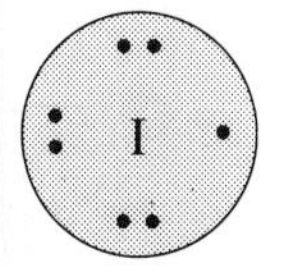

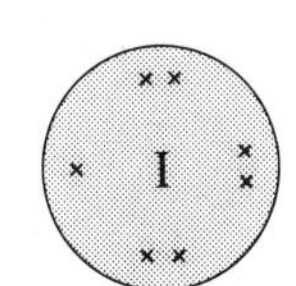

the iodine molecule

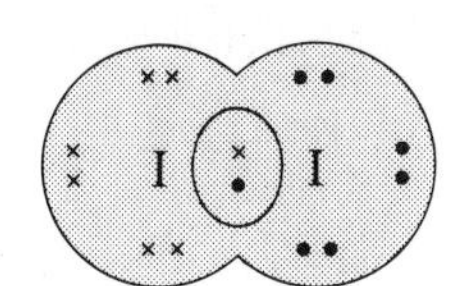

Sometimes, the covalent bond is simply drawn as a single line between the atoms. The line represents the bonding electrons; the other non-bonding electrons are left out, so that the molecule of iodine is drawn like this: I – I.

Properties of molecular solids

Facts	*Theory*
Molecular solids are not very hard or strong. They have low m.p. and b.p. values. They dissolve readily in many non-polar solvents.	The particles present in the lattices of molecular solids do not attract one another strongly—they very probably have full outer shells of electrons.
They do not conduct electricity.	The particles are uncharged.
	The particles are called molecules. Molecules contain atoms covalently bonded together.

10.2 Macromolecular solids

Carbon (diamond): a macromolecular solid

Just as iodine was chosen as an example of a molecular solid, we shall take diamond as our example of a macromolecular solid. This is because diamond is an element—it is one form of the element carbon—so there is only one type of atom to think about. Diamond has all the properties typical of a macromolecular solid:

it is very hard indeed
it has very high fixed points (m.p. 3500°C and b.p. 4800°C)
it is a very poor conductor of electricity and heat
it does not dissolve in any solvents.

When we apply the same bonding ideas to carbon atoms as we did to iodine atoms, a striking difference can be seen:

Element	*Separate atoms*		*Bonded atoms*
Iodine	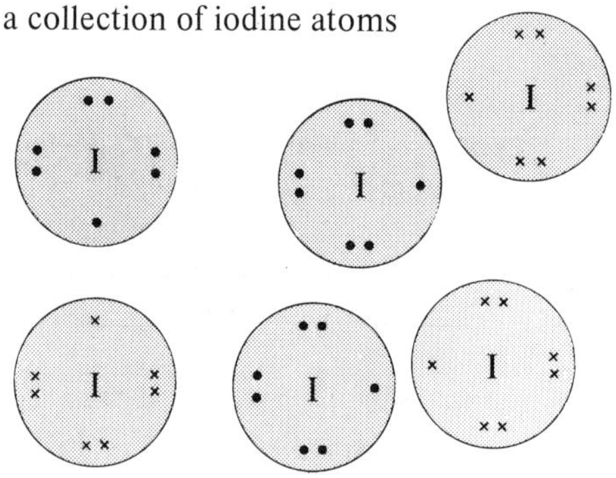 a collection of iodine atoms	→	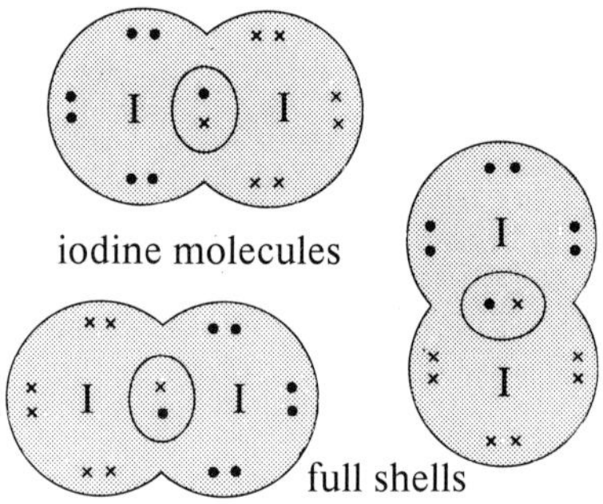 iodine molecules full shells
Carbon	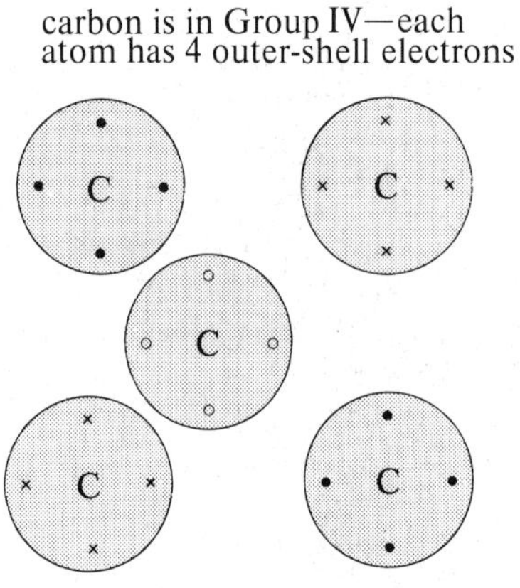 carbon is in Group IV—each atom has 4 outer-shell electrons	→	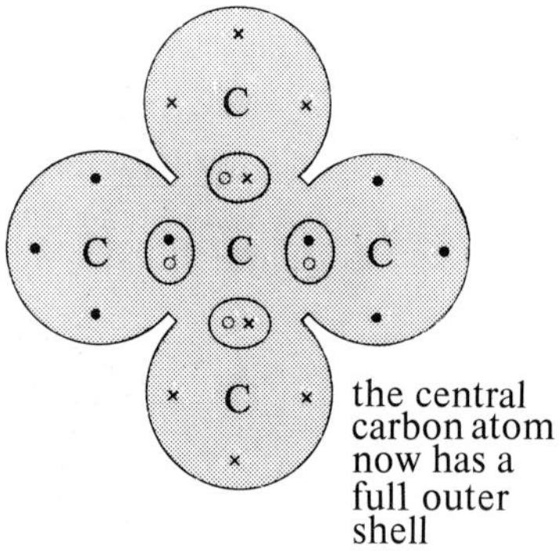 the central carbon atom now has a full outer shell

Look back at the atoms in the iodine molecules: each has a full shell. Now look at the five bonded carbon atoms: only the centre one has a full shell. It has managed this by forming covalent bonds with four other atoms. To reach a full shell, each of the other carbon atoms must also form covalent bonds with four more. When every atom does this, a giant network of carbon atoms is produced:

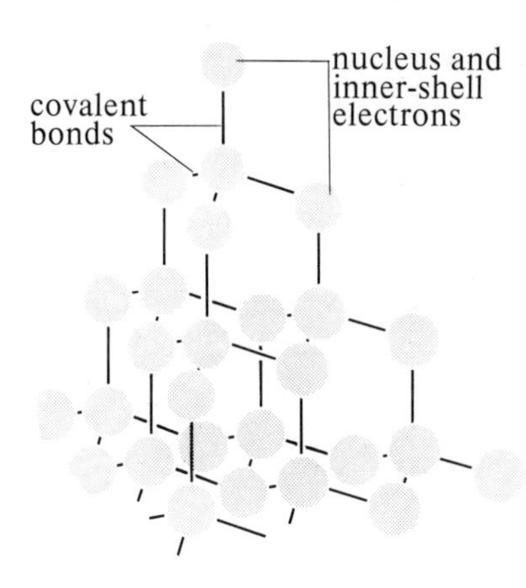

Fig. 1
The diamond lattice

Remember that every atom bonds to four other atoms. The result is an incredibly large molecule. The number of atoms in it is so large that the molecule is called a **macromolecule**. (In Greek, *makros* means great.)

Macromolecules and covalency

In the molecular iodine lattice, each iodine atom forms one covalent bond. In the macromolecular carbon lattice, each carbon atom forms four covalent bonds.

The number of covalent bonds formed by an atom is called its covalency. This word is often shortened to valency.

So iodine has a valency of one while carbon has a valency of four.

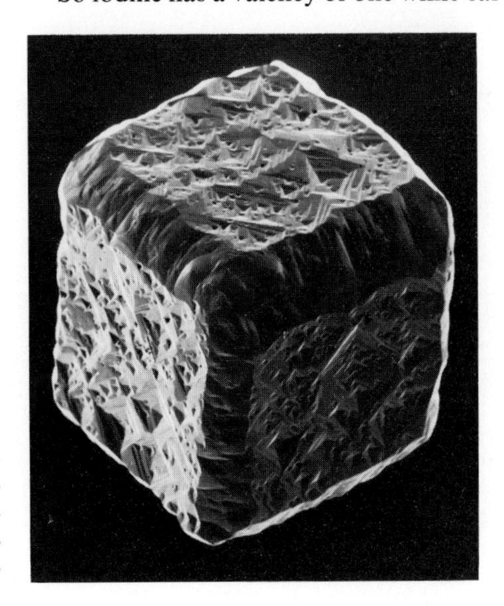

A macromolecule: a natural diamond that has not yet been cut or polished. Here it is magnified × 40

Properties of macromolecular solids

All the properties of macromolecular solids suggest that there are very strong forces between the particles in their lattices. We think that these forces are covalent bonds between atoms. A useful comparison can be made between the properties of carbon and iodine:

Facts	*Theory*	
	Behaviour	*Particle structure*
Iodine solid is soft and flaky with a low m.p. and b.p. 	An iodine molecule itself is held together quite strongly. The forces of attraction between molecules is weak, however—they all have full shells—so the lattice of molecules is easily broken up.	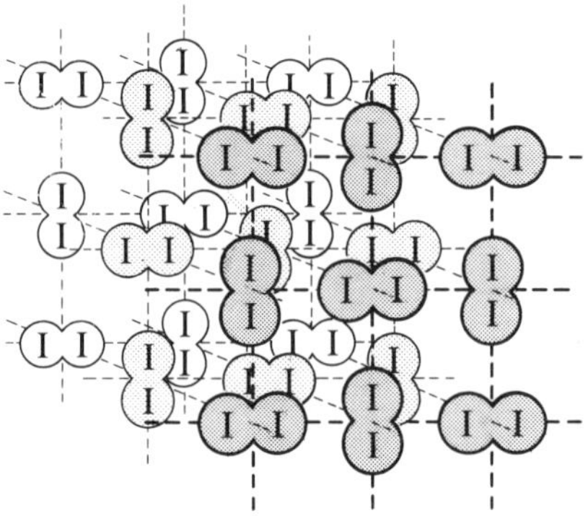 lattice of iodine molecules
Diamonds are very hard and have very high m.p. and b.p. values. 	The whole lattice is held together by covalent bonds. These are strong forces of attraction, so a considerable amount of energy is needed to break up the lattice.	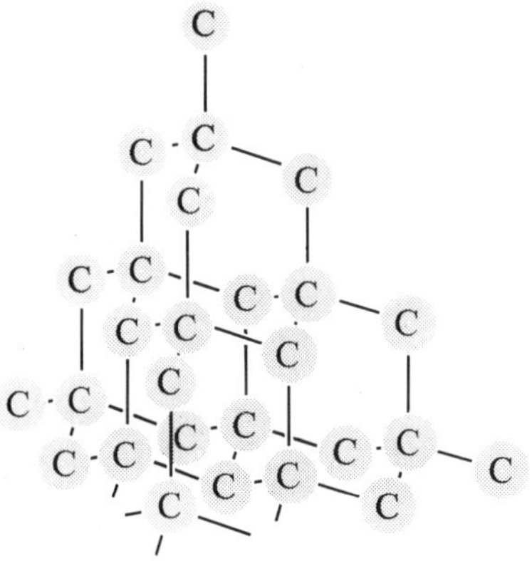

10.3 Metallic solids

Sodium: a metallic solid

We have chosen sodium as our example because it is a metal that we have already discussed. It shows the typical properties of a metal:

it is easily bent and shaped
it conducts heat and electricity very well
it has a shiny surface (when freshly scraped, however, it tarnishes very quickly).

Sodium is a Group I element, which means it has only one electron in its outer (third) shell. When we try to apply the same bonding ideas to sodium atoms that we did to iodine and carbon atoms, it is difficult to see how full shells can be produced:

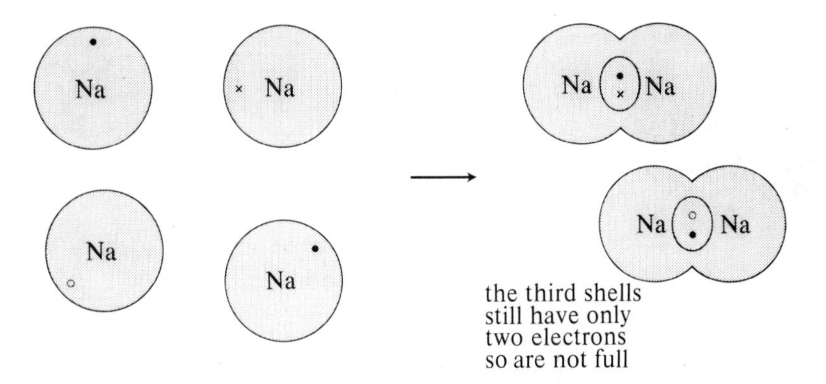

As shown above, a sodium atom cannot reach a full outer shell by sharing its one outer-shell electron. The only way a full shell can be produced is by *losing* the outer-shell electron. When this electron is removed, the complete shell underneath becomes the outer shell:

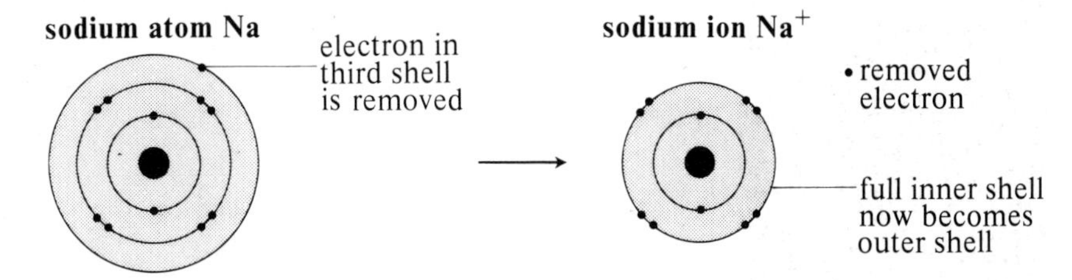

11 protons, 11 electrons ∴ neutral   11 protons, 10 electrons ∴ ⊕ charged

Metallic bonding

If all the sodium atoms lose their third-shell electrons, as shown in the diagram, the result is a large number of free electrons in the lattice.

Notice that the particle left behind after the electron loss now has one more proton than electron (11 +, 10 −). It therefore has a charge of 1 +. This charged particle is called an **ion**.

An ion is a particle which is charged because the number of protons in it does not equal the number of electrons in it.

The lattice that results from this loss of electrons is shown on the next page:

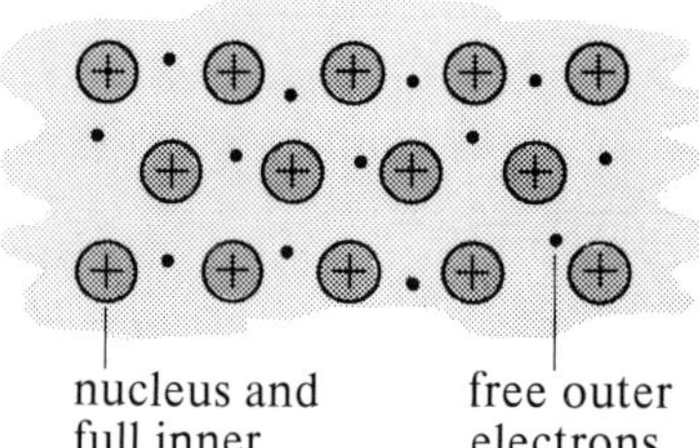

Fig. 2
A metal lattice

The positive ions in the lattice repel each other (like charges repel). Yet they are held together by their attraction to the free electrons between them—the attractive forces outweigh the repulsive forces.

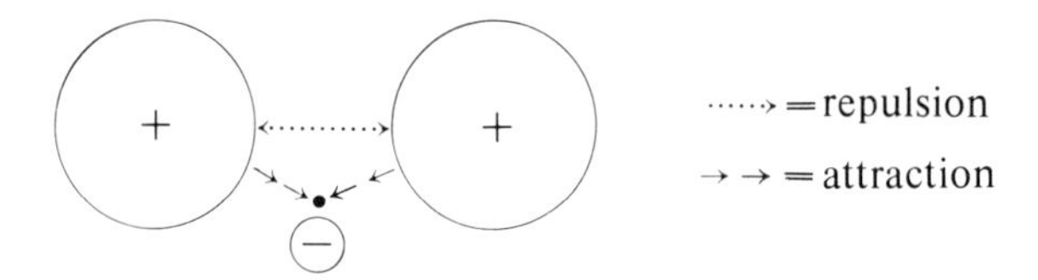

The electrons are free to move anywhere in the lattice. They are said to be **delocalised** (not in one place).

The bonding which results from the attraction of positive ions to delocalised electrons in a lattice is called **metallic bonding**.

The metallic bond is the force of attraction between two positive metal ions and the delocalised electrons between them.

Metal lattices

All metal lattices have the same type of structure as the one in Fig. 2. They all consist of positive ions ⊕ and delocalised electrons ⊖. But there are some differences: some metal atoms must lose more than one electron each, to give ions with full outer shells. Take aluminium, for example, Al(2,8,3). It has three outer-shell electrons:

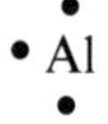

Each aluminium atom must lose three electrons to obtain a full outer shell. The remaining ion therefore has a charge of 3+.

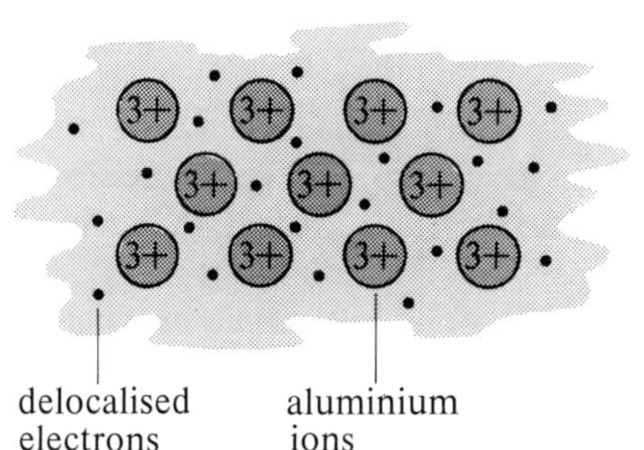

Fig. 3
The aluminium lattice

Properties of metallic solids Using these ideas about metal lattices, we can explain the properties of metallic solids.

Facts	*Theory*
Metals are malleable and ductile: they can be bent, shaped and drawn into wires.	The layers in the lattice can slide past each other without breaking the metallic bond. 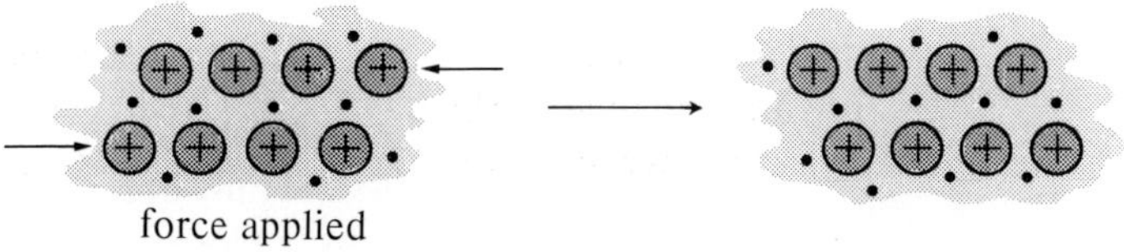
Metals conduct electricity well.	Electric current is a flow of electrons. 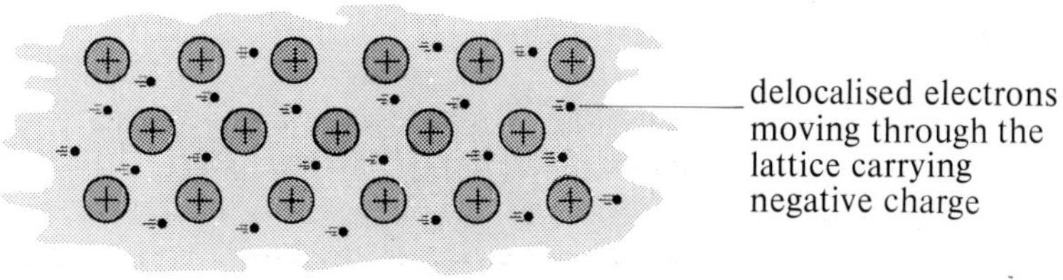
Metals conduct heat well.	Heat travels through the lattice as a result of the rapid random movement of electrons. 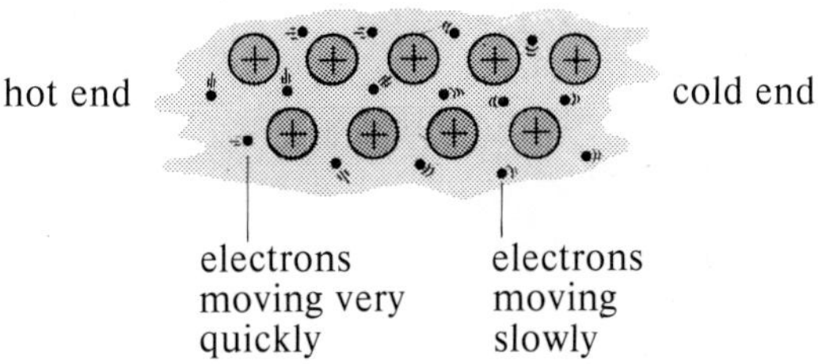

10.4 Ionic solids

Salt: an ionic solid

There are no elements that have the properties of an ionic solid: all ionic solids are compounds. The most common of these compounds is ordinary salt, sodium chloride. It has all the typical properties:

- it is hard but shatters when it is hit
- it has high fixed points (m.p. 801°C and b.p. 1465°C)
- it does not conduct heat or electricity well
- it dissolves in a polar solvent, water, to give a solution that conducts electricity.

Sodium is in Group I; it has one outer-shell electron. Chlorine is in Group VII; it has seven outer-shell electrons. When sodium and chlorine atoms bond together there is only one way in which all the atoms can reach full shells. A sodium atom can lose its outer-shell electron, leaving a full shell:

$Na \rightarrow Na^{+} + e^{-}$

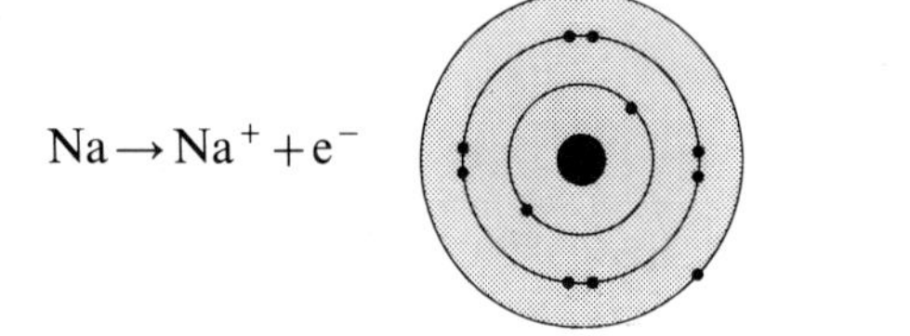

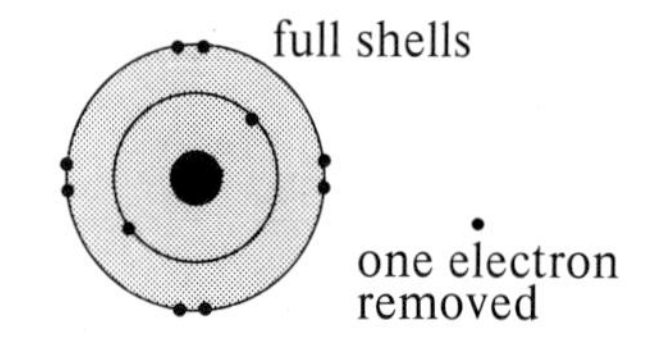

and a chlorine atom can accept this electron, to gain a full outer shell:

$Cl + e^{-} \rightarrow Cl^{-}$

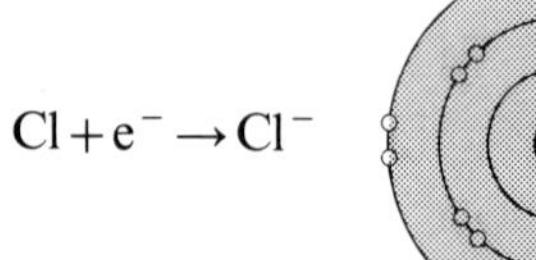

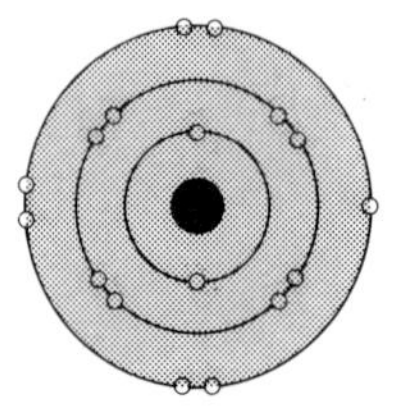

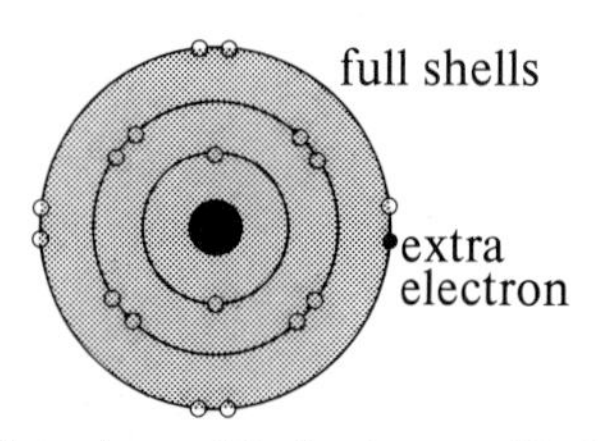

Ionic bonding

We can describe the whole process above as a **transfer** of electrons.

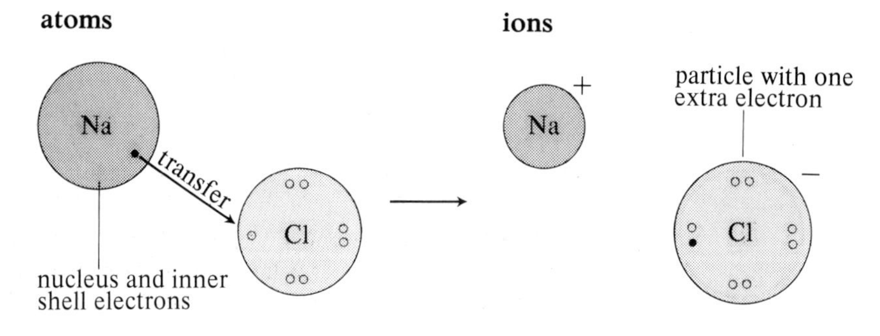

The sodium atoms each lose an electron to become positive ions, called **cations**. The chlorine atoms each gain an electron to become negative ions, called **anions**. The oppositely-charged ions attract one another electrostatically in the ionic lattice. In fact each ion attracts opposite ions all around it, like this:

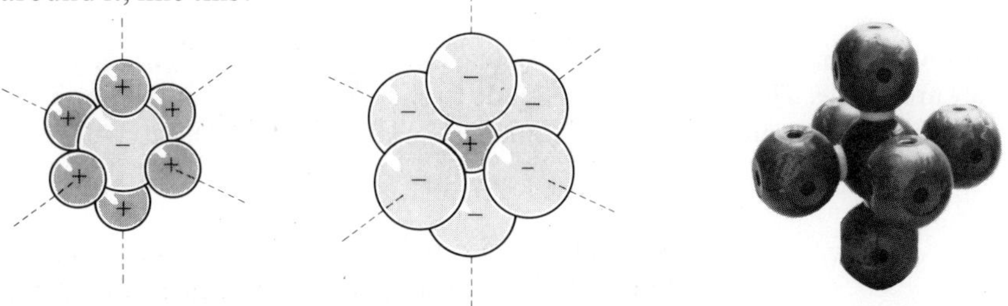

and these collect together into an ionic lattice, like this:

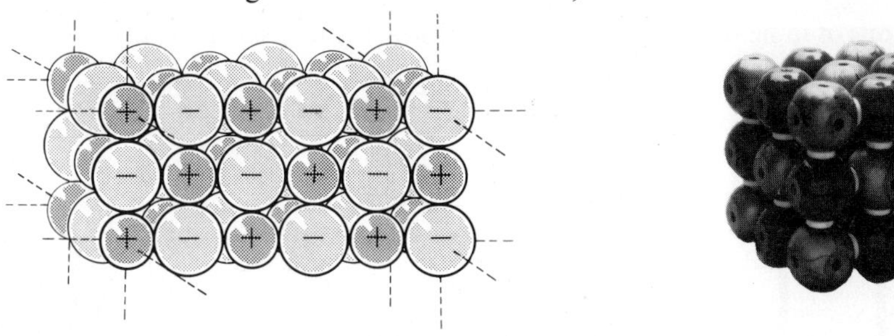

The bond between ions is called an **ionic bond**.
An ionic bond is the force of attraction between two oppositely-charged ions, which have been formed by electron transfer.
An ionic bond is sometimes called an **electrovalent bond**.

Ions and electrovalency

Sometimes more than one electron must be transferred, to produce a cation or an anion with a full shell. For example:

Metal atom →	*cation*	*Non-metal atom* →	*anion*
Mg →	Mg^{2+} + $2e^-$	O + $2e^-$ →	O^{2-}
2,8,2	2,8	2,6	2,8
Al →	Al^{3+} + $3e^-$	N + $3e^-$ →	N^{3-}
2,8,3	2,8	2,5	2,8

The number of electrons lost or gained by an atom is called its electrovalency. This word is also often shortened to valency.
Sodium atoms have valency 1; nitrogen atoms have valency 3.

Properties of ionic solids

Using these ideas about ionic bonding, the properties of ionic solids can be explained:

Fact	*Theory*
Although ionic solids are hard, they are brittle. They shatter when hit by a hammer.	Strong electrostatic forces hold the ions together. If one layer is made to slide, particles of the same charge move next to each other. The layers then repel each other. 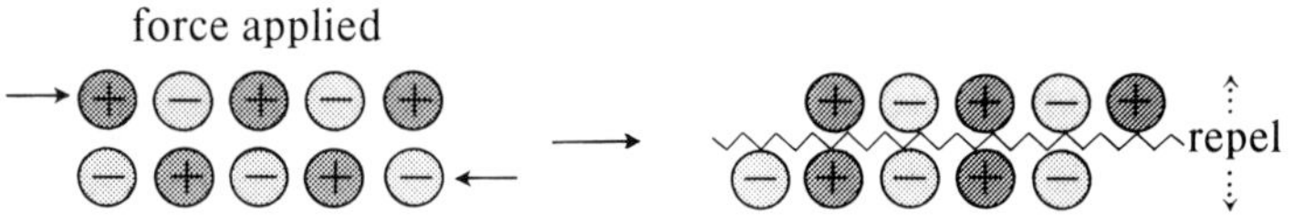
Solutions of ionic solids in water conduct electricity. 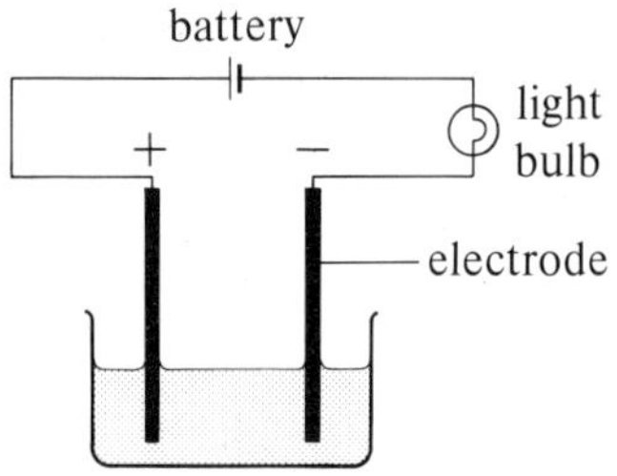	In solution, the ions have become free to move. 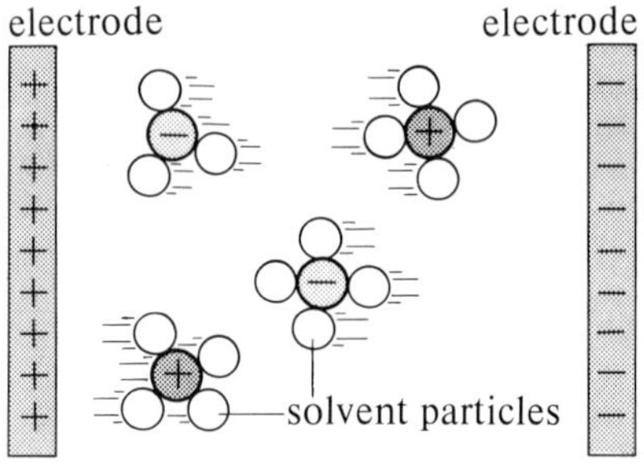 Ions, surrounded by solvent particles, move in opposite directions through the solution carrying charge.

Summary

The properties of a solid substance depend on the type of particle in the lattice and the way in which the particles in the lattice attract each other.

Atoms can bond together in different ways:
- by sharing electrons between two atoms—covalent bonding;
- by sharing electrons among all the atoms in the lattice—metallic bonding;
- by transferring electrons from one atom to another—ionic bonding.

In all these cases, the atoms achieve full electron shells.

When atoms bond covalently together, the particle they form is called a molecule.

The number of covalent bonds formed by an atom is called its covalency, or just valency.

The number of electrons lost or gained by an atom, in forming an ionic bond, is called its electrovalency, or just valency.

An ion is a particle which has gained or lost electrons, so is charged.

A cation is a positive ion, and an anion is a negative ion.

Questions

1 Explain the meaning of the term *covalent bond.* Both molecular and macromolecular solids contain covalent bonds.
However, molecular solids are soft, while macromolecular solids are hard. Try to explain why this is so.

2 (a) Carefully draw a diamond lattice, showing clearly the covalent bonds. How many covalent bonds does each carbon atom in the middle of the lattice form?
What is the covalency of each carbon atom?
Is the lattice made up of atoms or molecules?
(b) Repeat part (a) for iodine instead of diamond.

3 (a) Explain the meaning of the term *ionic bond.*
(b) How are these bonds formed?
(c) What happens to some electrons during the forming of these bonds?

4 Using simple diagrams, try to explain these facts about ionic substances.
(a) They shatter when hit.
(b) They conduct electricity when dissolved in water.

5 (a) Carefully draw a sodium chloride lattice.
(b) What is the charge on each ion?
(c) How many sodium ions are there around each chloride ion?
(d) How many chloride ions are there around each sodium ion?

6 (a) Explain the meaning of the term *metallic bond.*
Using clear simple diagrams, try to explain these facts.
(a) Metals conduct electricity.
(b) Metals conduct heat.
(c) Metals are ductile.

7 (a) Carefully draw part of a copper lattice, making sure that you label the ions, the inner shell electrons, and the delocalised electrons.

8 (a) Draw some circles like the ones below.

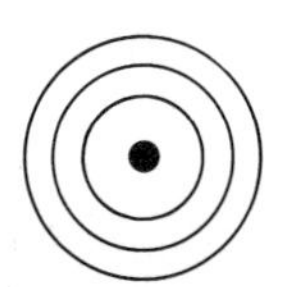

Sodium
($Z = 11$)

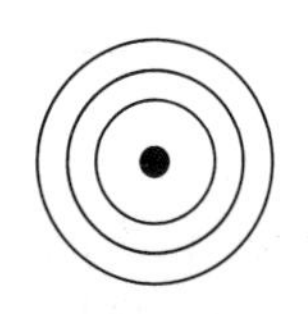

Magnesium
($Z = 12$)

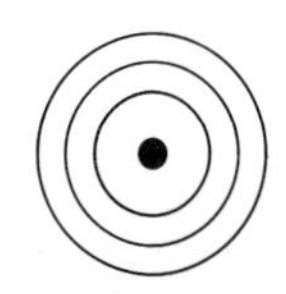

Aluminium
($Z = 13$)

(b) By filling in the right number of dots in each case, use them to show the electronic structures of sodium, magnesium, and aluminium.
(c) Under each diagram, write the number of outer-shell electrons the atom has.
(d) If these outer-shell electrons were lost to the lattice, the rest of the atom would become charged. It would become an ion. Under each diagram write down the sign and the size of the charge on the ion. Then write down whether it is an anion or a cation.

9 Here are diagrams of sodium and chlorine atoms.

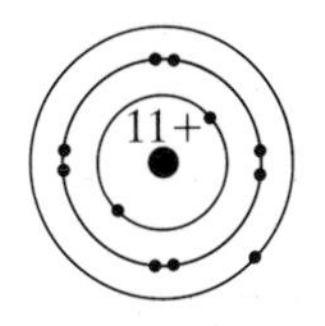

sodium

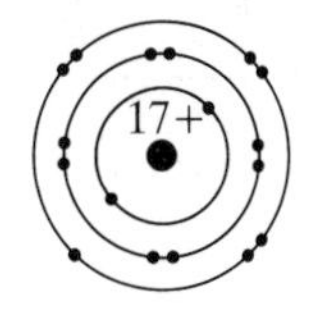

chlorine

(a) How many outer electrons has each atom got?
(b) What is the charge on each of these atoms?
(c) Suppose the sodium atom loses one electron to the chlorine atom. Now what is the charge on these particles?
(d) These particles have different charges. What are particles with these different charges called?
(e) What effect will these two charged particles have on each other?
(f) How many outer-shell electrons has each charged particle got?

11 Atoms, molecules, and ions: a review

This is the last chapter of Section 1. In it we will reinforce some of the ideas of the previous chapter, and account for some of the facts we missed last time. For example, the noble gases all have full shells of electrons. The theory so far suggests that their atoms should not be able to bond. But the noble gases can be made to solidify by cooling them down. What holds their atoms together?

11.1 Atomic lattices

The noble gases

The noble gases is the name given to the Group VIII elements. In Chapter 7, their properties were described; we explained these in Chapter 8 by assuming that the atoms all have full shells of electrons. Particles with full shells of electrons show no tendency to form covalent, ionic or metallic bonds, and have little force of attraction for one another. However, it is possible to obtain the Group VIII elements in the solid state:

Facts	*Theory*
Solid neon exists at temperatures below −249°C. Solid argon exists at temperatures below −189°C.	Group VIII atoms do attract one another, but the forces are so weak that they are very easily overcome. The lattice is made of atoms.

Because the lattices of these solid Group VIII elements are made of atoms, they are called **atomic lattices**.
An atomic lattice is an ordered arrangement of atoms in three dimensions.

Model of an atomic lattice

Van der Waals' forces

Forces of attraction between uncharged particles with full electron shells are called Van der Waals' forces. These weak forces are named after the Dutch scientist who studied them, Van der Waals. The strength of the forces gets greater as the total number of electrons in the particles gets larger:

Facts	*Theory*
Both neon and argon solids are soft and flaky. Neon melts at a lower temperature than argon.	The atomic lattice of a solid noble gas is held together by weak Van der Waals' forces. Ne atoms have 10 electrons; Ar atoms have 18 electrons. Van der Waals' forces are therefore stronger between Ar atoms.

Although we have mentioned them here for the sake of completeness, an explanation of Van der Waals' forces is beyond the scope of this book. However, you will find them explained in most good A-level chemistry books.

Macromolecular solids

The particles in a macromolecular lattice are atoms, so this type of lattice is also an atomic lattice. A quick glance at diamond's structure will remind you that it is an ordered arrangement of carbon atoms in three dimensions:

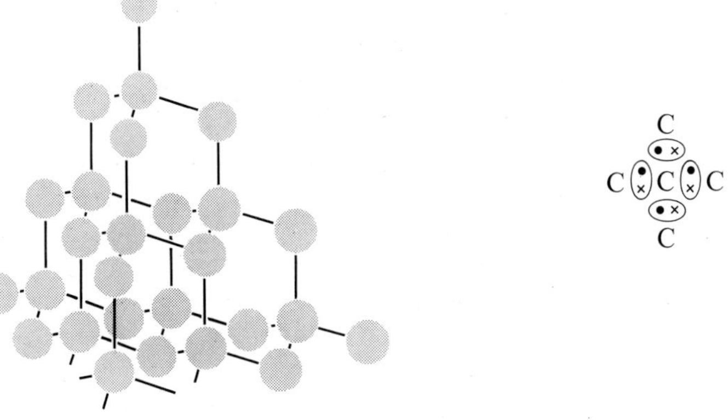

Fig. 1
The diamond lattice

Each atom shares four pairs of electrons with four other atoms. However, this atomic lattice is not quite the same as the argon lattice above. The carbon atoms are all held together by strong covalent bonds, but the argon atoms are only held in their lattice by weak Van der Waals' forces.

Metallic solids

Although the outer-shell electrons of metal atoms are delocalised in a metal lattice, the structure is still an ordered arrangement of metal atoms in three dimensions. Therefore a metallic solid is also a type of atomic lattice. Solid magnesium, for example, is a lattice of charged magnesium atoms—magnesium ions, Mg^{2+}—in a cloud of delocalised electrons.

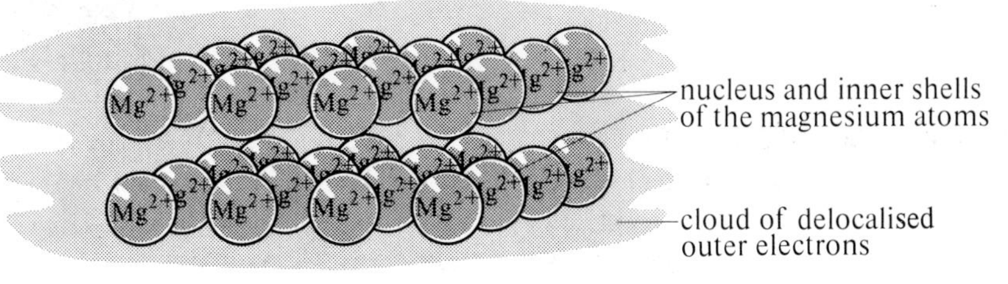

Fig. 2
The magnesium metal lattice

A comparison of atomic lattices

Table 11.1 Comparing the three different kinds of atomic lattice

Lattice	*Behaviour of atoms in the lattice*
Solid argon	The outer-shell electrons are localised within each atom. Van der Waals' forces hold the atoms together.
Solid diamond	The outer-shell electrons are grouped in pairs between the atoms, making covalent bonds. These strong bonds hold the lattice together firmly.
Solid magnesium	The outer-shell electrons are delocalised through the whole lattice. Strong electrostatic forces hold the metal cations to the delocalised outer-shell electrons.

11.2 Molecular lattices

A molecular lattice is an ordered arrangement of molecules in three dimensions.

Read page 89 again to remind yourself of the meaning of the word **molecule**. Molecules are made of atoms that are bonded covalently to one another. They are uncharged particles. All the atoms in a molecule have reached full outer shells by sharing electrons.

The bonds inside molecules

So far we have only looked at molecules of elements, e.g. iodine:

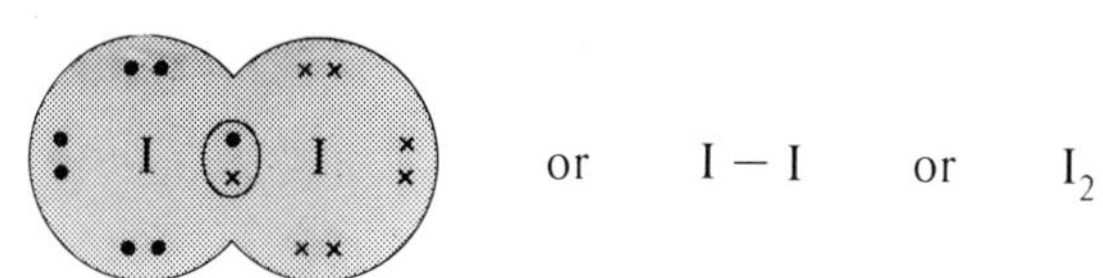

Many compounds are molecular—they contain small groups of atoms covalently bonded together into molecules, with full outer shells of electrons. Here are two examples:

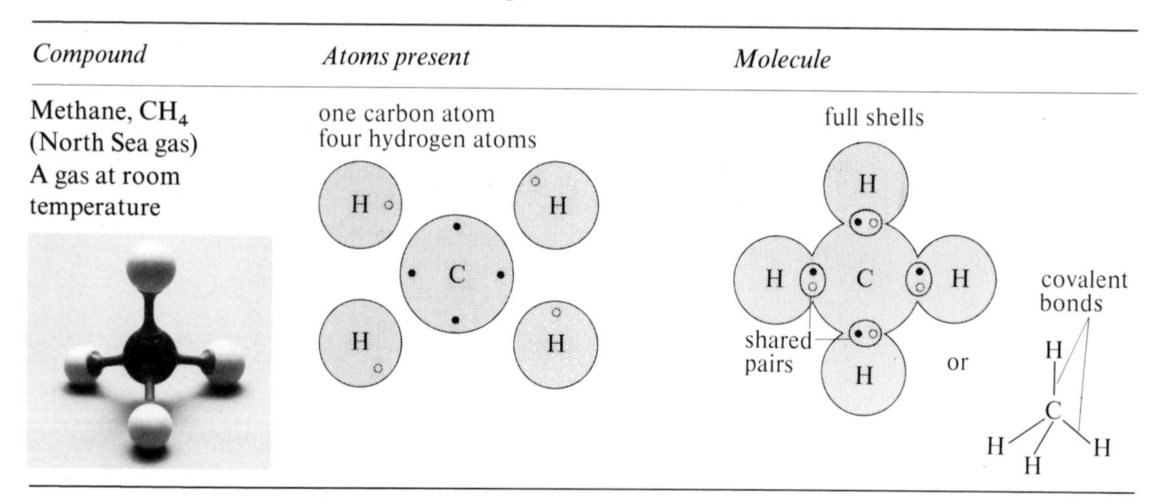

Compound	*Atoms present*	*Molecule*
Methane, CH_4 (North Sea gas) A gas at room temperature	one carbon atom four hydrogen atoms	full shells

Compound	*Atoms present*	*Molecule*
Ammonia, NH_3 A gas at room temperature 	one nitrogen atom three hydrogen atoms 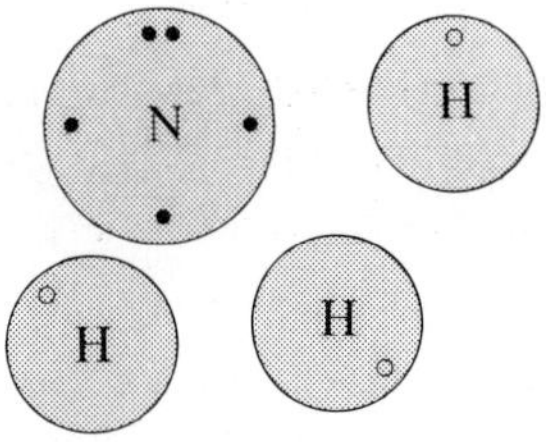	full shells; 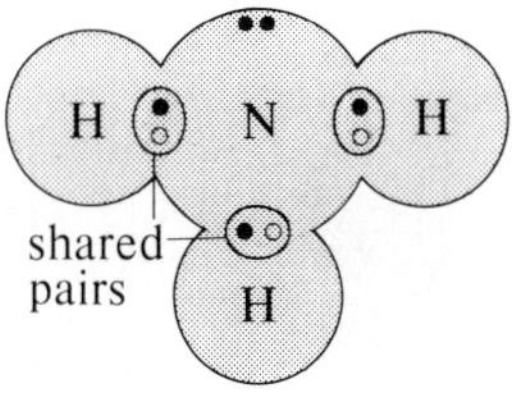 shared pairs; or; lone pair; 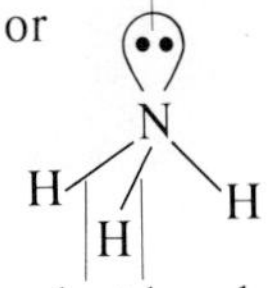 covalent bonds

Notice that there is an important difference between the molecules of methane and ammonia. All the outer-shell electrons of the atoms in the methane molecule are sharing in covalent bonds. In the ammonia molecule, the nitrogen atom has one pair of electrons that does not take part in bonding. It is called a **non-bonding pair** or **lone pair**.
A lone pair is a pair of outer-shell electrons that is under the control of the nucleus of one atom only.
In ammonia, the lone pair is under the control of the nitrogen nucleus only.

Two or more bonds between atoms

Some of the most common molecular substances are those that make up the air around us. The temperature of the atmosphere is far too high for their solid or liquid states to exist, so they are gases:

Table 11.2

Gas	*Amount in air*	*Formula of molecule*
Nitrogen	78 %	N_2
Oxygen	21 %	O_2
Carbon dioxide	0·04%	CO_2

The bonds inside the molecules of these gases are not quite the same as those inside the iodine, methane, and ammonia molecules. The atoms in N_2, O_2, and CO_2 must all have full shells; but the only way they can achieve this is by forming more than one covalent bond between them. Their bonding is shown on the next page.

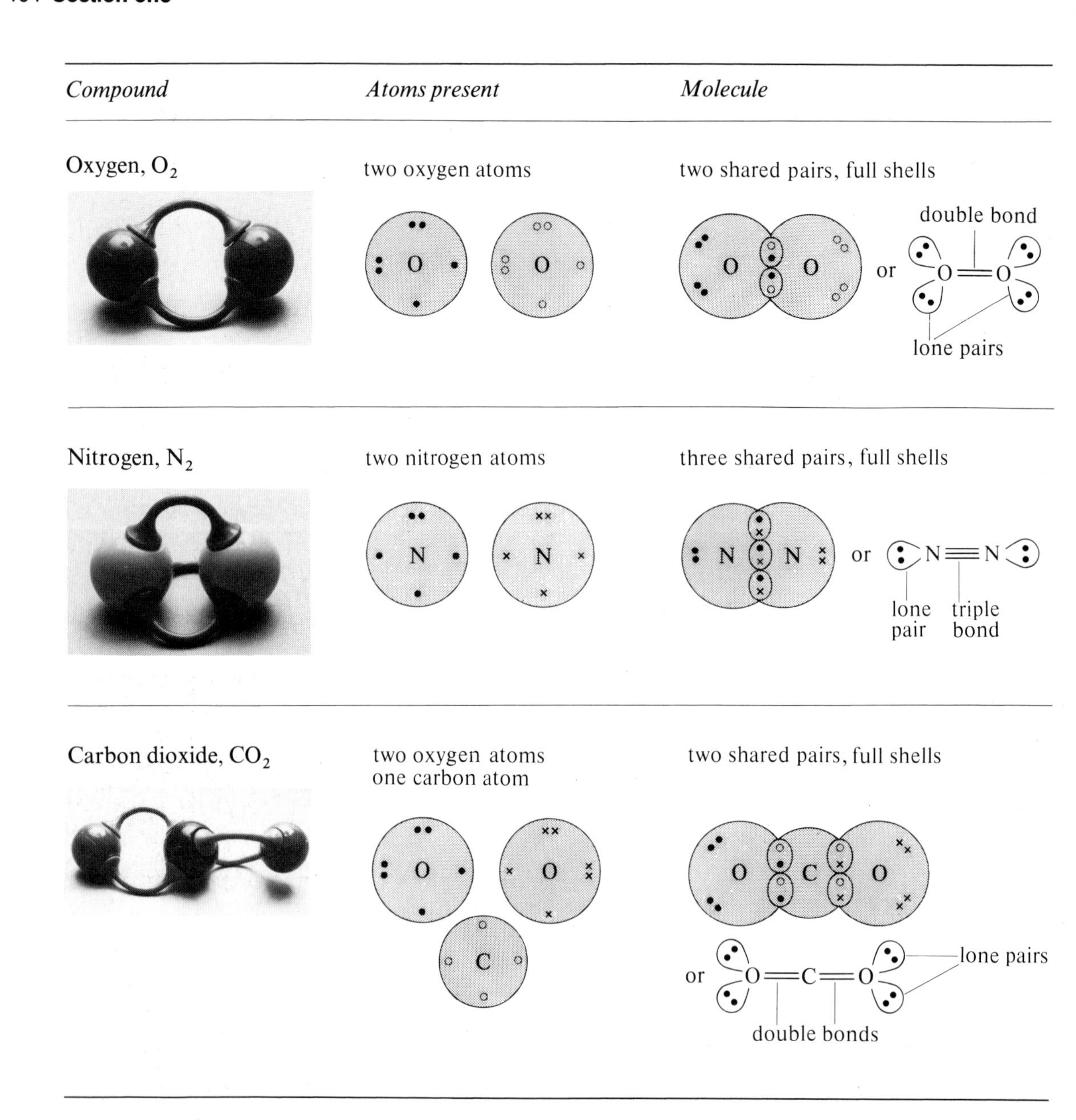

From the diagrams above, you can see that all the atoms in the three molecules have managed to gain full outer shells, by sharing the right number of electrons.

A single covalent bond is the force of attraction between a shared pair of electrons and the nuclei of two atoms.

A double covalent bond is the force of attraction between two shared pairs of electrons and the nuclei of two atoms.

A triple covalent bond is the force of attraction between three shared pairs of electrons and the nuclei of two atoms.

Table 11.3 The bonding within some common molecules

Molecule	*Electronic structure*	*Molecular structure*	*Model*
Hydrogen, H_2	 H (•×) H	H—H	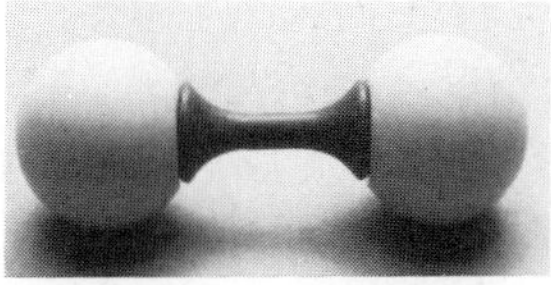
Chlorine, Cl_2	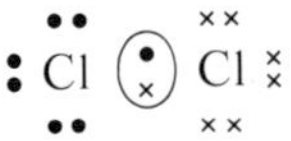 Cl (•×) Cl	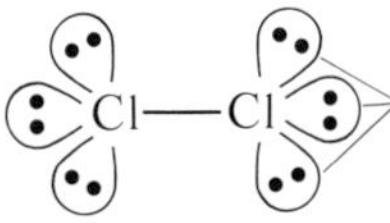 Cl—Cl; lone pairs	
Water, H_2O	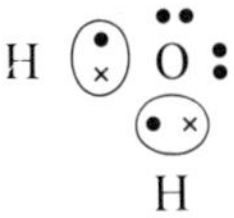 H (•×) O (•×) H	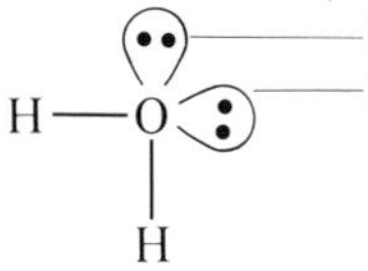 H—O—H; lone pairs	
Ethane, C_2H_6	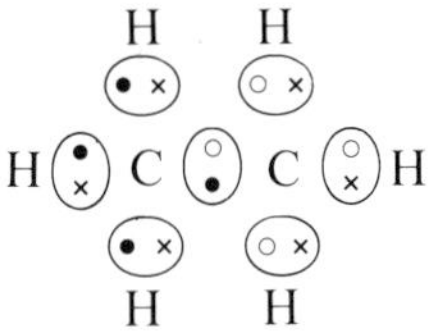	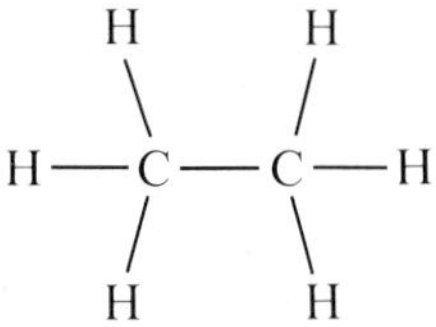 H₃C—CH₃	
Ethene, C_2H_4	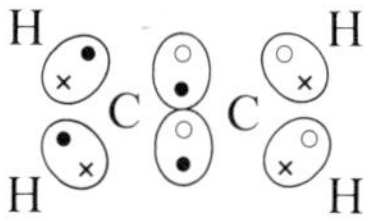	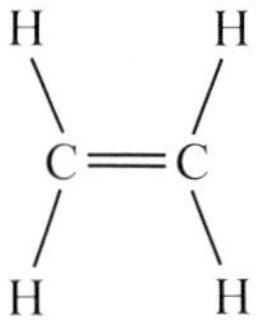 C=C	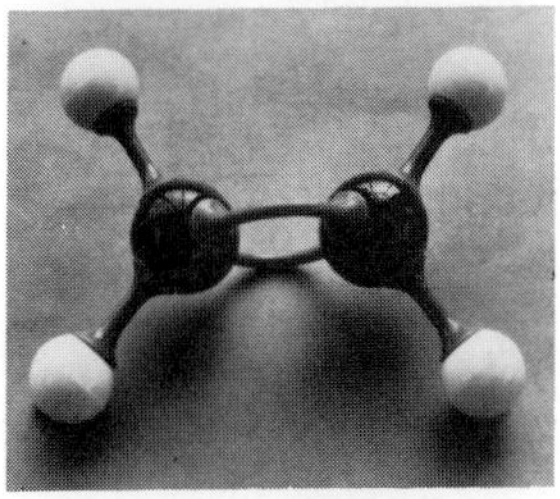
Tetrachloro-methane, CCl_4	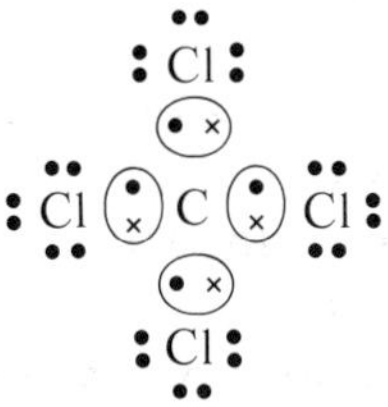	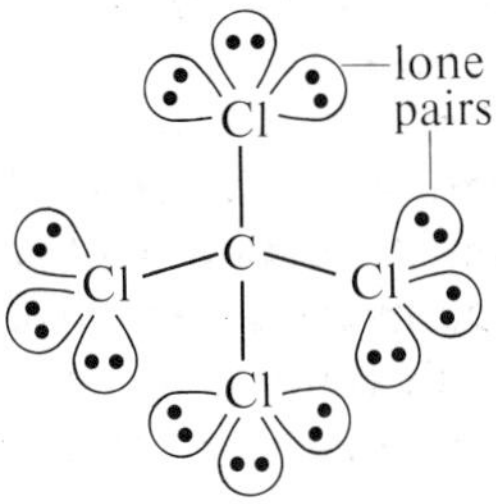 lone pairs	

Forces between molecules

Like the atoms of the noble gases, the atoms in molecules all have full outer shells of electrons. Also, like the noble gas atoms, they are not charged. Particles with full outer shells will not form strong bonds with other particles; but, as we saw earlier, they do have weak forces of attraction for one another.

The forces between molecules are of exactly the same sort as the forces between the uncharged noble gas atoms. They are also called Van der Waals' forces. The boiling-points of molecular systems give a good measure of the strength of the Van der Waals' forces.

Table 11.4 How boiling-points change with Van der Waals' forces

Molecule	*Number of electrons in each molecule*	*b.p. of liquid system/°C*	*Strength of Van der Waals' forces*
H_2	2	−253	
N_2	14	−196	
O_2	16 (increasing)	−183 (increasing)	increasing
Cl_2	34	−35	
CCl_4	74	+77	
I_2	126	+185	

The larger the number of electrons in each molecule, the stronger the forces *between* the molecules. The stronger forces mean that the system is held together more tightly and therefore the b.p. value is higher.

To summarise, molecules have:

strong forces within them, holding the atoms together;
weak forces of attraction between one another.

11.3 Ionic solids

Types of ion

On page 96, we saw that two different types of ion can be produced, by electron transfer to or from an atom. If an atom gains electrons, it becomes a negative ion—an anion. If an atom loses electrons, it becomes a positive ion—a cation. The resulting ions are usually particles with full outer shells of electrons.

This electron transfer need not always happen between single atoms. A group of covalently-bonded atoms can also gain or lose electrons to form a **molecular ion**. Once again this produces a particle containing atoms with full shells.

Atoms that bond covalently	*Electron transfer*	*Molecular ion produced*	*Charge*
one nitrogen atom four hydrogen atoms 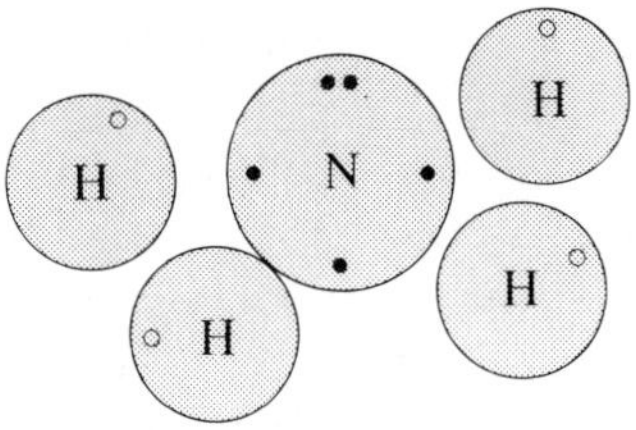	lose one electron •	molecular cation, full shells 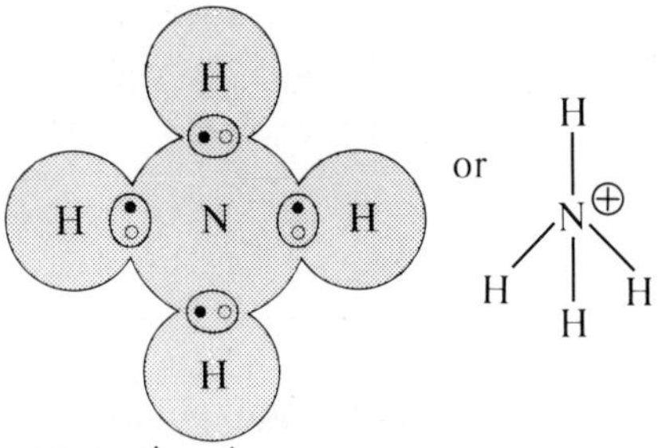 ammonium ion	1+ NH_4^+
one oxygen atom one hydrogen atom 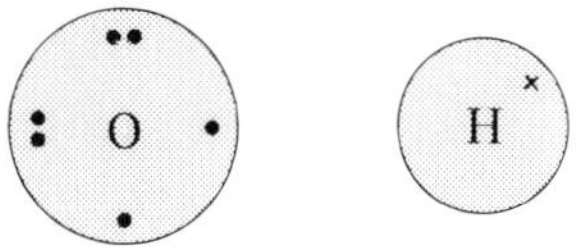	gain one electron ○	molecular anion, full shells 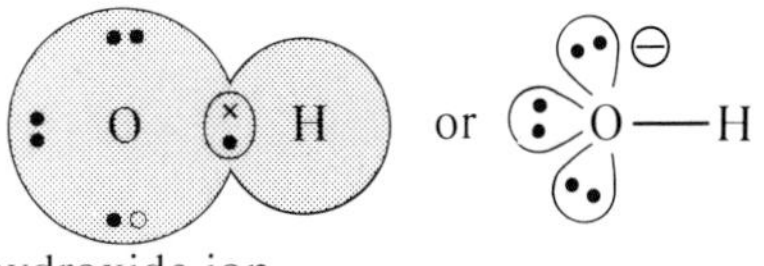 hydroxide ion	1− OH^-
one carbon atom three oxygen atoms 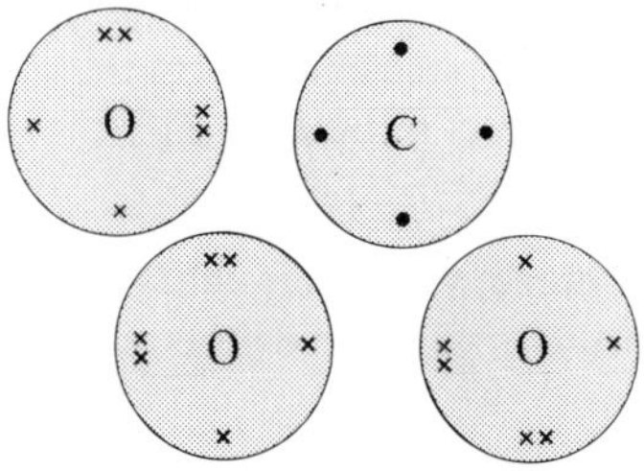	gain two electrons ○○	molecular anion, full shells 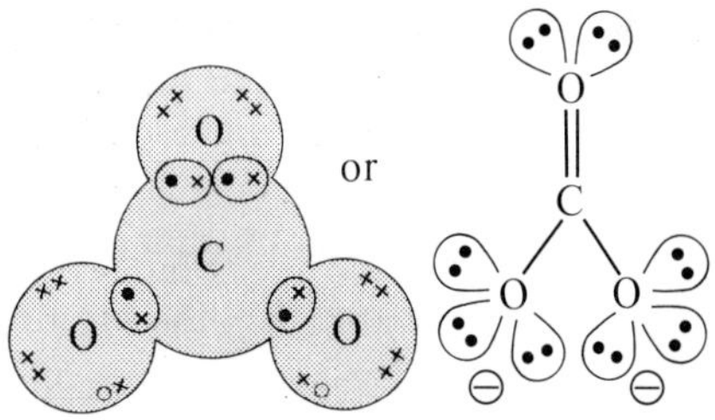 carbonate ion	2− CO_3^{2-}

A molecular ion has atoms covalently bonded within its structure. It is a charged particle because it does not have equal numbers of electrons and protons.

The bonding inside most molecular ions gets very complicated and need not concern us here. Table 11.5, on the following page, lists the names, formulas, and charges of the most common molecular ions. For comparison, we have included the most common atomic ions.

Table 11.5 The most common atomic and molecular ions

	Cations	*Formula*	*Anions*	*Formula*
Atomic ions	Sodium	Na^{+}	Chloride	Cl^{-}
	Potassium	K^{+}	Bromide	Br^{-}
			Iodide	I^{-}
	Magnesium	Mg^{2+}		
	Calcium	Ca^{2+}	Oxide	O^{2-}
	Copper(II)	Cu^{2+}	Sulphide	S^{2-}
	Zinc	Zn^{2+}	Nitride	N^{3-}
	Iron(II)	Fe^{2+}		
	Iron(III)	Fe^{3+}		
	Aluminium	Al^{3+}		
Molecular ions	Ammonium	NH_4^{+}	Hydroxide	OH^{-}
	Hydroxonium	H_3O^{+}	Hydrogen carbonate	HCO_3^{-}
			Hydrogen sulphate(VI)	HSO_4^{-}
			Nitrate(V)	NO_3^{-}
			Manganate(VII)	MnO_4^{-}
			Carbonate	CO_3^{2-}
			Sulphate(IV)	SO_3^{2-}
			Sulphate(VI)	SO_4^{2-}
			Dichromate(VI)	$Cr_2O_7^{2-}$

The formulas of ionic compounds

An ionic compound can be made from two atomic ions, two molecular ions, or one atomic and one molecular ion. When you know its name, you can work out the formula of an ionic compound, by balancing the charges of the ions making it up. The compound itself has no charge—the charges on the ions cancel each other out. So you must have the correct number of each type of ion. You can do this in the following way:

1. Write down the formulas of the ions in the substance, with their charges.
2. Adjust the number of one or both ions, so that the total charge is zero.
3. Collect up the ions and write the formula.
4. Put brackets around the formula of a molecular ion where necessary.

For example:

Compound	*1 Ions present*	*2 Balance*	*3 Formula*
Potassium chloride	K^{+} and Cl^{-}	K^{+} Cl^{-}	KCl
Magnesium hydroxide	Mg^{2+} and OH^{-}	Mg^{2+} $OH^{-}OH^{-}$	$Mg(OH)_2$
Ammonium sulphate(VI)	NH_4^{+} and SO_4^{2-}	NH_4^{+} NH_4^{+} SO_4^{2-}	$(NH_4)_2SO_4$
Aluminium nitrate(V)	Al^{3+} and NO_3^{-}	Al^{3+} NO_3^{-} NO_3^{-} NO_3^{-}	$Al(NO_3)_3$

Summary

The table below summarises what you have learned about atomic, molecular, and ionic lattices:

	Atomic lattice		*Molecular lattice*	*Ionic lattice*
	Metals	*Macromolecules*		
Examples	Magnesium	Diamond	Carbon dioxide	Sodium chloride
Lattice	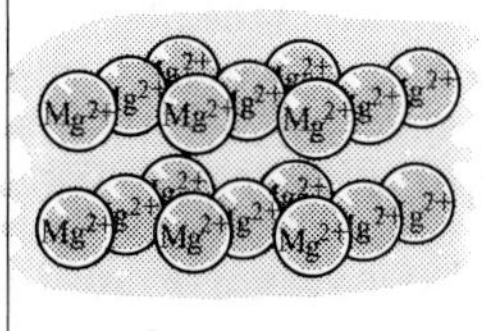	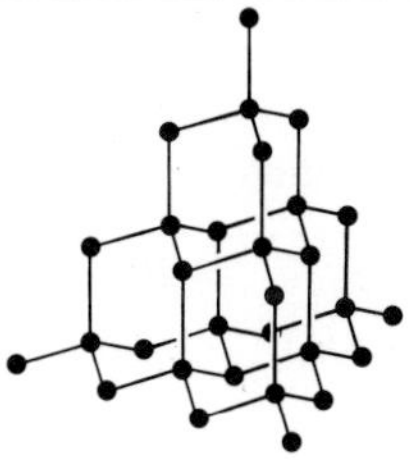	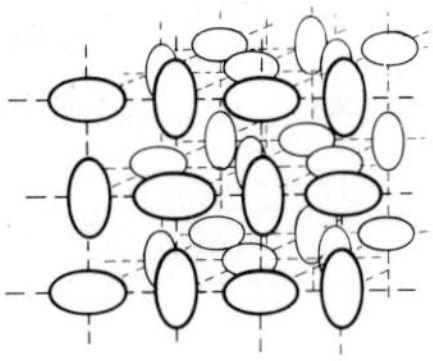	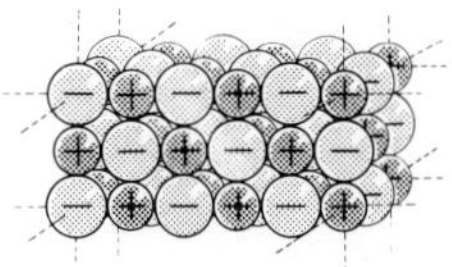
Lattice consists of	Cations and delocalised electrons	Atoms and shared electrons	Molecules	Ions resulting from electron transfer
Bonding	Metallic	Covalent	Covalent inside the molecules, but Van der Waals' forces between molecules	Ionic
Occurs between	Metal atoms	Non-metal atoms	Non-metal atoms	Metal atoms, and non-metal atoms
Physical properties of solid	Malleable and ductile; conducts heat and electricity well.	Very hard. Very high m.p. Insulator.	Soft and weak. Low m.p. Insulator.	Hard but brittle. High m.p. Solids are insulators but liquid phases conduct.

Questions

1 Molecular solids are not very hard. They melt easily and they do not conduct electricity. Try to think of some everyday substances with these properties. Write down the names of three of them, and say how their use depends on their properties.

2 Macromolecular solids are very, very hard. They are difficult to melt, and they do not conduct electricity. Granite is an example. Suggest two everyday uses for substances with these properties, explaining how the use you choose depends on a particular property.

3 Copy down this list of solids and their melting points:

Solid	*m.p./°C*
Phosphorus	44
Silicon	1410
Sulphur	119
Germanium	937
Arsenic	817
Silicon dioxide	1700

Decide which are molecular solids and which are macromolecular. Beside each solid write which sort it is.

4 Draw dot and cross diagrams to show the bonding in the following:
CH_3Cl $HOCl$ Br_2 N_2H_4 O_2 N_2 NH_3 HCl H_2O

5 Draw dot and cross diagrams to show the bonding in the following:
OH^- NH_4^+ CO_3^{2-} H_3O^+

6 (a) Explain the meaning of the term *Van der Waals' forces.*
(b) Between what sorts of particles are they found?
(c) Under what physical conditions do they become important?
(d) Reread page 106, and then explain what property of the molecule controls the strength of the Van der Waals' forces.

7 Here is a list of some molecules and the melting points of the solids they form.

Substance	*Molecule*	*m.p./°C*	*No. of electrons*
Phosphorus	P_4	44	60
Iodine	I_2	114	106
Sulphur	S_8	119	128

(a) What are the forces *inside* the molecules called?
(b) What are the forces of attraction *between* the molecules called?
(c) When molecular solids melt, which type of force is overcome?
(d) In which of the above solids are the molecules held to each other most strongly? What is the evidence for this?
(e) Explain the connection between the melting-points of these elements and the number of electrons they have.

8 Copy this table and write the correct answer in each space.

	Copper	*Diamond*	*Iodine*	*Potassium chloride*
Lattice consists of				
Type of bonding				
Will solid conduct?				
Will solid dissolve in water?				
Will solid melt easily?				
Will solid be brittle?				

9 Using Table 11.5, and the instructions on page 108, work out the correct formulas for:
(a) sodium carbonate
(b) calcium hydroxide
(c) calcium hydrogen carbonate
(d) aluminium sulphate (VI)
(e) potassium dichromate (VI)

10 By looking up the charges on the ions and then adjusting the number of each ion so that the plus and minus charges balance, work out the formulas of the following compounds: sodium, bromide; potassium oxide; calcium chloride; magnesium oxide; aluminium chloride; iron(III) sulphate; copper(II) nitrate.

Section two

Water and aqueous solutions

12 Water

Water is the most common compound found on Earth. The table below shows how water can be found almost anywhere.

System	*Water content by % mass*
seas, lakes, and rivers	more than 95%
humans	about 70%
trees	up to 60%
certain rocks (e.g. gypsum)	up to 20%
the air on a summer's day	about 1%

12.1 The water cycle

Water never stays in one place for long. The energy from the sun warms the seas, lakes, and rivers, and water evaporates into the atmosphere. When the air cools down, the water vapour condenses into very small droplets which make up clouds. When the droplets are big enough, they fall as rain or snow. The water drains back into the rivers, lakes, and seas, completing a cycle called **the water cycle**.

If all the water in the atmosphere at any one time rained down, the shower would only last a few hours. However, the air never runs out of water because it is always being replaced by more water evaporating from the seas. The time that water spends in each part of the water cycle is called the **residence time**. The residence time for water in the atmosphere is therefore very short.

But if all the water in the ice-caps suddenly melted, it would keep all the rivers flowing for a thousand years! The residence time for water in the polar ice-caps is very long. In between these two extremes are the residence times for the water in rivers and springs and for the water in the lakes and seas.

12.2 The uses of water

Apart from being essential to life, water plays a very important part in the economy of a country. For example,

Need	*Use*
domestic water supply	drinking, cooking, and washing
waste disposal	sewage, household waste
power production	cooling, hydroelectricity
industrial needs	cooling, waste removal
food production	irrigation, fishing
transport	shipping of goods, travel
recreation	water sports

The amount of water used by a community depends on its state of economic development. For bare survival, a man needs about two and a half litres a day for himself, as well as water to irrigate his crops. In a developed country like Britain, each person uses about a hundred and forty litres per day. This is made up in the following way:

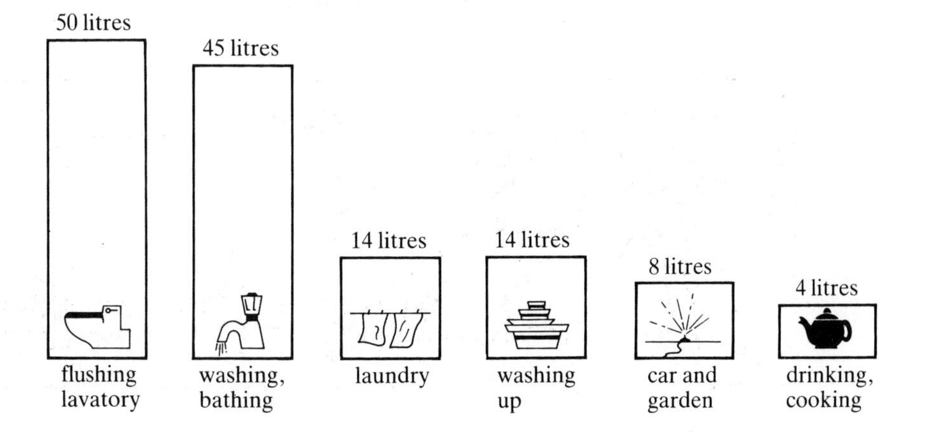

The amount of water used by each individual is tiny compared with the amount used in food production and in industry. For example, about two million litres of water are used every day to prepare the food for an average family of four—it is just as well that there is a water cycle or we would soon run out of water!

12.3 Water pollution

Although nearly all the causes of pollution are the result of human activity, natural events such as storms also cause pollution. The commonest result of water pollution is a drop in the amount of dissolved oxygen in the water. Dissolved oxygen is important because the useful bacteria in water, which break down impurities, are **aerobic bacteria**. 'Aerobic' means that they use up oxygen in the process. (Aerobic exercises are those which use up a lot of oxygen.) As the dissolved oxygen level in water falls, **anaerobic bacteria** begin to take over. These do not need oxygen for survival, and make the water dark and foul smelling. Eight of the more common sources of pollution are illustrated and discussed below.

Sewage Domestic sewage and waste from food-producing industries decrease the amount of dissolved oxygen in the water. The high concentration of organic (carbon-based) compounds in sewage causes an increase in the rate at which plant and animal life decays. The result is dark coloured water which smells foul.

Disease carrying organisms These get into the water from human and animal sewage, from certain industries like tanning, and from drainage off refuse tips. Most reservoirs are fenced off, and the water in them is carefully purified before being supplied to homes and factories.

Fertilizers Some of the fertilizers used by farmers gets washed off into rivers and lakes. High concentrations of nitrates and phosphates in the water cause increased growth algae and weeds. These use up the dissolved oxygen in the water, choking other forms of life and making the water smell and taste bad.

Sediments The rain washes mud and silt sediments into the water supplies. These can cause problems in purification plants by blocking channels and pipes.

Organic chemicals The two main sources of organic (carbon-based) compounds in water are detergents and pesticides. By causing water to foam, detergent prevents oxygen from easily dissolving in it. It can also upset the balance of essential fluids in the cells of some organisms. Pesticides accumulate in the body and so even small amounts can threaten fish. Dead and decaying fish add to the pollution problem as all organic matter uses up dissolved oxygen. Some detergents and pesticides can be broken down into less harmful substances by bacteria. Pollution problems can be lessened if more of these **biodegradable** products are used.

Inorganic chemicals Soluble minerals, as well as acids and alkalis, find their way from factories and mines into the water supply. Some of these pollutants (like the oxides of sulphur and nitrogen) get into the atmosphere from factory chimneys. Rain washes them out of the air into rivers and lakes. Other pollutants are introduced directly into rivers as they are washed out of the factory in cooling or cleaning water. These are often toxic to animals and plants, and cause corrosion in water treatment plants.

Radioactive materials The discharge of these materials is supposed to be restricted by law, but recently, increased radioactivity has been detected in the vicinity of some industrial plants, such as the one at Sellafield. The radiation given off by radioactive materials can cause cancers or mutation in a wide variety of organisms, including humans.

Heat Power stations use large amounts of water for cooling. They return the water pure, but slightly hotter. For example, the average temperature of the water next to a power station is often as high as 25°C in summer. This increase in temperature has two bad effects. Firstly, less oxygen dissolves in hot water than in cold, and secondly, the increased temperature increases the rate of decay of any material in the water. Decay nearly always uses up oxygen, so the amount of dissolved oxygen is further decreased.

Water often contains other impurities, which, because they are not harmful, are not thought of as pollutants. For example, the dissolved salts of calcium and magnesium make water **hard** without causing pollution (see page 298). Local authorities sometimes add substances to water for health reasons. One example of this is the addition of tiny amounts of sodium fluoride to reduce tooth decay in children.

12.4 Water treatment

Water supplied to the public must be free from impurities. Impurities can be insoluble (such as clay, silt, and micro-organisms) or soluble (such as industrial or agricultural pollutants, and dissolved organic compounds). All these impurities must be reduced to safe amounts before the water is piped to homes and factories.

Water works There is no universal way of purifying water. The method used depends on the local situation, but most waterworks carry out the jobs illustrated below.

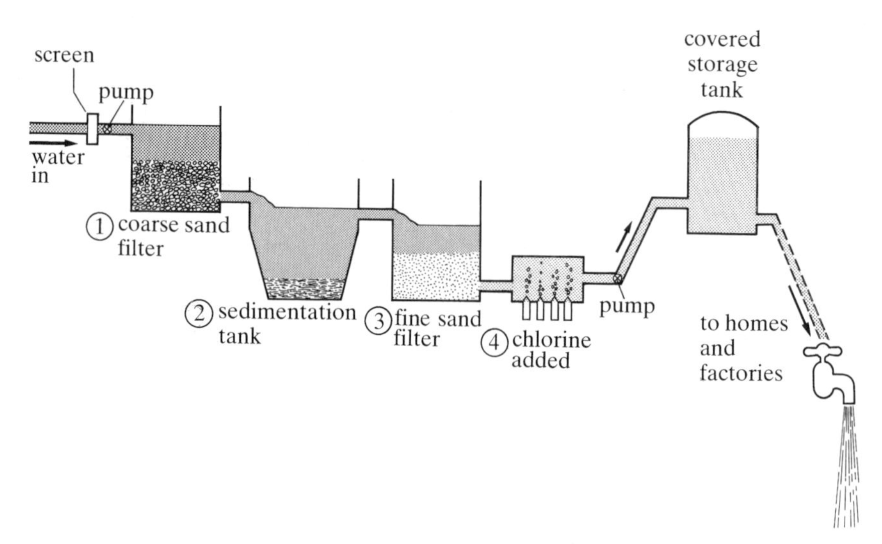

Screening This removes everything from tree branches to particles of grit.

Storage While water is stored in a reservoir, fine particles settle out. Different sources of water get well mixed up so that the taste and quality are consistent. Also, during storage, the ultraviolet radiation in sunlight destroys some of the germs in the water. However, there are problems with storing water in reservoirs. Fall-out from the atmosphere adds to the pollution of the water and, if the water is stored for more than ten days, algae begin to grow in it.

Aeration To keep up the levels of dissolved oxygen in the water, it is often pumped over small waterfalls. This stirs up the water and increases the amount coming into contact with the air.

Filtration The water is filtered through beds of sand and gravel. These filter beds contain an upper layer of fine quartz sand (about two metres deep) on top of a layer of coarse sand and one of gravel. After a few days, the upper layer traps and grows enough algae and plankton to decompose organic waste and to use up nitrates and phosphates. In doing this, the plankton and algae release oxygen. In the lower layer, other bacteria further break down any organic waste remaining.

Sterilization Even with all this treatment, the water still contains some bacteria. These are now killed by passing either chlorine or ozone into the water. A concentration of only five parts per million is enough to sterilize the water.

Sewage Plant The final stage in the treatment cycle of water is the sewage plant. Sewage contains all the waste from the baths, sinks, and lavatories of homes and factories. A sewage plant makes the water safe to be passed back into the waterworks. These are the jobs it carries out.

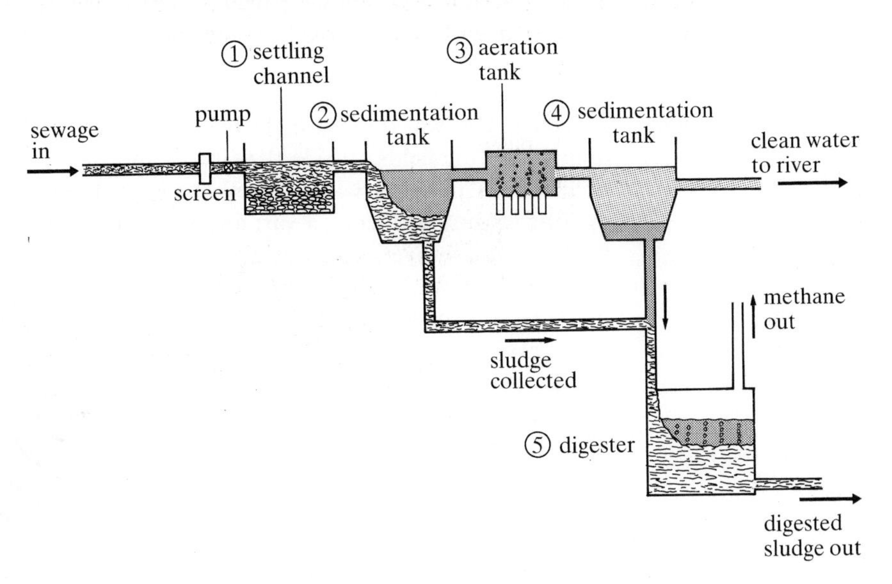

Screening This removes the biggest pieces of waste.

Settling The sewage goes into settling tanks where the larger particles sink to the bottom.

Sedimentation In the sedimentation tank, even smaller particles sink to the bottom and form a sludge.

Aeration From the sedimentation tank, the liquid is run off and is sprinkled over circular beds of stones. Bacteria on the stones digest all the harmful things in the water. Sometimes further treatment is needed, and the water goes into a tank in which air is blown through it.

Second sedimentation The sludge from the second sedimentation tank is run off into another tank called a digester, where more bacteria feeds on it. This process makes methane gas, which can be used as a fuel.

Sludge disposal Finally, all the unwanted sludge is removed to be dumped at sea, or is dried, and burnt to ash.

12.5 The River Thames: A case study

History

Since Roman times, the Thames has been a convenient place to dump London's rubbish. Sewage and household waste was thrown into the streets or into open drains. It rotted there until rain washed it into the river, where it gradually decayed on its way to the sea. This journey often took a long time because the Thames is 'tidal' at London. This means that the flow of water to the sea is affected by the coming and going of the tide at its estuary. For example, it takes about eighty days for something dropped into the river from London Bridge to travel the forty miles to the sea.

Growing problem

As the population of London grew, it became necessary for sewage to be collected. This was done at night by 'night-carts', and the sewage was dumped on the fields outside the city. In the 1800s, the popular invention of Thomas Crapper's 'water-closet' quickly led to the overloading of the existing sewers. Suddenly, a vast amount of untreated sewage was being flushed into the river. This was also accompanied by the great growth of London as a port. The huge amount of shipping traffic added to the pollution in the river. A crisis was reached by the 1850s. Fourteen thousand had died in the cholera epidemic of 1848/9, and conditions had got so bad that the hot dry year of 1856 was known as the 'year of the great stink'.

A massive new sewage system was built in the 1860s with outfalls into the Thames at Beckton and Crossness. This improved the situation a little in central London, but really only moved the problem downstream. The state of the river got steadily worse, but little more was done about it. The bombing in the Second World War resulted in damage to several of the Victorian pumping stations so that by the 1950s the water in the London reach of the Thames was black and dead. The graphs below show how much dissolved oxygen there was in the riverwater near London in 1900, 1935, and 1955.

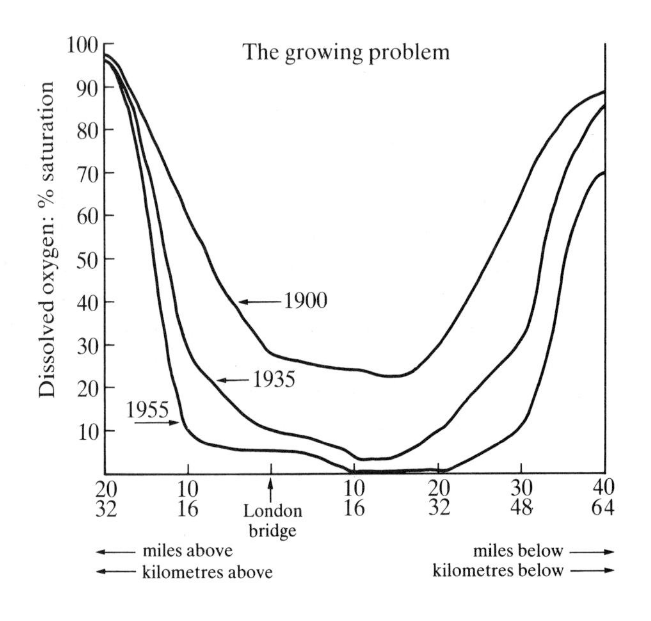

Water supply and main drainage of London

By the 1950s, long stretches of the Thames in London smelt foul. There were floating islands of rubbish, and only a few hardy eels could survive in the water. The invention of modern detergents after the war further increased the problem. In the dry summer of 1959, moored boats at Woolwich were completely covered with great banks of dirty foam. Not only did these detergents cut down the amount of oxygen which could dissolve in the river, but they poisoned the fish in the estuary and dissolved the natural oils on the feathers of the waterbirds.

The great clean-up

Public pressure finally forced the Government to provide the money to tackle the problem. A special committee under the Water Pollution Research Board identified the major causes of pollution in the Thames.

Cause of pollution	*% of total pollution*
domestic and industrial sewage	79
industrial discharges	12
waste from the upper Thames	4
from tributaries in London	3
from storm water	2

The need to improve the treatment of sewage was obvious, and so the rebuilding of the sewage works at Crossness and Beckton was started. The first was finished in 1964 and the second in 1974. These two plants, costing seventy-four million pounds, are able to treat 60% of London's sewage. They have replaced many older and smaller works, and now discharge almost pure water into the river. However, no sewage works can prevent storms from washing some untreated sewage into the Thames. In 1973, a bad storm left the river filled with dead fish as the dissolved oxygen fell to below 10%. In the seventies, there was not a big enough reserve of dissolved oxygen in the river to cope with sudden emergencies.

New, stricter laws were passed to prevent anyone from discharging effluent into the river without a licence. This licensing check-up found that a particular problem was being caused by the six big paperworks on the banks of the river. They pumped waste water into the river containing almost no dissolved oxygen but a lot of dissolved paper pulp. Two solutions were found: air was pumped into the waste water to raise its oxygen level; and a 'coagulation' plant was designed to precipitate most of the pulp out of the waste water. This extra pulp, which had been wasted, is now used to make cartons. Tighter control was also applied to detergent manufacturers. Their products must now be 'biodegradable' which means that they decompose when left in waste water for a month or so. The problem of foam has gradually got better as a result.

The effect of all these measures can be seen in the improving 'oxygen sag' curves below. Not only do they show that the level of dissolved oxygen is rising, but that the stretch of polluted water in the river is getting shorter.

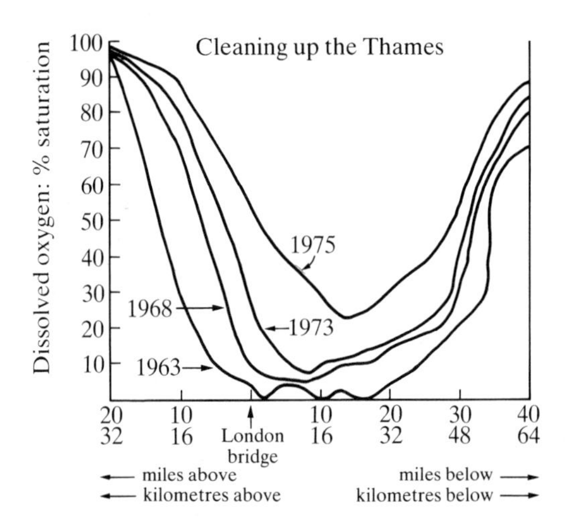

Today Unfortunately, not all the effects of the clean up have been welcome. There has been a rapid growth of weed in the cleaner water, both on the bottom of boats and on dock walls. Swarms of shrimps have often blocked the intakes of power stations, and a type of toredo worm, new to London, has been boring holes into wooden piles and boat planks. But, in 1957, there were practically no fish at all in the Thames at London, but, by the end of 1975, 86 different species had been found in the river! No salmon had been found in the Thames since 1860, but in 1974 one was found in the intake of a power station at West Thurrock. This fact is particularly important, because salmon do not survive in water with less than 30% dissolved oxygen.

Summary

Water is the most common substance known, filling rivers, lakes, and seas. It is also present in all forms of life, as well as in some rocks.

During the water cycle, water evaporates from the sea, forms clouds, falls as rain or snow and then returns to the sea in streams, rivers, and springs.

Water pollution is a danger which society must control. The main sources of pollution are: sewage, industrial pollutants, and fertilizers.

Sewage plants treat sewage to make it safe to be passed onto the waterworks.

The waterworks makes water safe to be piped into homes and factories. Recently, a campaign of concern for the water in the Thames has seen the pollution levels drop enough to allow salmon back into the river.

Questions

1 (a) Draw a diagram of the water cycle.
(b) In which steps do the following changes take place:
(i) evaporation
(ii) condensation
(iii) solution
(c) In which part of the cycle is most of the water found?
(d) What is the meaning of the phrase *residence time?*
(e) Explain why the residence time of water in the atmosphere is so short.
(f) A lot of pollution is pumped into the atmosphere. Do you think the pollution of the atmosphere is helped or made worse by the short residence time of water in the atmosphere? Explain
(g) In which part of the cycle is the residence time longest?

2 Water has many uses in everyday life.
(a) List five different types of use for water.
(b) Explain why the amount of water needed by a person in a developing country is so much less than used by a person in a developed country like Britain.
(c) What processes use really large amounts of water?
(d) With so much water being used everyday, why don't we run out of it?

3 (a) List the main causes of water pollution.
(b) Divide these causes up into those caused by domestic users, and those caused by industrial users.
(c) Explain why heat can be thought of as a form of pollution.

4 (a) Bacteria from sewage are a form of pollution, because they are a threat to health. Can you think of any good bacteria?
(b) Explain what these 'good' bacteria can do.
(c) What is the difference between *aerobic* and *anaerobic* bacteria?

5 Water pollutants are substances that get into the water unintentionally, as the result of some human activity.
(a) Write down as many impurities you can think of, that get into the water through natural processes. Explain how they get into the water.
(b) Give some examples of substances that are deliberately added to the water. Explain why they are added.

6 (a) Draw a diagram of a water treatment plant, and label each part.
(b) List the steps in the water purification process, and explain the purpose of each step.
(c) Now list the steps in two columns, one headed physical separations, and the other chemical treatment.

7 (a) Draw a diagram of a sewage treatment plant.
(b) List the steps in the sewage treatment process, and explain the purpose of each step.
(c) Try to list some uses that sewage might be put to.

8 Many large cities like London are built on rivers.
(a) Why does it take so long for the rubbish thrown into the Thames in London to reach the sea?
(b) How long does it actually take for this rubbish to reach the sea?
(c) Why did pollution suddenly get so much worse in London in the 1880s?
(d) What were the main effects of this pollution?
(e) How was it cured?

9 (a) When was pollution in the Thames at its worst?
(b) What were the major causes of pollution in London?
(c) What steps were taken to tackle the problem?
(d) Did the discovery of detergents make the problem of pollution better or worse?
(e) What were the signs that indicated that the Thames was becoming cleaner again?
(f) Are there any problems caused by the clean-up of the Thames? Explain.

13 The chemistry of water

Remember once again the aim of the chemist:

(i) to know the facts about chemical systems;
(ii) to find a theory that explains the facts.

In the first part of this chapter, section 13.1, we will look at the chemical properties of water. You have met some of these already. Then in section 13.2 we will try to find theories to explain these properties.

13.1 Facts

Occurrence As we have seen in Chapter 12, water is the most common compound on earth. Almost all forms of life contain high proportions of water, and it is even found trapped in the structures of some rocks.

Physical properties Compare the properties of water with those of other hydrogen and non-metal compounds, such as methane (CH_4) and ammonia (NH_3):

Table 13.2 The physical properties of some hydrogen/non-metal compounds

Compound	*Formula*	*m.p. /°C*	*b.p. /°C*	*Temperature range of liquid state/°C*	*Density at 20°C /g dm^{-3}*	*Smell*
Water	H_2O	**0**	**100**	**100**	**998**	**None**
Methane	CH_4	−182	−160	22	0·67	None
Ammonia	NH_3	−78	−33	45	0·71	Pungent and choking
Hydrogen fluoride	HF	−83	19	102	0·75	Pungent and choking
Hydrogen chloride	HCl	−114	−85	29	1·52	Pungent and choking
Hydrogen sulphide	H_2S	−83	−62	21	1·42	Bad eggs

Table 13.1 shows that water has:

higher fixed points than other non-metal hydrogen compounds;
a greater range of temperature during which it is in the liquid state (except for hydrogen fluoride);
a higher density at 20°C—it is a liquid at room temperature, while the others are gases.

Electrical properties

We can look at the electrical properties of water using a conductivity meter. You may have seen one of these in the laboratory. By connecting the substance into a circuit wired to the meter, you can measure how well that substance conducts.
In Table 13.2 we compare the result with the result for copper. Copper metal is a good conductor; as you probably know it is used in nearly all electrical wiring.

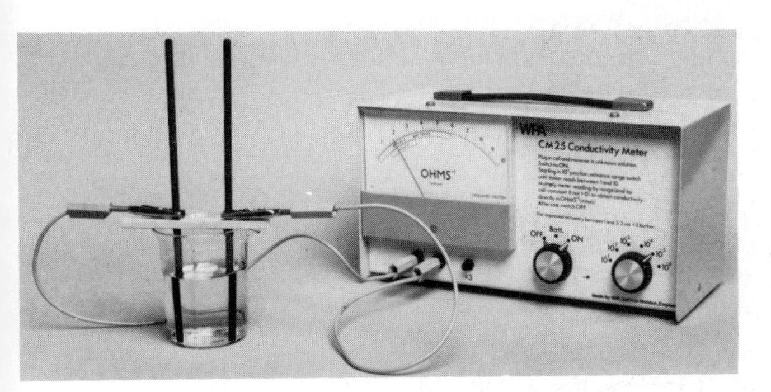

Above, apparatus for testing the conductivity of water

Table 13.2

Substance	*Conductivity*/Ω^{-1} *cm*$^{-1}$
Pure water	1×10^{-7} or 0·0000001
Tap water	1×10^{-3} or 0·001
Copper	64100

So pure water hardly conducts at all! Tap water conducts a little because it contains many dissolved impurities. But both conduct very badly compared with copper.

Solvent properties

Water is a polar solvent; it has the particular property of being able to dissolve many ionic solids. The solutions that are produced are called **aqueous solutions.** (We first met these ideas in Chapter 9; read the whole of the chapter again to remind yourself of their meaning.)
Ionic solids have these properties:

they are hard but brittle;
they have high m.p.'s;
they do not conduct electricity in the solid state;
they do conduct electricity in the liquid state;
when dissolved in water, they give solutions that do conduct electricity.

A polar solvent has these properties:

in a fine stream, it is attracted towards a charged plastic ruler;
it dissolves many ionic solids.

As well as water's ability to dissolve ionic solids, it also dissolves a number of molecular substances, for example sugar, and alcohol.

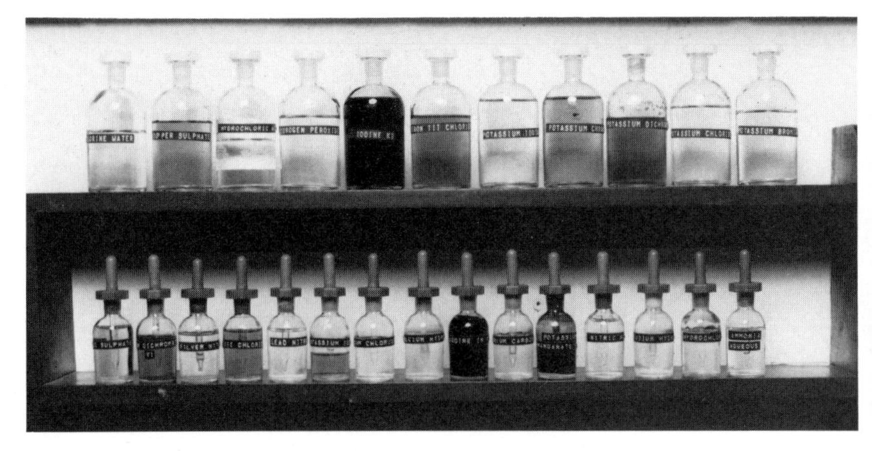

Some aqueous solutions

Aqueous solutions

Most laboratories have rows of bottles that contain substances dissolved in water—aqueous solutions. When certain types of aqueous solutions are mixed, a chemical change occurs. When one solution is poured into the other, a solid suddenly appears in the mixture. This type of chemical change is called **precipitation**.

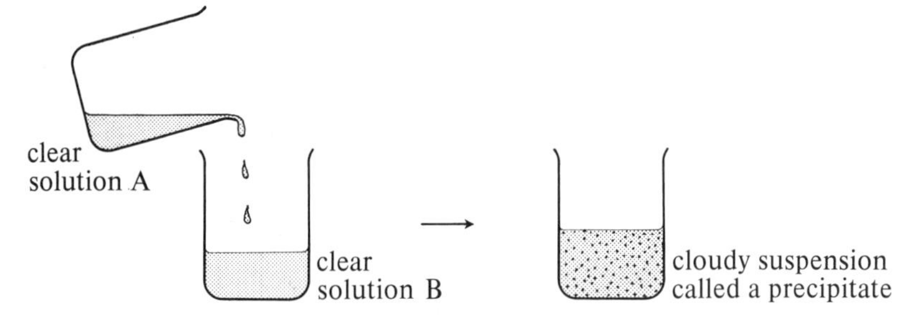

Fig. 1
Precipitation

Precipitation is a chemical reaction between solutes in solution, during which an insoluble product is formed.

Table 13.3 shows some solutions which give precipitates when mixed.

Table 13.3

Aqueous solution A	*Aqueous solution B*	*Product*
$AgNO_3$ Silver nitrate(V)	Any metal chloride e.g. $NaCl$, $CaCl_2$, $AlCl_3$	A curdy, white precipitate forms.
$BaCl_2$ Barium chloride	Any metal sulphate(VI) e.g. Na_2SO_4, $CuSO_4$	A fine, powdery, white precipitate forms.
NaOH Sodium hydroxide	A compound of any metal, except those of Group I	A solid is precipitated which is often coloured and rather gelatinous (jelly-like).

In a system equation, we show precipitates by giving them a downward arrow $\downarrow$. This shows the movement of the insoluble product.

We can add silver nitrate(V) solution to a number of different metal chloride solutions, like those in Table 13.3. The *same* white curdy precipitate forms every time. This precipitate is silver chloride. So we can write some system equations:

Solution A	*+solution B*	$\longrightarrow$ *precipitate*	*+solution*
$AgNO_3$	$+NaCl$	$\longrightarrow AgCl\downarrow$	$+NaNO_3$
$2AgNO_3$	$+CaCl_2$	$\longrightarrow 2AgCl\downarrow$	$+Ca(NO_3)_2$
$3AgNO_3$	$+AlCl_3$	$\longrightarrow 3AgCl\downarrow$	$+Al(NO_3)_3$

The same thing happens if we try barium chloride with different metal sulphates—see Table 13.3 again. Barium sulphate is insoluble and white:

$$BaCl_2 + Na_2SO_4 \longrightarrow BaSO_4\downarrow + 2NaCl$$
$$BaCl_2 + CuSO_4 \longrightarrow BaSO_4\downarrow + CuCl_2$$

Metal hydroxides except those of Group I (like KOH and NaOH) are insoluble. Calcium and barium hydroxides are just slightly soluble. Many hydroxides are coloured, so form coloured precipitates:

$$2NaOH + FeSO_4 \longrightarrow \underset{\text{(green)}}{Fe(OH)_2\downarrow} + Na_2SO_4$$

$$2NaOH + CuCl_2 \longrightarrow \underset{\text{(blue)}}{Cu(OH)_2\downarrow} + 2NaCl$$

Precipitation, like crystallisation, is the reverse of dissolving. If a solid comes out of solution slowly, a regular solid lattice has time to form and crystals are produced. We call this process crystallisation. But if the solid is formed very quickly, there is no time for a large regular lattice to grow; instead many tiny particles form in the liquid. This we call precipitation.

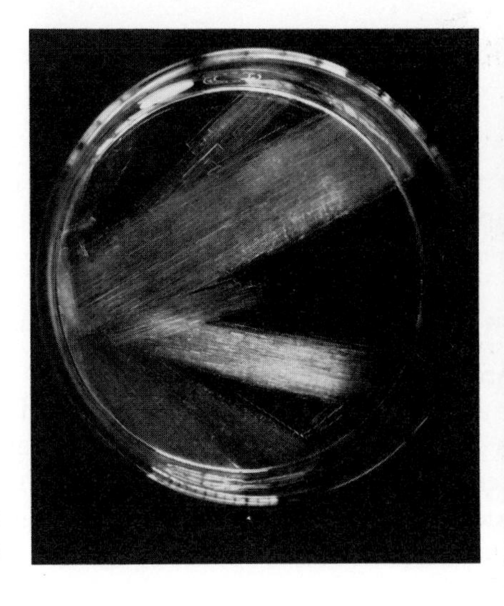

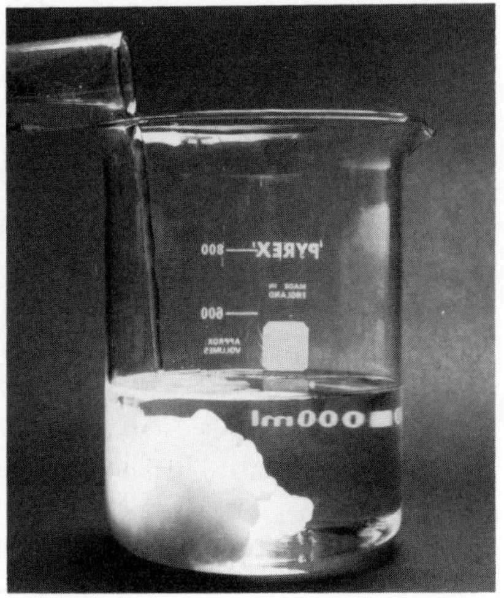

Left: crystals growing slowly
Right: the rapid formation of a precipitate

Making insoluble salts by precipitation

If the compound you want to make is insoluble, you can make it using a precipitation reaction. For example, to make insoluble lead sulphate you could follow these steps:

1 Choose a soluble lead compound, say lead nitrate. Choose a soluble sulphate, say sodium sulphate. Make aqueous solutions of each of them, in equal amounts of water. (If you are working carefully, you can calculate how much of each compound to use, from the equation for the reaction.)

2 Mix the aqueous solutions together. A precipitation reaction occurs, producing the insoluble product lead sulphate.

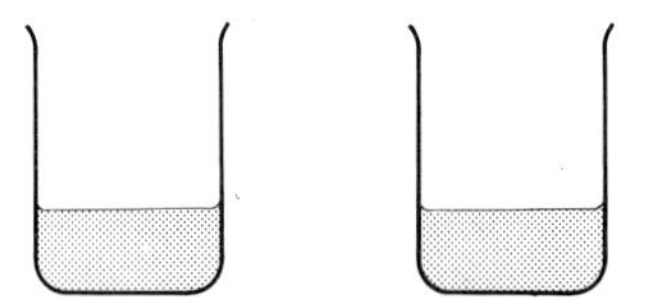

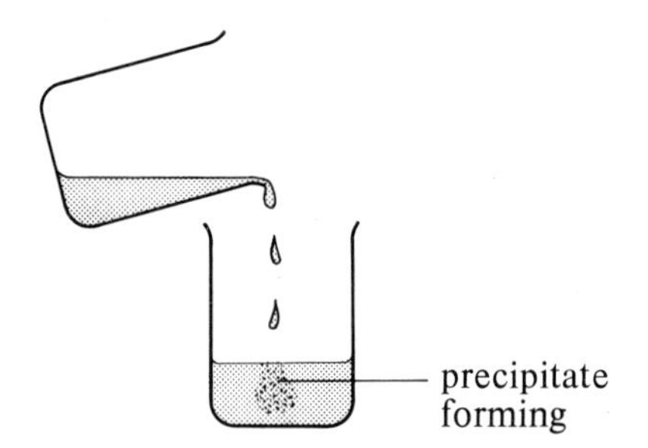

3 As lead sulphate is the only insoluble product, it can be filtered off.

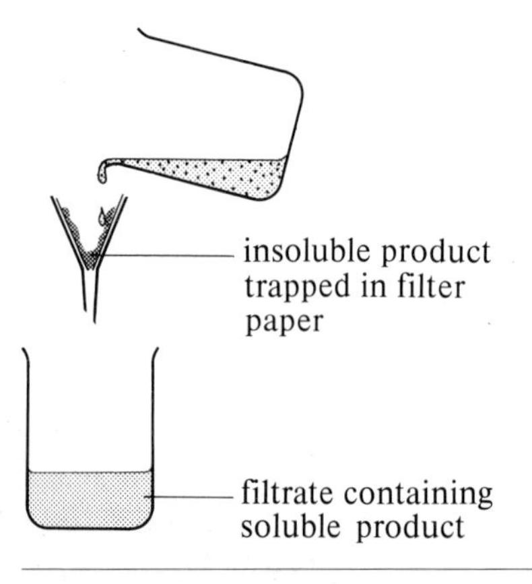

4 The lead sulphate is washed with distilled water. This rinses any soluble substances out of it.

5 The filter paper is then dried—usually in a warm oven—and the lead sulphate scraped off it.

The list below is useful in deciding which soluble compounds to use.

(i) All nitrates are soluble.
(ii) All chlorides are soluble, except those of silver, lead, and mercury.
(iii) All sulphates are soluble, except those of barium, calcium, and lead.
(iv) All carbonates are insoluble, except those of Group I metals.
(v) Most hydroxides are insoluble except those of Group I metals. However the Group II metals calcium and barium have slightly soluble hydroxides.

13.2 Theory

The water molecule Water is a compound of hydrogen and oxygen. Hydrogen atoms and oxygen atoms bond covalently to produce water molecules.

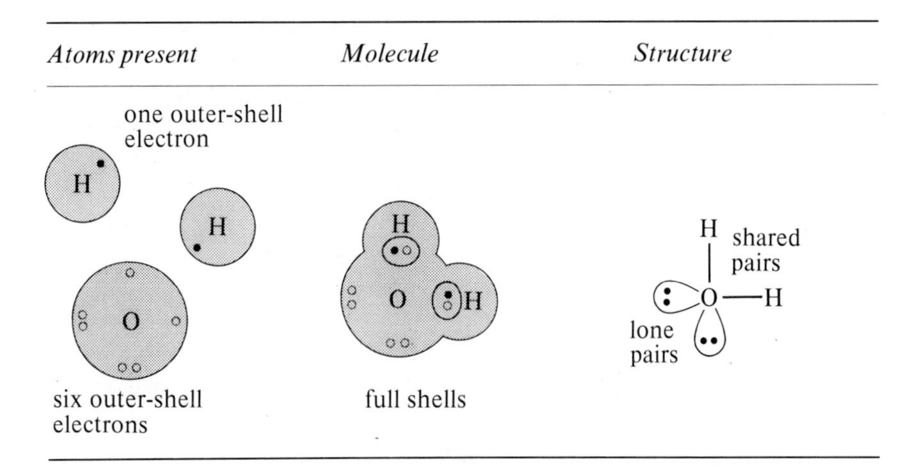

Note the two different types of electron pair in the water molecule. Two pairs are shared between hydrogen and oxygen, to bond the atoms together. The other two pairs, the lone pairs, belong only to the oxygen.

Water molecule	*Lone pairs*	*Shared pairs*
O, H, H	O	O, H, H

Core charge and the polarity of molecules The shared pairs in the water molecule are *not shared equally*. The attraction of the oxygen nucleus is stronger than that of the hydrogen nucleus:

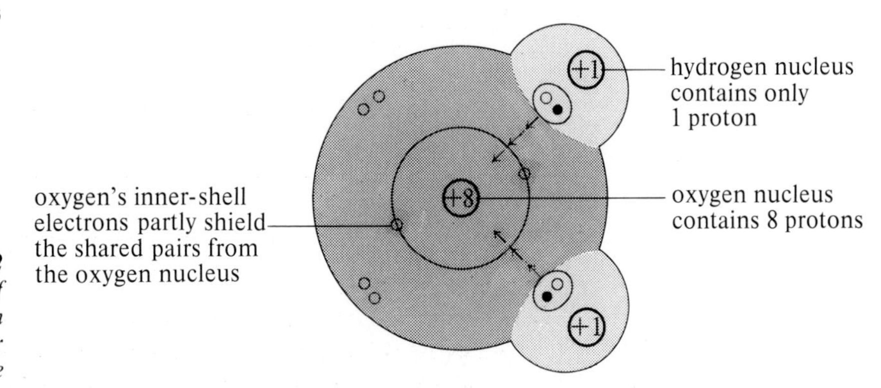

Fig. 2
Unequal sharing of electrons between oxygen and hydrogen in the water molecule

The pull of the oxygen nucleus with its eight protons is stronger than that of the hydrogen nucleus with only one proton. The shared pairs do not feel the full attraction of the oxygen nucleus, however, because oxygen's inner shell of electrons acts as a shield.

Sometimes it is useful to work out how strongly an atom attracts electrons in a bond. This measure of attraction is called the **core charge** of the atom. **An atom's core charge is a measure of the control it has over its outer-shell electrons:**
Core charge = nuclear charge − number of inner-shell electrons

Table 13.4

Atom	*Nuclear charge*	*Number of inner-shell electrons*	*Core charge*
Oxygen	$^{16}_{8}O \longrightarrow +8$	Only one inner shell. It contains 2 electrons.	$+8-2 = +6$
Hydrogen	$^{1}_{1}H \longrightarrow +1$	No inner shell, so no shielding electrons	$+1-0 = +1$

From the table above, you can see that oxygen has a higher core charge than hydrogen. So in the water molecule the oxygen attracts the shared pairs towards itself—and away from the hydrogens. This means that the oxygen atom becomes slightly negatively charged, and the hydrogen atoms become slightly positively charged. Chemists show this on diagrams by using the Greek letter δ (delta), which is taken to mean 'a little'. So $\delta+$ means slightly positive and $\delta-$ means slightly negative.

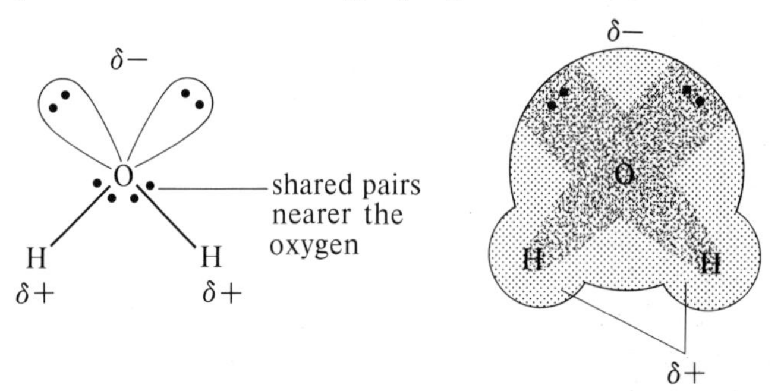

Fig. 3
Water, a polar molecule

Molecules that have an uneven spread of charge are called **polar** molecules. As you see from the diagrams above, water molecules are polar. This property helps us to explain a number of facts, as shown on the next page.

Fact	*Theory*
Water (hydrogen oxide) has unusually high fixed points, 0°C and 100°C. Compare with hydrogen sulphide which melts at −83°C and boils at −62°C.	The more positive end of one water molecule attracts the more negative end of another water molecule strongly. This attraction is repeated throughout the solid and liquid. These forces are called **hydrogen bonds**. H O H $\delta+$ $\delta-$ O H H Because of hydrogen bonding, more heat energy is needed to make ice melt and water boil. So water's fixed points are high.
A fine stream of water, running from a burette, is attracted towards a charged rod. The same effect is seen, whether the rod is charged positively or negatively.	+ charged rod; water molecules with their $\delta-$ ends attracted towards the charged rod; deflected stream; or; − oppositely charged rod; water molecules with their $\delta+$ ends attracted towards the charged rod; deflected stream

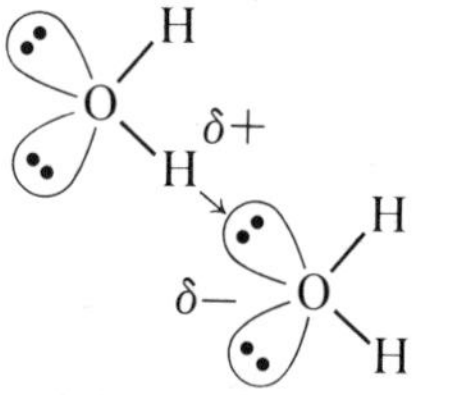

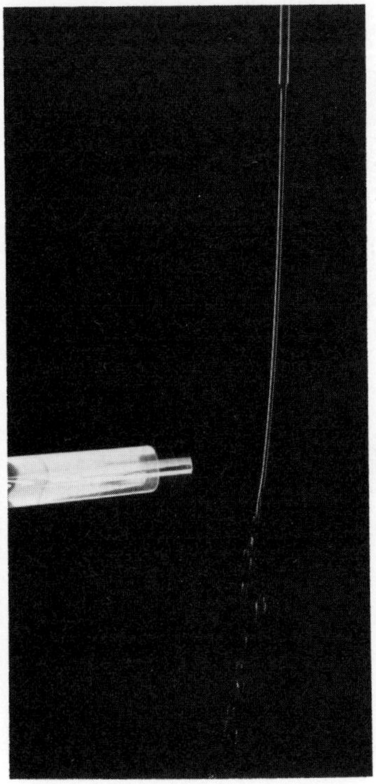

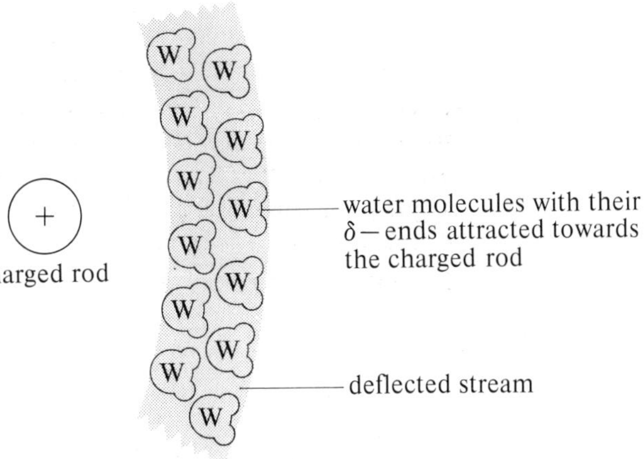

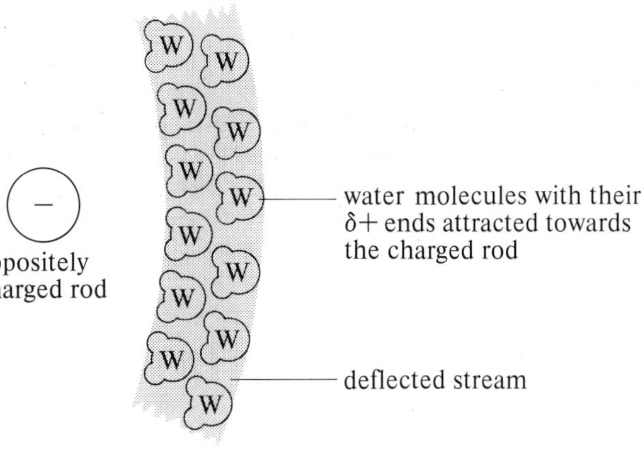

The effect of water molecules on ionic lattices

Ionic solids contain cations and anions arranged in an ordered way in three dimensions. This arrangement is called an **ionic lattice**. Read page 97 again to remind yourself of ionic lattices.

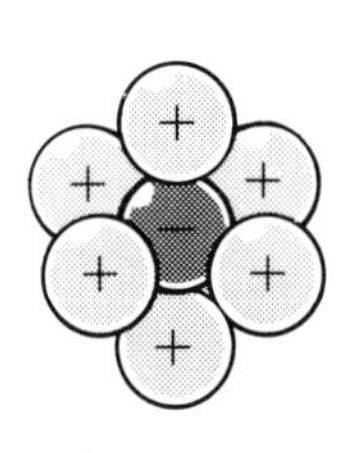

Each ion is surrounded by six other ions in the lattice.

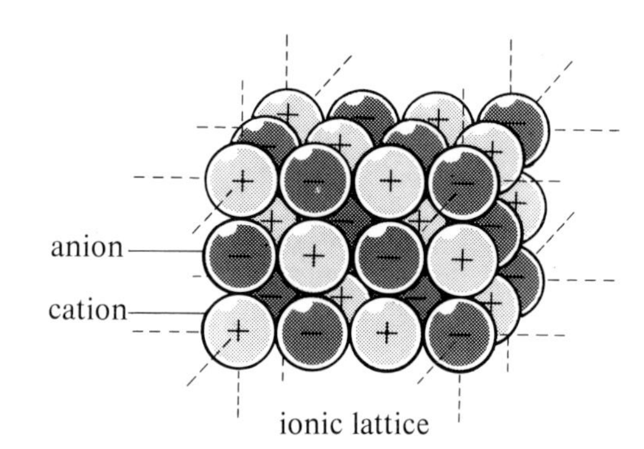
anion
cation
ionic lattice

The charged ions in the lattice have the same effect on water molecules as charged rods do. The ions attract the water molecules and the forces holding the lattice together are weakened. Single ions break free from the lattice, each surrounded by a cluster of water molecules:

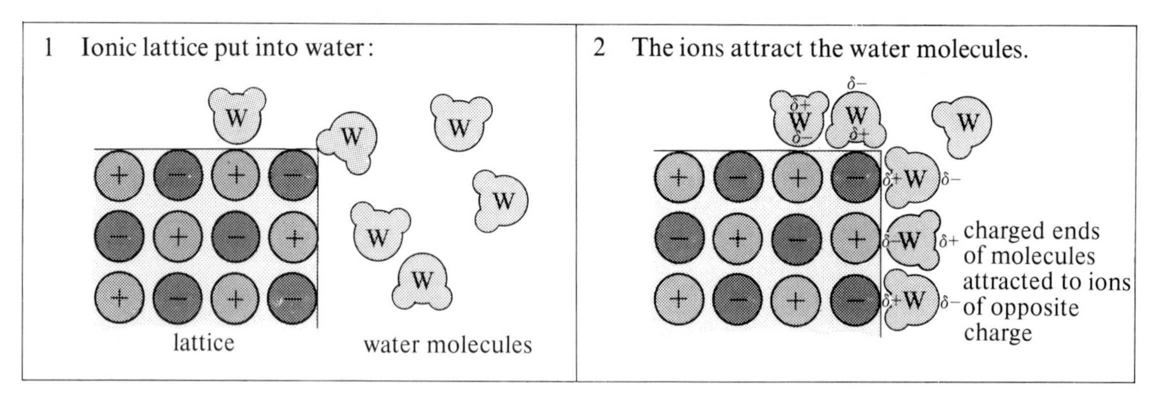
1 Ionic lattice put into water:
lattice
water molecules
2 The ions attract the water molecules.
charged ends of molecules attracted to ions of opposite charge

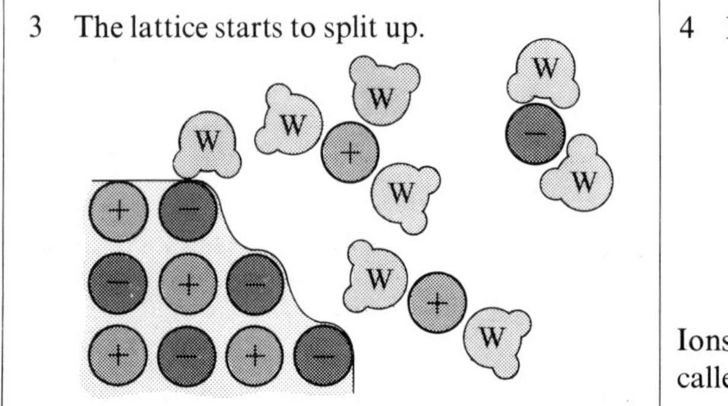
3 The lattice starts to split up.

4 Free ions go into solution.

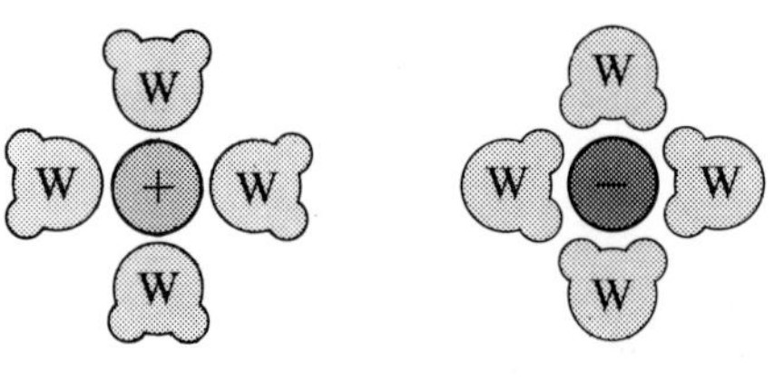

Ions surrounded by water molecules are called **aquo-ions**.

A special symbol is used to represent an ion surrounded by water molecules: **(aq)**; (aq) stands for 'aqueous'. We can write an equation to show a typical ionic solid dissolving in water, for example common salt, $Na^+ Cl^-$:

$$NaCl_{(s)} \longrightarrow Na^+_{(aq)} + Cl^-_{(aq)}$$

$Na^+_{(aq)}$ means a sodium ion surrounded by water molecules.
Compare a sodium ion in solid sodium chloride with a sodium ion in salt solution:

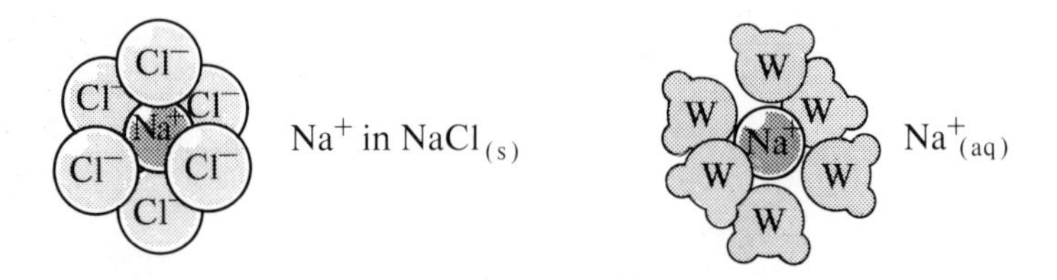

When the lattice dissolves in water, the ions separate from each other; this means they can behave quite independently of each other. This is a very important idea. Think of solutions of copper sulphate and of copper chloride, for example: both will have certain properties in common, because both contain copper aquo-ions. It is these that make both of them blue. However, they will have some different properties as well, because one contains sulphate aquo-ions while the other has chloride aquo-ions:

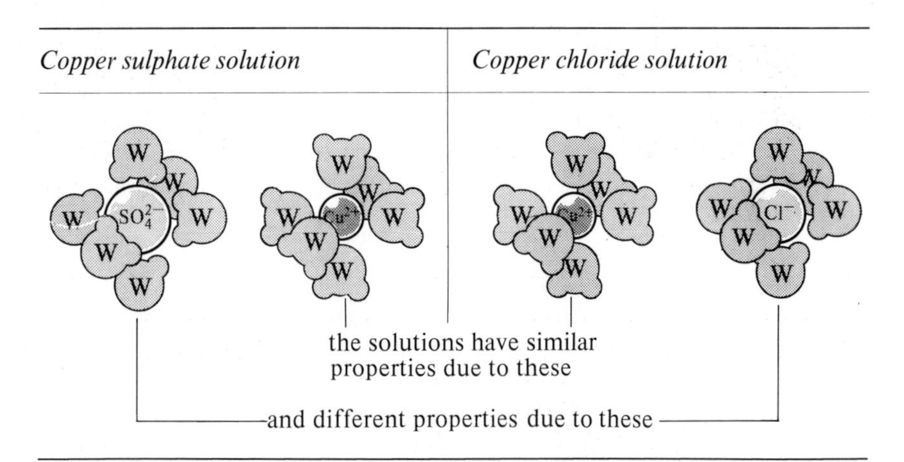

Precipitation and spectator ions

One very common reaction between aquo-ions is precipitation. You read the facts about it in section 13.1. It is the reverse of the dissolving process. During precipitation, oppositely charged aquo-ions attract one another. As they move together their clusters of water molecules are forced away. A solid lattice forms. For example, when a solution of silver nitrate is mixed with a solution of sodium chloride, solid particles of silver chloride are produced:

$$\underset{\text{solution}}{AgNO_3} + \underset{\text{solution}}{NaCl} \longrightarrow \underset{\text{solid}}{AgCl} + \underset{\text{solution}}{NaNO_3}$$

Earlier in the chapter you saw that *any* metal chloride will produce silver chloride solid when added to silver nitrate solution. This is another example of ions behaving independently in solution. The silver ions react with the chloride ions to form solid silver chloride. None of the other ions present are involved at all in the reaction:

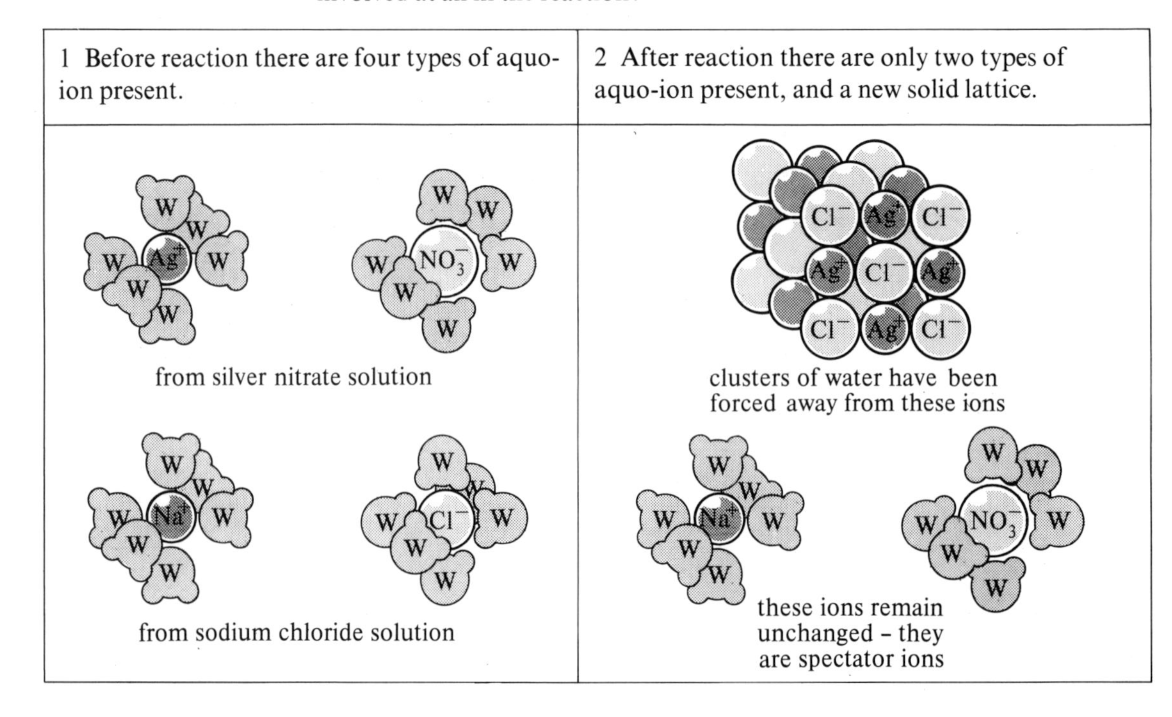
1 Before reaction there are four types of aquo-ion present.
2 After reaction there are only two types of aquo-ion present, and a new solid lattice.
W
Ag^+
NO_3^-
from silver nitrate solution
Na^+
Cl^-
from sodium chloride solution
clusters of water have been forced away from these ions
these ions remain unchanged - they are spectator ions

Since the only ions that change are the silver and chloride ions, these are the only ones usually written in the equation. The equation becomes:

$$Ag^+_{(aq)} + Cl^-_{(aq)} \longrightarrow AgCl_{(s)}$$

This is a particle equation, showing only the particles that have changed during the reaction. Since the sodium and nitrate ions take no part in the reaction they are called **spectator ions** and do not appear in the equation.

Similarly the reaction between barium chloride solution and copper sulphate solution is described as:

$$Ba^{2+}_{(aq)} + SO_4{}^{2-}_{(aq)} \longrightarrow BaSO_{4\,(s)}$$

and the precipitation of a metal hydroxide is shown, for example, as:

$$Fe^{3+}_{(aq)} + 3OH^-_{(aq)} \longrightarrow Fe(OH)_{3\,(s)} \quad \text{or}$$

$$Cu^{2+}_{(aq)} + 2OH^-_{(aq)} \longrightarrow Cu(OH)_{2\,(s)}$$

Summary

Water, or hydrogen oxide, is a common solvent.

Water is polar and has unusually high fixed points.

It is polar because the oxygen atom in the water molecule attracts the shared electron pairs away from the hydrogen atoms.

It has unusually high fixed points because there is an extra force of attraction between water molecules, called hydrogen bonding.

Water tends to dissolve ionic solids, because of its polarity.

Solutions of substances in water are called aqueous solutions.

Precipitation is a chemical reaction between solutes in solution, during which an insoluble product is formed.

The core-charge of an atom is the charge of its nucleus minus the number of inner-shell electrons, which act as a shield.

Questions

1 (a) Draw a water molecule and show how the charge is spread in it.
(b) Draw two water molecules next to each other and the same way round. Mark in any forces of attraction or repulsion between the two molecules.
(c) Draw a clear diagram showing how water molecules arrange themselves around a positively-charged ion.
(d) Repeat part (c) for a negatively-charged ion.
(e) How do you think water molecules are arranged in ice? Draw a diagram to show this.

2 (a) Explain what the term *polar solvent* means.
(b) Why is water a polar solvent?
(c) How would you demonstrate that water is polar?
(d) Name two other polar solvents.

3 Water is a liquid at room temperature, which is unusual for a non-metal hydrogen compound. It is also a very bad conductor.
What do these facts tell you about the nature of water particles and the bonding in water?

4 Describe with the help of diagrams how you would prepare a pure sample of silver chloride.

5 (a) Explain what the symbol $_{(aq)}$ written after an ion or molecule means.

6 Think about two solutions, one of copper chloride, the other of copper sulphate.
(a) What properties would they have in common?
(b) What makes them have these common properties?
(c) How could you tell them apart?
Describe a simple test you could do to distinguish them from each other.

7 Sodium chloride is a hard solid with a melting point of 801°C. This means that it is difficult to melt it in even the best of bunsen flames. However, when solid sodium chloride is put into water the crystal falls apart, and the sodium chloride dissolves. Explain these facts.

8 Draw a table like this one:

Solution	*Added to solution of:*		
	$CuSO_4$	KCl	$FeCl_3$
NaOH			
$AgNO_3$			
$BaCl_2$			

9 (a) Draw dot and cross diagrams showing the arrangement of electrons in the following molecules: water, H_2O; ammonia, NH_3; methane, CH_4.
(b) What do they all have in common?
(c) How does the methane molecule differ from the other two?
(d) In which of these three substances do you think the molecules would be most strongly attracted to neighbouring molecules?
(e) Look up the table of physical properties on page 119, and see if there is any data there that might help you in the answer to (d) above. What is it?

10 (a) When salt dissolves in water, the ions that are together in the lattice become separated in the solution. Write an equation for the dissolving of salt in water.
(b) Silver chloride is made by a precipitation reaction in which silver ions and chloride ions in solution join together. Write an equation for the precipitation of silver chloride.
(c) How are these two processes of dissolving and precipitation related to each other?

14 Acids and bases

14.1 Ammonia and hydrogen chloride gases

Ammonia (NH_3) and hydrogen chloride (HCl) gases are both extremely soluble in water. They both can be used in the 'fountain experiment':

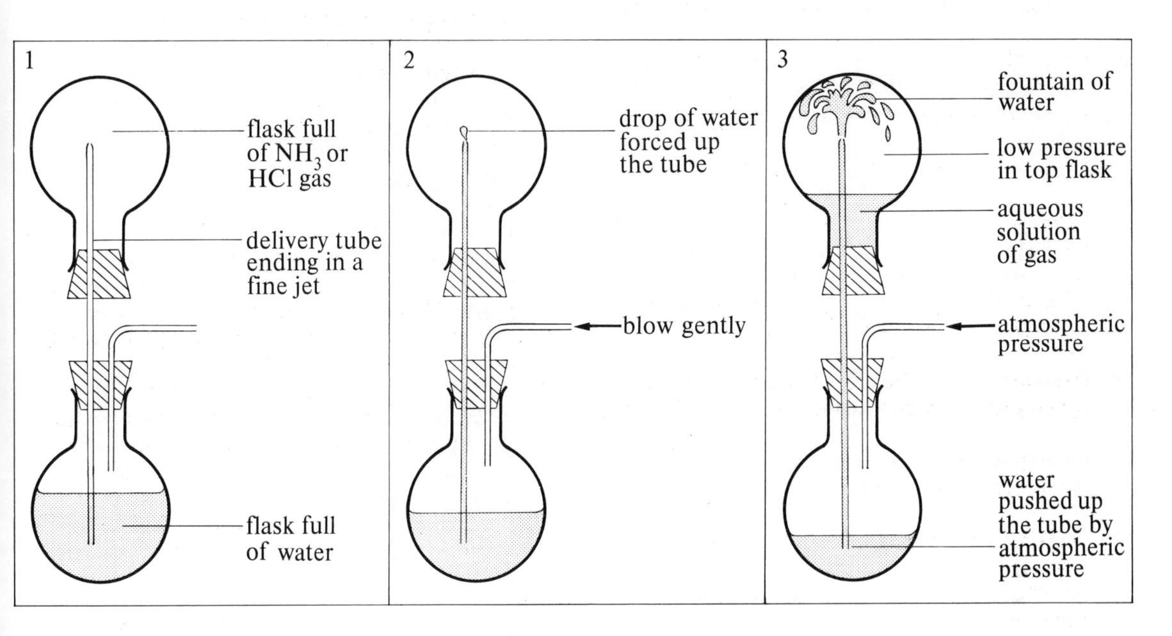

By blowing very gently into the side arm of the apparatus above, you can increase the pressure on the water in the bottom flask. So water is forced up the tube towards the top flask (step 2 in diagram). When the first drop of water reaches the top flask, stop blowing. If the gas in the top flask is very soluble, like ammonia or hydrogen chloride, almost all of it will dissolve in this one drop. The pressure in the top flask will therefore fall to a very low value. The pressure in the bottom flask will be greater than that in the top flask; the water will rush up the tube and into the top flask like a fountain.

Using this experiment you can show that ammonia and hydrogen chloride are both much more soluble in water than most other gases.
Ammonia is over 40000 times more soluble than air in water at 0°C.
Hydrogen chloride is 17000 times more soluble than air in water at 0°C.

Properties of the aqueous solutions

The aqueous solutions of the two gases:

change the colours of substances called indicators;
conduct electricity much better than pure water.

Indicators are coloured dyes usually obtained from plants. Some common indicators are litmus, phenolphthalein, and methyl orange. Universal indicator, another common one, is a mixture of several indicators.

Table 14.1 Comparing ammonia and hydrogen chloride solutions with pure water

Property	*Pure water*	*Ammonia solution*	*Hydrogen chloride solution*
Effect on litmus	No change	Blue	Red
Effect on universal indicator	Green	Purple-blue	Red
Conductivity (using a meter and electrodes—see page 158)	Almost zero	3000 times that of pure water	350000 times that of pure water

Compounds which dissolve in water and give solutions which:

turn litmus and universal indicator red are acids.
turn litmus and universal indicator blue are alkalis.

So hydrogen chloride is an acid. Ammonia is an alkali.

Reaction between the aqueous solutions

When aqueous solutions of ammonia and hydrogen chloride are mixed, some changes can be observed. These are listed in the table below.

Steps	1 Take 25 cm^3 of ammonia solution.	2 Take 25 cm^3 of hydrogen chloride solution.	3 Pour one into the other.
Temperature/°C	16·0	16·0	After mixing, 16·7
Colour of added universal indicator	Blue	Red	After mixing, green
Conductivity	About 3000 times greater than pure water.	About 350000 times greater than pure water.	About 40000 times greater than pure water.

The reaction has all the properties of a chemical change.

A new substance is produced:	it turns indicator green; its solution conducts better than water.
An energy exchange happens:	the temperature goes up; the reaction is exothermic.
Fixed amounts are needed:	if a little extra hydrogen chloride solution is added the indicator will end up red; if too much ammonia solution is added the indicator will end up blue.

The new product is neither an acid nor an alkali. It is **neutral. Neutral compounds do not affect litmus, and leave universal indicator green.**

14.2 Acids and alkalis

Indicators and pH

Table 14.2 How some aqueous solutions affect indicators

Substance in aqueous solution		*With litmus*	*With universal indicator*
Hydrogen chloride	HCl	Red	Red
Hydrogen sulphate(VI)	H_2SO_4		
Hydrogen nitrate(V)	HNO_3		
Sugar (glucose)	$C_6H_{12}O_6$	Unchanged	Green
Salt (sodium chloride)	$NaCl$		
Ammonia	NH_3	Blue	Blue
Sodium hydroxide	$NaOH$		
Sodium carbonate	Na_2CO_3		

You can therefore rearrange the substances in the list above according to their effect on the two indicators:

Acids	*Neutral substances*	*Alkalis*
Hydrogen chloride	Sugar	Ammonia
Hydrogen sulphate(VI)	Salt	Sodium hydroxide
Hydrogen nitrate(V)		Sodium carbonate

Special names are given to the aqueous solutions of the acidic substances:

Compound dissolved in water		*Name of solution*
Hydrogen chloride	HCl	Hydrochloric acid
Hydrogen sulphate(VI)	H_2SO_4	Sulphuric acid
Hydrogen nitrate(V)	HNO_3	Nitric acid

We sometimes use a special scale of numbers as a way of saying whether a solution is acidic, neutral, or alkaline. The scale is called the **pH scale**. The numbers usually range from 0 to 14:

if a solution has a pH below 7, it is acidic;
if a solution has a pH of 7, it is neutral;
if a solution has a pH above 7, it is alkaline.

Examples:

Hydrochloric acid	pH = 1
Water	pH = 7
Ammonia	pH = 12
Sodium hydroxide	pH = 14

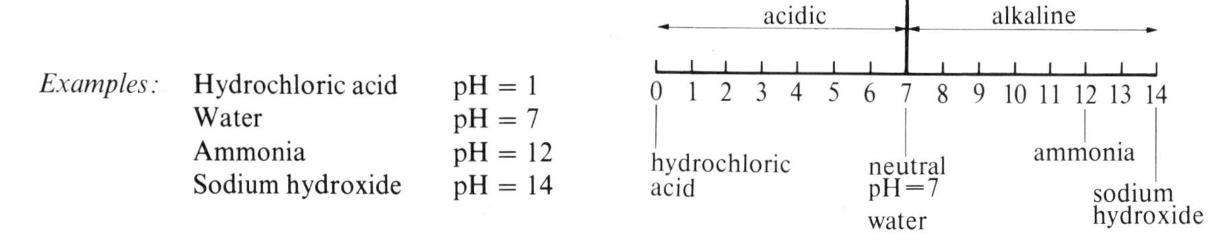

The colour an indicator goes, when you add it to a solution, depends on the pH of the solution. So you can use indicators to give you an idea of the pH of any solution. Table 14.3 shows the different colours.

Table 14.3 How indicators change colour with pH

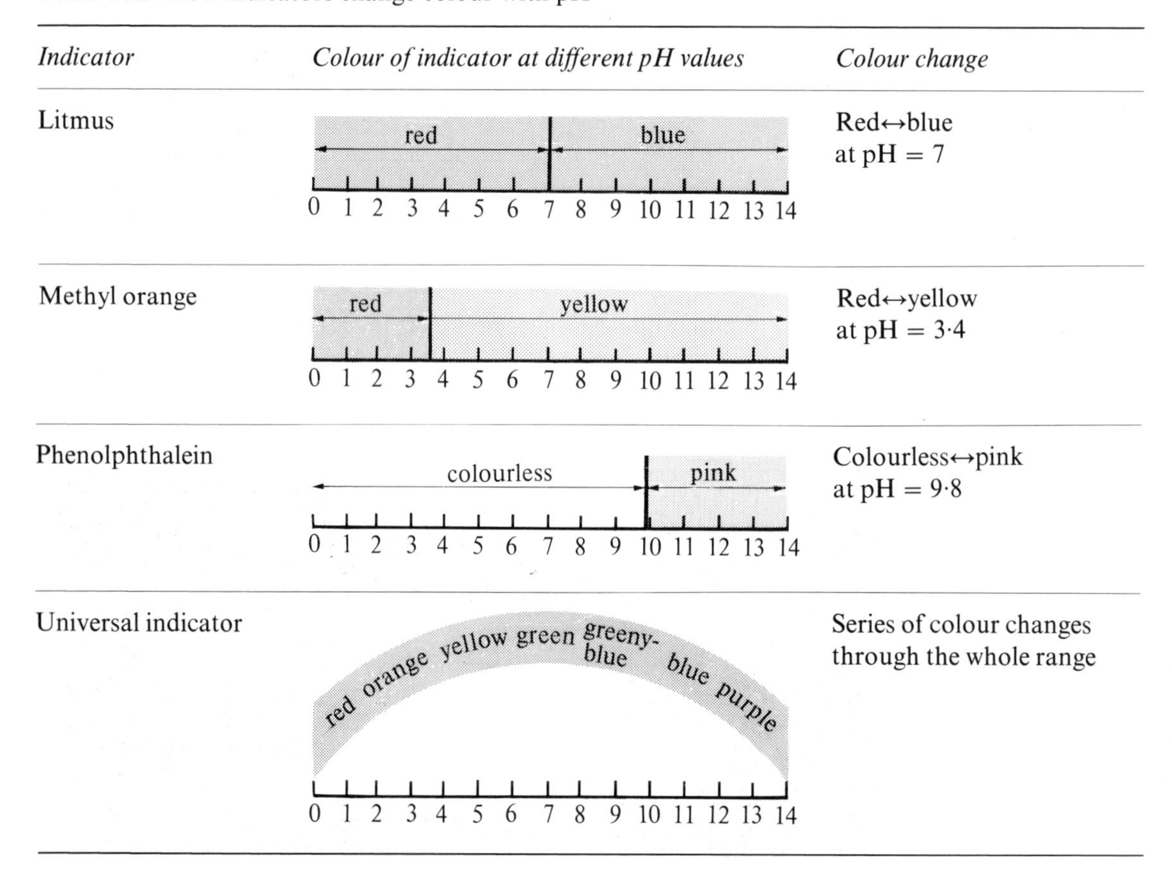

Indicator	*Colour of indicator at different pH values*	*Colour change*
Litmus	red (0–7) / blue (7–14)	Red↔blue at pH = 7
Methyl orange	red (0–3·4) / yellow (3·4–14)	Red↔yellow at pH = 3·4
Phenolphthalein	colourless (0–9·8) / pink (9·8–14)	Colourless↔pink at pH = 9·8
Universal indicator	red orange yellow green greeny-blue blue purple (0–14)	Series of colour changes through the whole range

14.3 Neutralization

Neutralisation and salts

At the beginning of the chapter, we saw the effect of adding an *acidic* solution (hydrochloric acid) to an *alkaline* solution (ammonia). A *neutral* solution is produced by the chemical change. We say that ammonia **neutralises** hydrochloric acid.

Acidic solution	*plus*	*alkaline solution*	*produces*	*neutral solution*
Hydrochloric acid	+	ammonia	⟶	The neutral substance is called a **salt**

When an acid is neutralised by an alkali, the products are a salt and water.
Common salt, sodium chloride, is made by neutralising the acid hydrochloric acid by the alkali sodium hydroxide.
Hydrochloric acid solution + sodium hydroxide solution ⟶
neutral sodium chloride solution
In everyday life, we rely on neutralisation in a number of important ways. For example:

Agriculture — The soil often contains acidic solutions absorbed in it. Plants can best take up most of the minerals they need when the soil is very slightly alkaline. So farmers neutralise the acids in the soil by using the alkali calcium hydroxide, 'lime'.

Medicine — Blood has a pH of 7·3 which means it is slightly alkaline. Injections of drugs or fluids must contain liquids of almost exactly the same pH. Changing the pH of a person's blood by one pH unit would kill him.

Digestion — The fluids in human stomachs are acidic, pH 1 or 2. Too much acidity, however, causes indigestion. One can neutralise the extra acidity by swallowing an alkaline substance, for example, sodium bicarbonate.

Left: this antacid powder is a mixture of the bases calcium carbonate and magnesium carbonate. For bases and neutralisation see next page.
Right: lime being spread on acid soil.

Bases

You have seen that alkalis are soluble substances that can neutralise acids. But alkalis are not the only substances that can neutralise acids. *All* the oxides, hydroxides and carbonates of metals can do it, even though many of them are insoluble in water. The examples below use insoluble compounds.

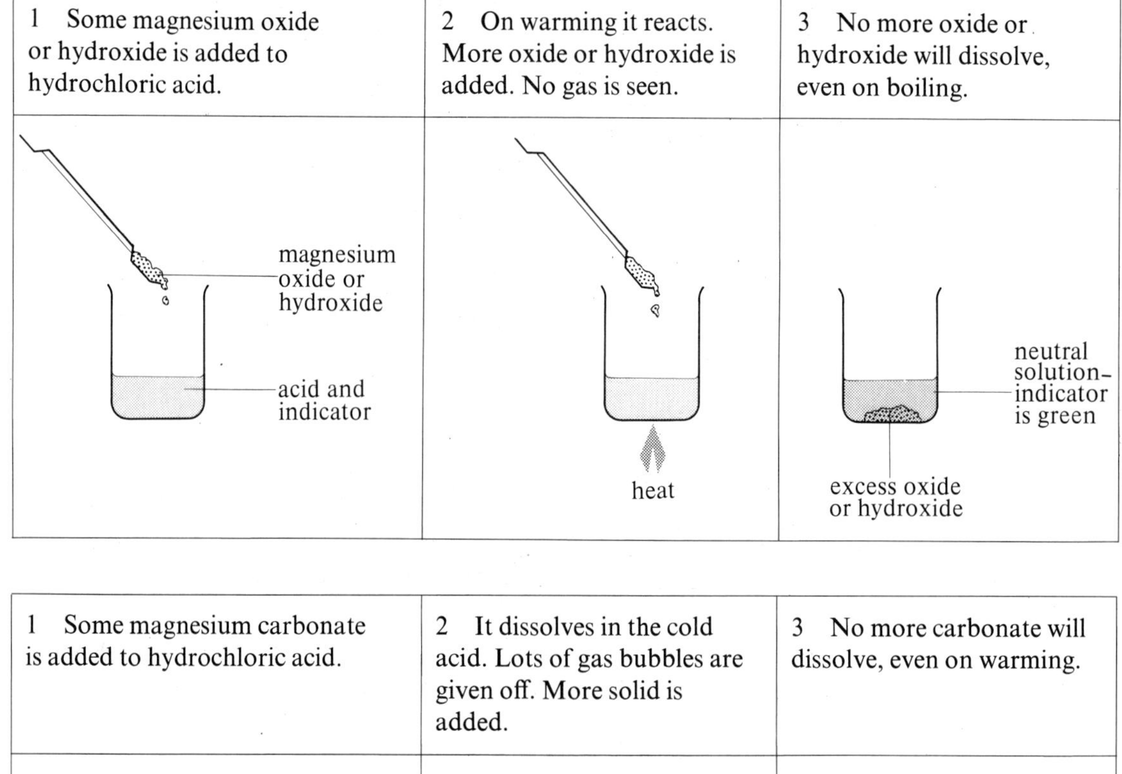

A general name is given to all substances that can neutralise acids. They are called **bases**.

A base is any substance that can neutralise an acid to produce a salt.

The bases are:

metal oxides
metal hydroxides
metal carbonates.

In the experiments on page 140, magnesium oxide, magnesium hydroxide, and magnesium carbonate were used. The neutral solutions all contain the same salt, magnesium chloride, $MgCl_2$. We can show this by filtering the excess solid off each solution and then boiling the solution to remove most of the water. When the remaining solution cools, white crystals are obtained in each case. The crystals all melt at 714°C—the melting-point of magnesium chloride. So we can write these equations for the reactions:

$$MgO + 2HCl \longrightarrow MgCl_2 + H_2O$$
$$Mg(OH)_2 + 2HCl \longrightarrow MgCl_2 + 2H_2O$$
$$MgCO_3 + 2HCl \longrightarrow MgCl_2 + H_2O + CO_2$$

All these experiments can be repeated using a different acid or a different metal compound. The results always follow the same pattern:

Zinc oxide	ZnO	neutralise sulphuric acid, H_2SO_4, to give the salt zinc sulphate, $ZnSO_4$
Zinc hydroxide	$Zn(OH)_2$	
Zinc carbonate	$ZnCO_3$	
Copper oxide	CuO	neutralise nitric acid, HNO_3, to give copper nitrate, $Cu(NO_3)_2$
Copper hydroxide	$Cu(OH)_2$	
Copper carbonate	$CuCO_3$	

Can you see any similarities in these different neutralisations? In each one, the hydrogen part of the acid is replaced by a metal, to give a salt:

$$2HCl \longrightarrow MgCl_2$$
$$H_2SO_4 \longrightarrow ZnSO_4$$
$$2HNO_3 \longrightarrow Cu(NO_3)_2$$

HCl and HNO_3 are called **monobasic** acids because they each have only one H to replace. H_2SO_4 is a **dibasic** acid because there are two H's to replace.

Summary To summarise, here are some useful rules to remember about neutralising acids:

acid + metal oxide $\longrightarrow$ salt + water
acid + metal hydroxide $\longrightarrow$ salt + water
acid + metal carbonate $\longrightarrow$ salt + water + carbon dioxide gas

Acid salts When a dibasic acid is completely neutralised, both hydrogens are replaced:

$$H_2SO_4 + 2NaOH \longrightarrow \underset{\text{sodium sulphate}}{Na_2SO_4} + 2H_2O$$

$$H_2CO_3 + 2KOH \longrightarrow \underset{\text{potassium carbonate}}{K_2CO_3} + 2H_2O$$

but it is also possible to replace only one hydrogen:

$$H_2SO_4 + NaOH \longrightarrow \underset{\text{sodium hydrogen sulphate}}{NaHSO_4} + H_2O$$

$$H_2CO_3 + KOH \longrightarrow \underset{\text{potassium hydrogen carbonate}}{KHCO_3} + H_2O$$

In this case the salts produced are **acid salts** because they still have a replaceable hydrogen. In solution they are more acidic than the normal salt but less acidic than the acid:

Solution	Na_2SO_4	$NaHSO_4$	H_2SO_4
pH	7	3	1

14.4 Making pure salts

A pure salt is usually made by neutralising the correct acid by a base:

Salt needed	*Acid to use*	*Example*
A chloride	Hydrochloric acid	$ZnO + 2HCl \longrightarrow ZnCl_2 + H_2O$
A sulphate(VI)	Sulphuric acid	$CuO + H_2SO_4 \longrightarrow CuSO_4 + H_2O$
A nitrate(V)	Nitric acid	$CaCO_3 + 2HNO_3 \longrightarrow Ca(NO_3)_2 + H_2O + CO_2\uparrow$

Nearly all bases are insoluble in water. Those that do dissolve are called **alkalis** and give **alkaline solutions**.
The common alkaline solutions found in the laboratory are:

sodium hydroxide solution
potassium hydroxide solution
solution of ammonia in water—aqueous ammonia
calcium hydroxide solution
barium hydroxide solution

We use two main methods for making salts, and these are described on the following pages. Method A is for using with soluble bases (alkaline solutions) and method B is for insoluble bases. Both methods have one aim: to get exactly the right amounts of acid and base to react. An excess of one reactant makes the final product—the salt—impure.

Method A To make a pure salt, starting with a soluble base (alkali)

Steps		*Example*
1 Measure out a volume of alkaline solution, say 20 cm^3. Add some phenolphthalein indicator, then run acid into the flask of alkali from a burette.	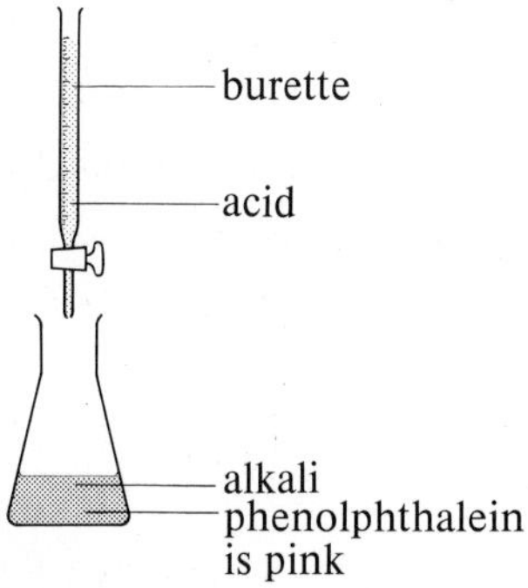 burette; acid; alkali; phenolphthalein is pink	Alkali: sodium hydroxide Acid: hydrochloric acid
2 Continue adding acid until one more drop of acid is *just* enough to turn the solution from pink to colourless. The base has now been neutralised.	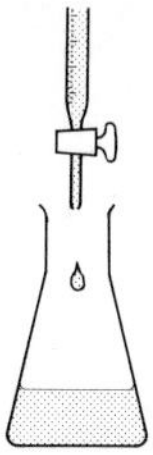	Acid + alkali → salt + water $HCl + NaOH \rightarrow NaCl + H_2O$
3 Measure the volume of acid needed for the neutralisation, from the reading on the burette.	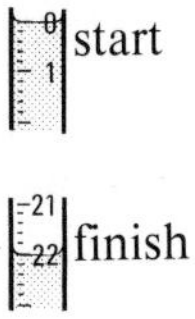 0; 1; start; 21; 22; finish	Start: 0 cm^3 Finish: 22 cm^3 therefore 22 cm^3 used
4 Repeat the process, this time without adding indicator. Run the measured volume of acid from the burette, into the 20 cm^3 of alkali in the flask.	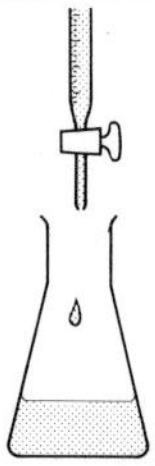	Add 22 cm^3 of acid to 20 cm^3 of alkali
5 Evaporate off most of the water from the solution, over a water bath. Leave the remaining solution to cool, so that the salt can crystallise out. Dry off the crystals on filter paper.	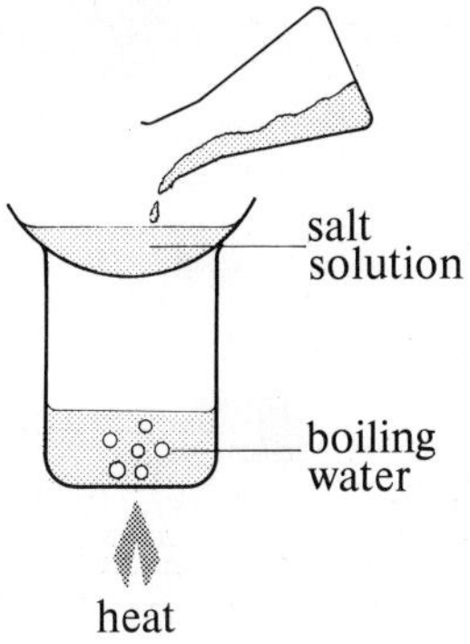 salt solution; boiling water; heat	

Method B To make a pure salt, starting with an insoluble base

Steps		*Example*
1 Measure out a volume of acid solution. Add a spatula-full of solid—metal oxide, metal hydroxide or metal carbonate—and warm the beaker.	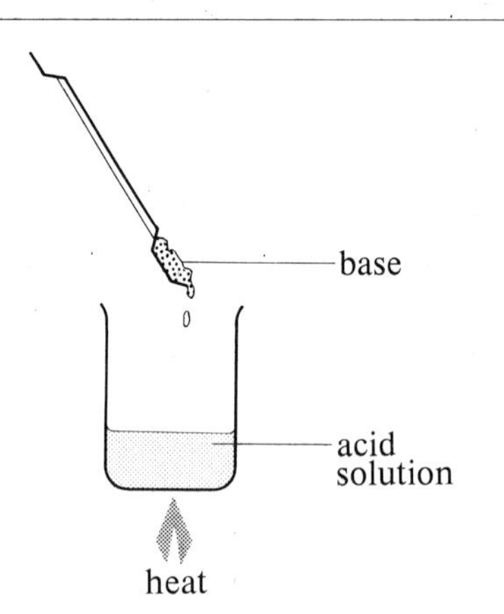	Base: copper(II) oxide Acid: sulphuric acid
2 Continue adding solid reactant to the solution until no more dissolves. Warm and stir the solution to make sure the reaction is complete.	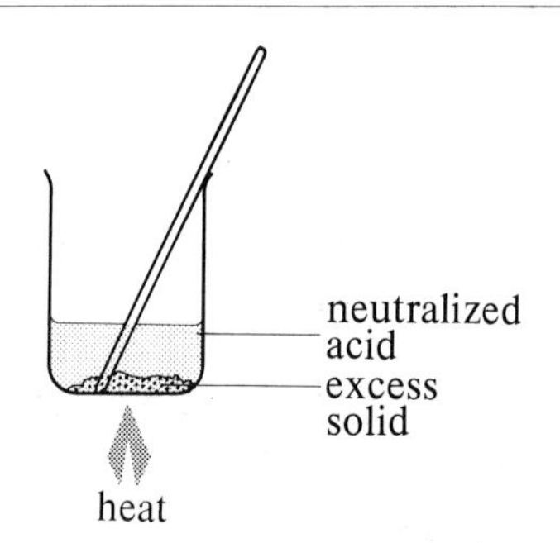	base + acid → salt + water $CuO + H_2SO_4 \rightarrow CuSO_4 + H_2O$
3 Filter off the excess solid and collect the filtrate which is a pure salt solution.	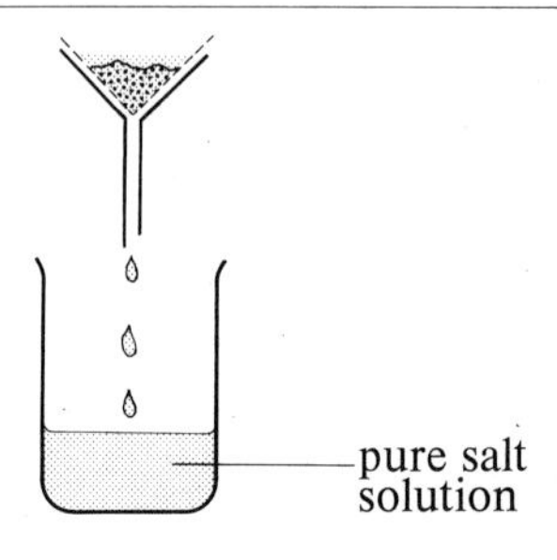	
4 Evaporate off most of the water from the solution over a warm bath. Leave the remaining solution to cool, so that the salt can crystallise out. Dry off the crystals on filter paper.	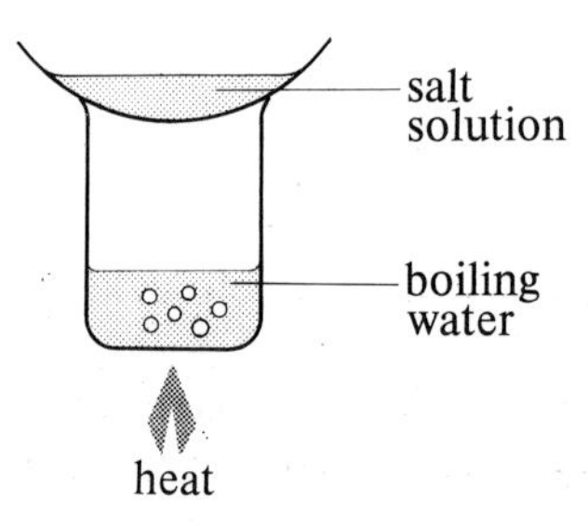	

14.5 Proton transfer: the theory of acids and bases

We have already met a number of facts about acids, bases, neutralisation and salts. We must now explain them. We shall start with the two gases hydrogen chloride and ammonia, dissolving in water—see page 135 again. Pure hydrogen chloride and pure ammonia are both molecular substances; so too is pure water.

Polar molecules You may remember from Chapter 13 that the bonding electron pairs in water are unevenly shared. They are attracted more towards the oxygen atom. This results in the hydrogen atoms being slightly positive and the oxygen atom slightly negative so that the water molecule is polar.

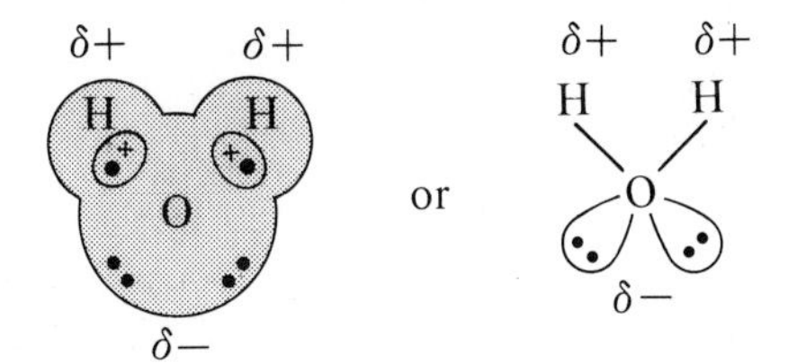

For exactly the same reasons both hydrogen chloride and ammonia molecules are polar. Chlorine and nitrogen atoms both have more protons than the hydrogen atom, therefore both have a greater core charge than it. So in HCl and NH_3 the hydrogen atoms are slightly positive, $\delta+$, and the chlorine and nitrogen atoms are slightly negative, $\delta-$.

Hydrogen chloride	*Water*	*Ammonia*
δ+ H Cl δ− lone pairs	δ+ δ+ H H O δ− lone pairs	δ+ H δ+ H N H δ+ δ− lone pair
or H Cl	or H H O	or H H H N

Molecules to ions We do not expect molecular liquids to conduct electricity, because they do not contain charged particles free to move through the system. We expect only solutions of ions, or ions in the liquid state, to conduct electricity. And yet, from page 136, you can see that both hydrogen chloride solution and ammonia solution *do conduct*, much better than pure water. Somehow, uncharged particles must have changed to give free, charged particles—molecules must have changed to ions.

Proton transfer: acids

We must find a theory to explain how molecules can change into ions. Let us think about hydrogen chloride in water first, and the way in which the hydrogen chloride and water molecules might collide with each other. Since both molecules are polar, it is likely that the more negative end of one (a lone pair) will collide with the more positive end of the other.

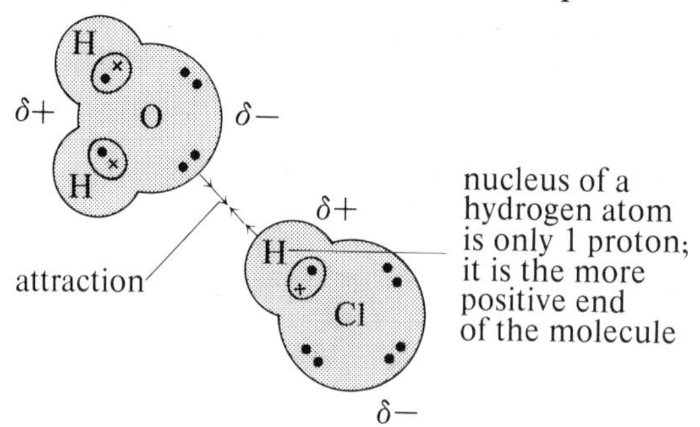

If the attractive force of the oxygen lone pair is strong enough, it can split the hydrogen chloride molecule in two. The nucleus of the hydrogen atom—which is just a proton—is torn away from the molecule.

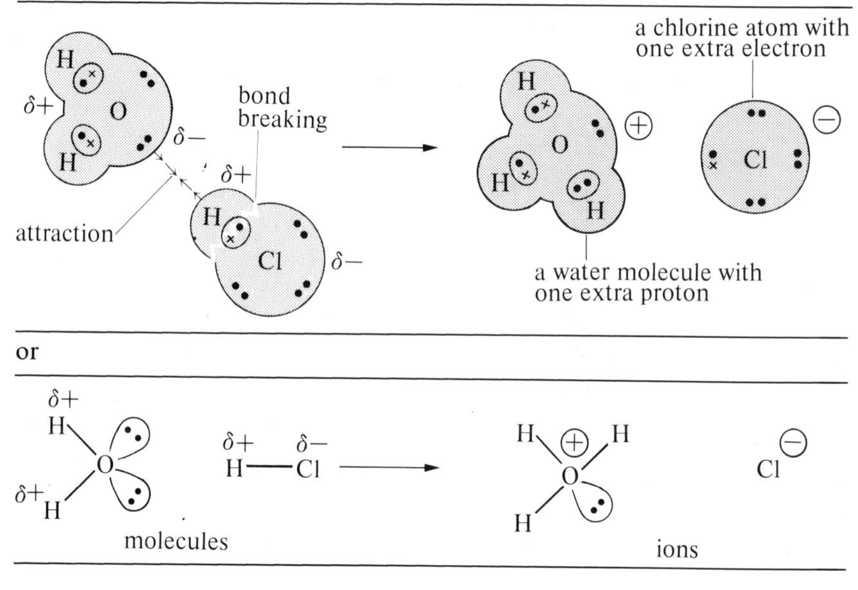

You can see from the above diagram that:

(i) the water molecule has gained a hydrogen nucleus (a proton);

(ii) the hydrogen chloride molecule has lost a hydrogen nucleus but kept its bonding electron;

(iii) molecules have been changed into ions.

In other words, a proton has been transferred from hydrogen chloride to water. We say that water has been **protonated**.
A simple equation, which represents all these changes, is:

$$H_2O_{(l)} + HCl_{(g)} \longrightarrow H_3O^+_{(aq)} + Cl^-_{(aq)}$$

molecules ions

The protonated water molecule, whose formula is $H_3O^+_{(aq)}$, is called the **hydroxonium ion.**
Protonated water molecules (hydroxonium ions) cause litmus and universal indicator to turn red.
All acids cause litmus and universal indicator to turn red. So all acids must react with water to produce hydroxonium ions, $H_3O^+_{(aq)}$. Compare the three acids below:

hydrochloric acid $H_2O_{(l)} + HCl_{(g)} \longrightarrow H_3O^+_{(aq)} + Cl^-_{(aq)}$

nitric acid $H_2O_{(l)} + HNO_{3\,(l)} \longrightarrow H_3O^+_{(aq)} + NO_3{}^-_{(aq)}$

sulphuric acid $2H_2O_{(l)} + H_2SO_{4\,(l)} \longrightarrow 2H_3O^+_{(aq)} + SO_4{}^{2-}_{(aq)}$

Notice that each H_2SO_4 molecule can produce two $H_3O^+_{(aq)}$ ions.

Proton transfer: alkalis

Ammonia solution turns universal indicator blue. This shows that hydroxonium ions are *not* produced when ammonia molecules meet water molecules. However, just as a water molecule can be protonated, it can also be **deprotonated**. In other words, it can have a proton removed from it. Look at this diagram:

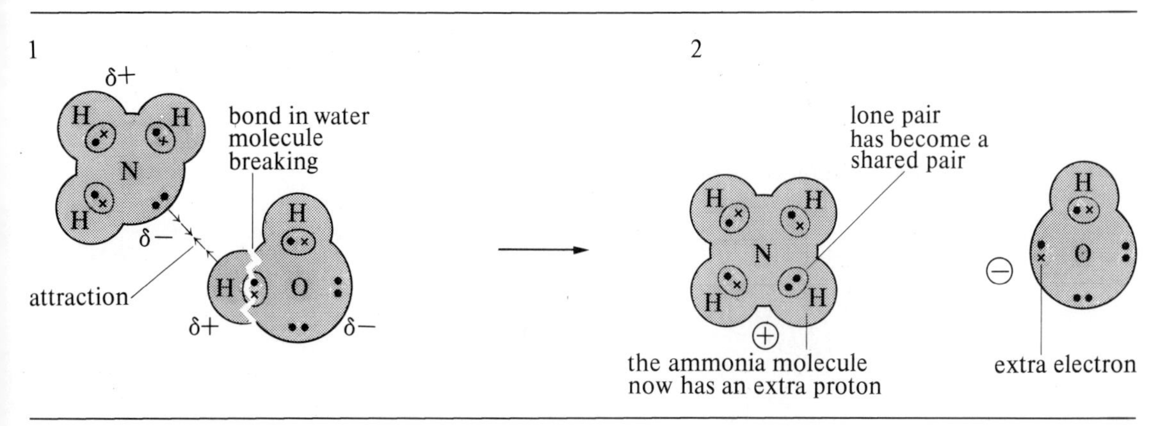

or

From the diagram you can see that:

(i) the water molecule has lost a hydrogen nucleus (a proton), but kept its bonding electron;
(ii) the ammonia molecule has gained a hydrogen nucleus;
(iii) molecules have again been made into ions.

This time, a proton has been transferred from water to ammonia. The water has been deprotonated. A simple equation for these changes is:

$$\underbrace{NH_{3\,(g)} + H_2O_{(l)}}_{\text{molecules}} \longrightarrow \underbrace{NH^+_{4(aq)} + OH^-_{(aq)}}_{\text{ions}}$$

Ammonia solution conducts about a hundred times less well than hydrogen chloride solution. This suggests that not all the molecules are turned into ions in an ammonia solution.
The deprotonated water molecule, whose formula is $OH^-_{(aq)}$, is called the **hydroxide ion.**
Deprotonated water molecules (hydroxide ions) cause litmus and universal indicator to turn blue.
All alkalis cause litmus and universal indicator to turn blue. So all alkalis must produce a solution of hydroxide ions, $OH^-_{(aq)}$ in water. Compare the three alkalis below:

ammonia	$NH_{3\,(g)} + H_2O_{(l)} \rightarrow NH^+_{4(aq)} + OH^-_{(aq)}$
sodium hydroxide	$NaOH_{(s)} + H_2O_{(l)} \rightarrow Na^+_{(aq)} + OH^-_{(aq)}$
calcium hydroxide (lime water)	$Ca(OH)_2 + H_2O_{(l)} \rightarrow Ca^{2+}_{(aq)} + 2OH^-_{(aq)}$

Neutralisation: alkalis

You have now seen three forms of the water particle, two of them produced by proton transfer:

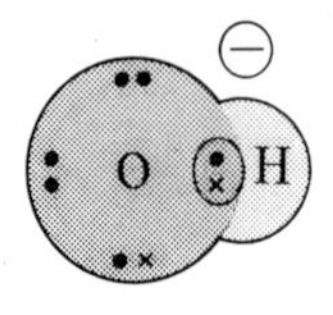

hydroxide ion–deprotonated water

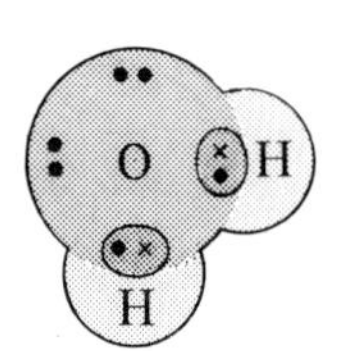

neutral water molecule

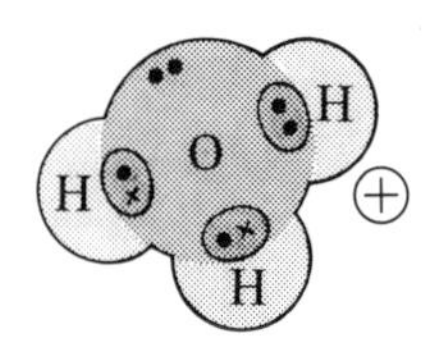

hydroxonium ion–protonated water

The 'proton transfer' theory explains the final set of facts that we met on page 136. When ammonia and hydrogen chloride solutions are mixed in the experiment, the universal indicator turns green. Green is also its colour in pure water. What happens is that the hydroxide ions of the ammonia solution collide with the hydroxonium ions of the hydrogen chloride solution, producing water.

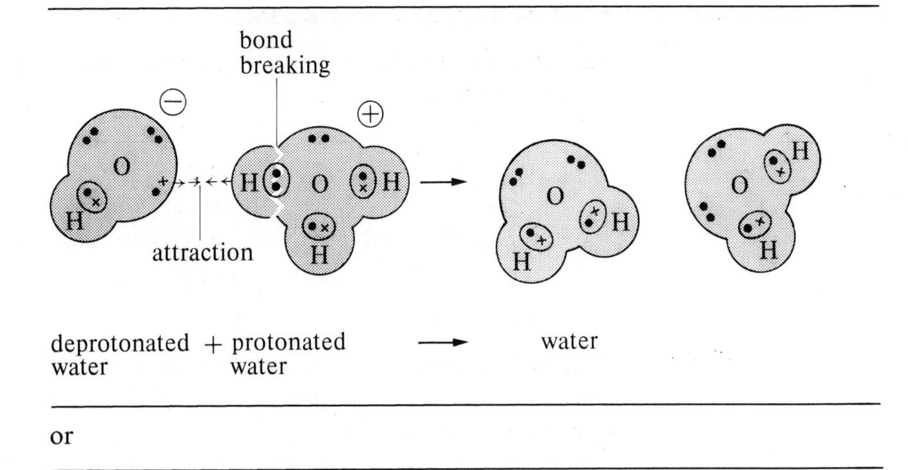

or

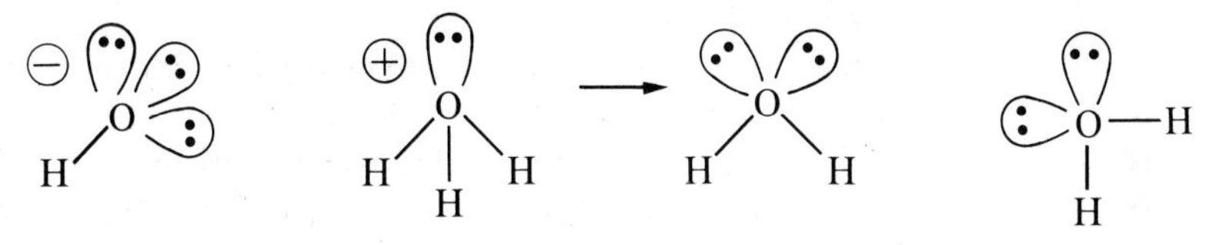

It is this proton transfer which causes neutralisation. The neutralisation of an acid by an alkali can therefore be simply described by the equation:

$OH^-_{(aq)} + H_3O^+_{(aq)} \longrightarrow 2H_2O_{(l)}$

The green or **neutral** colour of the indicator clearly shows that neither hydroxide ions nor hydroxonium ions are present in excess. The solution conducts, however, because it contains ammonium ions (NH_4^+) and chloride ions (Cl^-). These ions have taken no part in the reaction, and do not affect the colour of indicators.

The universal indicator colour green shows a neutral solution—neither hydroxide ions nor hydroxonium ions are present in excess.

Neutralisation: bases Unlike alkalis, most metal oxides and metal carbonates are insoluble. Therefore they do not produce hydroxide ions in solution. However, they are still able to neutralise acids. This means that there are particles in their lattices that can attract protons away from H_3O^+ ions. The hydroxonium ions of the acid solution lose their protons to these particles and become neutral water molecules.

Metal oxides Metal oxides contain the oxide ion O^{2-}. It has a charge of 2 –. Watch what it does in this diagram:

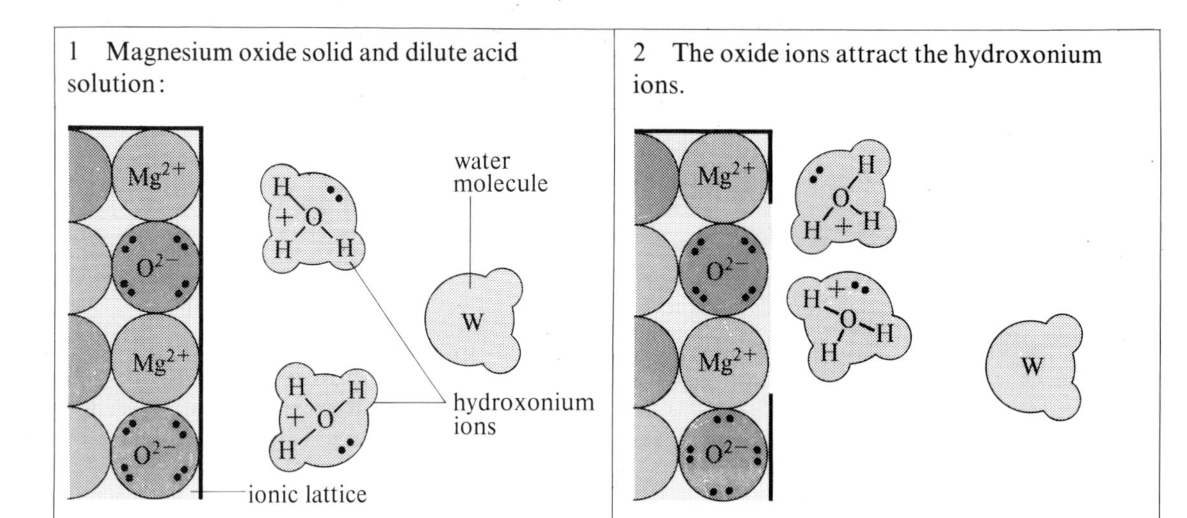

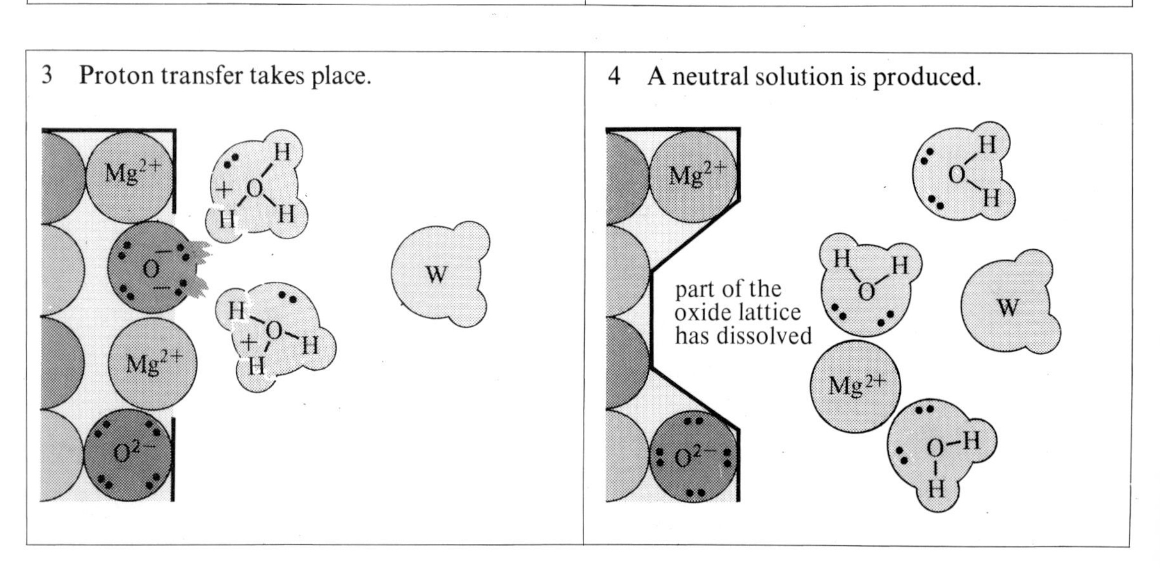

It is the oxide ions in the metal oxide lattice that deprotonate the hydroxonium ions. We can show all the changes in a single equation:

$$MgO_{(s)} + 2H_3O^+_{(aq)} \longrightarrow Mg^{2+}_{(aq)} + 3H_2O_{(l)}$$

Metal carbonates Metal carbonates neutralise acid and carbon dioxide is given off. Just as in the metal oxide lattice, the anions in the metal carbonate lattice can be protonated. The hydroxonium ions lose their extra protons to these anions and become neutral water molecules.

Metal carbonates, like magnesium carbonate, contain the molecular anion CO_3^{2-}:

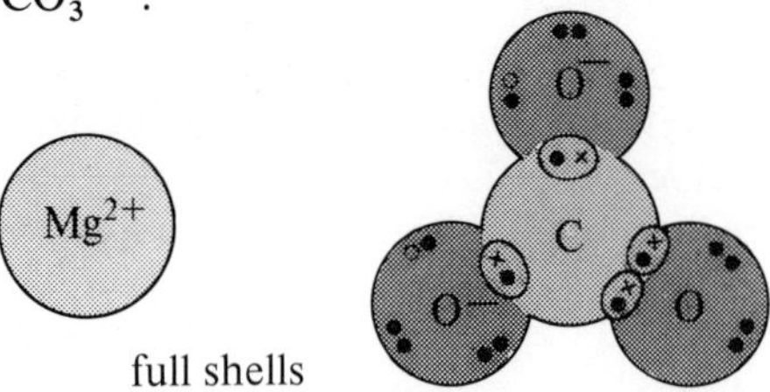

These anions are protonated in the same way as the oxide ions (page 150):

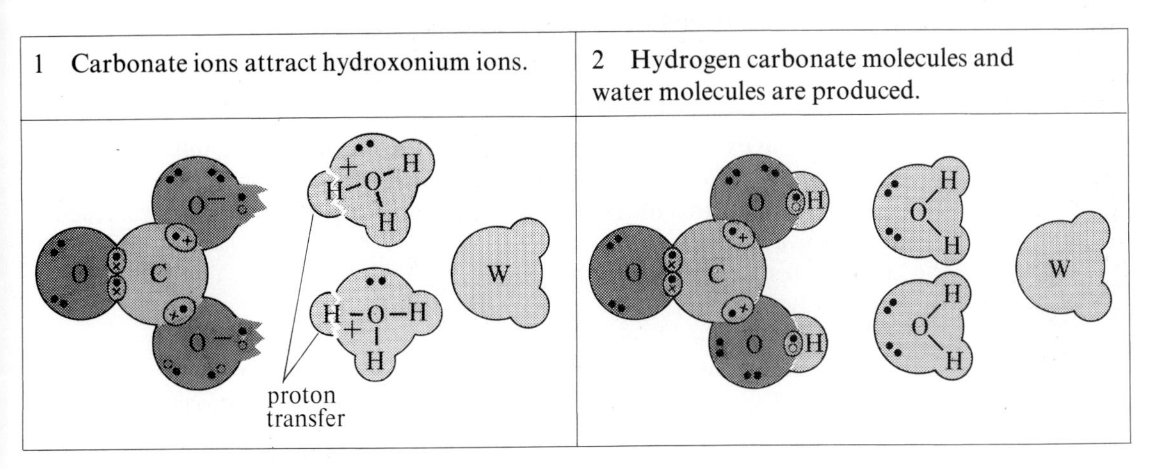

1 Carbonate ions attract hydroxonium ions.

2 Hydrogen carbonate molecules and water molecules are produced.

Hydrogen carbonate molecules are unstable and tend to split in two. Carbon dioxide molecules and water molecules are produced.

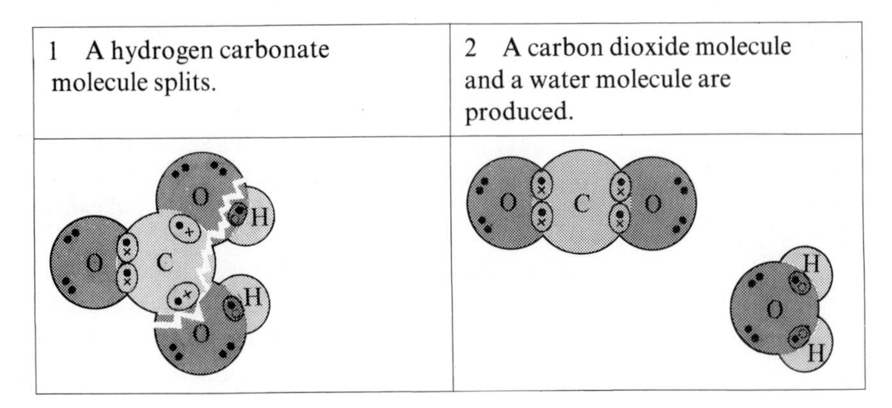

1 A hydrogen carbonate molecule splits.

2 A carbon dioxide molecule and a water molecule are produced.

All these changes can be shown in a single equation. For magnesium carbonate and acid it is:

$$MgCO_{3\,(s)} + 2H_3O^+_{(aq)} \longrightarrow Mg^{2+}_{(aq)} + CO_{2\,(g)} + 3H_2O_{(l)}$$

Summary

The facts and theory of acids, alkalis, bases, and salts are summarized in the two tables below:

Acids and bases

Facts	*Theory*
Acids dissolve in water to give solutions of pH less than 7. The solutions turn litmus and universal indicator red.	Acidic particles are proton donors. They protonate water molecules to give hydroxonium ions, $H_3O^+_{(aq)}$.
Alkalis dissolve in water to give solutions of pH greater than 7. The solutions turn litmus and universal indicator blue.	Alkaline solutions contain deprotonated water molecules, called hydroxide ions, $OH^-_{(aq)}$.
Bases neutralise acids. Common bases are metal oxides, metal hydroxides, and metal carbonates. Most of them do not dissolve in water. The alkalis are the bases that do dissolve in water.	Basic particles are proton acceptors. They can accept protons from hydroxonium ions. Neutral water molecules are produced as a result.
The product of a neutralisation between an acid and a base is called a salt.	The ions that remain in solution after the neutralisation can come together (after evaporation) to form a solid salt lattice.

Testing for acidity

Facts	*Theory*
Acidic solutions: (i) have pH values less than seven; (ii) turn litmus and universal indicator red.	Hydroxonium ions, $H_3O^+_{(aq)}$, affect the colour of the dyes present in the indicators.
Acidic solutions: (i) react with oxides and hydroxides of metals, so the compounds dissolve; (ii) react with metal carbonates to give off carbon dioxide gas.	Hydroxonium ions protonate: (i) oxide or hydroxide ions in their lattices; (ii) carbonate anions in metal carbonate lattices.

Questions

1 Carefully draw diagrams of the water, ammonia, and hydrogen chloride molecules, marking in the polar charges on them.

2 Ammonia and hydrogen chloride both dissolve very well in water.
(a) Write equations for their reactions with water.
(b) What similarities are there in these two reactions?
(c) What differences are there between these two reactions?

3 On page 146 you can see a reaction in which the hydrogen chloride molecule *loses* a proton to a water molecule. On page 147 you can see an ammonia molecule *gaining* a proton from a water molecule.
Write an equation for a reaction between an ammonia and a hydrogen chloride molecule.

4 (a) What is the pH of pure water?
(b) When ammonia is bubbled into water does it become acidic or alkaline? Does the pH of the solution increase or decrease? What colour would the solution be if it contained universal indicator?
(c) When hydrogen chloride is bubbled into water does it become acidic or alkaline? Does the pH of the solution increase or decrease? What happens to the pH if the solution is added to aqueous ammonia solution?

5 What starting materials could be used to make the following salts?
(a) Calcium chloride
(b) Aluminium nitrate(V)
(c) Magnesium sulphate(VI)
(d) Zinc carbonate

6 Write equations for the reactions between your starting materials for question 4.

7 Given some pure iron(II) oxide powder, describe, with diagrams, how you would prepare a pure sample of iron(II) sulphate(VI). Give the relevant equations.

8 Explain carefully, with examples, the meaning of the following terms:
(a) acid (b) alkali (c) base (d) indicator
(e) neutralization

9 On pages 143 and 144 two methods of making salts are illustrated.
(a) Why is an indicator needed in method A, but not method B?
(b) How do you know when the neutralization reaction is complete in method B?
(c) Why is it necessary to repeat the process of running acid into the alkali in method A?
(d) Why, in both methods, is the evaporation carried out on a water bath, instead of heating the salt solution directly?

10 Make a revision sheet comparing the three methods of making salts shown on pages 126, 144, 143.

15 Reactions of metals with water and acids: the electrochemical series

15.1 Reactions of pure metals

Metals and cold water You might find it surprising that some metals react with cold water! We will look at four metals: sodium, magnesium, calcium, and zinc.

Table 15.1 Behaviour of some metals in water

Metal put into beaker of pure water	*Effect on the metal*	*Effect on the water*	*Electrical conductivity of the mixture*
(i) Sodium	Dissolves very violently. Becomes so hot that it melts. Shoots randomly across the surface.	Bubbles of explosive gas are given off. Added indicator turns from green to blue.	About 300000 times better than pure water.
(ii) Magnesium	No observable effects.	If left for a long time, added indicator shows faintly blue near the metal.	Almost the same as pure water.
(iii) Calcium	Dissolves very quickly. A white suspension is produced instead of a solution. (Compare sodium.)	Bubbles of explosive gas are given off. Added indicator turns from green to blue.	About 10000 times better than pure water.
(iv) Zinc	No observable effects.	No observable effects.	The same as pure water.

You can see that sodium is the most **reactive** of these metals. Zinc is the least reactive.

Chemical reactions often go faster when the temperature is higher, so we can try the experiments again using steam instead of cold water. We need only try zinc and magnesium with steam—sodium and calcium react violently with cold water so will react dangerously with steam.

Metals and steam The apparatus used for this experiment is shown below.

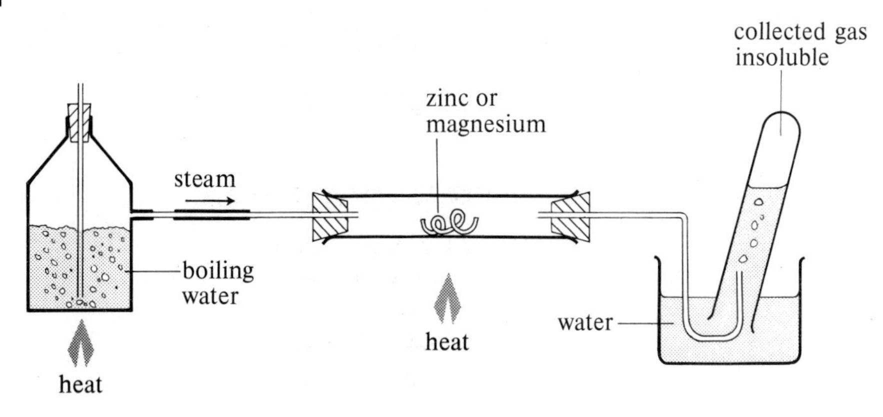

Table 15.2 The behaviour of metals in steam

Metal	*Effect on metal*	*Effect on steam*
Magnesium	Magnesium ribbon glows brightly. A white ash remains at the end.	An insoluble gas can be collected. The gas explodes when released in the air near a lighted splint. The reaction goes quickly once started.
Zinc	Zinc granules slowly becomes coated with a white crust.	A little insoluble gas appears. Most of the steam condenses in the beaker of water.

Conclusion

(i) The four metals can be put in this order of reactivity with water, starting with the most reactive: Na, Ca, Mg, Zn.

(ii) The explosive gas produced is hydrogen.

(iii) The other product each time is a metal oxide or metal hydroxide—a base.

(iv) These oxides or hydroxides are soluble in water (alkalis) only when a Group I metal is used.

Metals and dilute acids Many metals that do not react with either cold water or steam will dissolve in dilute acids. Zinc, which hardly reacted at all in steam, reacts quickly in cold dilute acid. The same explosive gas—hydrogen—is produced.
The equation for the reaction is:

$$Zn + 2HCl \longrightarrow ZnCl_2 + H_2\uparrow$$

The steps in the experiment are shown at the top of the next page.
If you look back to page 144 you will see that this reaction is almost identical to the reaction of a base with an acid. In both cases, the acid is neutralised and a salt solution is produced. Therefore you could use Method B to make a pure salt starting with a reactive metal and an acid.

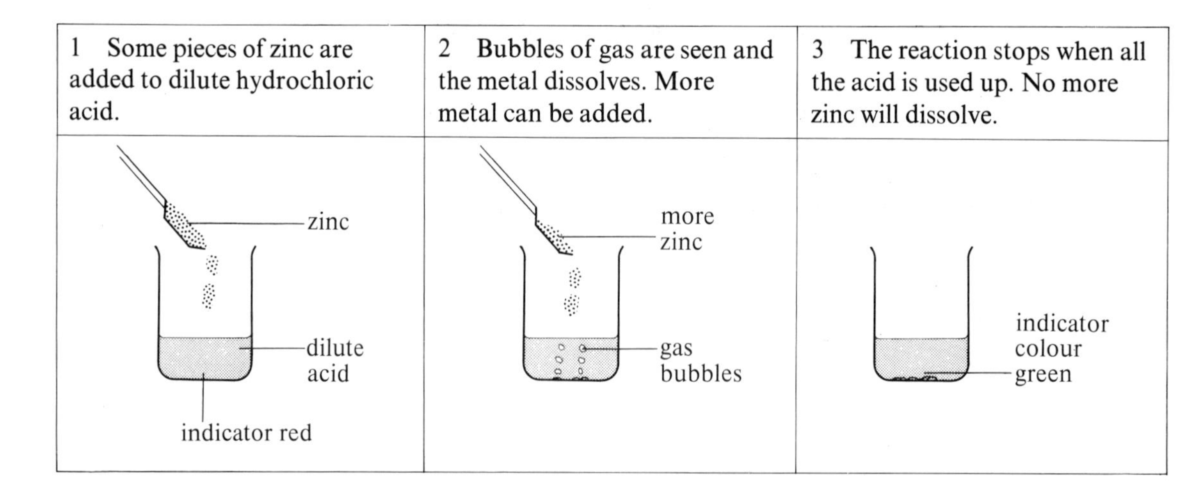

By adding different metals to acids and watching their reactions, even the fairly unreactive ones can be put into a reactivity list.

Metals and concentrated acids

Copper and silver do not react with either water or dilute acid. Dilute acids are made by adding oily liquids such as hydrogen sulphate(VI) and hydrogen nitrate(V) to water. These unpleasant and dangerous liquids are often called concentrated acids. We can try the effect of two of them, concentrated sulphuric acid and concentrated nitric acid, on our unreactive copper:

Table 15.3 The effect of concentrated acids on copper

Reactants	*Observations*	*Tests of gases produced*
Copper and concentrated sulphuric acid	On warming, the copper starts to dissolve. A pungent, throat-catching gas is given off. Adding water (carefully!) produces a blue solution.	(i) Turns damp blue litmus paper red. (ii) Turns purple acidified potassium manganate(VII) paper colourless.
Copper and concentrated nitric acid	The copper immediately starts to dissolve. Clouds of choking brown gas are given off. Adding water again gives a blue solution.	Turns damp blue litmus paper red.

These reactions are obviously not as simple as the other metal reactions! We can show that the two gases are in fact the dioxides of sulphur and nitrogen. The blue solutions are aqueous solutions of copper salts: nearly all copper salts dissolve to give blue solutions. The equations for the reactions are:

$$Cu + 2H_2SO_4 \longrightarrow CuSO_4 + 2H_2O + SO_2\uparrow$$
(blue solution) (pungent, throat-catching gas)

$$Cu + 4HNO_3 \longrightarrow Cu(NO_3)_2 + 2H_2O + 2NO_2\uparrow$$
(blue solution) (brown, choking gas)

Summary Similar experiments, using water and acids, can be tried for all the common metals. Table 15.4 summarises the results.

Table 15.4 Summary of metal reactions with water and acids

Metal	*Group*	*Reaction with water*	*Reaction with dilute acids*	*Reaction with concentrated acids*
Potassium Sodium	I I	Very rapid, producing an alkaline solution and hydrogen gas.	Violent.	
Calcium	II	Fast in cold water, producing an insoluble suspension and hydrogen gas.		
Magnesium	II	Very slow in cold water, fast in steam, producing an insoluble white solid and hydrogen gas.	All react producing salt solutions and hydrogen. Reactivity increases up the list.	All react, even copper and silver. The reactions are very complicated.
Aluminium	III	No reaction.		
Zinc Iron	Transition block	No visible reaction in pure water.		
Lead	IV		Very slow.	
Copper Silver	Transition block	No reaction.	No reaction.	

All the reactions with water and dilute acids are called **displacement reactions**. In each case, a metal has taken the place of (displaced) a hydrogen atom in the water or dilute acid:

$$2Na + 2HOH \longrightarrow 2NaOH + H_2\uparrow$$
$$Mg + 2HCl \longrightarrow MgCl_2 + H_2\uparrow$$

Copper and silver do not displace hydrogen from either water or dilute acids.

15.2 The electrochemical series

Reactivity of metals

Table 15.4 is a good guide to the reactivities of the metals. We can make from it a list of the metals in order of the ease with which they displace hydrogen. This list is called the **reactivity series** or the **electrochemical series**.

Hydrogen itself is fitted into the list:

below metals that are able to displace it from water and acid
above metals that are unable to displace it.

Table 15.5 The electrochemical series (E.C.S.)

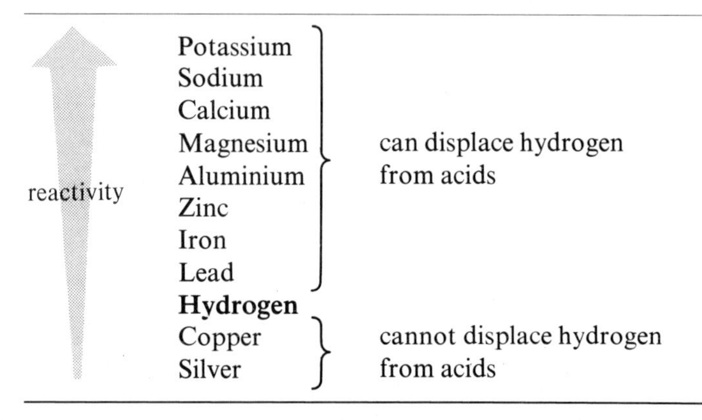

Reactivity (increasing upwards)	Element	
	Potassium	can displace hydrogen from acids
	Sodium	
	Calcium	
	Magnesium	
	Aluminium	
	Zinc	
	Iron	
	Lead	
	Hydrogen	
	Copper	cannot displace hydrogen from acids
	Silver	

Displacement reactions

Look again at the two metal reactions at the bottom of the opposite page. These reactions are called displacement reactions because the metal has displaced hydrogen from its compounds. But hydrogen is not the only element that can be displaced from a compound. If you put an iron nail into a solution of copper sulphate for example, it will dissolve. At the same time, copper metal is produced:

1 Iron nail and blue copper sulphate solution:

2 Pale green solution and copper are produced.

Iron has displaced copper from copper sulphate in solution.
The equation for the reaction is:

$$Fe + CuSO_4 \longrightarrow FeSO_4 + Cu$$

The experiment can be repeated with different metals and solutions.

Table 15.6 Some typical results

Pure metal	*Added to solutions of the compounds*			
	$FeSO_4$	$Pb(NO_3)_2$	$CuSO_4$	$AgNO_3$
Magnesium Mg	Displacement	Displacement	Displacement	Displacement
Iron Fe	No reaction	Displacement	Displacement	Displacement
Lead Pb	No reaction	No reaction	Displacement	Displacement
Copper Cu	No reaction	No reaction	No reaction	Displacement

The metals in Table 15.6 are arranged in the order in which they appear in the electrochemical series. A simple rule, that works for all the experiments, can be discovered from the results:
Any metal in the electrochemical series can displace a metal lower in the series from solutions of its compounds.

Reactivity of compounds

Another interesting idea comes out of Table 15.6 also. Look at the row of iron reactions. You can see that iron is quite a reactive metal, displacing three other metals from their solutions. Now look at the column of iron sulphate reactions. It shows that a solution of iron sulphate is fairly unreactive—only magnesium can displace the iron from it.
The results for copper and its compound are the reverse of this. Copper metal is unreactive, displacing only silver from solution. However a solution of copper sulphate is quite a reactive one—three metals can displace copper from it.
The more reactive a metal is, the more unreactive are its compounds in solution. Its ions will be difficult to displace.
Conversely, an unreactive metal will have reactive ions which are easy to displace from solution. This is summarized in Table 15.7.

Table 15.7 The reactivity series for metals and their ions

Metals		*Metal ions*	
Potassium	increasingly reactive (arrow pointing up)	$K^{+}_{(aq)}$	increasingly reactive (arrow pointing down)
Sodium		$Na^{+}_{(aq)}$	
Calcium		$Ca^{2+}_{(aq)}$	
Magnesium		$Mg^{2+}_{(aq)}$	
Aluminium		$Al^{3+}_{(aq)}$	
Zinc		$Zn^{2+}_{(aq)}$	
Iron		$Fe^{2+}_{(aq)}$	
Lead		$Pb^{2+}_{(aq)}$	
Hydrogen		$H_3O^{+}_{(aq)}$	
Copper		$Cu^{2+}_{(aq)}$	
Silver		$Ag^{+}_{(aq)}$	

15.3 Metal reactions with water and acids: the theory

In the last two sections we met a number of facts about metals, water, and acids:

some metals react with water;
less reactive metals can react with dilute acid;
some metals are more reactive than others—the metals can be put in a reactivity series.

Metal atoms

In suggesting theories to explain these facts, we will take the metals potassium, calcium, sodium, and magnesium as our examples.

Fact	*Theory*
Potassium and calcium react with water. Hydrogen gas bubbles off and escapes. The products that are left can conduct electricity much better than pure water.	The products that are left must contain ions. These ions, free to move in solution, are able to carry charge, i.e. conduct electricity. The metal lattice is changing to metal ions in solution.

The metal lattice is therefore losing electrons, and breaking up to form cations in solution. We can decide how easily this happens by looking at the core charge of a single metal atom. You may remember that the core charge of an atom is a measure of the control it has over its outer-shell electrons.

Core charge = nuclear charge − number of inner-shell electrons.

Metal	Potassium	Calcium
Structure	nucleus with 19 protons 2,8,8,1	nucleus with 20 protons 2,8,8,2
Core charge	$+19-18 = +\mathbf{1}$	$+20-18 = +\mathbf{2}$

Fact	*Theory*
Potassium is more reactive than calcium.	Electrons can leave a potassium lattice more easily than a calcium lattice, because the potassium nucleus exerts less attraction over its outer-shell electrons—its core charge is lower.

Water molecules

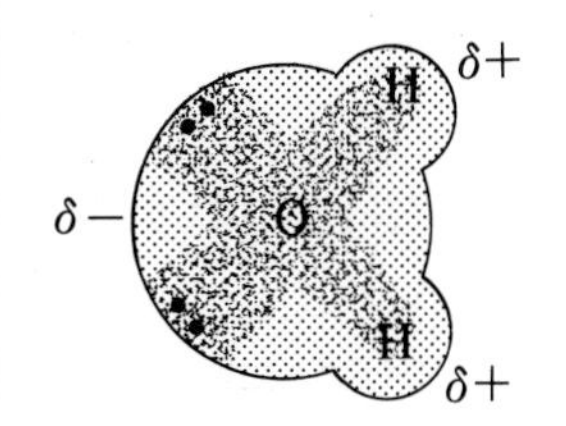

We have already discussed water molecules and their unusual properties in some detail, in Chapter 13. You will remember that a water molecule is polar with one end slightly positive and the other end slightly negative. A water molecule can change in two ways:

(i) its more negative end (the lone pairs) can attract positively charged particles, like protons (see Chapter 14, page 146),
(ii) its more positive end can attract negatively charged particles, like electrons.

It is this second possibility that we must now look at more closely.

A water molecule cannot fit in any more electrons: the shells in its bonded atoms are full. But suppose two water molecules attract a pair of electrons between them: if they then lose a proton each, a molecule of hydrogen and two deprotonated water molecules are produced.

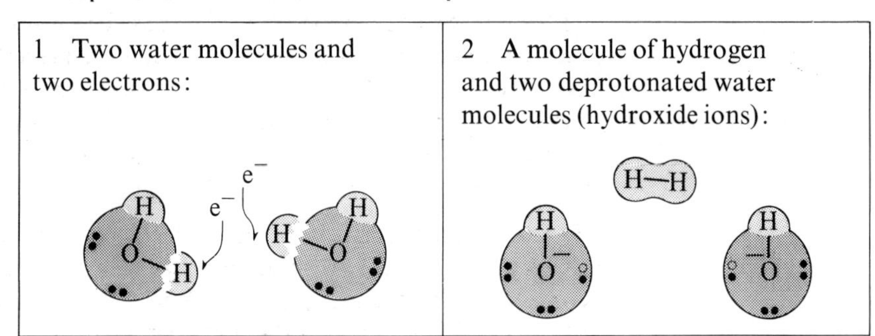

1 Two water molecules and two electrons:

2 A molecule of hydrogen and two deprotonated water molecules (hydroxide ions):

Metal lattices and water

Now look at the reaction of a metal with water:

Facts	*Theory*
Many metals react with water or steam.	Metal lattices consist of positive metal ions and electrons. The lattice can lose its electrons to water molecules.
The products are metal hydroxides (or oxides), and hydrogen gas.	The water molecules become hydroxide ions and hydrogen molecules. The metal ions are no longer held together by the electron cloud, so the lattice breaks up.
Example: $2Na + 2H_2O \longrightarrow 2NaOH + H_2\uparrow$	$2Na_{(s)} \longrightarrow 2Na^+_{(aq)} + 2e^-$ $2e^- + 2H_2O \longrightarrow 2OH^-_{(aq)} + H_{2\,(g)}$

The reaction above is shown below in picture form, to help you imagine it:

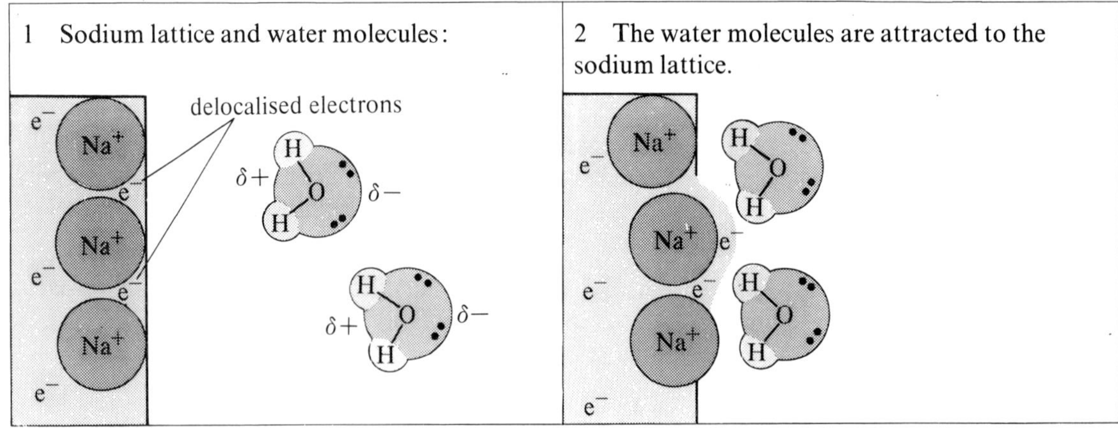

1 Sodium lattice and water molecules:

2 The water molecules are attracted to the sodium lattice.

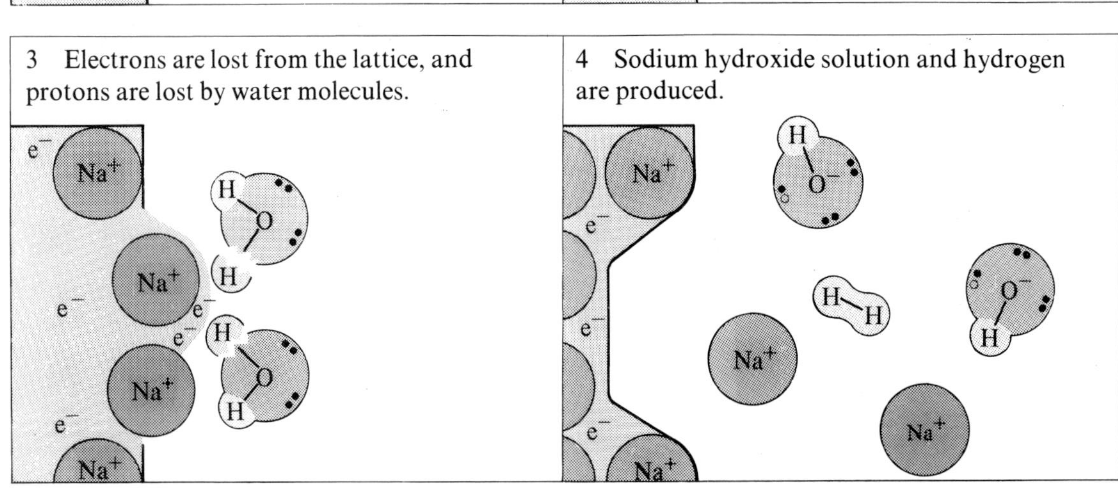

3 Electrons are lost from the lattice, and protons are lost by water molecules.

4 Sodium hydroxide solution and hydrogen are produced.

Metal lattices and acids

If the metal is not very reactive, it must be holding its electron cloud more firmly in the lattice. It will not react with the protons in water very easily. But if the water molecules are given extra protons—that is, if you add an acid—then even some less reactive metals can dissolve. The diagram that follows shows what happens.

You will see that it is very like the last diagram. The important difference is that the metal is being attacked by hydroxonium ions, H_3O^+, instead of water molecules, H_2O. Because the hydroxonium ions are positive, they are more attractive to the electrons than water molecules are.

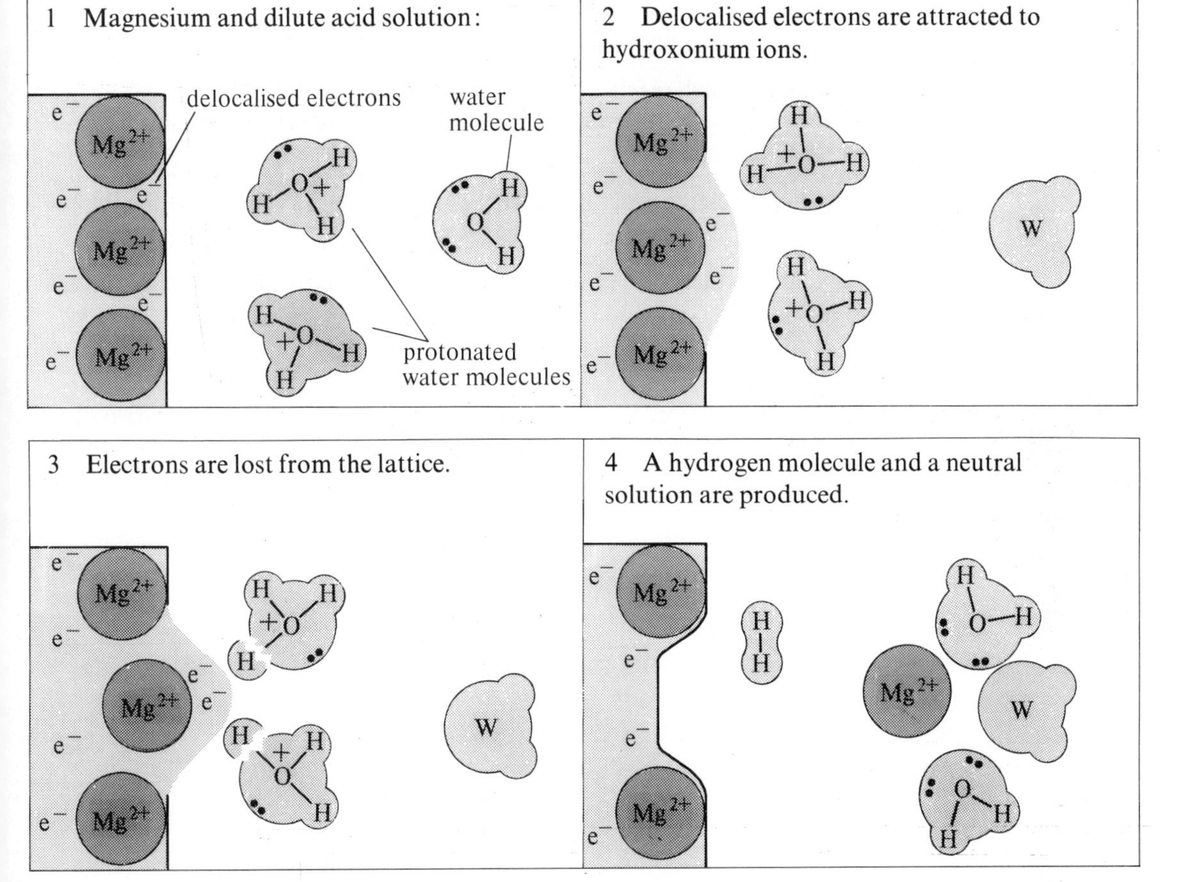
1 Magnesium and dilute acid solution:
delocalised electrons
water molecule
protonated water molecules
2 Delocalised electrons are attracted to hydroxonium ions.
3 Electrons are lost from the lattice.
4 A hydrogen molecule and a neutral solution are produced.

Again we can show the whole change in a single equation:

$$Mg_{(s)} + 2H_3O^+_{(aq)} \longrightarrow Mg^{2+}_{(aq)} + 2H_2O_{(l)} + H_{2\,(g)}$$

Summary

Some metals react with water. The Group I metals do so, producing soluble bases (alkalis); the Group II metals do so, and produce almost insoluble bases. Both reactions produce hydrogen.

Less reactive metals can react with acids, producing salt solutions and hydrogen.

The reactions of metals with water or acid are called displacement reactions, because hydrogen is displaced from the water or acid.

The metals can be arranged in a list, in order of decreasing reactivity. When hydrogen is included, the list is called the electrochemical series.

Any metal which reacts easily with water or acid is difficult to displace again from solution. In other words, if a metal is reactive, its ions in solution are unreactive.

So you can tell the order of reactivity of metal ions in solution, using the electrochemical series. It is the reverse of the order of metal reactivity.

Any metal will displace a metal below it in the electrochemical series, from a solution of its compounds.

Questions

1 Draw a table like the one below:

Reactions	*Similarities*	*Differences*
Sodium with water		
Calcium with water		

Fill in two ways in which the reactions are similar, and two ways in which the reactions are different.

2 Explain why we store sodium metal in jars of oil.

3 Draw another table like the one in question 1. This time, fill it in for the reactions of magnesium and zinc in steam.

4 Write an equation for the reaction of zinc with steam.

5 Rubidium is a metal in the same group as sodium (Group I). A small piece of rubidium is held beneath the surface of some water in a beaker, as shown in the diagram. **This experiment is dangerous. Do not try it for any Group I metals.**

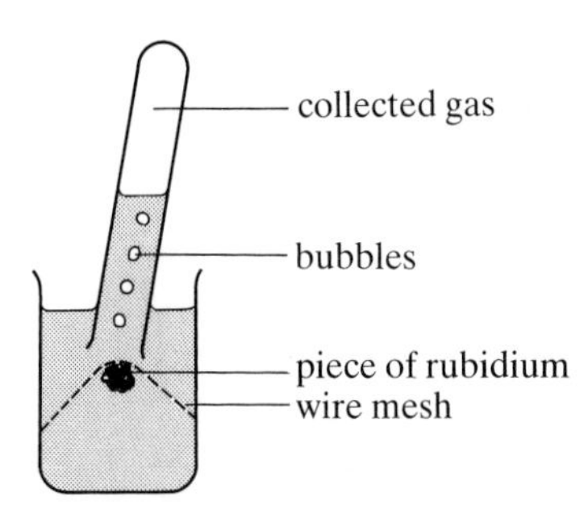

(a) What do you think the gas is?
(b) How could you test for this gas?
(c) What happens to the rubidium?
(d) If universal indicator was added to the solution in the beaker, what colour would it go?
(e) Why is the rubidium held under a wire mesh?

6 Look at this list of metals:
iron aluminium zinc magnesium calcium copper
(a) Which reacts most readily with water?
(b) Which reacts least readily with water?
(c) For which would you need concentrated acid to make the metal dissolve?

7 A piece of magnesium ribbon is dropped into the test-tube as shown in the diagram.

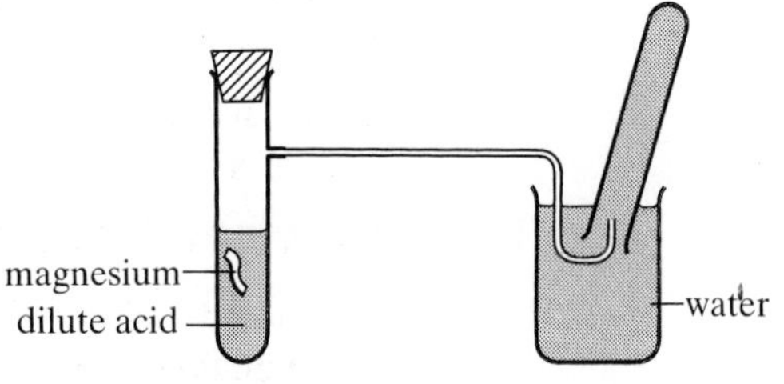

(a) What happens to the magnesium?
(b) What would you see happening to the liquid?
(c) What would happen to the water in the upside-down tube?
(d) Write an equation for this reaction.

8 (a) Copy out the table below:

Pure metal	*put into an aqueous solution of*		
	Hydrogen chloride	Zinc chloride	Copper(II) sulphate(VI)
Magnesium			
Iron			
Silver			
Lead			

If you think a reaction will take place when magnesium is put into an aqueous solution of hydrogen chloride, write *Yes* in the appropriate box. If not, write *No*. Repeat for all the metals and solutions.
(b) What do you call this type of reaction?
(c) Write an equation for each reaction that does take place.

9 Write in your own words what you think the electrochemical series is, and what patterns it shows.

10 Draw a simple diagram to show:
(a) the electronic structure of a water molecule
(b) the outer-shell electronic structure of a single magnesium atom
(c) the effect that a magnesium atom has on a water molecule during the reaction of magnesium and steam
(d) the electronic structure of the products of the reaction.

11 Men have known about and used gold, silver, copper, iron, and lead for thousands of years. Why have we only started using aluminium and titanium in the last hundred years?

16 Electrolysis and cells

16.1 Electrolysis

You have come across electrolysis several times by now. In Chapter 4 we met it as a method of decomposing a molten salt using electrical energy. In Chapter 10 it was used as evidence for the existence of charged particles called ions in molten and dissolved salts. In this chapter we are going to look at the process in greater detail.

The usual apparatus for electrolysis in the laboratory consists of two rods dipping into a liquid. The two rods are connected to the terminals of a battery:

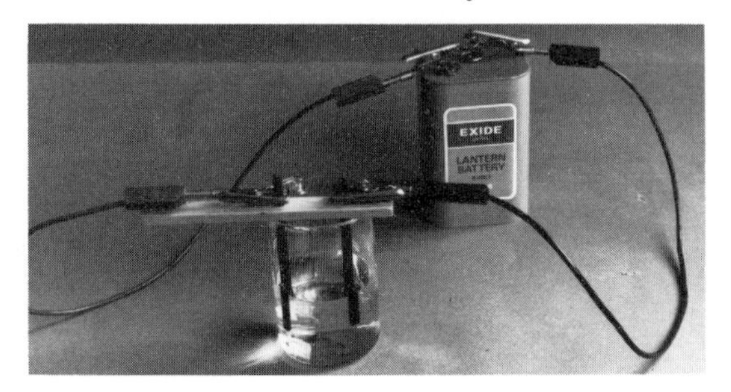

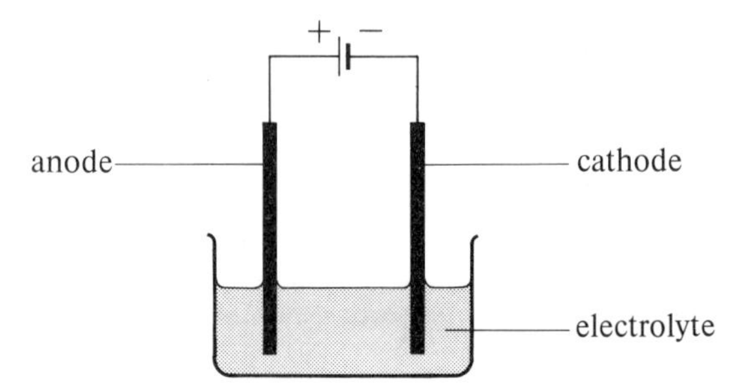

Fig. 1
Apparatus for electrolysis

The liquid being electrolysed is called the **electrolyte**.

The two rods dipping into the electrolyte are called **electrodes**. These are often made of graphite. The positive electrode is called the **anode** and the negative electrode is called the **cathode**. The whole set-up is called an **electrolytic cell**.

Electrolytes

A liquid is an electrolyte if it contains charged particles which can move through it carrying current.

Ionic solids produce electrolytes. When an ionic solid is melted, the ions are free to move and can carry a current. And when an ionic solid is dissolved in water, the aquo-ions are free to move and they can carry the current. For example:

$$NaCl_{(s)} \xrightarrow{\text{heat}} Na^+_{(l)} + Cl^-_{(l)}$$

solid lattice, ions fixed → liquid ions free to move

$$NaCl_{(s)} + H_2O_{(l)} \longrightarrow Na^+{}_{(aq)} + Cl^-{}_{(aq)}$$

solid lattice, ions fixed → aquo-ions free to move

The other main type of electrolyte is produced when a molecular substance reacts with water to form ions. For example:

$$\underset{\text{molecules}}{HCl_{(g)} + H_2O_{(l)}} \longrightarrow \underset{\text{ions}}{H_3O^+_{(aq)} + Cl^-_{(aq)}}$$

If nearly all the molecules react to form ions, the solution will conduct well. The substance is said to produce a **strong** electrolyte. Hydrogen chloride does this. So its solution in water—hydrochloric acid—is called a **strong acid.**

But if only a few molecules react, to form only a few ions, the solution will not conduct well. The substance is said to produce a **weak** electrolyte. Ammonia is like this, and is called a **weak base**:

$$\underset{\text{molecules}}{NH_{3(g)} + H_2O_{(l)}} \longrightarrow \underset{\text{few ions}}{NH^+_{4\,(aq)} + OH^-_{(aq)}}$$

All acids and bases are either strong or weak. Examples are:

Strong—exist mainly as ions		*Weak—exist mainly as molecules*	
Acids	*Bases*	*Acids*	*Bases*
Hydrochloric acid, HCl Sulphuric acid, H_2SO_4 Nitric acid, HNO_3	Sodium hydroxide, NaOH Potassium hydroxide, KOH	Carbonic acid, H_2CO_3 Ethanoic acid, CH_3COOH (acetic acid)	Ammonia, NH_3

Molecular substances may dissolve in solvents other than water; but the solutions will not conduct unless ions are produced. For example, hydrogen chloride dissolves in the molecular liquid toluene, but it remains as molecules in solution. No ions are produced and the solution is not an electrolyte.

To summarize, these substances are electrolytes:

molten salts
solutions of salts in water
solutions of acids in water
solutions of alkalis in water.

The movement of ions

When the battery is connected to an electrolytic cell, one electrode becomes positively charged and the other negatively charged. The ions in the liquid are attracted to the electrode with opposite charge and move through the solution. Ions with a negative charge are attracted to the positive anode; that is why they are called **anions**. Ions with a positive charge are attracted to the negative cathode; that is why they are called **cations**.

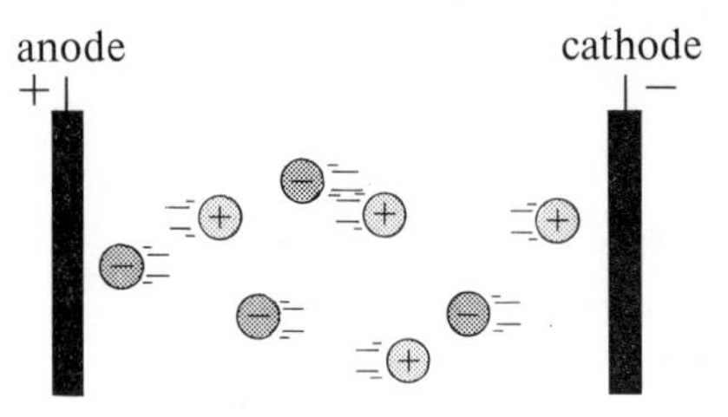

If a salt containing a coloured ion is used, the ions can actually be seen moving. A simple experiment is set up like this:

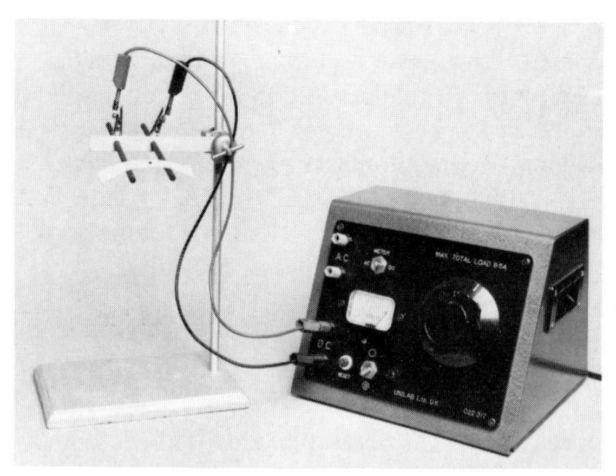

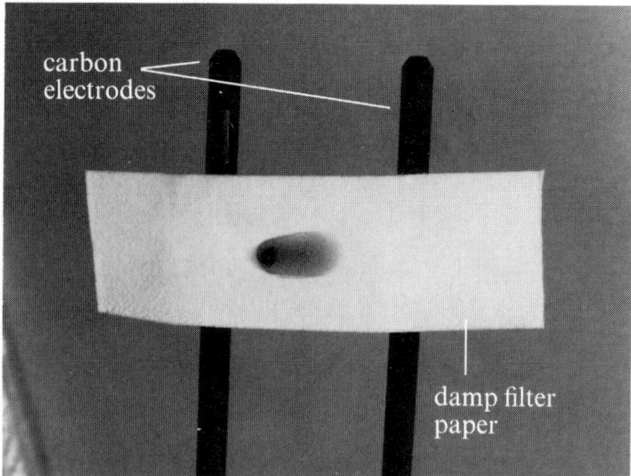

For the experiment, carbon electrodes are connected to a power pack. A close-up of the electrodes is shown on the right.

When a crystal of the coloured salt potassium manganate(VII) is put on the damp strip of filter paper, it dissolves in the solution. A patch of colour can soon be seen moving towards one electrode. In the case of the manganate(VII) ion, a purple patch moves towards the anode—the electrode on the right above.

In a molten salt the moving ions are quite simple. For example in molten NaCl:

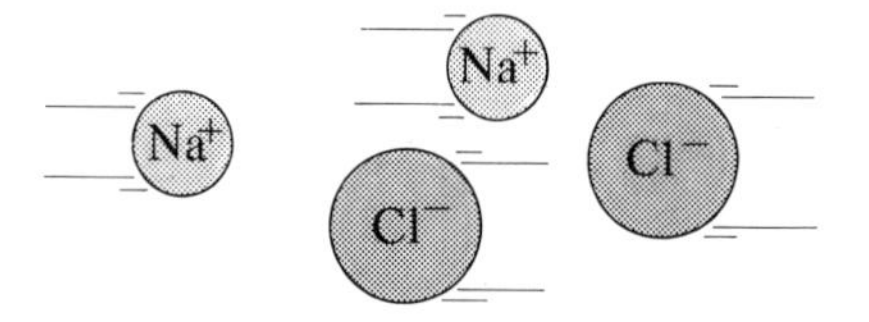

But in an electrolyte which is an aqueous solution, each moving ion drags along a cluster of water molecules. The moving particles in dissolved NaCl would look like this:

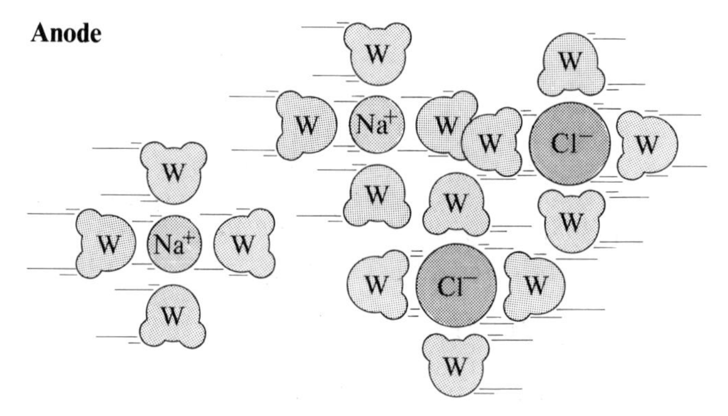

So in electrolytes that are aqueous solutions, the moving particles are aquo-ions made up of simple ions surrounded by water molecules.

Electrode reactions

When electrolysis takes place, no reaction can be seen in the middle of the electrolyte. It is only at the electrodes that you can see signs of chemical change. These signs are usually bubbles of gas, or changes in pH, or metals being deposited. The reactions at the electrodes are called **electrode reactions**, and they fall into two main classes: (i) reactions of molten salts and (ii) reactions of aqueous solutions.

(i) Molten salts

As you have seen, in molten salts the ions moving through the solution are simple ions. They will finish up clustered around the electrodes. It is here that they either gain or lose electrons.

At the cathode
Positive cations gain electrons from the negative cathode. They become neutral atoms. For example:

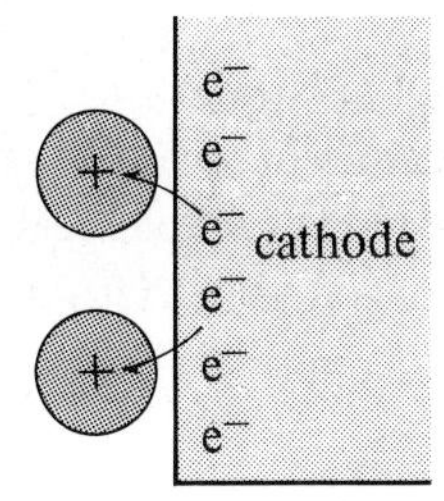

$$Na^+_{(l)} + e^- \longrightarrow Na_{(l)}$$

In molten salts, the cations are almost always metal ions. So the products of cathode reactions are usually metals.

At the anode
Negative anions lose electrons to the positive anode. They too become neutral atoms. For example:

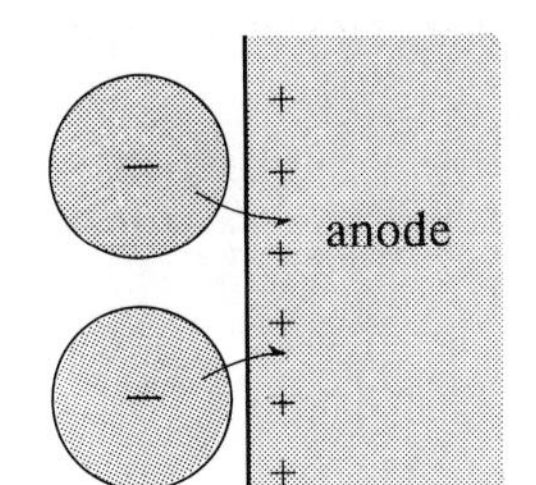

$$2Cl^-_{(l)} \longrightarrow Cl_{2\,(g)} + 2e^-$$

Anions are nearly always the ions of non-metal elements. So anode reactions usually produce non-metals.

For an example of the electrolysis of a molten salt, look back at the electrolysis of molten lead bromide, described on page 33.

(ii) Aqueous solutions

When an aqueous solution is electrolysed the picture is more complicated because the ions are not simple. Each electrode is surrounded by bulky aquo-ions.

At the cathode
Sometimes a metal is produced and sometimes hydrogen gas. The cathode gives up electrons, which can react with either metal ions (producing the metal) or water molecules (producing hydrogen gas). You can use the electrochemical series to predict which of these things will happen—remember that metals above hydrogen in the series displace hydrogen from water or acids (see page 158). Take sodium for example:

$$2Na_{(s)} + 2HOH_{(l)} \longrightarrow 2NaOH_{(aq)} + H_{2\,(g)}$$

Sodium is *more reactive* than hydrogen—it can displace hydrogen from water, giving sodium ions in solution. Therefore we can conclude that a sodium ion in solution is *less reactive* than a hydrogen atom in a water molecule. It is easier to displace the hydrogen atom from a water molecule than to displace a sodium ion from solution:

$$Na_{(s)} \underset{\text{difficult to displace ions from solution}}{\overset{\text{goes into solution easily}}{\rightleftharpoons}} Na^{+}_{(aq)}$$

$$H_{2(g)} \underset{\text{easy to displace hydrogen from water}}{\overset{\text{reacts less easily to form water}}{\rightleftharpoons}} H_2O_{(l)}$$

Now, back to the cathode reaction. Suppose the dissolved salt is sodium chloride. Then around the cathode will be clusters of sodium aquo-ions. It is easier to displace hydrogen atoms from water molecules in the aquo-ion than to displace the sodium ion from it. So electrons from the cathode react with the water molecules, and hydrogen is displaced:

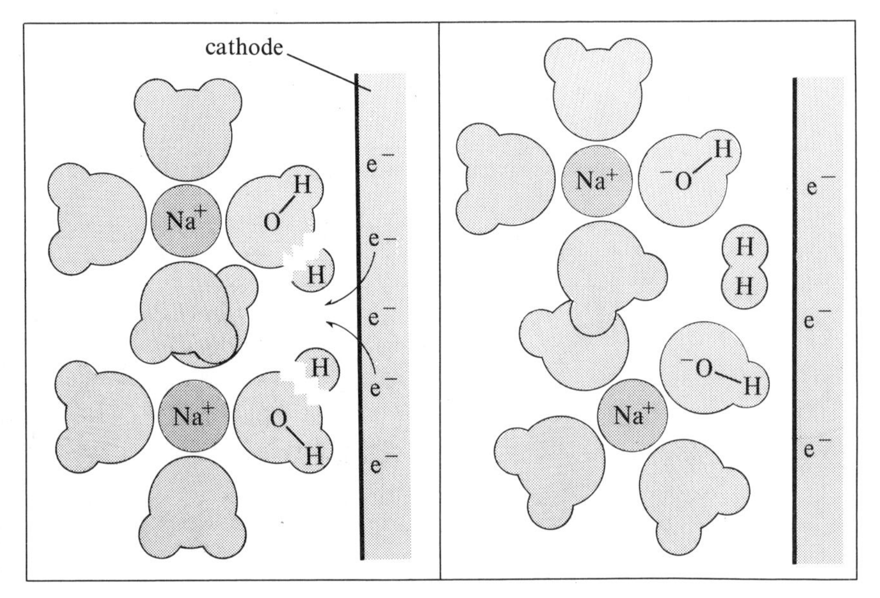

Fig. 2
Aquo-ions reacting at the cathode

The result is the decomposition of water molecules:

$$2\ HOH_{(l)} + 2e^- \longrightarrow H_{2(g)} + 2OH^-_{(aq)}$$

Hydroxide ions are formed so the solution becomes alkaline around the cathode.

The cathode reaction above can be summed up in this way:
Because sodium is above hydrogen in the electrochemical series, hydrogen is formed at the cathode instead of sodium metal.
This leads to a general rule:
If a metal is above hydrogen in the electrochemical series, electrolysis of solutions of its salts always produces hydrogen at the cathode.

A tiny part of the surface of a cathode, showing copper deposited by electrolysis, magnified × 3200. Can you explain why the pattern is regular?

If a metal is *below* hydrogen in the electrochemical series, however, then it is *less* reactive than hydrogen. Therefore its ions are *more* reactive than are the hydrogen atoms in water molecules. Think about the electrolysis of copper chloride, for example:

$$Cu_{(s)} \underset{\text{easy to displace ions from solution}}{\overset{\text{difficult to get into solution}}{\rightleftharpoons}} Cu^{2+}_{(aq)}$$

Copper aquo-ions will be clustered round the cathode. But in this case the metal ions, rather than the water molecules, will accept the electrons from the cathode. The water molecules are not affected.

$Cu^{2+}_{(aq)} + 2e^- \longrightarrow Cu_{(s)}$

So we have a second rule:

If a metal is below hydrogen in the electrochemical series, electrolysis of solutions of its salts always produces the metal at the cathode.

The two different types of cathode reaction can clearly be seen, by electrolysing sodium chloride solution and copper(II) chloride solution. In both of them, chlorine is formed at the anode.

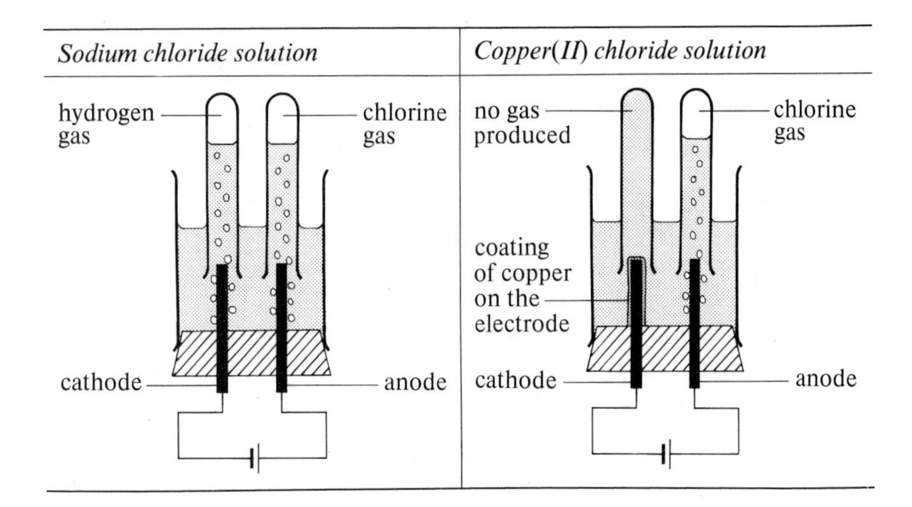

At the anode

Like the cathode reactions, these are also complicated for aqueous solutions, because of the different reactivities of non-metal ions compared with water. They are summarised below.

(i) If the dissolved salt is a chloride, chlorine gas is produced.

$2Cl^-_{(aq)} \longrightarrow Cl_{2\,(g)} + 2e^-$

(ii) If the dissolved salt is a sulphate or nitrate, oxygen gas is produced. This is the result of water in the aquo-ions being decomposed.

$$6H_2O_{(l)} \longrightarrow 2O_{2\,(g)} + 2H_3O^+_{(aq)} + 2e^-$$

The solution around the anode becomes acidic, because of the hydroxonium ions H_3O^+.

The electrolysis of water

Pure water cannot be electrolysed because it consists almost entirely of uncharged molecules. These cannot carry charge. However, if a little sulphuric acid is added, ions are produced:

$$H_2SO_{4\,(l)} + 2H_2O_{(l)} \longrightarrow 2H_3O^+_{(aq)} + SO^{2-}_{4\,(aq)}$$

The solution will now conduct and can be electrolysed. The product at the cathode is hydrogen, as you would expect, and at the anode oxygen is given off. Since the two reactions add up to the decomposition of water, the process is called **the electrolysis of water**. It is usually done in a **voltameter** which is the name of the apparatus below.

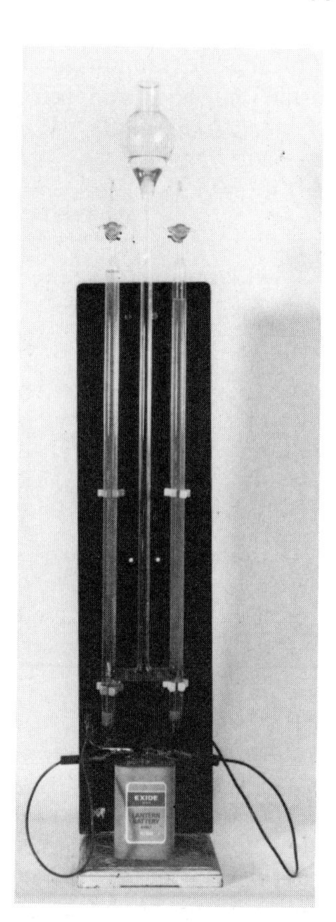

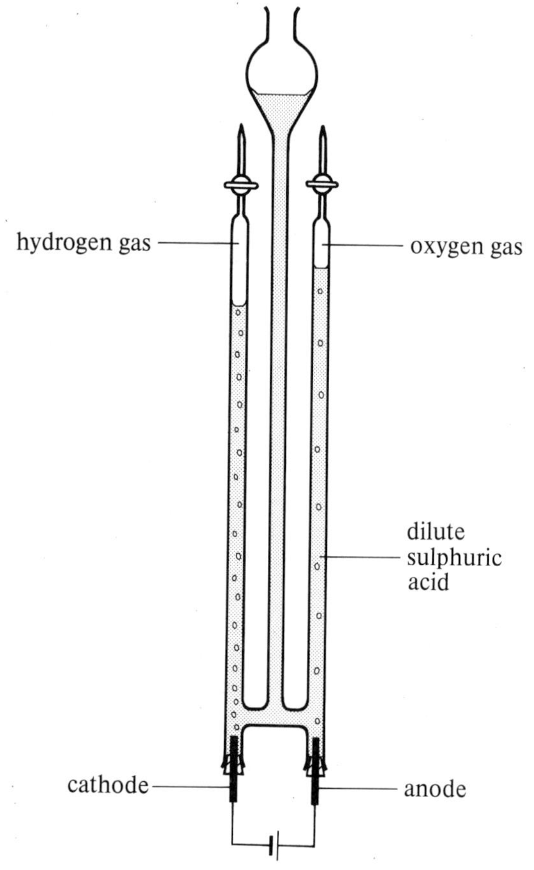

Fig. 3
A voltameter

Some common examples of electrolysis

Below are listed some molten salts and aqueous solutions that are often electrolysed. Before looking at the electrode reactions, see if you can work them out for yourself.

Molten salts

(i) Lead bromide

Cathode reaction: $Pb^{2+}_{(l)} + 2e^- \longrightarrow Pb_{(l)}$

Anode reaction: $2Br^-_{(l)} \longrightarrow Br_{2\,(g)} + 2e^-$

Observations: Globules of lead produced at cathode, bromine gas bubbles off the anode.

(ii) Sodium chloride

This electrolysis is difficult to do in the laboratory because of the high m.p. of NaCl.

Cathode reaction: $2Na^+_{(l)} + 2e^- \longrightarrow 2Na_{(l)}$

Anode reaction: $2Cl^-_{(l)} \longrightarrow Cl_{2\,(g)} + 2e^-$

Aqueous solutions

(i) Sodium chloride solution

Cathode reaction: $2H_2O_{(l)} + 2e^- \longrightarrow 2OH^-_{(aq)} + H_{2\,(g)}$

Anode reaction: $2Cl^-_{(aq)} \longrightarrow Cl_{2\,(g)} + 2e^-$

Observations: Gases seen at both electrodes; solution becomes alkaline.

(ii) Copper sulphate solution with graphite or platinum electrodes

Cathode reaction: $Cu^{2+}_{(aq)} + 2e^- \longrightarrow Cu_{(s)}$

Anode reaction: $3H_2O_{(aq)} \longrightarrow \frac{1}{2}O_{2\,(g)} + 2H_3O^+_{(aq)} + 2e^-$

Observations: Cathode becomes plated in copper;
oxygen given off at anode;
solution becomes acidic.

(iii) Copper sulphate solution with copper electrodes

Cathode reaction: $Cu^{2+}_{(aq)} + 2e^- \longrightarrow Cu_{(s)}$

Anode reaction: $Cu_{(s)} \longrightarrow Cu^{2+}_{(aq)} + 2e^-$

Observations: Nothing seen; but anode gets lighter and cathode gets heavier.

(iv) Water containing a little sulphuric acid

This is the electrolysis of water.

Cathode reaction: $2H_3O^+_{(aq)} + 2e^- \longrightarrow 2H_2O_{(l)} + H_{2\,(g)}$

Anode reaction: $3H_2O_{(l)} \longrightarrow \frac{1}{2}O_{2\,(g)} + 2H_3O^+_{(aq)} + 2e^-$

Observations: Hydrogen produced at the cathode, oxygen at the anode;
the gases are produced in the ratio 2 volumes of hydrogen to 1 volume of oxygen;
no change in pH.

In this process, like electrolysis, a car body is given an undercoat of paint. The body is charged and acts like an electrode, attracting the paint in the dip.

Summary

(i) Substances can be decomposed, by dipping electrodes into their solutions or liquids and connecting the electrodes to a battery.

(ii) The battery causes the decomposition to happen, by making the electrodes positive and negative.

(iii) The electrodes transfer electrons. They themselves usually remain unchanged. (But see electrolysis (iii) on page 173.)

16.2 Cells

In the last section we looked at chemical reactions brought about by using a battery. In this section we will look at the batteries themselves. The correct name for a single battery is a **cell**. In a cell, a chemical reaction is used to produce current.

If you dip zinc and copper into acid at the same time, you have a simple cell. If the metals are connected by wire, a current will flow between them. The current will cause a bulb to light up.

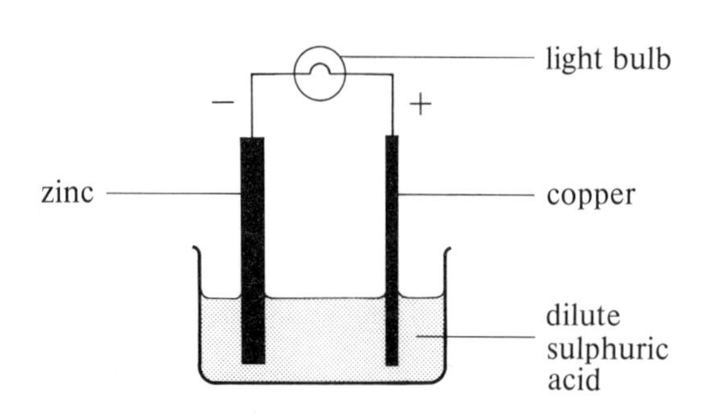

Fig. 4
A simple cell

Think back to the electrochemical series. Zinc is higher in the list than copper; this means that zinc forms ions in solution more easily than copper does. When the two metals are dipped into dilute acid, the zinc begins to dissolve.

$$Zn_{(s)} \longrightarrow Zn^{2+}_{(aq)} + 2e^-$$

The electrons freed by the reaction travel up the rod and into the wire. The zinc has become the negative terminal of the cell.

In principle, any two different metals dipped into an electrolyte will make a cell. The further apart the metals are in the electrochemical series, the larger the voltage of the cell. So a cell differs from electrolysis in these ways:

(i) A chemical reaction is causing a current to flow. This is the *opposite* of electrolysis.

(ii) The electrodes are usually made from two different materials. (In some cells, both electrodes are made from the same material. However at this level you do not need to understand how such cells work.)

(iii) At least one of the electrodes will be chemically changed during the reaction. In the example above, the zinc electrode dissolves.

In practice, the electrode metals are carefully chosen and the cells carefully designed to meet special needs. You are probably already familiar with some cells.

The battery or dry cell

A battery in a torch or transistor radio must give a steady current, it must work any way up and it must not leak. It is often called a **dry cell**. Here is a photograph and a cut-away drawing of one:

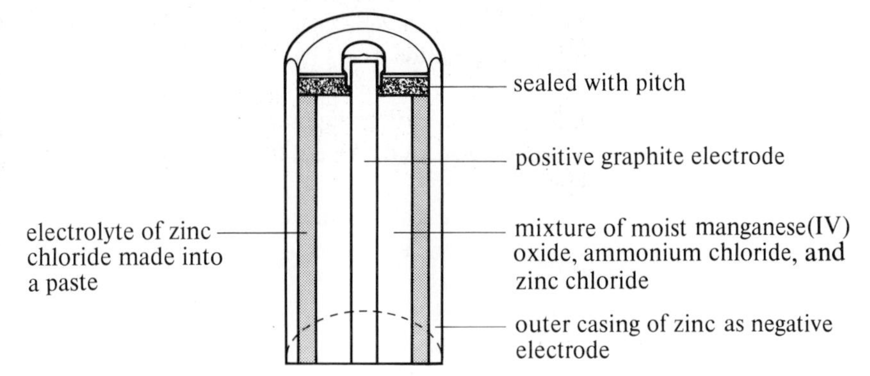

The reactions are complicated, but they can be represented as:

$$2MnO_2 + 2NH_4^+ + 2e^- \longrightarrow Mn_2O_3 + 2NH_3 + H_2O$$

$$Zn + 4NH_3 \longrightarrow Zn(NH_3)_4^{2+} + 2e^-$$

Notice that the outer case dissolves as the battery is used. You may have come across old batteries in torches: they are usually sticky, because the electrolyte paste leaks out as the case dissolves away.

The accumulator

In a car a cell is needed for giving bursts of energy—to start the engine or run the lights. Once the engine is going, energy from it is used to recharge the cell. Because it can save up or accumulate electrical energy this type of cell is often called an **accumulator**.

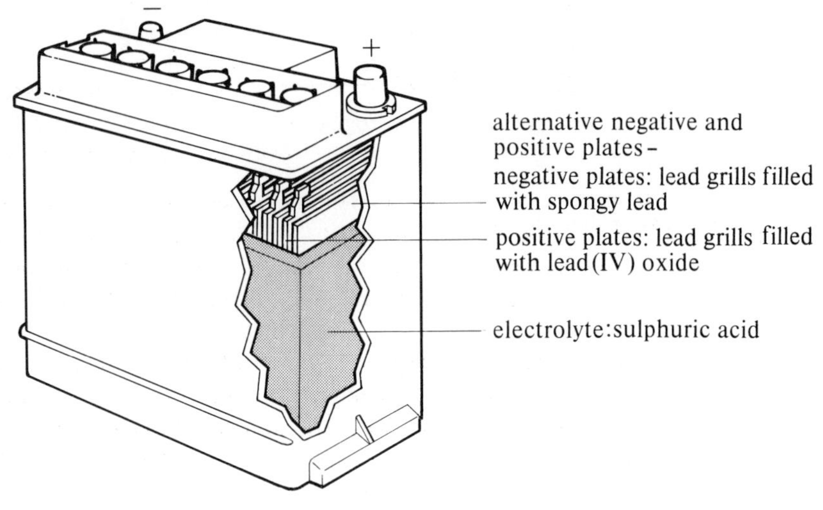

When the accumulator is being used, the reactions are

positive plates: $PbO_{2\,(s)} + 4H_3O^+_{(aq)} + 2e^- \longrightarrow Pb^{2+}_{(aq)} + 6H_2O_{(l)}$

negative plates: $Pb_{(s)} \longrightarrow Pb^{2+}_{(aq)} + 2e^-$

Notice that lead ions are produced at both plates. This is important because when the battery is being charged up the reverse reactions take place. If two different metal ions had been produced, there would be no way of making each sort go to its right plate for the reverse reaction. As they are all the same it does not matter.

The fuel cell

A lot of research is going into the design of a cell that will combine oxygen and hydrogen, and give out energy as electricity. The basic idea is a simple one.

Decomposing water is an endothermic reaction needing energy:

$$2H_2O_{(l)} \xrightarrow{\text{energy in}} 2H_{2\,(g)} + O_{2\,(g)}$$

Combining oxygen and hydrogen is therefore an exothermic reaction producing energy:

$$2H_{2\,(g)} + O_{2\,(g)} \xrightarrow{\text{energy out}} 2H_2O_{(l)}$$

The energy can be given out as electrical energy.

The idea is simple to demonstrate. Electrolyse some dilute sulphuric acid using graphite rods. The rods quickly become coated in bubbles of hydrogen and oxygen gas. Now disconnect the battery and connect a lamp to the rods. The lamp will light up.

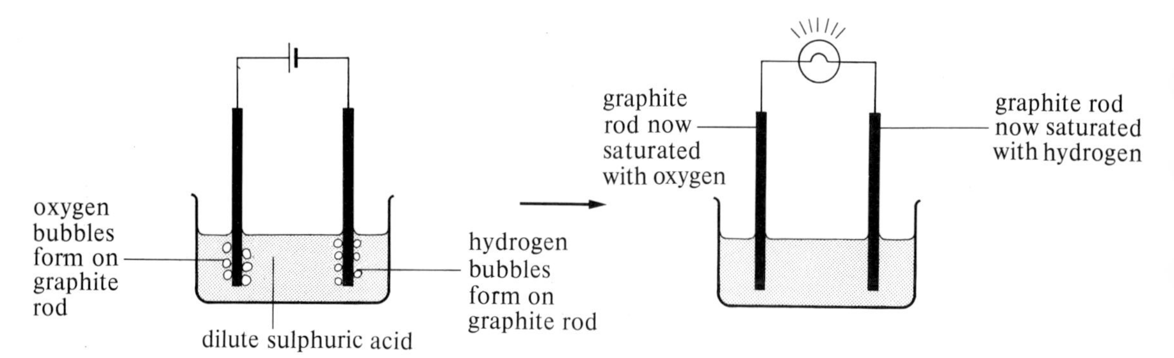

In a fuel cell, only the second stage will take place. The two gases, oxygen and hydrogen, are combined, giving water as the product.

Summary

Electrolysis is a decomposition reaction, brought about by electricity.

Electrolytes are liquids that can carry current.

There are two different types of electrolyte:
(i) molten salts
(ii) aqueous solutions of acids, alkalis, or salts.

In molten salts, the ions are simple ions. In aqueous solutions, the ions are aquo-ions.

The electrodes are the rods that carry current to and from the electrolyte. The positive electrode is called the anode. The negative electrode is called the cathode.

At the cathode, positive ions from the liquid gain electrons. At the anode, negative ions from the liquid lose electrons.

During electrolysis, the ease with which elements are displaced from solution is in the opposite order to the electrochemical series.

In a cell (or battery), a chemical reaction causes current to flow. So it is the opposite of electrolysis.

In most cells, the electrodes are two different materials, usually metals.

The further apart these metals are in the electrochemical series, the greater the voltage produced.

Questions

1 Suppose you set up the experiment given below.

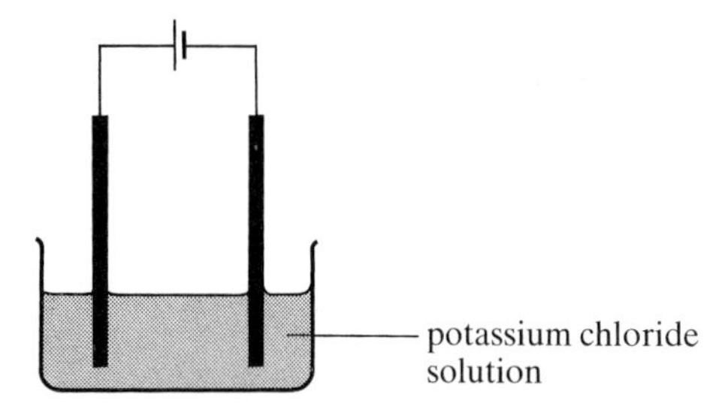

(a) What would you see at the cathode?
(b) What would you see at the anode?
(c) Would the pH change at the cathode? Explain.
(d) What would happen to indicator added just around the anode? Explain.
(e) What gases would be produced?

2 Repeat question 1, this time using silver nitrate solution as the electrolyte.

3 (a) How many mistakes are there in the diagram below?
(b) Draw another diagram, based on this one, but without any mistakes.

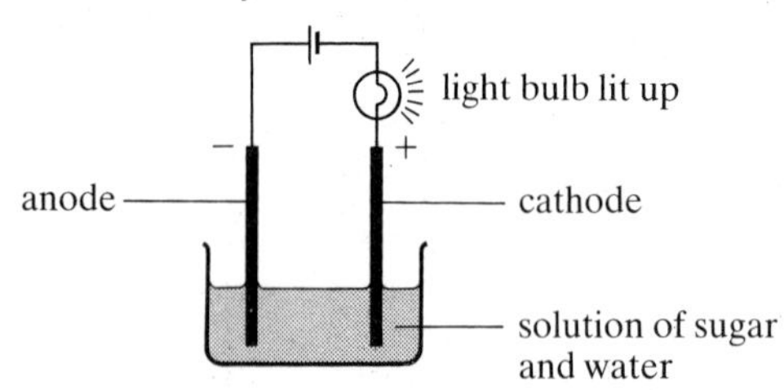

4 (a) Explain the meaning of the words *strong* and *weak* when they are applied to electrolytes.
(b) Give two examples of strong electrolytes, and two of weak ones.

5 Solid sodium chloride does not conduct electricity. Both molten and dissolved sodium chloride do conduct electricity.
(a) Draw a small portion of the sodium chloride lattice.
(b) Explain why the solid does not conduct.
(c) What particles are produced when the solid is melted?
(d) What happens to these particles during electrolysis?
(e) What particles are produced when the solid is dissolved in water?
(f) What happens to these during electrolysis?

6 Here are diagrams of three simple cells.

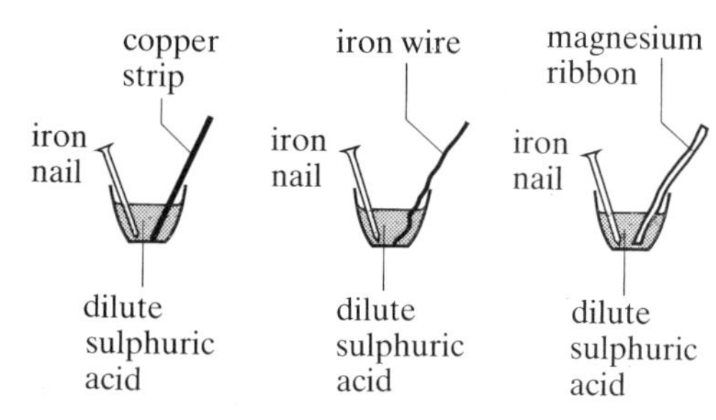

(a) Which of these cells does not produce electricity if the two metals were connected? Explain why.
(b) Which would produce the highest voltage? Explain why.
(c) In which cell is the iron the positive terminal?
(d) In which cell is the iron the negative terminal?

7 Study the diagrams of the two cells shown on page 171, and then by reading the explanation about them carefully, answer the following questions.
(a) Why is no sodium metal produced at the cathode of the sodium chloride cell?
(b) What is decomposed around the cathode and what will be the result?
(c) What happens to the cathode in the copper chloride cell?
(d) Why is no hydrogen gas made in the copper chloride cell?
(e) What would happen to the colour of the solution in the copper chloride cell as electrolysis went on?

8 When water containing a little dilute sulphuric acid is electrolysed, oxygen and hydrogen are made.
(a) Write down the cathode reaction for the process.
(b) Write down the anode reaction for the process.
(c) What will happen to the pH of the acid solution being electrolysed?
(d) What will be the ratio of the volume of oxygen to the volume of hydrogen produced? Explain your answer.

9 Explain the difference between a dry cell, like the batteries used in torches and radios, and an accumulator battery of the sort used in cars.

Section three

A closer look at chemical change

17 The reactivity of the elements

17.1 Reactivity of metals: the facts

The electrochemical series

Metals can be put in order of reactivity by comparing their reactions with water, with dilute acids, and with the solutions of salts of other metals. We compared these reactions of metals in Chapter 15. The result was the **electrochemical series.** This is a list of the metal atoms in decreasing order of their tendency to become positive ions. For example, the reaction

$$Zn_{(s)} \longrightarrow Zn^{2+}_{(aq)} + 2e^-$$

goes more readily than

$$Cu_{(s)} \longrightarrow Cu^{2+}_{(aq)} + 2e^-$$

When a reaction happens easily, the reverse reaction will not happen easily; therefore

$$Cu^{2+}_{(aq)} + 2e^- \longrightarrow Cu_{(s)}$$

goes more readily than

$$Zn^{2+}_{(aq)} + 2e^- \longrightarrow Zn_{(s)}$$

The last statement agrees with the observations in electrolysis, where cations become atoms at the cathode. Metals lowest down in the list are easiest to displace from solution.

Reactions with non-metals

The reactions of metals with pure non-metals also fit the same trend of reactivity. For example:

(i) Metals at the top of the electrochemical series, like potassium, sodium, calcium, and magnesium, burn brightly in a gas jar of oxygen or chlorine gas.

$$2Mg_{(s)} + O_{2(g)} \longrightarrow 2MgO_{(s)}$$

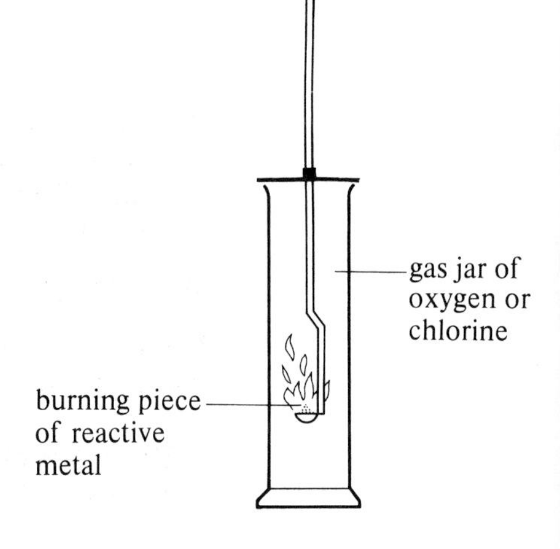

(ii) Metals in the middle of the electrochemical series, like aluminium, zinc, iron, and even copper, react when they are heated in a stream of the gas. Once the reaction starts the metal glows.

$$Zn_{(s)} + Cl_{2(g)} \longrightarrow ZnCl_{2(s)}$$

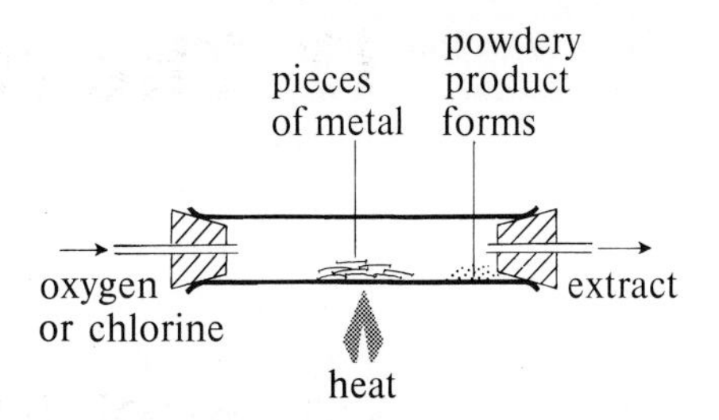

In both cases the metals react with the non-metals to form ionic products. The metal atoms become positive ions; the non-metals become negative ions.

17.2 Reactivity of non-metals: the facts

The reactions of metals with oxygen and chlorine tell us about more than just the reactivity of these metals. The reactions also tell us that oxygen and chlorine are very reactive. And on page 64 we described the reaction between iodine and aluminium: this had to be done in the fume cupboard.

It is safer to test the reactivity of non-metals by using displacement reactions in solution. Non-metals—often diluted in water—can be added to compounds of other non-metals in solution. For example:

(i) When chlorine in water is added to a solution of potassium bromide, bromine is displaced. The equation for the reaction is:

$$Cl_{2(aq)} + 2Br^-_{(aq)} \longrightarrow 2Cl^-_{(aq)} + Br_{2(aq)}$$

Chlorine is therefore more reactive than bromine.

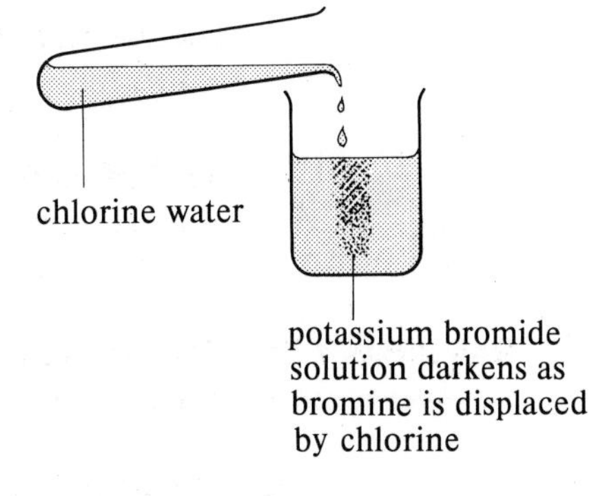

(ii) When a crystal of iodine is added to a solution of sodium sulphide, sulphur is displaced. The equation for the reaction is:

$$8I_{2(s)} + 8S^{2-}_{(aq)} \longrightarrow S_{8(s)} + 16I^-_{(aq)}$$

Iodine is therefore more reactive than sulphur.

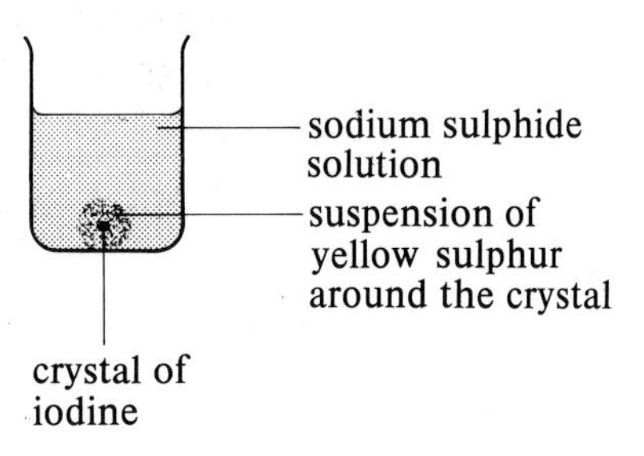

The results of a whole series of these reactions can be collected into a table like Table 17.1 which follows.

Table 17.1 Some typical results*

	Solutions of non-metal compounds				
Non-metals	*Potassium chloride*	*Potassium bromide*	*Potassium iodide*	*Water*	*Sodium sulphide*
Chlorine (chlorine water)		Displacement	Displacement	—	Displacement
Bromine (bromine water)			Displacement	—	Displacement
Iodine	—	—		—	Displacement
Oxygen (hydrogen peroxide)	—	—	Displacement		Displacement
Sulphur	—	—	—	—	

* A dash means that no reaction takes place.

By looking at the number of displacement reactions that each non-metal can do, you can see that the order of reactivity of the elements is:

reactivity (increasing upwards)

Chlorine
Oxygen } about the same
Bromine }
Iodine
Sulphur

The order of reactivity of the compounds is the reverse. You can see that sulphides react with all the other non-metals—they are the most reactive.

reactivity (increasing downwards)

Chlorides } about the same
Oxides }
Bromides
Iodides
Sulphides

17.3 Comparing the reactivities of the elements in the Periodic Table

If the trends in reactivity of the metals and non-metals are shown in the Periodic Table, an interesting pattern appears.

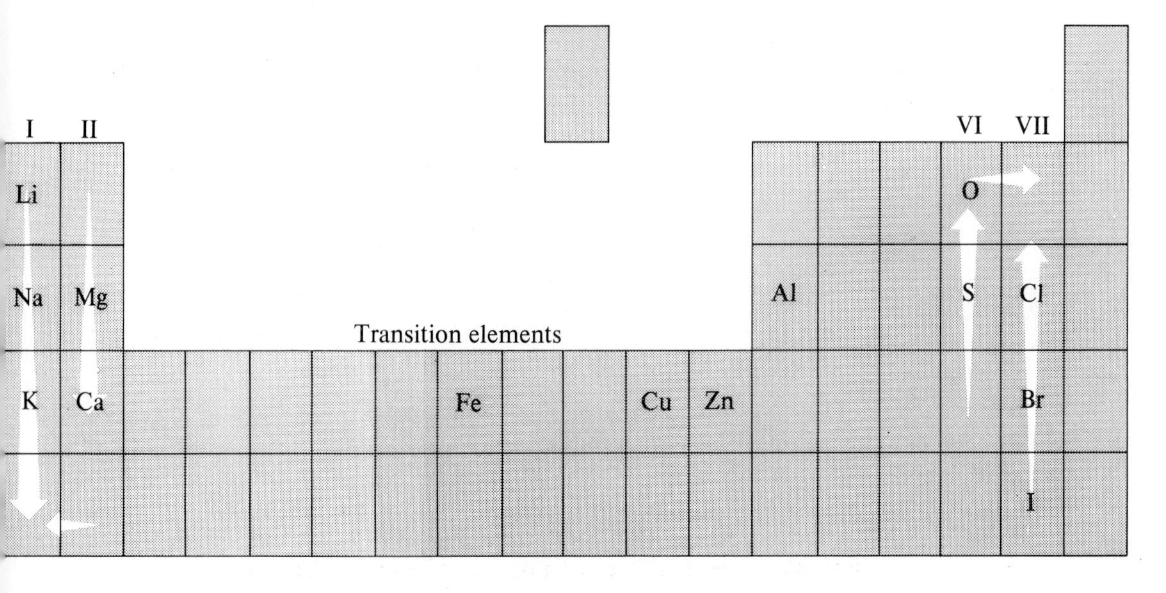

The arrows show increasing reactivity. The trends can be summarised as:

(i) an increase in reactivity going down Groups I and II;
(ii) a decrease in reactivity going from a Group I element to the Group II element in the same period;
(iii) an increase in reactivity going up Groups VI and VII;
(iv) an increase in reactivity going from a Group VI element to the Group VII element in the same period;
(v) no clear trend in reactivity in the transition elements.

So, in general, reactivity increases towards the top right and bottom left of the Periodic Table.

17.4 Reactivity and the structure of the atom: the theory

There are two sets of facts we must try to explain:

(i) there are trends in reactivity for metals and non-metals;
(ii) the trend for metals is in the opposite direction to the trend for non-metals.

The loss and gain of electrons If you look back at the reactions of metals that led us to the electrochemical series in Chapter 15, you will notice that they have one thing in common. In every case the metal lattice loses electrons, producing positive ions. Think about a single atom:

$$Na \longrightarrow Na^+ + e^-$$

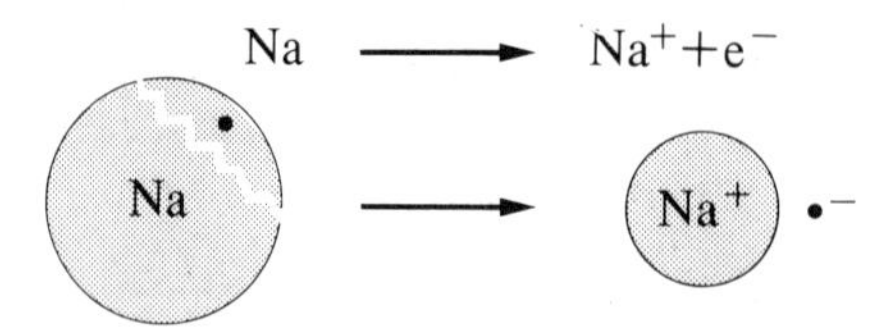

With the non-metal elements the opposite happens. Atoms of non-metals react to gain electrons and become negative ions:

$$Cl_2 + 2e^- \longrightarrow 2\,Cl^-$$

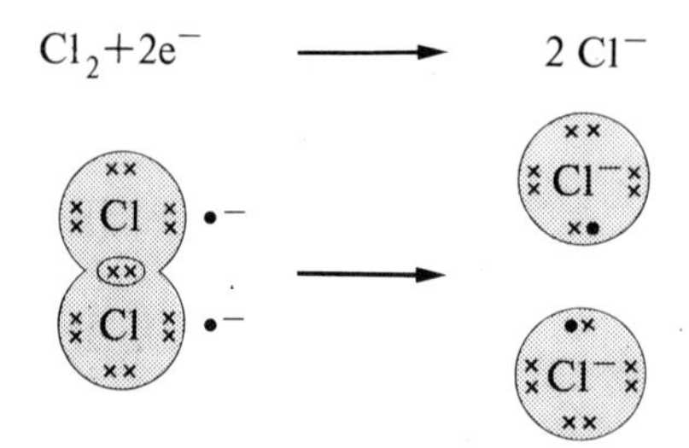

Losing electrons and gaining electrons are opposite processes, so it is not surprising that the reactivities of metals and non-metals show opposite trends.

Atomic size and structure There are two things to remember about the sizes and structures of atoms:

(i) As you move down a group in the Periodic Table, the atoms get larger. This is because more shells are being added. The outer shell is increasingly shielded from the nucleus by the inner shells.
(ii) As you move across a period the number of shells stays the same, but the number of protons in the nucleus increases. This means that the nucleus more strongly attracts the electrons in the atom and draws them closer to itself; the atoms get smaller.

Metal atoms How does this affect reactivity? Let us think first about the metal atoms of Groups I and II. Remember that they react to lose electrons. As the atoms get larger, the outer shells get further from and more shielded from the positive nucleus. The nucleus therefore attracts the outer-shell electrons less

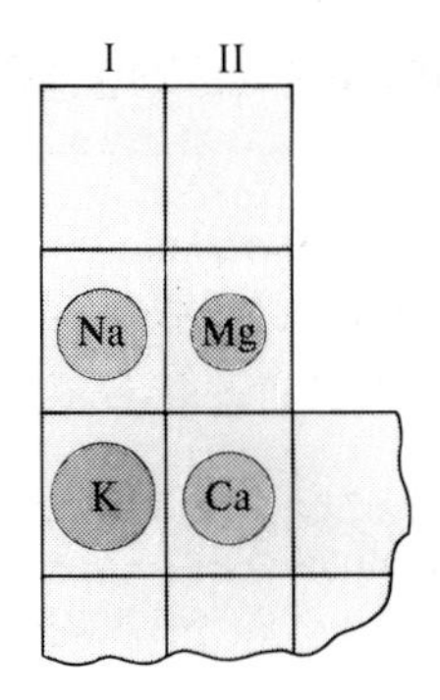

strongly; it becomes easier to remove them and form metal ions. So reactivity increases down Groups I and II. Going from Group I to Group II, the loss of the outer-shell electrons becomes less easy, for the reasons given in point (ii) above. So reactivity is lower in Group II than in Group I.

Non-metal atoms Now think about non-metal atoms. They react to gain electrons and become negative ions. As you go down a group, the outer shell gets further from the nucleus. It becomes more difficult to attract electrons into it to form ions. So reactivity decreases as you go down Groups VI and VII. In other words, reactivity increases as you go up Groups VI and VII.
Going from Group VI to Group VII, the atoms gain an extra proton and get smaller. So it becomes easier for the nucleus to attract an extra electron. Reactivity therefore increases from Group VI to Group VII.

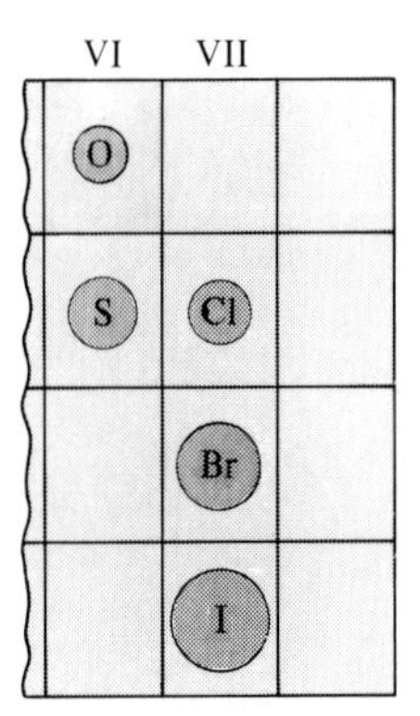

Summary

The electrochemical series gives the metals in order of decreasing reactivity.

The non-metals can also be arranged in order of decreasing reactivity.

The more reactive an element is, the less reactive its ionic compounds, and vice-versa.

Reactive metals contain atoms that lose electrons easily. Reactive non-metals contain atoms that gain electrons easily.

Reactivity is therefore connected with the size of the atom—the ease of losing and gaining electrons depends on the distance of the outer shell from the nucleus.

The sizes of atoms increase down any group of the Periodic Table, and decrease along any period of the Periodic Table.

Reactivity increases down Groups I and II.

Reactivity decreases down Groups VI and VII.

(Note: there is no clear trend of reactivity in the transition elements.)

Questions

1 Rewrite the following sentences, putting in the correct word from each pair in italics.
When metals react, their *atoms/molecules* react by *gaining/losing* electrons and so becoming *negatively/positively* charged. This means that the particles that form are *anions/cations*. The number of electrons *lost/gained* for elements in the main groups is *the same/not the same* as the group number, and the amount of charge on the ion is equal to the group number.

2 (a) Do the metals in Groups I and II get more or less reactive as you go down the group?
(b) When these metals react do they lose or gain electrons?
(c) Do the atoms of the elements in Groups I and II get bigger or smaller as you go down the group?
(d) Will the electrons in the outer shells of these atoms be more or less strongly held as you go down the group?
(d) Write a sentence relating the reactivity of these elements to the sizes of their atoms.

3 By going through the same sort of questions, think about the reactivity of the elements in Groups VI and VII.
Now write a sentence explaining why the reactivity in these groups change as you go down the group.

4 On page 183 there is an outline diagram of the Periodic Table.
(a) List the metals on it in order of reactivity, with the most reactive first.
(b) What is this sort of list called?
(c) Bracket together any elements in the same group.
(d) What pattern can you see relating group number to reactivity?

5 Look again at the diagram of the Periodic Table on page 183.
(a) Where in the Periodic Table would you find the most reactive metal?
(b) Where in the Periodic Table would you find the most reactive non-metal?

6 Read pages 180 and 181 carefully, and then explain, with an example of a metal and a non-metal, the meaning of the words *displacement reaction*.

7 Say which of the following pairs will react. For those that do, give the equation of the reaction.
(a) $Cu_{(s)} + Fe^{2+}_{(aq)}$ (b) $Cu^{2+}_{(aq)} + Mg_{(s)}$
(c) $Zn_{(s)} + Fe^{2+}_{(aq)}$ (d) $Zn_{(s)} + Ag^{+}_{(aq)}$
(e) $Fe^{2+}_{(aq)} + Mg_{(s)}$ (f) $Mg_{(s)} + Ag^{+}_{(aq)}$

8 Say which of the following pairs will react. For those that do, give the equation for the reaction.
(a) $I_{2(aq)} + Br^{-}_{(aq)}$ (b) $Cl_{2(aq)} + Br^{-}_{(aq)}$
(c) $Br_{2(aq)} + Br^{-}_{(aq)}$ (d) $Cl^{-}_{(aq)} + Br_{2(aq)}$
(e) $Br_{2(aq)} + I^{-}_{(aq)}$ (f) $Cl_{2(aq)} + I^{-}_{(aq)}$

9 Describe two pieces of evidence that indicate that:
(a) magnesium metal is more reactive than iron
(b) copper is less reactive than iron
(c) iodine is less reactive than chlorine

10 Put the following in order of reactivity with the most reactive first:
(a) NaCl NaI NaBr
(b) $CuSO_4$ $ZnSO_4$ Na_2SO_4 $FeSO_4$

11 Describe how you would make a pure sample of anhydrous iron(II)chloride in the lab. Sketch the apparatus.

18 Oxidation and reduction

18.1 The transfer of electrons

All the reactions you have seen in the last three chapters have involved the loss and gain of electrons.

Metals When a metal reacts, the lattice loses electrons, forming positive ions. For a single metal atom, the reaction is:

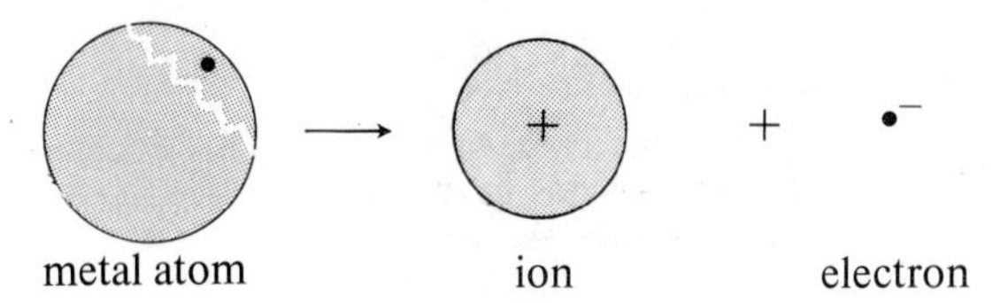

The electrochemical series tells you how easily this process happens when metals react with water, acids, or solutions of other metal salts. These electrons, lost from the metal atoms, may be gained by:

hydrogen atoms in water or acids
the ions of a less reactive metal
the atoms of a non-metal.

Non-metals When non-metals react, their atoms gain electrons to become negative ions:

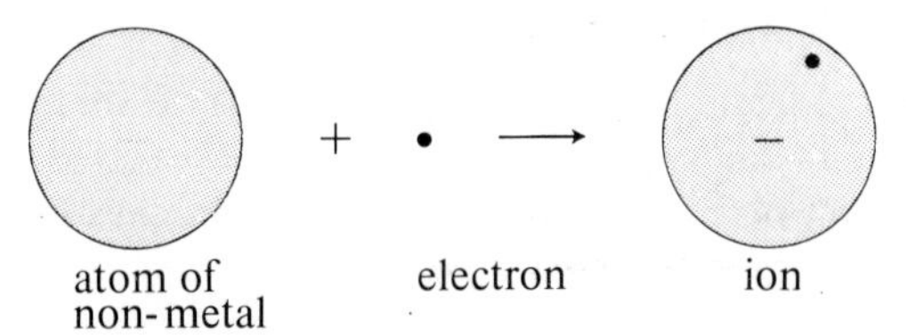

These electrons, gained by the non-metal atoms, have been lost from:

metal atoms
less reactive non-metal ions.

So all these reactions involve the transfer of electrons from the atoms or ions of one element to the atoms or ions of another element. When one particle loses electrons another particle gains them.

18.2 Oxidation and reduction

The processes of electron loss and gain are so common and important in the study of chemistry that special names are given to them.
The loss of electrons by an atom or ion is called oxidation.
The gain of electrons by an atom or ion is called reduction.

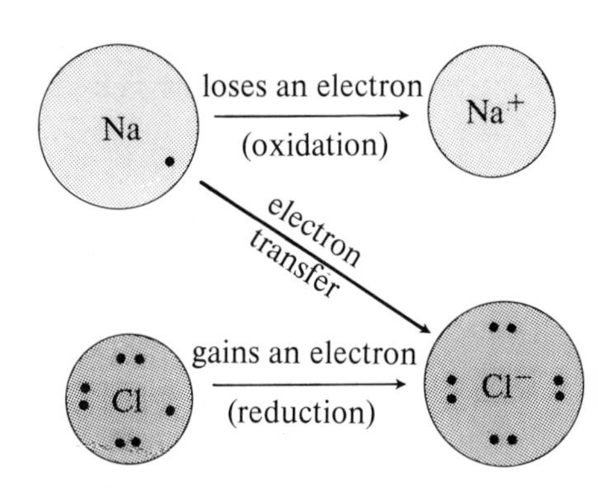

As we have seen, these two processes always take place at the same time. Electron loss and gain are just the results of the transfer of electrons. The diagram on the right summarises the processes that take place when sodium burns in chlorine.

Reactions involving electron transfer are often called **redox reactions**. The word *redox* is short for *reduction-oxidation.*

Sometimes one particular reaction is called an oxidation or a reduction. For example, the reaction

$$2Ca_{(s)} + O_{2(g)} \longrightarrow 2Ca^{2+}_{(s)} + 2O^{2-}_{(s)}$$

may be called the oxidation of calcium. This does not mean that reduction is *not* happening. It means that the chemist is concentrating on what happens to the atoms of calcium.

In general, when a metal reacts with oxygen or another reactive non-metal, the metal lattice loses electrons so the metal is being **oxidised**. In the reaction above, calcium is being oxidised. In the reaction below, magnesium is being oxidised:

$$Mg_{(s)} + Cl_{2\,(g)} \longrightarrow Mg^{2+}_{(s)} + 2Cl^{-}_{(s)}$$

Similarly a non-metal is being **reduced** when it reacts with a metal: the non-metal atoms gain electrons. In the last two examples oxygen and chlorine are being reduced.

Redox in industry

A very important redox reaction in industry is the one in which a metal compound—an ore from the ground—is converted into the metal. The word 'reduction' was first used to describe this process. It can be illustrated in the laboratory using copper(II) oxide:

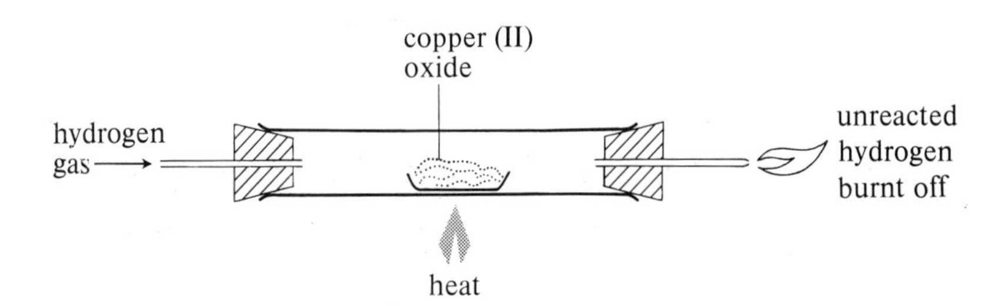

The copper(II) oxide slowly turns into copper.

$$CuO_{(s)} + H_{2(g)} \longrightarrow Cu_{(s)} + H_2O_{(g)}$$

You can see that oxygen is being removed from copper(II) oxide during the reaction. This led to the early definition of reduction: *reduction is the removal of oxygen*. So copper(II) oxide is being reduced. Using the modern broader definition, copper(II) oxide is being reduced because the copper ions in copper oxide are gaining electrons to become copper atoms.

In the same way, one early definition of oxidation was: *oxidation is the removal of hydrogen*. So in the reaction

$$H_2S_{(g)} + Cl_{2\,(g)} \longrightarrow 2HCl_{(g)} + S_{(s)}$$

hydrogen sulphide is being oxidised, because hydrogen is being removed from it. Using the modern definition, we say that hydrogen sulphide is being oxidised because the sulphur atoms in it are partly losing control of electrons, to become solid sulphur. How this happens will be clearer when you read section 18.3 below.

Redox and electrolysis In Chapter 16 we studied reactions at the electrodes during electrolysis. The electrolysis of molten lead bromide is a typical one. At the cathode, lead ions gain electrons and become lead atoms:

$$Pb^{2+}_{(l)} + 2e^- \longrightarrow Pb_{(l)}$$

At the cathode, therefore, a reduction reaction is taking place.

At the anode, bromide ions lose electrons and become bromine atoms:

$$2Br^-_{(l)} \longrightarrow Br_{2\,(g)} + 2e^-$$

An oxidation reaction is taking place at the anode.

So in electrolysis the processes of oxidation and reduction go on at the same time, but at opposite ends of the electrolytic cell. The electrons are transferred by the electric circuit. We can therefore include electrolysis in our definitions of oxidation and reduction—see below.

Summary Oxidation—the loss of electrons—occurs when

(i) an element reacts with oxygen or another reactive non-metal like chlorine
(ii) hydrogen is removed from a compound
(iii) an anode reaction takes place.

Reduction—the gain of electrons—occurs when

(i) an element reacts with a metal or hydrogen
(ii) oxygen is removed from a compound
(iii) a cathode reaction takes place.

18.3 The need for oxidation numbers

When two non-metal elements react together the loss or gain of electrons is not so clear. In this reaction:

carbon + oxygen ⟶ carbon dioxide

$C_{(s)} \quad + O_{2\,(g)} \qquad CO_{2\,(g)}$

carbon is obviously reacting with oxygen, so it is being oxidised. Logically the oxygen must be being reduced. But how can this be explained in terms of electrons? In carbon, oxygen, and carbon dioxide, the bonding is covalent—bonding electrons are shared between atoms:

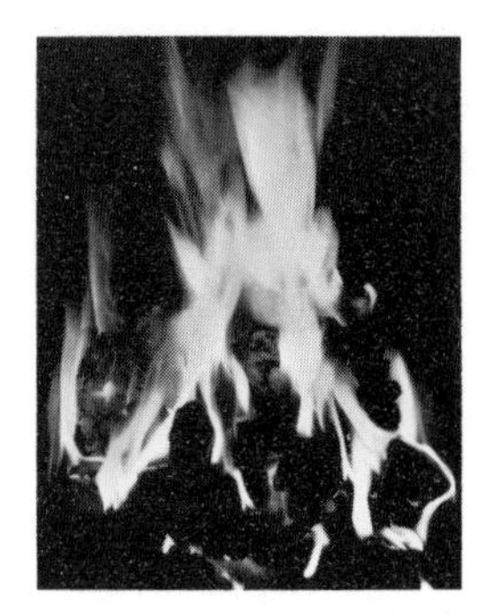

The reaction
$C + O_2 \longrightarrow CO_2$

carbon atoms sharing electrons + oxygen atoms sharing electrons ⟶ oxygen and carbon atoms sharing electrons

Both before and after this reaction, therefore, electrons are shared. There is no complete transfer of electrons and no ions are formed. But a closer look will show that there is *partial* loss and gain of electrons.

In the two elements, electrons are evenly shared: two carbon atoms share a pair of electrons equally; two oxygen atoms share two pairs of electrons equally. But in carbon dioxide there is uneven sharing because oxygen is better at attracting electrons than carbon is.

O C O leading to $\overset{\delta-}{O}=\overset{\delta+}{C}=\overset{\delta-}{O}$

bonding electrons nearer the oxygen atoms

So in the reaction with oxygen, carbon atoms lose partial control of the bonding electrons, therefore carbon is oxidised. The oxygen atoms gain partial control of the bonding electrons, and so oxygen is reduced.

Oxidation numbers are used to work out the control an atom has over the electrons around it. You may have already met these numbers in Chapter 5. Read pages 46–48 again.

18.4 Summary of ideas on oxidation numbers

1 Every element in every compound can be given an oxidation number. It is worked out according to a set of rules and tells us two things.

(i) The oxidation number itself is the same as the number of electrons each atom of the element uses, in forming bonds to atoms of other elements.
(ii) The sign of the oxidation number tells whether the atoms of the element have lost or gained control of bonding electrons.

2 Electrons are negative, so an atom that loses some control of the electrons around it becomes positive. The element will have a positive oxidation number. For example:

in Na^+, one electron has been lost, so the oxidation number is +I;
in Cu^{2+}, two electrons have been lost, so the oxidation number is +II.

3 Atoms that gain control of electrons become negative. The element will have a negative oxidation number:

in Cl^-, one electron has been gained; the oxidation number is −I.
in O^{2-}, two electrons have been gained; the oxidation number is −II.

4 In a pure element all the atoms are the same. Electrons are neither gained nor lost but shared evenly by the atoms:

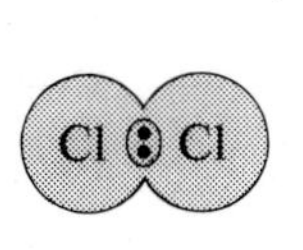

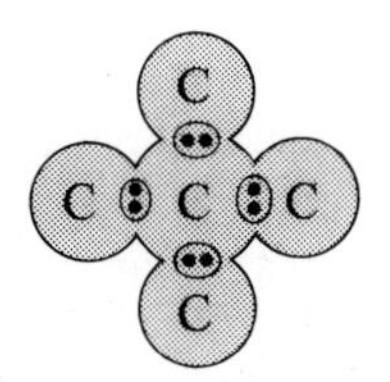

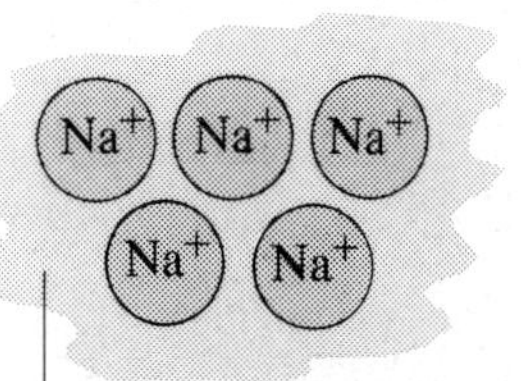

So the oxidation number of elements is always zero.

5 In an ionic compound the loss or gain of electrons is complete:

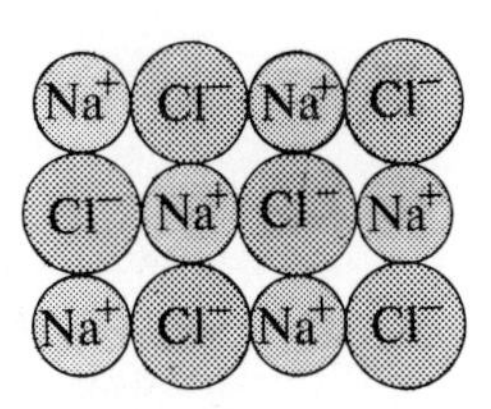

6 In a molecular compound the loss or gain of electrons is partial. This partial loss or gain is shown by the small delta sign, δ.

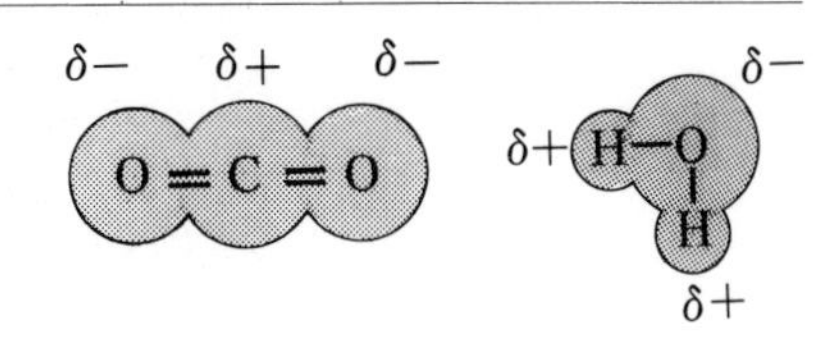

7 The oxidation number of an element often changes during a reaction:

Oxidation number change	*Name of change*
The oxidation number of an element in a substance goes down during a reaction. For example: $\overset{II}{Cu}O_{(s)} + H_{2(g)} \longrightarrow \overset{0}{Cu}_{(s)} + H_2O_{(g)}$	The substance is said to have been reduced. The reaction is a reduction for that substance. Copper oxide has been reduced.
The oxidation number of an element in a substance goes up during a reaction. For example: $2\overset{II}{Fe}Cl_{2(s)} + Cl_{2(g)} \longrightarrow 2\overset{III}{Fe}Cl_{3(s)}$	The substance is said to have been oxidised. The reaction is an oxidation for that substance. Iron(II) chloride has been oxidised.

Working out oxidation numbers These were given in Chapter 5 with some examples. If you have forgotten them, you will find it helpful to turn back and work through a couple of examples before going on.

Oxidation or reduction? Using oxidation numbers may seem confusing at first, but with a little practice you will find it quite easy. In the examples below, they are being used to find out which substance is being oxidised and which reduced. Work through these examples carefully.

Example 1 $2Na_{(s)} + Cl_{2\,(g)} \longrightarrow 2NaCl_{(s)}$

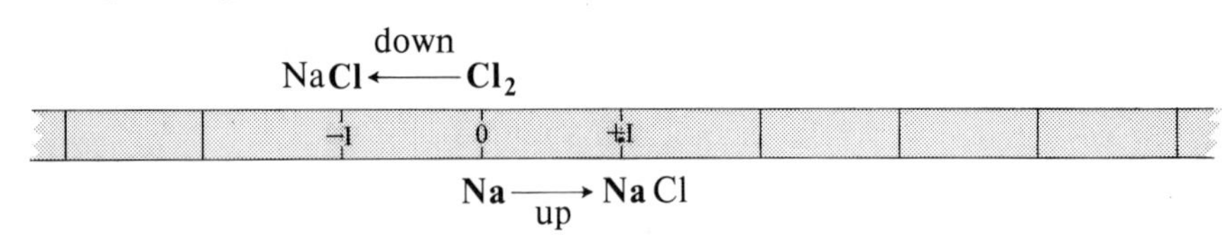

Chlorine's oxidation number has gone down from 0 to −I. Chlorine has been reduced.
Sodium's oxidation number has gone up from 0 to +I. Sodium has been oxidised.

Example 2 $CuO_{(s)} + H_{2\,(g)} \longrightarrow Cu_{(s)} + H_2O_{(g)}$

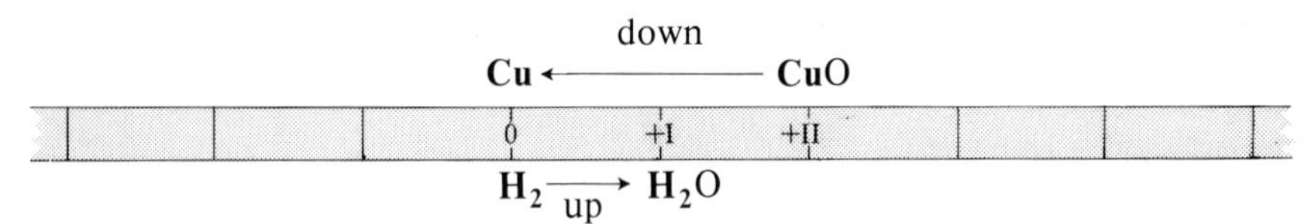

Copper's oxidation number has gone down from +II to 0. So copper oxide has been reduced.
Hydrogen's oxidation number has gone up from 0 to +I. Hydrogen has been oxidised.

Example 3 $2CuO_{(s)} + C_{(s)} \longrightarrow 2Cu_{(s)} + CO_{2\,(g)}$

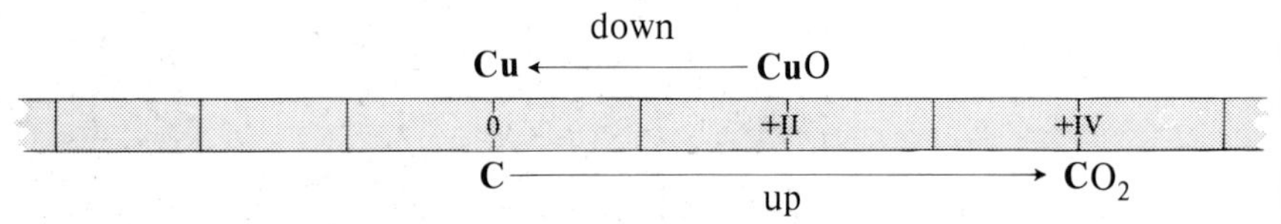

Copper's oxidation number has gone down from +II to 0. Copper oxide has been reduced.
Carbon's oxidation number has gone up from 0 to +IV. Carbon has been oxidised.

18.5 Recognising redox reactions

When the equation of a reaction is written down, you can tell whether it is a redox reaction or not by using oxidation numbers. All you have to do is work out the oxidation numbers for all the elements in the equation. If the oxidation number of any element has changed, going from reactant to product, the reaction is a redox one.

Example 1 $MnO_{2\,(s)} + 4HCl_{(aq)} \longrightarrow MnCl_{2\,(aq)} + 2H_2O_{(l)} + Cl_{2\,(g)}$

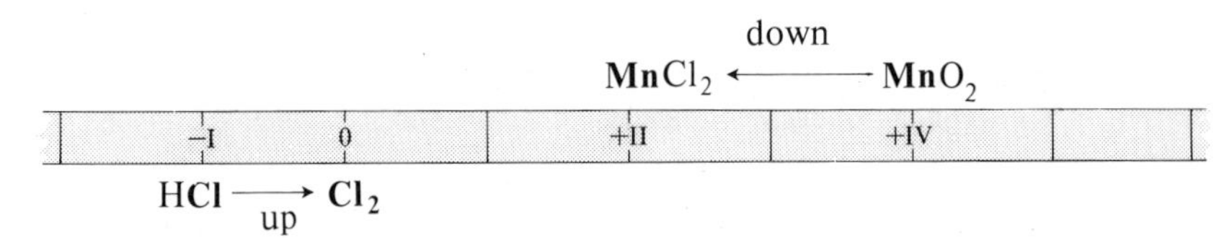

Conclusion:
The oxidation numbers of manganese and some of the chlorine have changed. This is a redox reaction.

Example 2 $\overset{II}{Cu}O_{(s)} + H_2\overset{VI}{S}O_{4\,(l)} \longrightarrow \overset{II}{Cu}\overset{VI}{S}O_{4\,(aq)} + H_2O_{(l)}$

no changes

Conclusion:
No oxidation numbers have changed. The reaction is not a redox one.

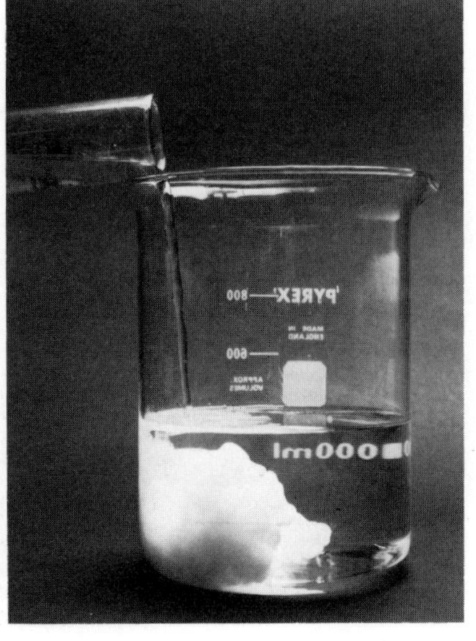

Left: rusting, a common problem. Find the equation for it on p. 311. Is it a redox reaction? What is being oxidised?

Right: a precipitate is formed when silver nitrate solution is added to sodium chloride solution. $AgNO_3 + NaCl \longrightarrow AgCl\downarrow + NaNO_3$ *Is this a redox reaction?*

18.6 Oxidising and reducing agents

An oxidising agent is a substance that causes the oxidation of another substance.
A reducing agent is a substance that causes the reduction of another substance.
Oxidising agents are sometimes called **oxidants**; reducing agents are called **reductants**.
With a little care, we can work out which substances are the oxidising and reducing agents, from the equation of a redox reaction.

Example 1 $MnO_{2\,(s)} + 4HCl_{(aq)} \longrightarrow MnCl_{2\,(aq)} + 2H_2O_{(l)} + Cl_{2\,(g)}$

down
$MnCl_2 \longleftarrow MnO_2$
−I 0 +II +IV
$HCl \longrightarrow Cl_2$
up

Conclusions:

(i) The oxidation number of manganese has gone down, so manganese(IV) oxide has been reduced. Therefore hydrogen chloride is acting as a reducing agent.

(ii) The oxidation number of chlorine has gone up, so hydrogen chloride has been oxidised. Therefore manganese(IV) oxide is acting as an oxidising agent.

Example 2 $C_{(s)} + 2H_2SO_{4\,(l)} \longrightarrow CO_{2\,(g)} + 2SO_{2\,(g)} + 2H_2O_{(l)}$

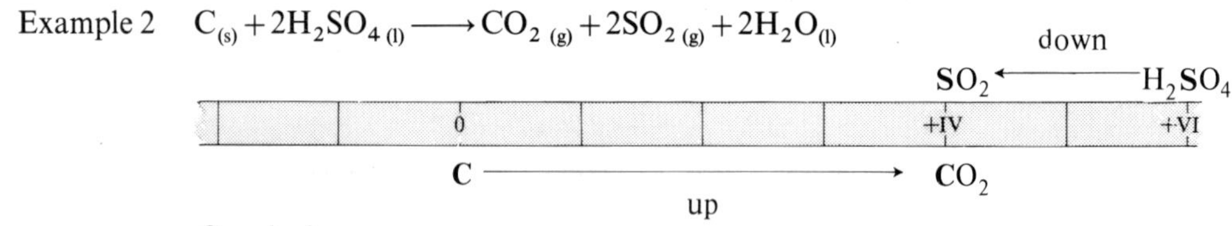

Conclusions:

(i) The oxidation number of carbon has gone up, therefore sulphuric acid is acting as an oxidising agent.

(ii) The oxidation number of sulphur has gone down, therefore carbon is acting as a reducing agent.

From the examples above, we can conclude that:
an oxidising agent is a substance that is easily reduced
a reducing agent is a substance that is easily oxidised
This follows logically from the original definitions of oxidation and reduction. During the oxidation of an element, its atoms lose some control of electrons. The atoms of a second element gain control of these electrons, so this second element is being reduced.

Table 20.1 shows some common oxidising and reducing agents, in order of their strengths. All the substances in this table are elements.

Table 18.1 Some oxidising and reducing elements

Reducing agents			*Oxidising agents*	
Potassium	$K_{(s)} \longrightarrow K^{+}_{(aq)} + e^{-}$	increasing strength (↑)	Oxygen	$O_{2(g)} + 4e^{-} \longrightarrow 2O^{2-}_{(aq)}$
Sodium	$Na_{(s)} \longrightarrow Na^{+}_{(aq)} + e^{-}$		Chlorine	$Cl_{2(g)} + 2e^{-} \longrightarrow 2Cl^{-}_{(aq)}$
Calcium	$Ca_{(s)} \longrightarrow Ca^{2+}_{(aq)} + 2e^{-}$		Bromine	$Br_{2(g)} + 2e^{-} \longrightarrow 2Br^{-}_{(aq)}$
Magnesium	$Mg_{(s)} \longrightarrow Mg^{2+}_{(aq)} + 2e^{-}$		Iodine	$I_{2(g)} + 2e^{-} \longrightarrow 2I^{-}_{(aq)}$
Aluminium	$Al_{(s)} \longrightarrow Al^{3+}_{(aq)} + 3e^{-}$		Sulphur	$S_{8(s)} + 16e^{-} \longrightarrow 8S^{2-}$
Zinc	$Zn \longrightarrow Zn^{2+}_{(aq)} + 2e^{-}$			
Hydrogen	$H_{2(g)} \longrightarrow 2H^{+}_{(aq)} + 2e^{-}$			
Carbon	$C_{(s)} \longrightarrow CO_{2(g)}$			

Right: when an iron nail is put into copper sulphate solution, it begins to dissolve. At the same time a coating of copper forms on it. Is this a redox reaction? What is the reducing agent?

Note that the metals in Table 18.1 are in the same order as in the electrochemical series (see page 158). The non-metals are in the same order as the activity list on page 182.
Many compounds can act as reductants or oxidants. Some common ones are listed in Table 18.2. Remember, they work because they are themselves easily oxidised or reduced.

Table 18.2 Some oxidising and reducing compounds

Reducing agents		
Carbon monoxide	It can be oxidised to carbon dioxide.	$\overset{II}{C}O_{(g)} \longrightarrow \overset{IV}{C}O_{2(g)}$
Hydrogen sulphide	It can be oxidised to sulphur.	$H_2\overset{-II}{S}_{(g)} \longrightarrow \overset{0}{S}_{8(s)}$
Sulphur(IV) oxide	It can be oxidised in water to sulphate(VI) salts.	$\overset{IV}{S}O_2 \longrightarrow \overset{VI}{S}O_4^{\ 2-}$
Oxidising agents		
Potassium manganate(VII)	It can be reduced to manganese(II) salts.	$K\overset{VII}{Mn}O_{4(s)} \longrightarrow \overset{II}{Mn}^{2+}_{(aq)}$
Potassium dichromate(VI)	It can be reduced to chromium(III) salts.	$K_2\overset{VI}{Cr}_2O_{7(s)} \longrightarrow \overset{III}{Cr}^{3+}_{(aq)}$
Sulphuric(VI) acid	It can be reduced to sulphur(IV) oxide.	$H_2\overset{VI}{S}O_4 \longrightarrow \overset{IV}{S}O_2$

Tests for oxidising and reducing agents

The ability of a substance to be an oxidant or a reductant is a very important part of its chemistry. Because of this, we have special tests to recognise these properties. Such tests must be easy to do and must involve a chemical change that is easy to see. The most obvious changes are colour changes—nearly all the tests involve a clearly visible colour change. Tables 18.3 and 18.4 list the common tests for oxidising and reducing agents in solution.

Table 18.3 Reactants used to test for oxidising agents

Reactant	*Product*	*The oxidation*
Iron(II) sulphate solution $FeSO_4$ **pale green**	Iron(III) sulphate solution $Fe_2(SO_4)_3$ **pale yellow-brown**	iron(II) ⟶ iron(III) $Fe^{2+}_{(aq)} \longrightarrow Fe^{3+}_{(aq)}$
Potassium iodide solution KI **colourless**	Iodine solution I_2 **brown**	iodide(−I) ⟶ iodine(O) $2I^-_{(aq)} \longrightarrow I_{2\,(aq)}$
Sodium sulphide solution Na_2S **colourless**	Sulphur suspension S_8 **pale yellow**	sulphide(−II) ⟶ sulphur(O) $8S^{2-}_{(aq)} \longrightarrow S_{8\,(s)}$

A substance which can cause any of these reactants to change colour as shown must be an oxidising agent.

Table 18.4 Reactants used to test for reducing agents

Reactants	*Product*	*The reduction*
Potassium manganate(VII) in dilute sulphuric acid solution $KMnO_4$ **purple**	Manganese(II) sulphate in solution $MnSO_4$ **colourless**	manganate(VII) ⟶ manganese(II) $MnO^-_{4(aq)} \longrightarrow Mn^{2+}_{(aq)}$
Potassium dichromate(VI) in dilute sulphuric acid solution $K_2Cr_2O_7$ **orange**	Chromium(III) sulphate in solution $Cr_2(SO_4)_3$ **green**	dichromate(VI) ⟶ chromium(III) $Cr_2O^{2-}_{7\,(aq)} \longrightarrow Cr^{3+}_{(aq)}$

A substance which can cause either of the colour changes shown above must be causing reduction—it must be a reducing agent.

Summary

Reactions involving the complete or partial transfer of electrons from the atoms of one element to the atoms of another element are called redox reactions. Redox is short for reduction-oxidation.

Reduction occurs when the atoms of an element gain more control over electrons around them. This happens when:

the element reacts with a reactive metal or hydrogen;
oxygen is removed from a compound containing the element;
a cathode reaction takes place.

During reduction the oxidation number of the element goes down.

Oxidation occurs when the atoms of an element lose control over the electrons around them. This happens when:

the element reacts with oxygen or another reactive non-metal;
hydrogen is removed from a compound containing the element;
an anode reaction takes place.

During oxidation, the oxidation number of the element goes up.

Oxidation numbers give an indication of how much control an atom has over the electrons it uses to form bonds.

Redox reactions can be recognised from their equations, because the elements change their oxidation numbers.

Oxidising agents cause other substances to be oxidised.

Reducing agents cause other substances to be reduced.

Oxidising and reducing agents are sometimes called oxidants and reductants.

Questions

1 Explain the meaning of the following:
(a) oxidation
(b) oxidising agent
(c) oxidation number

2 Explain:
(a) why oxidation and reduction always happen at the same time
(b) what happens to oxidation numbers during a redox reaction
(c) how you can tell from a written equation whether it is a redox equation

3 (a) List three compounds which are oxidising agents, of which two are solutions.
(b) List three compounds which are reducing agents.
(c) What type of elements are oxidising agents.
(d) What type of elements are reducing agents.
(e) Draw an outline of the Periodic Table, and mark in on it oxidising and reducing agents.

4 Look up two tests for reducing agents.
(a) Describe what you would do to carry out the tests.
(b) Describe what you would see if the test was positive.
(c) Write down the equations for the reactions that have taken place in the tests.

5 Look up two tests for oxidising agents.
(a) Describe what you would do to carry out the tests.
(b) Describe what you would see if the test was positive.
(c) Write down the equations for the reactions that have taken place in the tests.

6 Work out the oxidation numbers of the atoms underlined below:

$\underline{C}O$	$\underline{S}O_3$	$H\underline{Cl}$	$\underline{N}_2O$
$H\underline{N}O_3$	$H_2\underline{C}O_3$	$H_2\underline{S}O_3$	$H_2\underline{S}O_4$
$\underline{N}H_3$	$\underline{N}H_4OH$	$H_2\underline{O}_2$	$\underline{Cl}_2$

7 The equation for the reaction between magnesium and chlorine is:
$Mg + Cl_2 \rightarrow MgCl_2$
(a) Write down the oxidation numbers of the elements on both sides of the equation.
(b) Which oxidation number has gone up.
(c) Which oxidation number has gone down.
(d) What has been oxidised?
(e) What has been reduced?

8 For each reaction, say which substance is being oxidised and which is being reduced.
(a) $2CO + O_2 \rightarrow 2CO_2$
(b) $H_2 + S \rightarrow H_2S$
(c) $Mg + H_2SO_4 \rightarrow MgSO_4 + H_2$
(d) $PbO + CO \rightarrow Pb + CO_2$
(e) $4HCl + PbO_2 \rightarrow PbCl_2 + Cl_2 + 2H_2O$
(f) $PbS + 4H_2O_2 \rightarrow PbSO_4 + 4H_2O$
For each, write down the substance that acts as oxidising agent.

9 Which of the following equations describe redox changes?
(a) $NH_3 + HCl \rightarrow NH_4Cl$
(b) $3CuO + 2NH_3 \rightarrow 3Cu + 3H_2O + N_2$
(c) $MgO + H_2SO_4 \rightarrow MgSO_4 + H_2O$
(d) $Mg + H_2SO_4 \rightarrow MgSO_4 + H_2$
(e) $MnO_2 + 4HCl \rightarrow MnCl_2 + 2H_2O + Cl_2$
For each redox reaction, write down the oxidant and reductant.

10 From the list below, pick a substance that can be:
(a) an acid (b) an oxidant (c) a reductant
(d) a base
Give the reasons for your choice.
MnO_2 MgO H_2S $NaCl$ SO_2 HCl

11 Look back at page 173.
(a) Write down the cathode reaction for the lead bromide cell.
(b) What is happening to lead ions during this reaction?
(c) Is this process oxidation or reduction?
(d) Write down the reaction for the anode reaction in the same cell.
(e) What is happening to bromide ions during this reaction?
(f) Is this process oxidation or reduction?

19 Energy changes in chemical reactions

19.1 Exothermic and endothermic reactions

Chemical change

We first introduced the properties of chemical change in Chapter 4. The three major observations made about all chemical changes are:

(i) A new substance is produced. Reactants ⟶ products.
(ii) The amounts of reactants and products are in fixed ratio.
(iii) There is an energy exchange with the surroundings.

Read pages 31 and 32 again to remind yourself in more detail about the properties of chemical change.

Systems and surroundings

The coal on a fire reacts with the oxygen in the air and heats up the surroundings:

System and surroundings	*Energy exchange*
	From system to surroundings

This is an example of an **exothermic** reaction. 'Exo' comes from Greek and means 'out'. Energy is given out by the reacting system.

Trees grow by a chemical change. Carbon dioxide from the air combines with water to make new cells. This reaction is called **photosynthesis** and needs the energy from the sun to make it work:

System and surroundings	*Energy exchange*
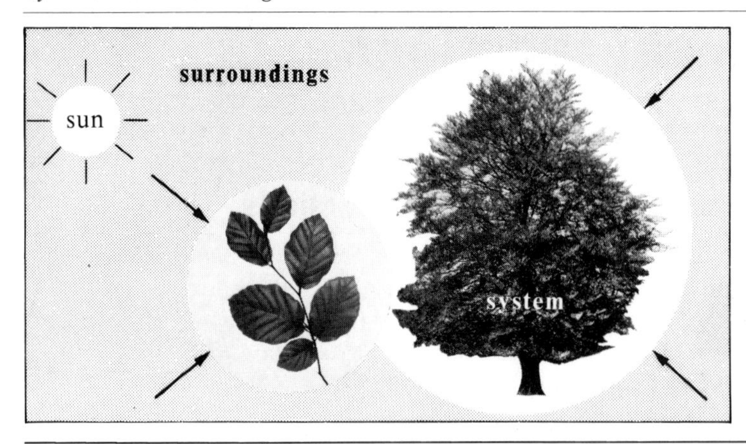	From surroundings to system

This is an example of an **endothermic** reaction. 'Endo' also comes from Greek and means 'in'. Energy is taken in by the system.

The energy taken in and given out can be in different forms:

heat energy, e.g.	taken in by a system warmed over a bunsen; given out by a coal fire.
electrical energy, e.g.	taken in during electrolysis; given out by the reaction in a battery.
light energy, e.g.	taken in during photosynthesis; given out when potassium metal reacts with water.

In the laboratory we are usually concerned with heat energy. Chemical reactions often occur between reactants dissolved in solution: we can tell whether they are exothermic or endothermic simply by using a thermometer. This is shown in the two experiments that follow.

Experiment 1

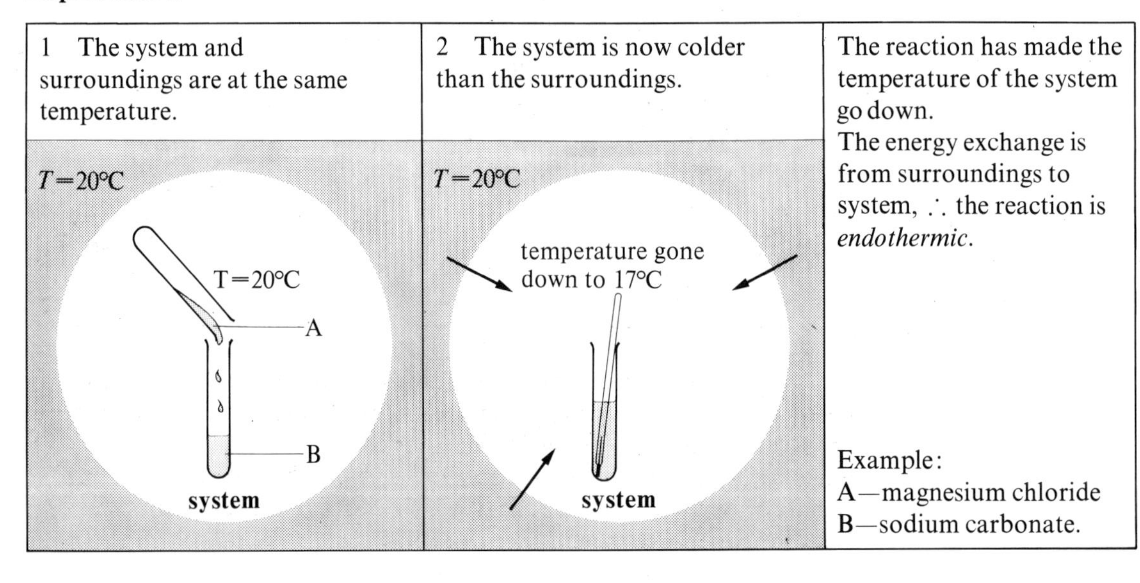

1 The system and surroundings are at the same temperature.
$T=20°C$
T=20°C
A
B
system
2 The system is now colder than the surroundings.
$T=20°C$
temperature gone down to 17°C
system
The reaction has made the temperature of the system go down.
The energy exchange is from surroundings to system, ∴ the reaction is *endothermic*.
Example:
A—magnesium chloride
B—sodium carbonate.

Experiment 2

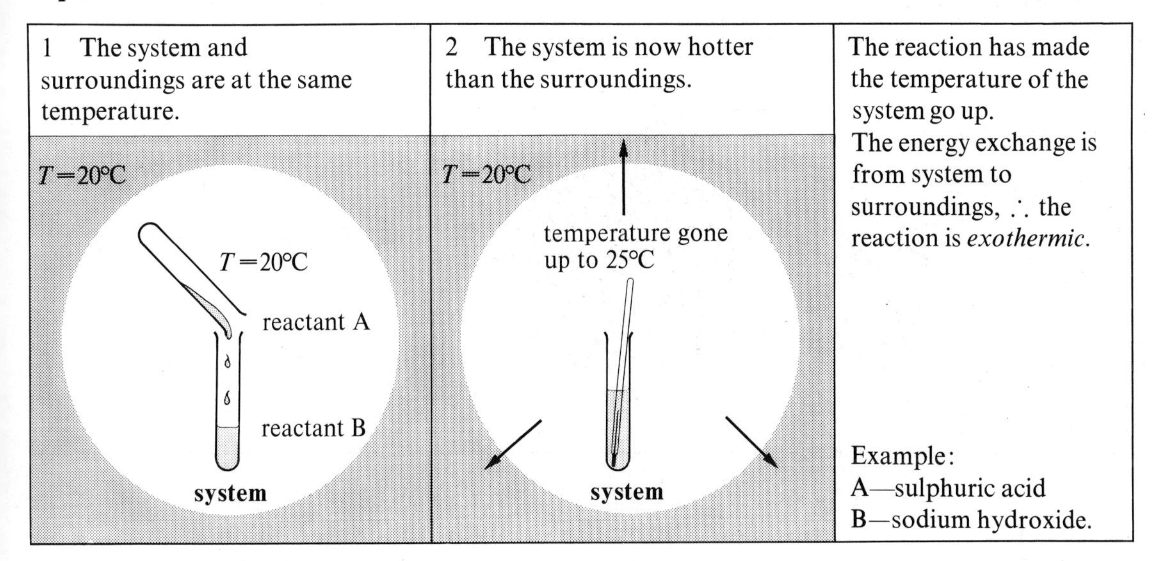

1 The system and surroundings are at the same temperature.
2 The system is now hotter than the surroundings.
The reaction has made the temperature of the system go up.
The energy exchange is from system to surroundings, ∴ the reaction is *exothermic*.
T=20°C
T=20°C
reactant A
reactant B
system
T=20°C
temperature gone up to 25°C
system
Example:
A—sulphuric acid
B—sodium hydroxide.

19.2 Reactants and products

Remember the two aims of chemists?

(i) To know the facts about chemical systems;

(ii) to find theories that explain these facts.

We now know a number of facts about endothermic and exothermic reactions. We must try to explain them using the particle theory.

We shall take a simple example of an exothermic reaction to introduce the theory slowly.

Fact	*Theory*	
Hydrogen and chlorine gas explode violently when put together in sunlight. Hydrogen chloride is formed. $H_{2(g)} + Cl_{2(g)} \longrightarrow 2HCl_{(g)}$	1 Bonds break.	2 Bonds form.
	Cl Cl H H	Cl H Cl H

It needs energy to break a bond. Think of two strong magnets attracting one another: it needs energy to pull them apart. In the same way energy must be supplied to pull two bonded atoms apart.

But what happens when bonds are made? Energy (like matter) cannot be created or destroyed. This means that *when a bond is made, exactly the same energy is given out as would be needed to break it.*

To prove this strange idea to yourself, consider the steps shown below:

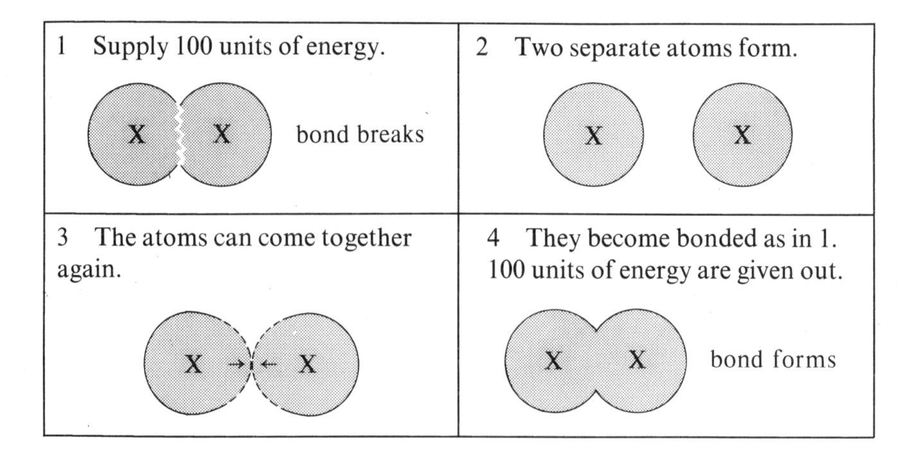

If we do not get 100 units of energy back in step 4, then we have found a way of destroying energy!

Bond energy

The amount of energy needed to separate two atoms is too small to measure directly. Instead, we use a unit called the mole, which we will discuss in later chapters. We always measure the energy needed to separate a mole of pairs of atoms.

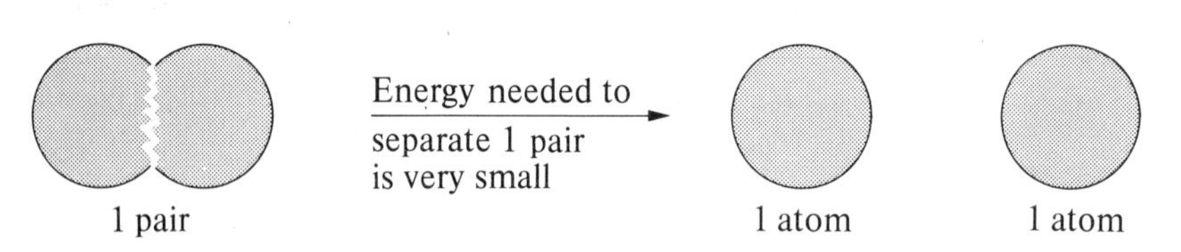

We scale the process up L (the Avogadro Constant) times:

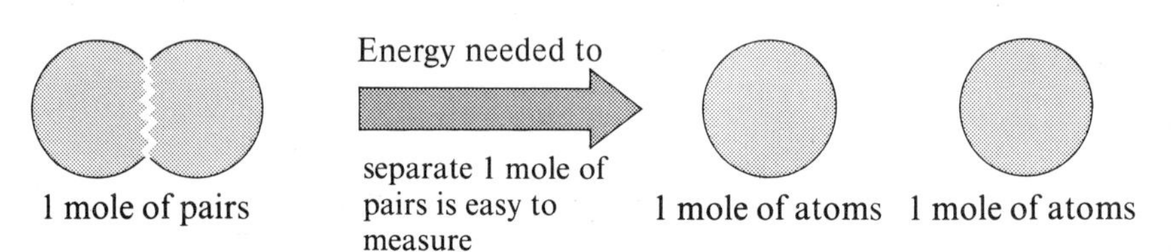

The energy needed to break one mole of bonds is called the bond energy.

When a mole of bonds is *made*, this bond energy is given out from the system to the surroundings.

Table 19.1 lists some common bond energies. They are measured in the normal units of energy—kilojoules per mole of bonds broken, written in short as $kJ\,mol^{-1}$.

Table 19.1 Some common bond energies

Bond	*Bond energy /kJ mol^{-1}*	*Bond*	*Bond energy /kJ mol^{-1}*
H–H	436	C–C	349
Cl–Cl	242	C–H	412
H–Cl	431	O–H	462

Using this list of bond energies, we can now explain why the hydrogen and chlorine reaction sketched on page 201 is exothermic:

Bonds broken	*Energy*	*Bonds made*	*Energy*
H–H	436 kJ mol^{-1}	H–Cl	431 kJ mol^{-1}
Cl–Cl	242 kJ mol^{-1}	H–Cl	431 kJ mol^{-1}
Total energy *in* = 678 kJ mol^{-1}		Total energy *out* = 862 kJ mol^{-1}	

We get more energy out from making the new bonds than was put in to break the old bonds. This extra energy given out is passed to the surroundings—the reaction is exothermic. In this case, the system gives out 184 kJ to the surroundings, for every two moles of hydrogen chloride made.

The energy given out or taken in by a reaction is often shown beside its equation:

$$H_{2(g)} + Cl_{2(g)} \longrightarrow 2HCl_{(g)} \qquad \Delta H = -184\,kJ\,mol^{-1}$$

ΔH, pronounced *delta H*, stands for the energy exchange between system and surroundings. The sign tells whether the system loses energy to the surroundings (−), or gains energy from the surroundings (+).
Exothermic reactions have negative ΔH values. Endothermic reactions have positive ΔH values.

$$C_{(s)} + O_{2(g)} \longrightarrow CO_{2(g)} \qquad \Delta H = -394\,kJ\,mol^{-1}\ \text{(exothermic)}$$

$$C_{(s)} + H_2O_{(g)} \longrightarrow CO_{(g)} + H_{2(g)} \qquad \Delta H = +137\,kJ\,mol^{-1}\ \text{(endothermic)}$$

19.3 Energy sources

The table below shows some energy sources and how we use them.

Source	*Example*	*Use*
the weather	sunshine	solar panels for water heaters
	wind	windmills for driving electrical dynamos
water	waterfalls	for generating hydroelectric power
	sea waves	for driving specially designed electrical dynamos
fuels to burn	coal	power stations and domestic fires
	oil	engines of all sorts e.g. cars, ships, and planes
	gas	cookers and central heating
nuclear fuels	uranium plutonium	nuclear power stations atomic bombs and warheads

The sun The sun is our most important source of energy. Plants and trees use its energy to combine water and carbon dioxide by photosynthesis (see page 335). All animals, including humans, live by eating plants or other animals and by converting the stored energy for their own growth. When a plant or an animal dies, it returns to the soil. The coal, oil, and gas shown above as 'fossil fuels', are the decayed remains of prehistoric plants and animals.

Waste Waste causes problems for two main reasons. Firstly, there is only a limited amount of fossil fuel beneath the earth and the seas. We are using it up both too quickly, and too wastefully. It has taken millions of years for the process of decay to provide us with these fuels. If we go burning them at the present rate, they will barely last into the next century. This is why so much effort is being put into finding alternative energy sources.

Secondly, once a fuel has been used, there are always waste products left behind. The disposal of these waste materials is a major headache. The burning of such massive quantities of fossil fuel is the main cause of atmospheric pollution. The waste products of nuclear fuels are even more dangerous. They are highly radioactive and give off radiation which can kill most forms of life. The nuclear industry reprocesses its waste as far as is possible, but there is quite a lot of radioactive material which is useless. This is sealed in special containers and buried deep in the ocean or in the ground.

The future The search is on for new methods of obtaining 'clean' electrical energy. Sunshine, wind, and water power already provide some. These sources of energy are clean because there are no harmful waste products to worry about, and the resources are not used up in the process. But the world's energy requirements cannot be satisfied by these sources alone. Various ideas are being researched at present. One of the most exciting is a process called **nuclear fusion**. This is the reaction which is happening within the sun. Scientists are developing the technology necessary for the reaction to occur on the Earth. As you can imagine, this is not very easy . . . the sun is extremely dense and is at a temperature of well over a million degrees Celsius! All the same, the researchers are quite confident of success by the turn of the century. While we wait, however, we ought to be cutting right down on the use of combustion engines for transport. New designs of battery and of battery-powered vehicles are constantly being developed but, so far, no really successful model has been designed. A few examples are shown in the photos below.

19.4 Activation energy

In all reactions, whether exothermic or endothermic, bonds must be broken before new bonds can be made. For this reason, many reactants need to be supplied with energy before the reaction actually begins. For example, you use a burning match to light a fire, and a spark to light a gas cooker.
If the reaction is exothermic, it will itself supply the energy necessary to keep going, once some bonds have been broken.

This energy needed to start a reaction is known as the **activation energy** for the reaction. It can be shown on an **energy diagram** as a hump between reactants and products.

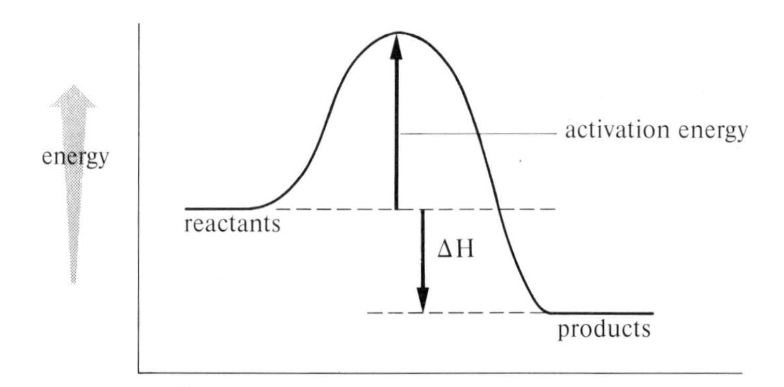

Summary

Exothermic reactions give energy out to the surroundings.

Endothermic reactions take energy in from the surroundings.

Reactions involve making and breaking bonds.

Energy is put in to break bonds and given out when bonds are made.

The energy needed to break one mole of bonds is called the bond energy. Bond energy is measured in kilojoules per mole, $kJ\ mol^{-1}$.

ΔH stands for the energy exchange between system and surroundings. If the sign of the ΔH value is $-$, the reaction is exothermic. If it is $+$, the reaction is endothermic.

The energy needed to start a reaction is called the activation energy.

We must use our supplies of coal, oil, and gas wisely; there are only limited amounts of these fossil fuels. Alternative sources of energy (e.g. from sunshine, wind, waterfalls, and waves) are very important.

Questions

1 Explain the meaning of the following:
(a) exothermic
(b) bond energy
(c) activation energy

2 Say whether the following processes are exothermic or endothermic:
(a) a gas fire burning
(b) cooking a cake
(c) photosynthesis
(d) the precipitation reaction between magnesium chloride and sodium carbonate
(e) the neutralisation of sulphuric acid by sodium hydroxide.

3 Energy is in great demand by society. We use it at home, we use it at work, and we use it when travelling.
(a) Write down two exothermic reactions which can be used to produce heat in the home.
(b) Write down two exothermic reactions that are used in industry, and explain what happens to the heat energy from them.

4 Write down two endothermic reactions that happen in the home.

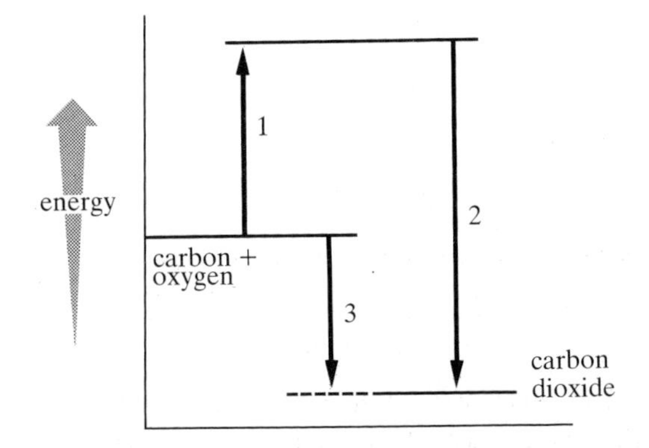

6 (a) What change does the arrow 1 represent? Is energy put in or is it given out?
(b) What change does arrow 2 represent? Is energy put in or is it given out?
(c) What does the arrow marked 3 represent?
(d) Is the reaction exothermic or endothermic? How can you tell?

For the reaction
$C + O_2 \rightarrow CO_2$ $\triangle H$ is $-394\,kJ\,mol^{-1}$
(a) What does the minus sign mean before the 394?
(b) What does $\triangle H$ mean?

7 To break one mole of hydrogen molecules into two moles of hydrogen atoms requires 436 kJ. To break half a mole of oxygen molecules into one mole of oxygen atoms requires 249 kJ. When one mole of hydrogen atoms bonds to one mole of oxygen atoms, 462 kJ of energy is given out.
(a) Make a larger copy of the energy diagram below and mark in these $\triangle H$ values in the

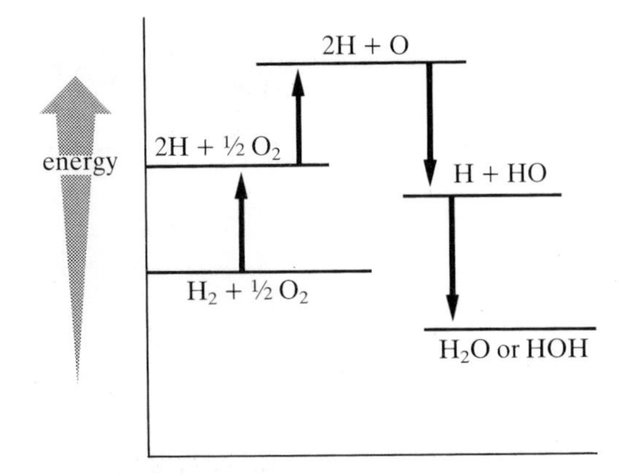

(b) Calculate the energy given out when one mole of water is made.

8 Calor gas is substance called propane, whose formula is C_3H_8. When 1 mole of propane burns, 2200 kJ of energy is given out.
(a) Calculate the mass of 1 mole of propane.
(b) Calculate the heat energy given out when 4·4 kg of propane is burned.
(c) What mass of propane must be burned to give out 17 600 kJ?

9 When 1 mole of ammonia (NH_3) is made from its elements, the energy change is $-46{\cdot}5$ kJ.
(a) When 1 mole of ammonia is decomposed, is energy taken in or given out? How much energy?
(b) What is the energy change when 34 g of ammonia is formed from its elements?

10 Explain the following:
(a) why cars need sparkplugs
(b) why plants need light to grow
(c) why fires make you hot

20 The rates of chemical reactions

20.1 Rate

The meaning of rate

Think about some old people who live at the top of a block of high-rise flats. When the lifts break down they have to walk up the stairs. It takes them twenty-five minutes to climb to their floor.

By stairs:
50 metres up in 25 minutes, or
2 metres per minute

By lift:
50 metres up in 1 minute, or
50 metres per minute.

It is faster to use the lift, slower to use the stairs.
The words *fast* and *slow* both describe the **rate** of going up.
Rate is a measure of the change that happens in a single unit of time.
The unit of time can be a second, a minute or an hour. So in our example:

> the rate of going up by stairs is 2 metres per minute;
> the rate of going up by lift is 50 metres per minute;
> it is 25 times faster to use the lift than the stairs.

Many different rates of change are used in everyday life. A few more examples are shown in Table 20.1. In every case, notice that rate is a measure of the change per unit time—per hour, per minute, or per second.

Table 20.1 Some everyday rates of change

Example	*Rate*
Going to work by bus	30 miles per hour (m.p.h.)
Going to work by train	60 miles per hour
Playing a long-playing record (L.P.)	$33\frac{1}{3}$ revolutions of the turntable per minute (r.p.m.)
Playing a single	45 revolutions of the turntable per minute
Pouring petrol from a spare can	50 cm^3 per second
Pumping petrol from a petrol pump	500 cm^3 per second

Rate of chemical change

In a chemical change reactants change to products. The rate of a chemical change is a measure of the amount of change occurring in a unit of time, e.g. per minute.
How can this rate be measured? To help answer the question, think about the two diagrams below:

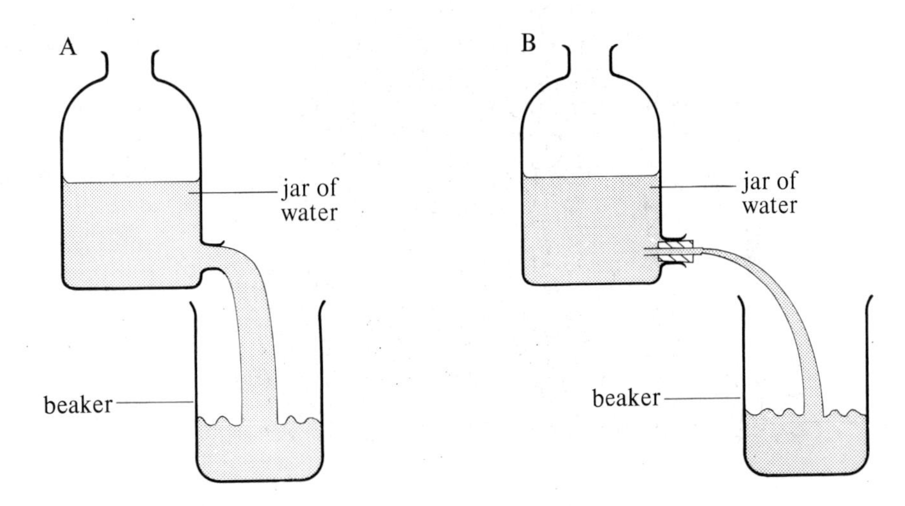

In both cases, water flows from the jar to the beaker. The rate of flow is clearly greater in A than in B. For both A and B, the rate can be found in two ways:

by measuring the volume of water lost by the jar per unit time
by measuring the volume of water gained by the beaker per unit time.

Similarly, in chemical change, the rate can be found by measuring:

the amount of reactant used up per unit time, or
the amount of product made per unit time.

For example, when magnesium dissolves in acid

$Mg + 2HCl \longrightarrow MgCl_2 + H_2\uparrow$

the rate can be calculated by measuring:

(i) the amount of magnesium or hydrochloric acid used up in a certain time, or
(ii) the amount of magnesium chloride or hydrogen formed in a certain time.

20.2 Measuring rates

Amounts of reactants and products

The next important question is: how can amounts of reactant or product be measured? This is often not easy to answer. In many reactions, reactants and products stay all mixed up, and in the same state. To separate one of them and measure the amount of it can be difficult.

In this chapter, as an introduction to the study of reaction rates, we shall look only at reactions that give a product in a different state from the reactants. Because the product is in a different state, the amount of it can usually be measured easily.

Gaseous products

Many reactions give a gas as one product. The rate of such a reaction can be found by one of two methods.
Method 1: Measure the volume of gas produced in a given time.
Method 2: Allow the gas to escape from the system. Measure the reduction in mass of the system after a given time. This works best for a heavy gas.

Method 1: Volume of gas

Reactive metal + acid ⟶ salt solution + hydrogen gas.
The reaction of a reactive metal with acid produces hydrogen, a light gas. So method 1 must be used to measure the reaction rate. The apparatus needed is shown below:

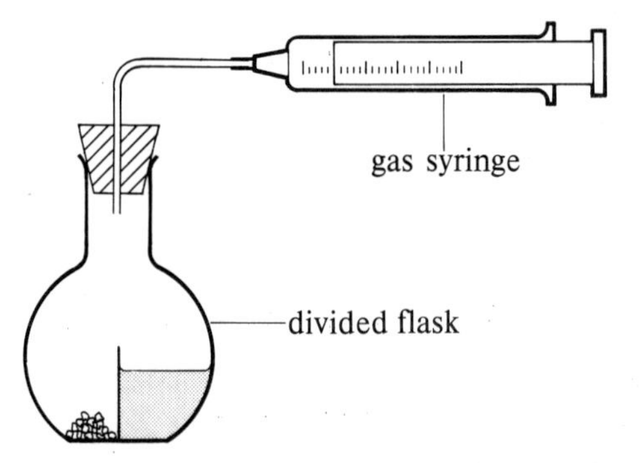

Fig. 1
Apparatus for measuring the rate of a reaction that produces a gas

The flask is divided into two compartments. The acid is put in one compartment, and the metal in the other. When the flask is tipped up the reactants mix and start to react, producing the gas. The gas pushes its way out into the syringe; the plunger of the syringe is forced back. Readings of volume are taken every minute, using the scale marked on the side of the syringe.
The results can be shown graphically, as in Table 20.2.

Table 20.2 Typical results for method 1

Time /min	*Volume /cm³*	*Plotted on a graph*
0	0	
1	12	
2	20	
3	26	
4	30	
5	33	
6	35	
7	36·5	
8	37·5	
9	38	
10	38	

The reaction is much *faster* at the start: 12 cm^3 are produced during the first minute, but only 3 cm^3 during the fifth minute (33 cm^3 – 30 cm^3 = 3 cm^3). Notice that the curve is steepest at the start: after nine minutes it has gone completely flat. That means the reaction is complete.

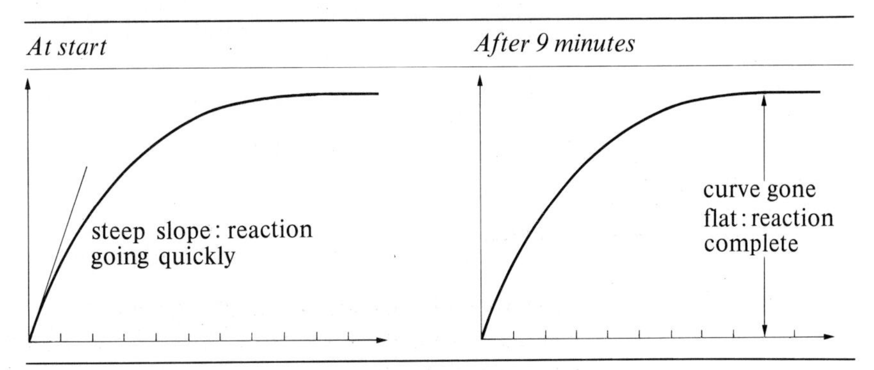

So the rate of reaction changes during the reaction. But you can find its rate at any given time by measuring the slope of the curve at that time. You do this by drawing a tangent to the curve.

Example. What is the rate of production of hydrogen, after two minutes? To find the answer, the steps are:

1. Draw a line up to the curve from the 2 min mark.
2. Carefully draw a tangent to the curve, at the point where the line touches it. The slope of the curve at time = 2 min is the same as the slope or gradient of this tangent.
3. Measure the gradient of the tangent. As you probably know from maths:

$$\text{gradient} = \frac{\text{change in vertical distance}}{\text{change in horizontal distance}}$$

These steps are shown on the diagram below:

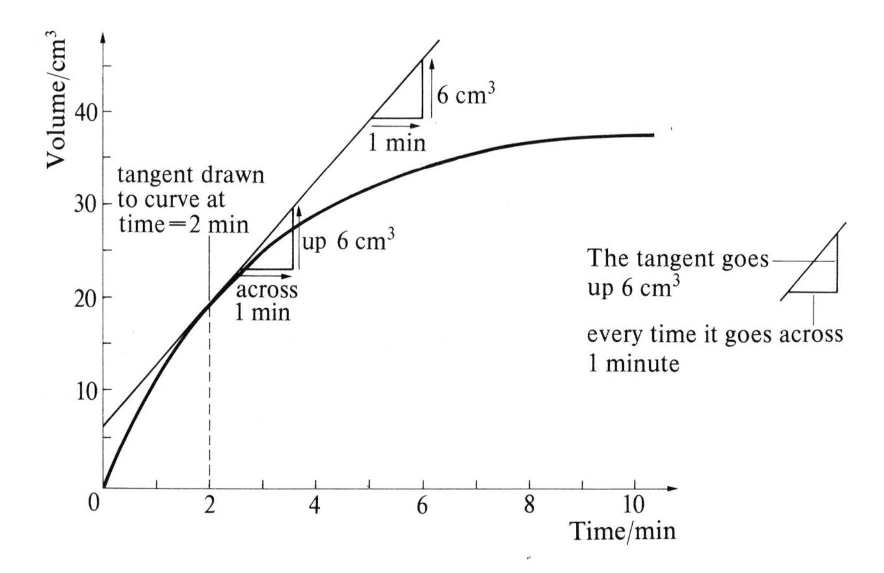

From the diagram, you can see that:

$$\text{gradient} = \frac{\text{change in vertical distance}}{\text{change in horizontal distance}} = \frac{\text{change in volume}}{\text{change in time}}$$

In other words, the gradient is the same as the rate. For the graph above:

$$\text{gradient} = \frac{6\ \text{cm}^3}{1\ \text{minute}}$$

$\therefore$ rate = 6 cm^3 per minute
= **6 cm^3 of hydrogen produced per minute**

After 9 minutes, when the curve has gone flat, the tangent to it is also flat:

$$\text{gradient} = \frac{0\ \text{cm}^3}{1\ \text{minute}}$$

$\therefore$ rate = **0 cm^3 of hydrogen produced per minute**
The reaction has stopped.

Method 2: Mass of system

This method works best for reactions producing heavy gases, like carbon dioxide. A suitable reaction would be:

Metal carbonate + acid ⟶ salt solution + carbon dioxide gas

The apparatus needed is shown below:

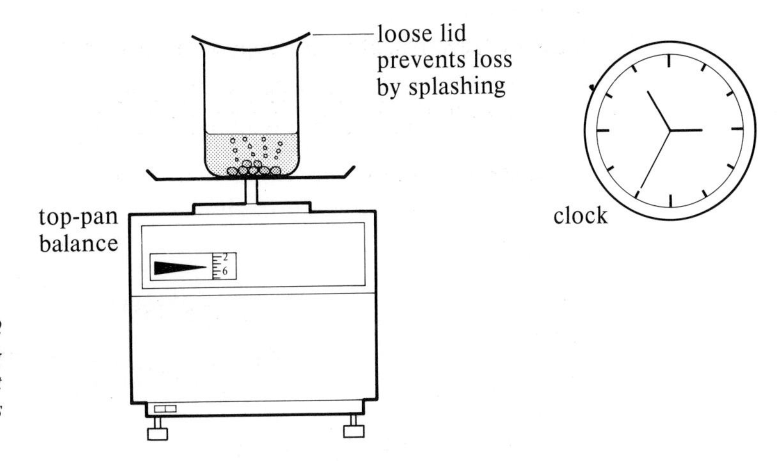

Fig. 2
Apparatus for measuring the rate of a reaction that produces a heavy gas

Because the lid is loose, the gas can bubble out from the beaker; the beaker therefore weighs less. The results are again shown graphically. This time the curve slopes down because the mass of the system decreases.

Example. What is the rate of reaction after 20 minutes?
Follow the same steps to find the rate as in method 1:

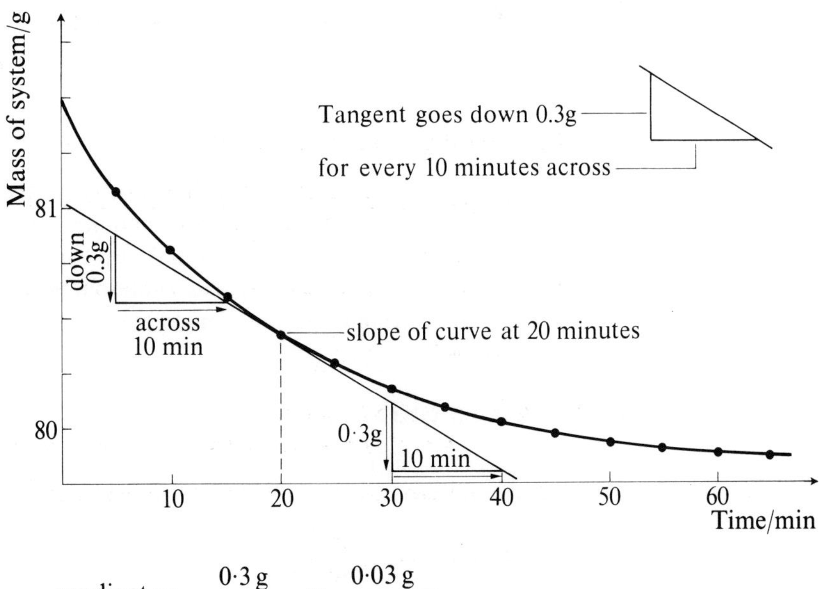

$$\text{gradient} = \frac{0{\cdot}3\text{ g}}{10\text{ minutes}} = \frac{0{\cdot}03\text{ g}}{1\text{ minute}}$$

∴ rate = **0·03 g of carbon dioxide produced per minute.**

Precipitated products

Reactions that occur in solution sometimes produce an insoluble product. This is precipitated as a cloudy suspension. An example is the action of dilute acid on sodium thiosulphate solution: a fine suspension of sulphur slowly forms. You can measure the rate of the reaction like this:

1. Pour a known volume of sodium thiosulphate, say 25 cm^3, into one beaker, and a known volume of dilute acid into another beaker.
2. Mark a cross on a piece of paper, with a pencil.
3. Put the paper under the thiosulphate beaker, so that you can see the cross through the liquid.
4. Quickly pour the acid into the sodium thiosulphate. As you pour it in, start a stop-clock.
5. As the cloudy suspension forms, it blots out the cross from view. Stop the clock the second the cross completely disappears.

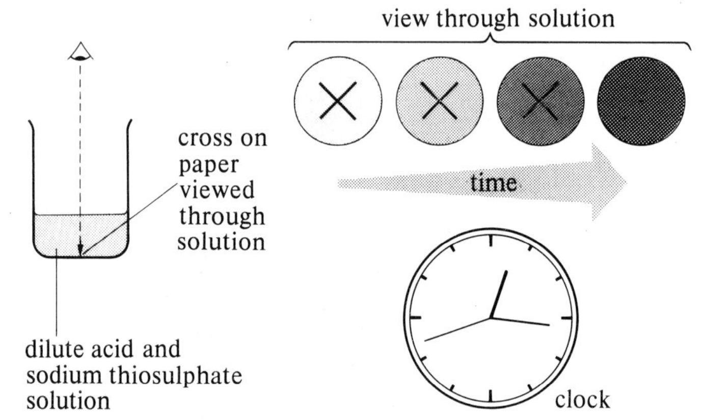

Fig. 3
Rate can also be followed by measuring the time taken for a mark to disappear

The faster the cross is blotted out, the faster the reaction is going.
You could repeat the experiment using different dilute acids, for example, to see how they affect the rate. If you always use the same volume of acid, and the same volume of thiosulphate solution, then you always have the same volume of liquid when you add them together. The cross will always disappear when a certain number of sulphur particles have formed in this volume. So you are really measuring the time it takes to produce a certain amount of sulphur. Using one acid solution it might take 100 seconds. Using a more concentrated solution it might take only 10 seconds.

20.3 Factors that affect rate

In this section you will see how these factors affect rate:

concentration
pressure
temperature
light
surface area
catalysis

Concentration The most convenient reaction for studying the effect of concentration on rate is the thiosulphate reaction described in the last section.

Concentration is the amount of reactant dissolved per unit volume of solution. If the amount of reactant dissolved in one unit of volume is large the concentration is high. Here is a way to find out how concentration affects rate:

1. Using a measuring cylinder, measure out five equal volumes of dilute hydrochloric acid (reactant A in the diagram below).
2. Now measure out five *different* volumes of sodium thiosulphate solution (reactant B). Add water to four of them, so that the final volume of each is the same. The *concentration* of each thiosulphate solution is now different.
3. Add one reactant A to one reactant B. Measure the time for the cross to disappear.
4. Repeat for the other four pairs of solutions.

Table 20.3 The experiment and its results

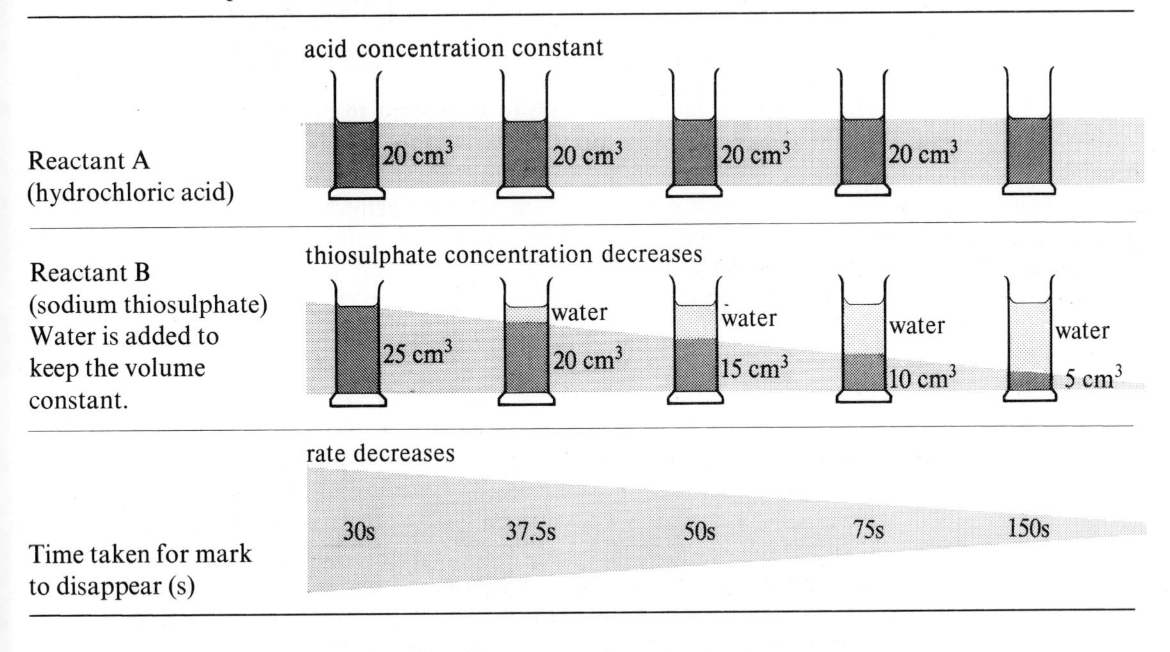

The longer it takes for the mark to disappear, the *slower* the rate.
As you can see from the times given in the diagram, the rate of the reaction is highest when the concentration of reactant B is highest. This makes sense: the more reactant there is present, the faster the reaction will go.
An increase in the concentration of reactants produces an increase in the reaction rate.
This rule holds true for most reactions. It also explains why the rate changes *during* a reaction. Look back to the graph in Table 20.2 (page 211), for example; you can see that the rate (slope of curve) decreases as the reaction goes on. The concentration of reactants is at its highest at the start of the reaction, before any has reacted. So the rate is highest at the start. The rate

decreases (the slope flattens) because the reactant concentration decreases: the reactants are being used up to form products. As time goes on, the rate gets lower and lower. Finally a reactant runs out and the reaction stops.

Pressure If the reactants are gases, you can push them into smaller volumes by putting pressure on them.

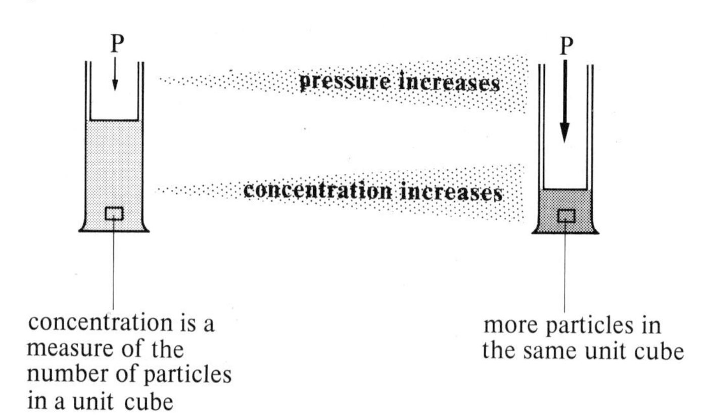

So an increase in pressure leads to an increase in concentration. This in turn leads to an increase in rate of reaction.

Temperature While the effect of concentration on rate is being studied, the temperature of the systems is always held constant. However, the hydrochloric acid/sodium thiosulphate experiment can be repeated, this time holding concentrations constant but varying the temperature.

Reactants (in water-baths)	ice A B	A B	A B	A B	constant concentration
Temperature	0°C	15°C	30°C	50°C	increase in temperature
Time for mark to disappear (s)	100s	40s	15s	3s	increase in rate

From the table above, you can see that:

The reaction rate increases as the temperature increases.

The rate of a typical reaction doubles for every 10°C rise in temperature.

Light Light, like heat, is a form of energy. Just as some reactions can be speeded up by heat, so some can be speeded up by light:

(i) A mixture of hydrogen and chlorine gases explodes when exposed to bright sunlight*:

$$H_2 + Cl_2 \xrightarrow{\text{light}} 2HCl$$

(ii) A precipitate of silver bromide or silver chloride darkens when light is shone on it; this is the basis of photographic film.

(iii) The process of photosynthesis (Page 335) goes faster in bright sunlight.

Reactions like these which are speeded up by light are called **photochemical reactions**. Photochemical reactions are not all that common, and most of the reactions you will see in the laboratory are unaffected by light.

Surface area The experiment using the divided flask (method 1 on page 210) can also be done with calcium carbonate and acid. In this case the gas pushing out the plunger is carbon dioxide:

$$CaCO_3 + 2HCl \longrightarrow CaCl_2 + H_2O + CO_2\uparrow$$

We can do the experiment twice, once using chips of calcium carbonate (marble chips) and once using powdered marble. Two sets of results are produced:

Marble chips

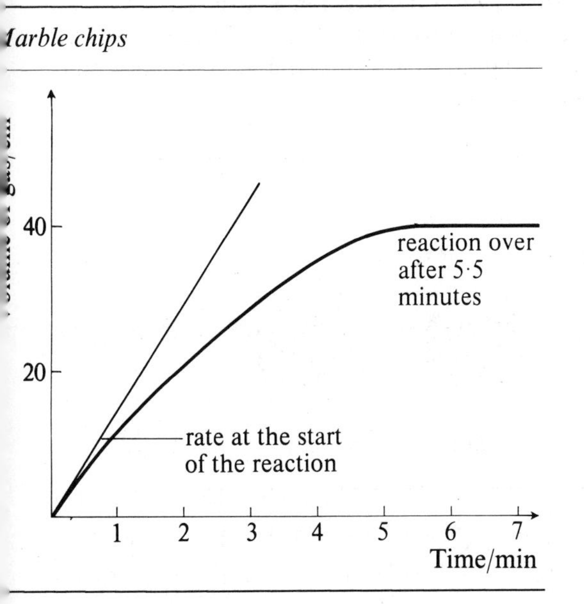

Powdered marble

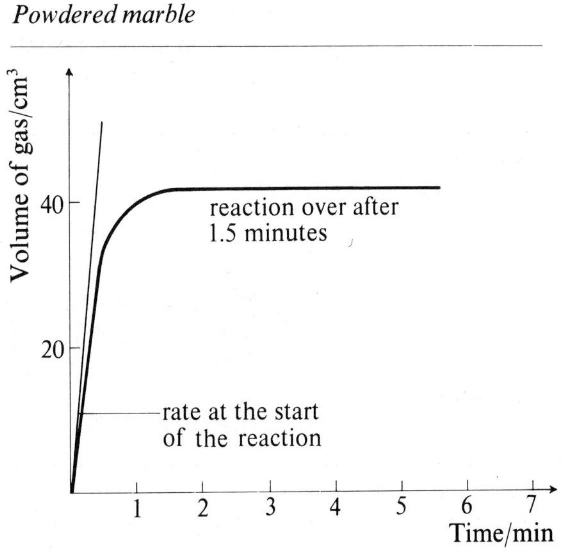

The reaction is much faster using the powdered marble. The powder has a larger **surface area** than the marble chips—most of the marble in a marble chip is inside the chip. The acid comes into contact with all the marble

* This reaction is dangerous and should not be attempted.

particles more quickly when it is powdered than when it is in chips.
The reaction rate increases when the surface area of the reactants is increased.

Catalysis

Wine gradually goes sour if left in the air. This is the result of a slow reaction between the alcohol in the wine and the oxygen in the air. An acid which is the major ingredient of vinegar is produced; the process takes several days.

Alcohol + oxygen $\longrightarrow$ acid

The same reaction happens much more quickly if a warmed spiral of platinum wire is held just over the surface of alcohol in a beaker. The energy given out by the reaction heats the spiral until it glows.

The alcohol is converted to acid in a few seconds: the platinum spiral remains unchanged at the end of the reaction.

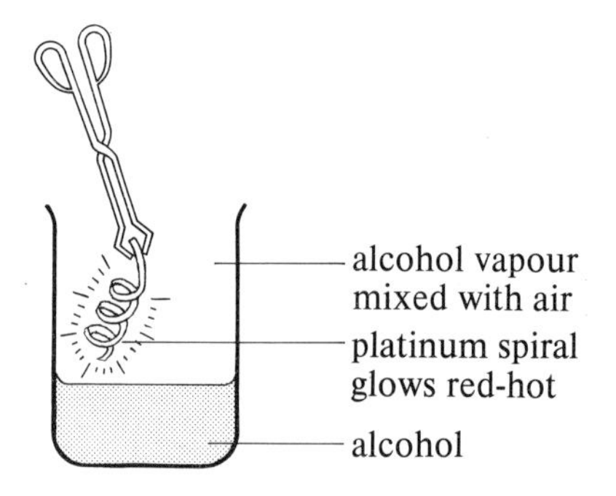

The platinum is said to **catalyze** the reaction. It is a **catalyst**.
A catalyst is a substance that changes the rate of a chemical reaction, but remains chemically unchanged at the end of the reaction.

Table 20.3 Some examples of catalysts

Reaction	*Catalyst*
Decomposition of hydrogen peroxide	Manganese(IV) oxide
Fermentation of sugar to alcohol	Yeast enzymes
Synthesis of ammonia	Iron
Oxidation of sulphur(IV) oxide to sulphur(VI) oxide	Vanadium (V) oxide

20.4 The collision theory

In the first three sections of this chapter, we have outlined the facts about rates of reaction. Now we must find theories to explain them.

The collision theory

The **collision theory** is the most common one used to explain the facts about rates. The two main statements of this theory are:

particles must collide before they can react
the colliding particles between them must contain enough energy to cause bonds to break.

We can illustrate the theory by the reaction of hydrogen with iodine:

1 A collision without enough energy:

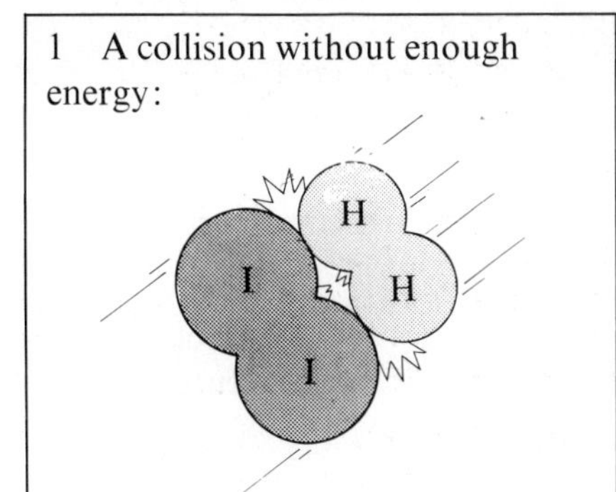

2 The molecules bounce apart. The collision has not been successful.

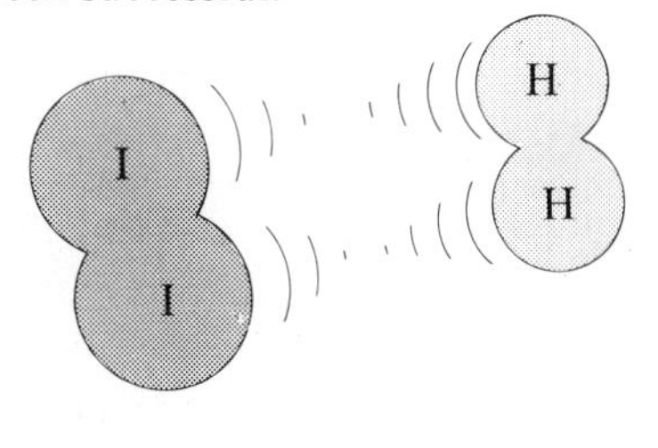

1 A collision with enough energy: bonds are broken and new bonds are formed.

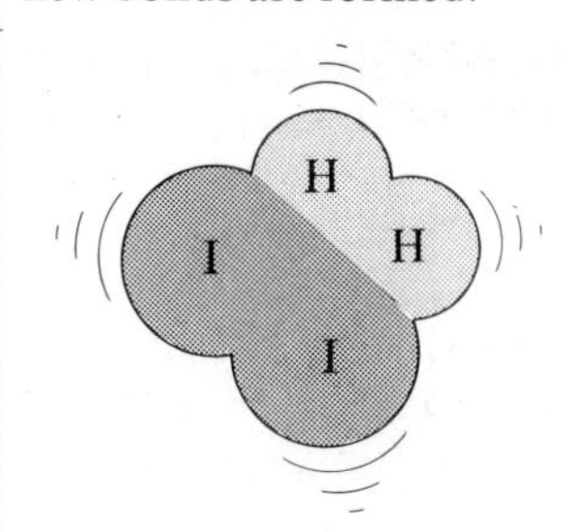

2 Hydrogen iodide molecules form. The collision has been successful.

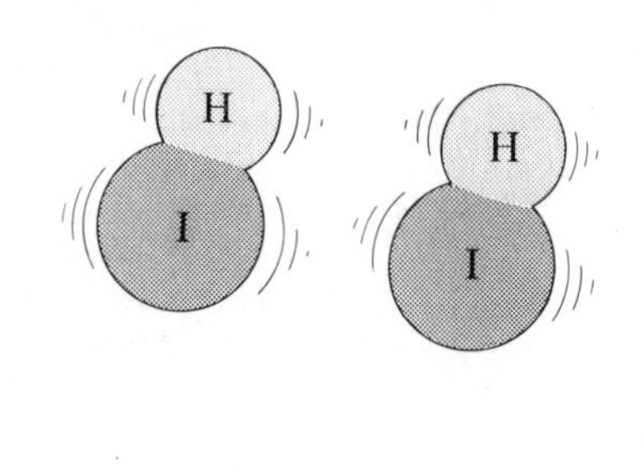

The collision theory explains rates

Facts	*Theory*
An increase in reactant concentration leads to an increase in reaction rate.	Higher concentration means a larger number of particles in a given volume. If there are more particles, there will be more collisions. More collisions means a larger number of successful collisions. Therefore the reaction rate will increase.
An increase in temperature leads to an increase in reaction rate.	The temperature of a system is a measure of the average kinetic energy of the particles in it. If the average kinetic energy of the particles increases, then more pairs of particles will have enough energy between them to collide successfully.
An increase in surface area leads to an increase in reaction rate.	A greater surface area means a larger number of particles are exposed. Therefore there will be a larger number of collisions between particles.
A catalyst can speed up the rate of a reaction.	The catalyst lowers the amount of energy needed for a successful collision. Therefore more collisions will be successful.

Summary

The rate of a chemical reaction measures the amount of reactant used up, or the amount of product made, per unit of time.

The unit of time could be a second, a minute, or an hour. It is usually a second.

The amount of reactant or product can be measured in any convenient way, for example as mass, volume, or concentration.

Rates of reaction depend on:
the concentration of reactants
the surface area of reactants
the temperature
the pressure
the presence of catalysts

A catalyst is a substance that speeds up the rate of a chemical reaction, but remains chemically unchanged at the end of the reaction.

Some reactions are affected by light energy; these are called photochemical reactions.

The collision theory explains the observed facts about the rates of chemical reactions.

Questions

1 The rate of travelling is always expressed as distance gone per unit time, miles per hour, kilometres per hour, and so on.
In what two ways can the rate of a chemical reaction be expressed.

2 (a) List four factors that affect the rate of a chemical reaction.
(b) Explain why a fire is always started with small sticks and paper rather than big logs.
(c) Why does a fire gradually burn more rapidly once it starts?
(d) Why does blowing air at it usually make it burn better?

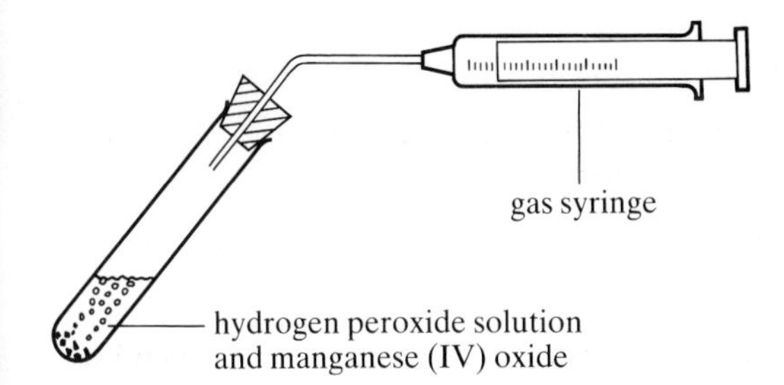

3 Some apparatus was set up as shown above. Readings of the volume of gas in the syringe were taken every minute. Here they are:

Time (min)	0	1	2	3	4	5	6	7	8
Volume of gas (cm^3)	0	40	70	92	107	115	120	120	120

(a) Explain the purpose of the manganese(IV) oxide.
(b) What gas is being produced?
(c) Carefully plot the readings on a graph, showing time along the *x*-axis.
(d) How long did it take for 50 cm^3 of gas to be produced?
(e) How long did it take for the reaction to be half completed?
(f) What was the volume of gas finally produced?
(g) Why was there no change in volume after 6 minutes?
(h) Did the reaction speed up or slow down as it proceeded?
(i) Explain why the rate of the reaction changed as it proceeded?
(j) Describe two ways of speeding up this reaction, without altering the amounts of chemicals involved.

4 To know something about the rate of a chemical reaction, you have to measure some property that changes as the reaction goes on. Describe three ways of doing this in the laboratory, illustrating your answers with diagrams of the apparatus or method.

5 Some reactions speed up as they proceed, like a fire burning. Others slow down, like a metal dissolving in acid. Why do these two examples behave in different ways?

6 (a) Why does the exhaust-pipe of a car rust more quickly than the rest of the body-work?
(b) Why does milk go sour more slowly in the fridge?
(c) Why do potatoes cook more quickly in a pressure-cooker?
(d) Why does glue like 'Araldite' take longer to set on a cold day?
(e) Why does the fine flour-dust in a flour mill sometimes explode?

7 A beaker containing some hydrochloric acid is placed on a direct-reading balance. Some chips of calcium carbonate (marble) are carefully added to the beaker. The weight of the beaker and the reacting system is recorded every minute.

Time/min	0	1	2	3	4	5
Mass/g	130·13	129·98	129·91	129·85	129·79	129·74
Time/min	6	7	8	9	10	11
Mass/g	129·69	129·66	129·63	129·61	129·60	129·60

(a) Plot these results on a graph, showing time on the horizontal axis.
(b) Why did the system lose weight?
(c) Write down three ways in which the reaction could have been speeded up.
(d) How long did it take for half the acid to be used up?
(e) Does the reaction slow down or speed up, as it proceeds?
(f) Give three reasons for the change in the rate of the reaction.
(g) On the same axes, sketch how the curve would look if:
  (i) one large lump of calcium carbonate was used
  (ii) powdered calcium carbonate was used.

21 The chemical industry

Since the industrial revolution of the mid-nineteenth century, man has increasingly reshaped the world in which he lives. Sprawling areas of tarmac now cover much of the open countryside and, in what is left, huge fields spread out in place of the old patchwork of pasture and hedgerow. The sloping factory roofs, the pylons, and the mining slag-heaps are new additions to the skyline, and industrial waste threatens to pollute the air, water, and soil. These are the darker sides of the chemical industries. However, the industry has brought about a dramatic increase in food production, material comfort, medical care, and life expectancy, as well as improved standards of working conditions, travel, and leisure activities. It would be wrong to ignore the bad things, but we should not forget how much we owe to the chemical industry.

21.1 The needs of man

The more developed a society is, the more **needs** it appears to have. We must look at a developing society to recognise man's most basic needs: for the starving, the primary needs go no further than food, water, and shelter. However, when these needs are fulfilled, others quickly come to take their place. For example, is there seed enough to prevent the pattern of starvation from repeating itself? Are there fertilizers and irrigation to make up for the barren soil? Is there machinery for farming and transport? Are there enough spare parts and fuel to keep the machinery running? As you can see, the questions come thick and fast. In general, we can group the essential needs of man into three broad classes: food, materials, and energy.

Food The world's population is growing at a frightening rate. Look at the graph beside the photographs below.

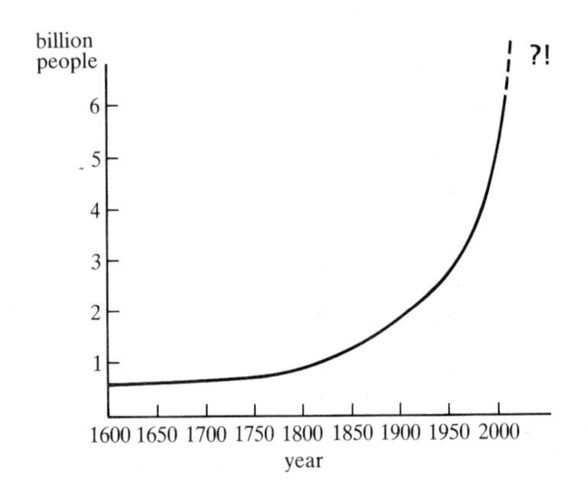

At the present rate of increase, there will be four times as many people alive in the year 2000 as there are today. The growing world population needs ever more food, and traditional methods of farming just cannot produce enough for everyone to eat. One answer lies in what is called **intensive farming**. Intensive farming increases the yield of farm produce by the large-scale use of:

modern machinery
artificial fertilizer
pesticides and weed killers
newly bred strains of plants.

In the developed world, intensive farming has been so successful that huge surpluses of food have been produced. The Common Market countries produce so much grain, meat, butter, and wine that there are literally mountains of food deep-frozen in storage. The reason for storing it is that food prices would tumble to such an extent that the farmers would be ruined if the surplus was put on the market. But there is something obscene about deep-freezing unwanted food while half the world starves to death. Unfortunately, it is not easy to get intensive farming methods going in developing countries. It costs a lot to set them up and the farmers need to know how to use the chemicals and machinery properly. The old methods are still used and these increasingly fail to meet the needs of a population which is growing much faster than that of the developed world. Hunger is the normal experience, and natural disasters like drought simply make the situation even more desperate.

Materials

The developed world needs a wide range of materials to maintain its present standard of living. These fall into three main classes:

metals like iron, aluminium, copper, and lead
building materials like brick, cement, plaster, glass, and wood
chemical feedstocks from which plastics, fertilizers, pesticides, paints, and drugs can be made.

Class	*Raw material*	*Used to make*
metals	bauxite	aluminium
	haematite	iron
	chalcocite	copper
building materials	clay	bricks
	limestone	cement
	gypsum	plaster
	sand	glass
chemical feedstocks	sulphur	sulphuric acid
	salt	chlorine, hydrogen, sodium hydroxide
	crude oil	plastics, solvents, fuels
	nitrogen (air)	ammonia, nitric acid

Many raw materials are produced by natural cycles such as the water cycle (page 112), the carbon cycle (page 336), the nitrogen cycle (page 359), or the rock cycle (shown below). If the cycle happens quickly enough, a raw material can be replaced and so the supply is never exhausted. A material of this sort is called a **renewable resource**. However, if the cycle is slow, there may be no chance of replacing the material as it is used up. Materials of this sort are called **finite resources**. Most minerals are finite resources because they are formed in the rock cycle which happens incredibly slowly. Fossil fuels are also finite resources. We should be very careful about the rate at which we use up finite resources. It is important to look for alternative resources and to recycle our existing ones as best we can. For example, waste paper, plastic, metal, and glass can all be recycled if they are collected separately.

The rock cycle

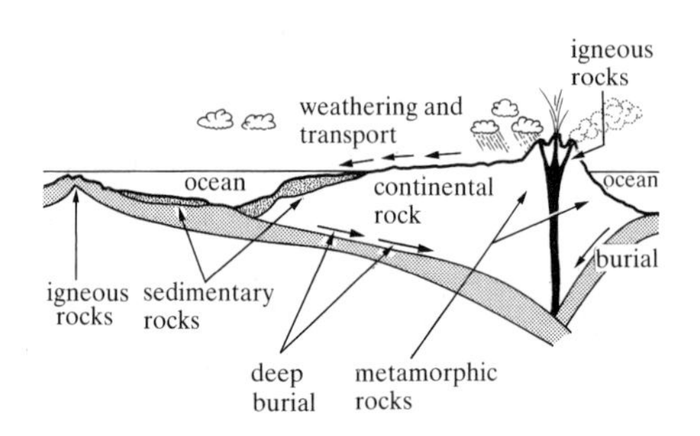

The rock near the centre of the Earth is very hot. It slowly reaches the surface as volcanic or igneous rock, pouring out during volcanic eruptions. Nearby rocks are melted by the heat of the eruption, and these are called metamorphic rocks. The weather wears these structures down into sedimentary rocks, carrying them via streams and rivers to be deposited in valleys and finally onto the seabed. The rocks on the seabed slowly sink back towards the centre of the Earth, restarting the cycle. The whole cycle takes hundreds of millions of years.

Energy

Some different energy sources are described on page 204. Industries, factories, and homes all need energy for heating and lighting, and for driving the machinery and transport vital to their existence. The more developed the country, the more energy it seems to need. Compare the energy used by the different countries in the graph shown below.

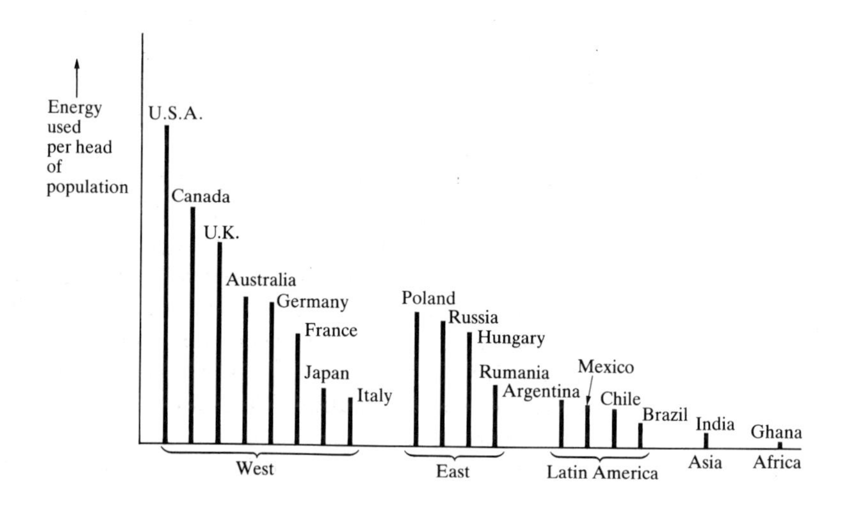

21.2 Setting up a chemical plant

The primary reason for setting up a chemical plant is to make a product to sell. If there is no market for the particular product, then the plant will quickly close. But, like any business, the plant must make the product as cheaply as possible. If a competitor comes up with a cheaper method of making it, the plant will again be threatened with closure. In the search for the cheapest methods, the company owning the plant must never forget safety. Many of the reactants and products in a chemical plant are toxic (poisonous). Precautions need to be taken to protect both workers, and those who live near the plant. If you were about to set up a chemical plant, here are five things you would need to look at carefully.

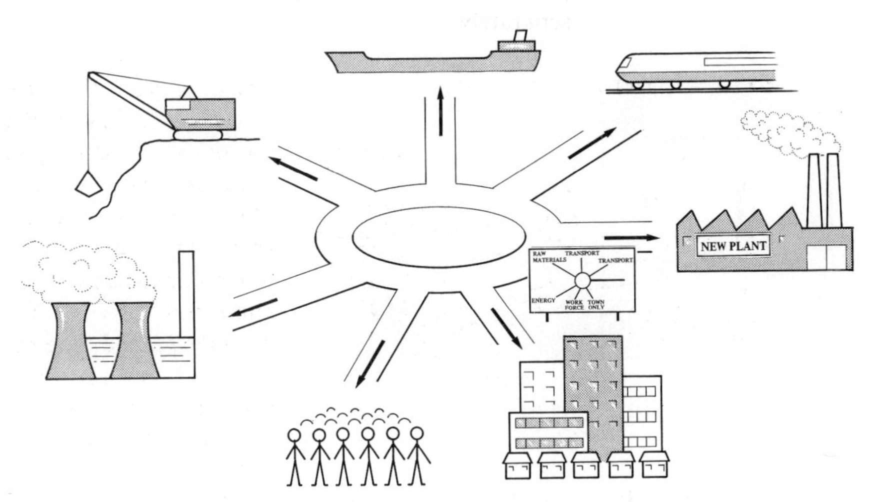

Raw materials These either occur naturally or are processed by another industry. The choice of raw materials depends on cost and availability.

Energy Every plant needs energy to start the reactions and to control the mixing of reactants and removal of products. Some plants need enormous amounts of energy (for example, the electrolysis cells shown on pages 174 and 175). Others not only use less energy but make their own as they run (for example, the sulphuric and nitric acid plants shown on pages 378 and 363).

Labour A skilled workforce is called for. Chemical engineers are needed to design and get the most out of the plant, computer experts to control the automatic parts of the plant, and production workers to man the machinery and to check that the plant is working safely and efficiently.

Transport The raw materials must be brought to the plant, and the finished products must be transported from the plant. The cost of this will depend on how close the plant is to a seaport, a rail link, or a motorway.

Environment Many chemical plants deal with toxic, explosive, or radioactive substances. It is not wise to build plants of this sort too close to areas of high population. The risk of harming those living in the locality becomes too great.

Consider these three case studies.

Steelworks

A steel plant was built in Corby, Northants, because it was near some iron ore beds and there was a skilled labour force available. All the other needs of the steelworks (particularly coke) were transported to Corby. Now that the richest deposits of iron ore have run out, the Corby steelworks have closed with the loss of a huge number of jobs. The Corby works could not produce steel as cheaply as its competitors once the iron ore had to be transported from outside. By contrast, the recently built steelworks at Llanwern are nowhere near any iron ore deposits. But they are right on the Severn estuary and next door to huge supplies of Welsh coal. Cheap foreign iron ore can be shipped straight to the plant and the energy costs to convert it to iron are as low as possible.

Aluminium

At Lynemouth, there is one of the biggest aluminium plants in Europe. Like Llanwern, it is not built beside a source of the mineral from which aluminium is made (bauxite). Bauxite comes from the Far East, and Lynemouth even pays for the purification of the mineral before shipping it all the way to the UK. By far the biggest cost is the energy required to run the plant. And this is the reason for building the aluminium works at Lynemouth. The plant is right on top of a coal seam and has its own private power station to provide the billions of watts of electrical energy needed. So the plant can afford to pay high transport costs and yet still stay competitive. Its major competitors are Canadian and Norwegian plants built beside hydroelectric power stations.

Nuclear fuels Plants which process nuclear fuel are extremely unpopular with the public and so are built as far away as possible from any centres of population. For example, one of the biggest plants is at Sellafield in Northumbria. Trains with specially designed railway containers are used to carry the dangerously radioactive materials to the plant. In a widely publicized experiment, one of these special trains was crashed into huge concrete blocks at 90 mph to test what would happen to the containers. None showed any signs of splitting. The UK nuclear industry is one of the world's leaders and would like to be able to process nuclear waste from other countries as well. However, many people are very concerned about the safety problems that would arise if the UK were to become the European centre of nuclear fuels.

21.3 Running a plant

Once a company has designed and built a chemical plant, the challenges are only just beginning! The plant must be run economically, but also safely. Safety must be guaranteed both for the workforce, and also for those living near the plant.

Workers In the early days of the chemical industry (at the turn of the century) not enough was known about safety. For example, many of the workers in the match-making industry of the time developed a particularly unpleasant form of phosphorus poisoning. Known to the doctors as 'phosphorus necrosis', or gangrene of the jaw, the workers called it phossy-jaw. It killed painfully.

More recently, the asbestos industry has had similar problems. Asbestos is a silicate compound that can be mined. It is very useful because it can be broken up into long thread-like fibres. These can be bound together or woven into a cloth-like material which is fireproof, and does not conduct electricity. Asbestos has been used as a building material for ceilings and fireproof cements, and in vehicles for brake-linings and clutch-plates. Many asbestos workers began to suffer from a lung disease called asbestosis, and many more developed lung cancers. There is evidence that almost half the population now have asbestos fibres in their lungs. This may well be one of the major causes of lung cancer. For this reason, asbestos is now banned as a building material, and is no longer used in the car industry. Where it is found in existing buildings, workers wearing special protective clothing to remove it.

Around the plant Pollution control is particularly important for today's chemical industry. The effects of pollution are described on pages 114 and 339. Unfortunately, accidents happen in all industries, and the chemical industry is no different. But an accident in the chemical industry can cause a major disaster. A recent example is described below to emphasize just how careful an industry must be about its safety precautions.

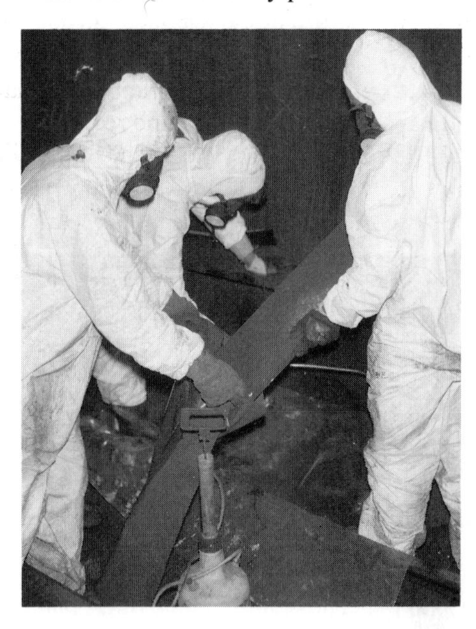

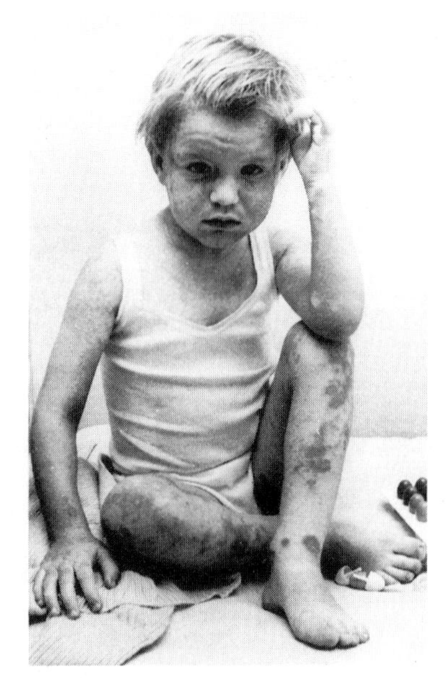

Seveso The Swiss owned company of Icmeas had a herbicide factory near the Italian town of Seveso. The main herbicide being produced was 2,4,5-trichlorophenoxyethanoic acid. This substance is like one of the natural hormones found in plants. When it is applied, the plant grows big stems without the roots or leaves to go with them. It soon dies. On the tenth of July 1976, there was an explosion at the plant that released a toxic substance called dioxin into the atmosphere. The chemical cloud that was released was blown over part of the nearby town. Only four kilograms of dioxin were released in the explosion, but these began to cause immediate problems. The townspeople began to suffer from stomach and liver poisoning, acute chloracne (huge blackheads giving off a sour smell), and from excessive growth of thick black hair. These symptoms were only the beginning. Pregnant women miscarried at twice the national rate, and checks revealed that almost every child living in the area next to the factory had a low blood count of lymphocytes (the white blood cells). Roche, the Swiss parent company, removed all the top soil to a depth of eight inches, burning all the vegetation and killing all the domestic animals and wildlife in the area.

Treating pollution

Pollution control is usually an expensive process. This is particularly true when an existing plant needs to be modified to reduce the amount of pollution it is causing. New plants are designed with an eye already on pollution levels. Many pollutants are expensive and if they are not thrown away the whole process is more economical.

There are three main ways of treating pollution.

Chemical treatment The pollutant is converted into something safer. For example, the oxides of nitrogen and carbon monoxide in the exhaust gases of a car are converted to nitrogen and carbon dioxide using a catalyst (see page 363).

Recycling Many valuable substances are thrown away instead of being recycled for use again. More and more local authorities are installing recycling plants to sort out domestic rubbish so that the waste metal, glass, plastic, and paper can be separated and recycled.

Dispersal It is safe for some pollutants to be diluted and spread more thinly over a wider area. For example, tall chimneys disperse ash and smoke higher into the atmosphere so that it is spread wider before it comes back to earth. Radioactive wastes are dumped into the sea, so that when their containers finally rust through, the sea water will dilute everything. Spills from tankers at sea are dispersed by pouring detergents onto them.

Dispersal is often the cheapest method of treating pollution. But it can cause even more problems than the pollutants themselves would have caused. For example, the detergent put onto an oil slick may remove the oil so that bird life is no longer visibly affected. But, at the same time, it can kill all the plankton in the sea. This causes long-term disruption of food chains, which affects many species including eventually the seabirds.

Summary

The chemical industry tries to satisfy the needs of society.

There are three main classes of need: food, materials, and energy.

The industry uses whatever raw materials are available to supply:
- fertilizers, pesticides, and herbicides to improve yields of grain
- drugs and feedstocks for animals to improve meat production
- materials such as metals, brick, cement, plastic, glass, and paper
- energy by processing fuels such as coal, petrol, gas, and uranium.

In addition, the chemical industry supplies the fabrics industry with artificial fibres and the pharmaceutical companies with the raw materials to make their drugs and cosmetics.

The chemical industry must be careful that its activities do not harm the environment. Three main causes for concern are:
- the rate at which finite resources like minerals and oil are used
- the safety of all who either work in a plant or live near it
- the control of the pollution of air, earth, and water.

Questions

1 (a) What are man's most basic needs?
(b) Why do the needs of a developing country appear to be different from the needs of a developed country?

2 Look at the population curve shown on page 222.
(a) How many people were alive in 1800?
(b) At what date did the world's population reach 3 billion?
(c) When might the world's population reach 6 billion?
(d) Why is it worrying that the world's population is growing?

3 (a) What is meant by 'intensive farming'?
(b) Make a table with the headings:
Advantages of intensive farming;
Disadvantages of intensive farming.
Fill the table in as fully as you can.

4 (a) List three classes of material which are important to the way of life of a developed country.
(b) What raw materials are used up to provide these materials? Give two examples for each class.

5 (a) What is a 'finite resource'?
(b) Give two examples of finite resources.
(c) What is a 'renewable resource'?
(d) Give two examples of renewable resources.
(e) What advice would you give to an industry which could choose between using a finite resource or a renewable resource?

6 (a) What is meant by 'recycling'?
(b) Try and think of a few everyday substances which could be recycled. Make a list of them.
(c) How many of the substances on your list are carefully recycled by society?
(d) If there are some which are not being recycled, try to say why you think that this is. What can be done about it?

7 (a) Draw diagrams of the rock cycle and of the water cycle side by side.
(b) What is the same about the two cycles, and what is different about them?

8 (a) What is the point of setting up a chemical plant?
(b) List and explain five factors that you would need to take notice of if you decided to set up a new chemical plant.

9 (a) What do you understand by the term pollution?
(b) Chemical plants must be careful not to cause pollution. Give an example of the way in which a plant might be in danger of polluting:
the atmosphere
the rivers
the soil.

10 (a) Once a chemical plant is set up and running, what are the sorts of things the management might be concerned about?
(b) List some possible reasons why a chemical plant might become unworkable.

11 (a) Why was it such a surprise when a salmon was caught in the Thames just upstream from London?
(b) Is this encouraging for the chemical industry?

12 (a) There are three main ways of treating pollution; list them.
(b) What are the advantages and disadvantages of each method?

Section four

Calculations in chemistry

22 The amount of each element in a compound

22.1 Masses of atoms

Standards

A **standard** is something with which other things can be compared. We often use standards in everyday life without even thinking about it. Look at this bus journey, for example:

Event	*Use of standard*	*Comparison*
You are charged 50p for your fare home.	Last week the fare was 25p.	It is now twice as expensive.
You find all the seats taken. A large man is completely filling one double seat.	A double seat should take two people.	He is twice as big as you. You have to stand!
Because of a traffic jam, the journey home takes two hours.	The timetable shows that it should take one hour.	The journey has taken twice as long as it should.

If you had not used standards you would not have known what to expect. You would probably have been much happier with the whole journey.

A carbon atom is standard

You already know that atoms are so light that their masses are not measured in grams. Instead, each atom is compared to a standard atom. The atom chosen as standard is the most common isotope of carbon, $^{12}_{6}C$. It is often called **carbon-12**. Compare it with a magnesium atom:

Atom	*Standard*	*Comparison*
Mg a magnesium atom	C a carbon-12 atom	A magnesium atom is twice as heavy as a carbon atom.

Relative atomic mass A_r

The carbon-12 atom has a mass of exactly 12 units. We can show this on a number scale like the one that follows. Because magnesium atoms are twice as heavy as carbon-12 atoms, magnesium is marked on the scale at 24.

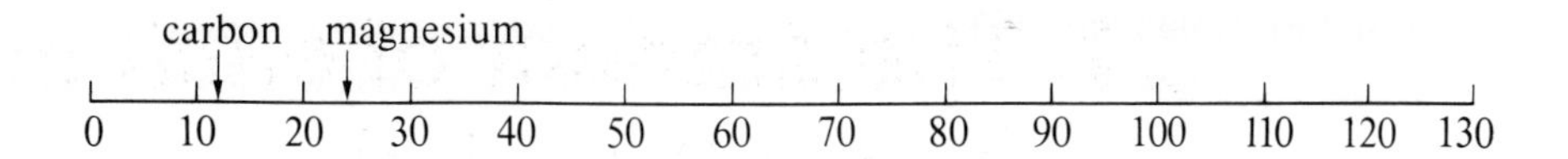

The scale does not start at carbon because some atoms are lighter than carbon—hydrogen and helium, for example.

The number scale above is called the **relative atomic mass scale** when used to compare the masses of atoms.

The relative atomic mass (A_r) of an element is the number of times heavier an average atom of the element is, than $\frac{1}{12}$ of the mass of a carbon-12 atom.

Atom	*Standard*	*Comparison*	A_r
Ag	C-12	An atom of silver is nine times as heavy as an atom of carbon-12.	$9 \times 12 =$ **108**

carbon magnesium silver

0 10 20 30 40 50 60 70 80 90 100 110 120 130

Here are some common A_r values, in order of increasing mass:

Table 22.1 A_r for some common elements

Element	*Atomic symbol*	A_r	*Element*	*Atomic symbol*	A_r
Hydrogen	H	1	Potassium	K	39
Helium	He	4	Calcium	Ca	40
Lithium	Li	7	Chromium	Cr	52
Carbon	C	12	Manganese	Mn	55
Nitrogen	N	14	Iron	Fe	56
Oxygen	O	16	Copper	Cu	63·5
Fluorine	F	19	Zinc	Zn	65
Sodium	Na	23	Bromine	Br	80
Magnesium	Mg	24	Silver	Ag	108
Aluminium	Al	27	Tin	Sn	119
Silicon	Si	28	Iodine	I	127
Phosphorus	P	31	Barium	Ba	137
Sulphur	S	32	Mercury	Hg	201
Chlorine	Cl	35·5	Lead	Pb	207

Relative molecular mass (M_r)

The same scale can be used to find the masses of *molecules*. The carbon-12 atom is again taken as standard, but this time the scale is called the **relative molecular mass scale (M_r).**

The relative molecular mass (M_r) of a substance is the number of times heavier an average molecule of the substance is than $\frac{1}{12}$ of the mass of a carbon-12 atom.

You can work out M_r values if you know the formula of the compound and the A_r of each atom in the formula. To find the total mass of the molecule, add up the A_r of all the atoms in it. Like this:

Compound	*Shape of molecule*	A_r	M_r	
HCl Hydrogen chloride	Cl H	H = 1 Cl = 35·5	1 H 1 Cl	1 × 1 = 1 1 × 35·5 = 35·5 **36·5**
H_2O Water	H O H	H = 1 O = 16	2 H 1 O	2 × 1 = 2 1 × 16 = 16 **18**

And here are two more M_r values worked out:

Compound	*Formula*	*Adding up A_r's*	
Carbon dioxide	CO_2	1 C 2 O	1 × 12 = 12 2 × 16 = 32 **44** = M_r
Sodium carbonate	Na_2CO_3	2 Na 1 C 3 O	2 × 23 = 46 1 × 12 = 12 3 × 16 = 48 **106** = M_r

22.2 Using ratios

What is a ratio?

A **ratio** is the simplest and most exact way to compare the amounts of two systems. Compare the systems below, for example:

System 1	*System 2*
A road 20 miles long	A road 2 miles long
A 50-kilo bag of sugar	A 5-kilo bag of sugar
A wallet with £25 in it	A wallet with £2.50 in it

The amount of each system 1 is ten times the amount of the corresponding system 2. So the systems in each pair are in a ratio of ten to one. This is usually written as 10:1.

In what ratio are the systems in each pair below?

System 1	*System 2*
A road 30 miles long	A road 20 miles long
A 6-kilo bag of sugar	A 4-kilo bag of sugar
A wallet with £15 in it	A wallet with £10 in it

Did you find that these systems are all in a ratio of three to two (3:2)?

Number ratios

A compound contains atoms of different elements. The numbers of each type of atom in the compound are often given as a ratio. For example:

Compound	*Formula*	*Number ratio*
Sulphur dioxide	SO_2	S:O = 1:2
Methane	CH_4	C:H = 1:4
Ethane	C_2H_6	C:H = 2:6 = 1:3

Mass ratios

A number ratio is not the only way of saying how much of each element there is in a compound. You can also write the masses of the elements present as a ratio. For example:

Compound	*Shape of molecule*	*Mass of each element present*	*Mass ratio*
Sulphur dioxide SO_2	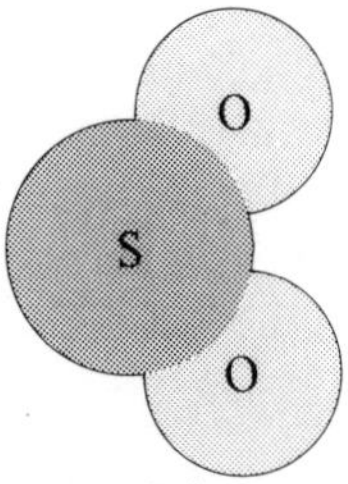 O S O	S $1 \times 32 = 32$ O $2 \times 16 = 32$	S:O = 32:32 = 1:1 There are equal masses of sulphur and oxygen in sulphur dioxide.
Methane CH_4	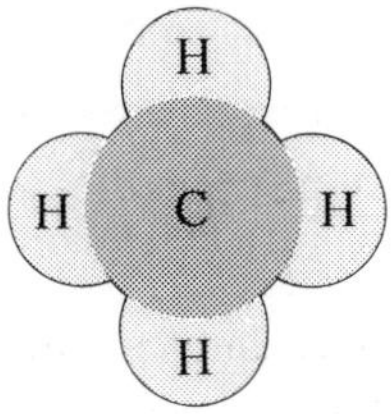 H H C H H	C $1 \times 12 = 12$ H $4 \times 1 = 4$	C:H = 12:4 = 3:1 The mass of carbon is three times the mass of hydrogen, in methane.

Mass ratios as percentages

You can easily express a mass ratio as percentages as long as you know exactly what a ratio means. Suppose the mass ratio of X to Y is 3:2.

This means	3 parts of the total mass is due to X; 2 parts of the total mass is due to Y; The total mass is 5 parts.
As fractions	X weighs $\frac{3}{5}$ of the total; Y weighs $\frac{2}{5}$ of the total.
As percentages	$\frac{3}{5}$ is $(\frac{3}{5} \times 100)\%$ or 60%; $\frac{2}{5}$ is $(\frac{2}{5} \times 100)\%$ or 40%. X weighs 60% of the total; Y weighs 40% of the total.

The ratio of X to Y as percentages is therefore 60%:40% or 60:40.

Comparing mass ratios and number ratios

The number ratio of the atoms of each element in a compound is hardly ever the same as the mass ratio of the elements in that compound. For example, look at sulphur dioxide and methane again:

Compound	*Number ratio*	*Mass ratio*
Sulphur dioxide, SO_2	S:O = 1:2	S:O = 1:1
Methane, CH_4	C:H = 1:4	C:H = 3:1

It is not surprising that these two ratios are different for the same compound. It is because each atom has a different mass. It takes the mass of two oxygen atoms to equal the mass of one sulphur atom; twelve hydrogen atoms are needed to balance the mass of one carbon atom.

We can find out the mass ratios for compounds by experiment. Number ratios must be worked out from the mass ratio. See examples on page 238.

Molecular and empirical formulas

The **molecular formula** shows the *actual number* of each atom in a molecule of the compound. The molecular formula for ethane is C_2H_6. So a molecule of ethane contains 2 atoms of carbon and 6 atoms of hydrogen.

An **empirical formula** is written in the same way, but the figures show the *simplest number ratio* of the atoms in the molecule. In ethane, the number ratio for C:H is 2:6. More simply it is 1:3.

Ethane: molecular formula C_2H_6; empirical formula CH_3

Sometimes the number ratio cannot be made any simpler than the actual numbers of atoms in the molecule. Then the two formulas are the same:

Compound	*Molecular formula*	*Number ratio*	*Empirical formula*
Hydrogen peroxide	H_2O_2	H:O = 2:2 or 1:1	HO
Water	H_2O	H:O = 2:1	H_2O

22.3 Worked examples

Mass ratios from number ratios

You can calculate the mass ratio of the elements in a compound, if you know the formula for the compound (a number ratio). To calculate the mass ratio in percentage form, the steps are:

1. Write down the formula for the compound.
2. Work out the mass of each element present, and the M_r
3. Write the mass of each element present as a fraction of the M_r.
4. Multiply each fraction by 100, to give a percentage.

Example. Calculate the percentage mass of each element in the compound.
(a) water (H_2O)
(b) carbon dioxide (CO_2)
(c) calcium carbonate ($CaCO_3$)

The working is shown below.

Compound	*Mass of each element present*	*Mass of each element as fraction of total*	*Mass of each element as % of total*
(a) Water H_2O	2 atoms H = 2 1 atom O = 16 Total = 18	H = $\frac{2}{18}$ O = $\frac{16}{18}$	H = $(\frac{2}{18} \times 100)\%$ = **11·1%** O = $(\frac{16}{18} \times 100)\%$ = **88·9%**
(b) Carbon dioxide CO_2	1 atom C = 12 2 atoms O = 32 Total = 44	C = $\frac{12}{44}$ O = $\frac{32}{44}$	C = $(\frac{12}{44} \times 100)\%$ = **27·3%** O = $(\frac{32}{44} \times 100)\%$ = **72·7%**
(c) Calcium carbonate $CaCO_3$	1 atom Ca = 40 1 atom C = 12 3 atoms O = 48 Total = 100	Ca = $\frac{40}{100}$ C = $\frac{12}{100}$ O = $\frac{48}{100}$	Ca = $(\frac{40}{100} \times 100)\%$ = **40%** C = $(\frac{12}{100} \times 100)\%$ = **12%** O = $(\frac{48}{100} \times 100)\%$ = **48%**

Number ratios from mass ratios

You can calculate the number ratio (empirical formula) of elements in a compound if you know the mass ratio. However, it is more difficult to turn mass ratios into number ratios than the other way round. Suppose the number ratio of carbon atoms to hydrogen atoms in a compound is 1:4. There is 1 carbon atom to 4 hydrogen atoms.

Mass of a carbon atom = 12
Mass of a hydrogen atom = 1
Mass of 4 hydrogen atoms = 4

So the mass ratio of carbon to hydrogen in the compound is 12:4 or 3:1.

In this example, the number ratio has been turned into a mass ratio, by multiplying each number by the correct A_r. The reverse of this process must be used to turn a mass ratio into a number ratio. The mass ratio is usually given as percentages (see page 236). The steps are:

1 Write down the symbols of the elements.
2 Write down their mass ratios, as percentages.
3 Divide the percentages by their A_r's—this will give a series of numbers.
4 Divide by the smallest number to give the simplest number ratio.

Example 1 Calculate the empirical formulas of the compounds whose mass ratios are:
(a) 60 % magnesium : 40 % oxygen
(b) 59 % sodium : 41 % sulphur
(c) 52·1 % carbon : 13·0 % hydrogen : 34·9 % oxygen

The working out is shown below.

Composition	*Mass ratio*	*Number ratio (by division)*	*Empirical formula*
(a) 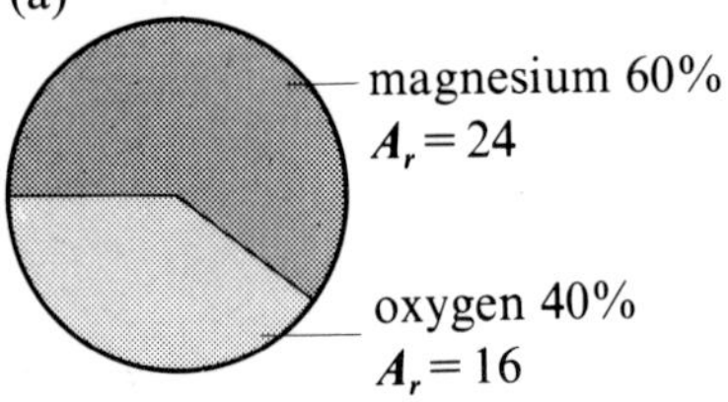 magnesium 60% $A_r = 24$; oxygen 40% $A_r = 16$	Mg:O 60 % : 40 %	$Mg:O = \frac{60}{24}:\frac{40}{16}$ $= 2{\cdot}5:2{\cdot}5$ or Mg:O = 1:1	**MgO**
(b) 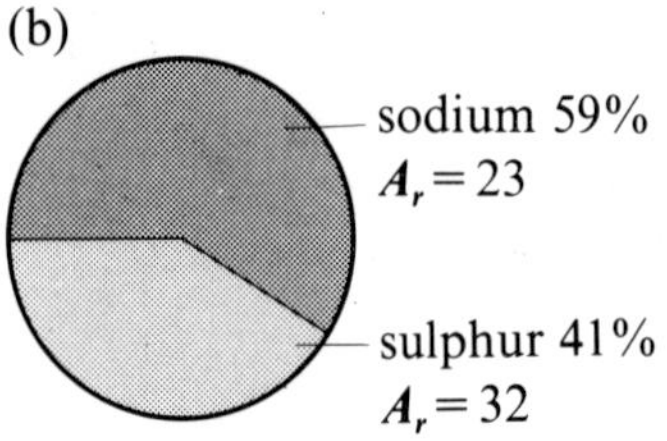 sodium 59% $A_r = 23$; sulphur 41% $A_r = 32$	Na:S 59 % : 41 %	$Na:S = \frac{59}{23}:\frac{41}{32}$ $= 2{\cdot}56:1{\cdot}28$ Divide by smallest number, 1·28 Na:S = 2:1	**Na_2S**
(c) 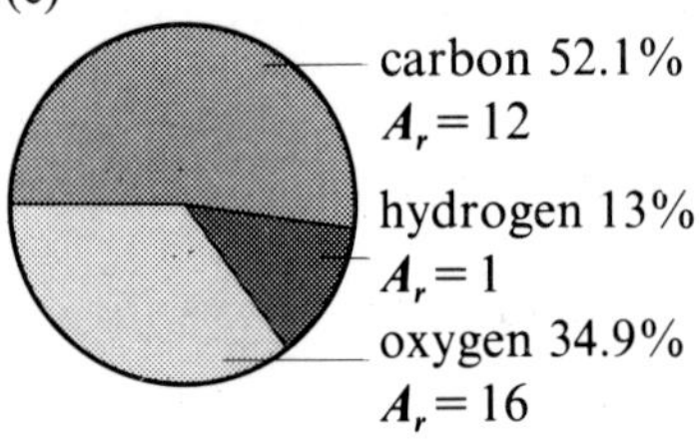 carbon 52.1% $A_r = 12$; hydrogen 13% $A_r = 1$; oxygen 34.9% $A_r = 16$	C:H:O 52·1 % : 13·0 % : 34·9 %	$C:H:O = \frac{52{\cdot}1}{12}:\frac{13{\cdot}0}{1}:\frac{34{\cdot}9}{16}$ $= 4{\cdot}35:13{\cdot}0:2{\cdot}17$ Divide by smallest number, 2·17 C:H:O = 2:6:1	**C_2H_6O**

Example 2 Calculate the empirical formulas of the compounds whose mass ratios are:
(a) 66.3% chlorine: 26.2% nitrogen: 7.5% hydrogen
(b) 62.2% iron: 35.6% oxygen: 2.2% hydrogen

Composition	*Mass ratio*	*Number ratio (by division)*	*Empirical formula*
(a) nitrogen 26.2% $A_r = 14$ hydrogen 7.5% $A_r = 1$ chlorine 66.3% $A_r = 35.5$	N:H:Cl 26.2% : 7.5% : 66.3%	$N:H:Cl = \frac{26 \cdot 2}{14} : \frac{7 \cdot 5}{1} : \frac{66 \cdot 3}{35 \cdot 5}$ $= 1.87 : 7.5 : 1.8$ Divide by smallest number, 1.87 N:H:Cl = 1:4:1	**NH_4Cl**
(b) oxygen 35.6% $A_r = 16$ hydrogen 2.2% $A_r = 1$ iron 62.2% $A_r = 56$	Fe : O : H 62.2% : 35.6% : 2.2%	$Fe:O:H = \frac{66 \cdot 2}{56} : \frac{35 \cdot 6}{16} : \frac{2 \cdot 2}{1}$ $= 1.11 : 2.22 : 2.2$ 2 Divide by smallest number, 1.11 Fe:O:H = 1:1:2	**FeO_2H_2** or **$Fe(OH)_2$**

Summary

Atoms and molecules are far too light to be weighed individually. Instead, the masses of atoms and molecules are compared to the mass of a standard atom.

The standard atom is carbon-12, ${}^{12}_{6}C$.

Using this standard, the mass of any atom is called its relative atomic mass (A_r); the mass of any molecule is called its relative molecular mass (M_r).

There are two kinds of formula for a compound:

the molecular formula, which shows the actual numbers of atoms in the molecules;

the empirical formula, which shows the simplest number ratio of the atoms present.

The empirical formula of a compound can be worked out from the mass ratio of the elements present in it.

Questions

1 The carbon-12 atom, $^{12}_{6}C$, is used as a standard of mass. In your own words, explain what this means.
Why is a single isotope of carbon used, instead of taking a sample of ordinary carbon?

2 (a) Explain what the term *relative atomic mass* means.
(b) Explain what the term *relative molecular mass* means.

3 Write down the $\boldsymbol{A_r}$ of each element, using only the information given below:
(a) An atom of silver is nine times as heavy as an atom of carbon-12.
(b) An atom of magnesium is twice as heavy as an atom of carbon-12.
(c) An atom of hydrogen is one-twelfth the mass of an atom of carbon-12.
(d) An atom of silicon is $2\frac{1}{3}$ times as heavy as a carbon-12 atom.

4 Work out the $\boldsymbol{M_r}$ for each compound. Use the table of $\boldsymbol{M_r}$'s on page 233 to help you.
(a) Methane, CH_4
(b) Sodium chloride, NaCl
(c) Sodium hydroxide, NaOH
(d) Aluminium chloride, $AlCl_3$
(e) Hydrogen nitrate, HNO_3

5 Write down the masses of the elements in each compound as a ratio. Make each ratio as simple as possible.
(a) Hydrogen fluoride, HF
(b) Nitrogen(IV) oxide, N_2O_4
(c) Ethyne, C_2H_2
(d) Butane, C_4H_{20}

6 Write down the mass of each element in each compound as a percentage. Use the examples on page 237 to help you. Where the answer is not a whole number, give it to one decimal place.
(a) Hydrogen fluoride, HF
(b) Aluminium hydride, AlH_3
(c) Calcium oxide, CaO
(d) Silicon hydride, SiH_4
(e) Calcium carbonate, $CaCO_3$
(f) Water, H_2O
(g) Methane, CH_4

7 Calculate the percentage of each element by mass in the following compounds. Give answers to one decimal place.
(a) FeO (b) Fe_2O_3 (c) Fe_3O_4
(d) $NaNO_2$ (e) $NaNO_3$ (f) $FeSO_4$

8 Calculate the percentage by mass of each element in the following compound. Give the answers to one decimal place.
(a) $Ca(OH)_2$ (b) $CuSO_4.5H_2O$
(c) NH_4NO_3

9 Write down the empirical formula for each compound.
(a) C_2H_6 (b) C_2H_4 (c) C_2H_2
(d) C_6H_6 (e) CuO (f) Cu_2O
(g) C_4H_{10} (h) H_2S (i) C_8H_{10}

10 Calculate the empirical formulas of the compounds which have the following compositions.
(a) Si 87.5%; H 12.5%
(b) Cu 79.8%; S 20.2%
(c) C 42.8%; O 57.2%
(d) C 27.3%; O 72.2%
(e) Mg 60.0%; O 40.0%
(f) K 54.9%; O 45.1%

11 Work out the empirical formulas of the compounds which have the following % composition.
(a) Ca 40.0%; C 12.0%; O 48.0%
(b) H 1.94%; Cl 67.94%; O 30.62%
(c) H 1.46%; Cl 51.82%; O 46.72%
(d) H 1.18%; Cl 42.0%; O 56.82%
(e) H 0.99%; Cl 35.4%; O 63.61%

12 Find the empirical formulas of the compounds which have the following % composition.
(a) K 38.6%; N 13.9%; O 47.5%
(b) Na 33.3%; N 20.3%; O 46.4%
(c) Na 27.4%; H 1.2%; C 14.3%; O 57.1%
(d) K 69.6%; O 28.6%; H 1.8%
(e) Ca 54.1%; O 43.2%; H 2.7%

23 The mole

23.1 Mass ratios

The mass ratio of two objects

Look at the two types of brick below. Let us say that one has four times the mass of the other.

Objects		*Mass ratio*
1 kg	4 kg	Brick : breeze block = 1 : 4

The mass ratio of equal numbers of the objects

The photograph below shows two piles of these bricks. There are a hundred in each pile.
The pile of ordinary bricks has a mass of 100 kg (100 × 1 kg).
The pile of breeze blocks has a mass of 400 kg (100 × 4 kg).

Systems		*Mass ratio*
100 kg	400 kg	Bricks : breeze blocks = 100 : 400 or 1 : 4

So the mass ratio of the two piles of bricks, 1 : 4, is the same as the mass ratio of the two single bricks. *That is true only because there is the same number of bricks in each pile.*

The Avogadro Constant

These ideas can also be applied to atoms:

C Mg mass ratio = 1:2

$A_r = 12$ $A_r = 24$

If a magnesium atom has twice the mass of a carbon atom, then a million magnesium atoms have twice the mass of a million carbon atoms. But atoms are so small that even a million of them cannot be weighed easily.

Bricks are sold by the hundred because this is a convenient number for builders. In the same way chemists need a convenient number of atoms; it must be a big enough number of atoms to be easily weighed. The number they use is an enormous one, called **the Avogadro Constant**. It is $6{\cdot}02 \times 10^{23}$. Written out in full it looks like this:

602000000000000000000000

The Avogadro Constant, $6{\cdot}02 \times 10^{23}$, is the number of atoms of carbon in exactly 12g of carbon-12.

To save writing this number out every time, we use a special symbol for it: *L* stands for the Avogadro Constant. How do chemists use *L*? Let us look again at carbon and magnesium:

C $A_r = 12$ Mg $A_r = 24$

1 carbon atom	1 magnesium atom	mass ratio = 1:2
1 million carbon atoms	1 million magnesium atoms	mass ratio = 1:2
L carbon atoms	*L* magnesium atoms	mass ratio = 1:2

L carbon atoms weigh 12 g, therefore *L* magnesium atoms must weight 24 g. Atoms are too small to count, but if we weigh out exactly 24 g of magnesium, we know it contains *L* atoms. Now look at carbon and silver:

C $A_r = 12$ Ag $A_r = 108$

1 carbon atom	1 silver atom	mass ratio = 1:9
1 million carbon atoms	1 million silver atoms	mass ratio = 1:9
L carbon atoms	*L* silver atoms	mass ratio = 1:9

L carbon atoms weigh 12 g, so *L* silver atoms must weigh 108 g. A pure silver bracelet that weighs exactly 108 g contains exactly *L* atoms.

The same argument works for all atoms. If we weigh out a number of grams of an element equal to its A_r, we know how many atoms it contains: L atoms, or $6{\cdot}02 \times 10^{23}$ atoms. And the argument also works for molecules:

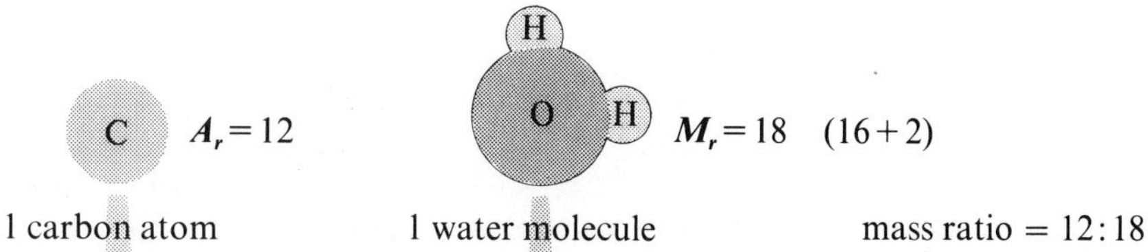

1 carbon atom — 1 water molecule — mass ratio = 12:18

L carbon atoms — L water molecules — mass ratio = 12:18

L carbon atoms weigh 12 g, therefore L water molecules must weigh 18 g.

An important rule comes out of all this:
When the number of grams of a substance equals its A_r or M_r, there are L particles of the substance present.

The mole

Builders think of bricks by the hundred. We think of eggs by the dozen. Chemists think of substances by the **mole**.
In a mole of a substance, the number of particles making up the substance equals the Avogadro Constant.
A mole of any substance therefore contains $6{\cdot}02 \times 10^{23}$ particles. 108 g of silver atoms is one mole; 18 g of water molecules is a mole; 32 g of oxygen molecules is a mole. So a mole is a very convenient amount because it can be weighed out easily. Just look up the A_r of the element, or the M_r of the compound, and weigh out that many grams.

Because a mole of substance always contains L particles, we must say what sort of particles we are thinking about. For example, a mole of atomic sulphur contains L atoms of sulphur and has a mass of 32 g. A mole of molecular sulphur (S_8) contains L sulphur molecules; but each of these is made of 8 sulphur atoms. So a mole of molecular sulphur contains $8 \times L$ sulphur atoms, and has a mass of $8 \times 32 = 256$ g. A mole of sodium contains L sodium ions and has a mass of 23 g. (Ions have practically the same mass as atoms.) A mole of electronic charge is L electrons, and weighs hardly anything!

Left to right: a mole of carbon, copper, zinc, sulphur, water, and calcium carbonate

23.2 Molar measurements

In this section we will look at four sets of measurements in which the idea of a mole is often used:

the mass of a system (solid, liquid, or gas)
the volume of a gas
the volume and concentration of a solution
the charge of an electric current

The mass of a system

You now know that the amount of a system that contains L particles is one mole of that system and weighs the appropriate A_r or M_r in grams. You can easily calculate the mass of a mole of any pure substance—solid, liquid, or gas—if you know its formula and the A_r values of the elements in it. Some examples are shown in Table 23.1.

Table 23.1 Calculating the mass of a mole

System	*Particle*	*Mass of a mole*
Carbon (solid)	M_r =12	C $\xrightarrow[\text{weigh}]{L \text{ of these}}$ 12 grams
Salt, sodium chloride (solid)	M_r =58.5	NaCl $\xrightarrow[\text{weigh}]{L \text{ of these}}$ 58·5 grams
Hydrogen (gas)	M_r =2	H_2 $\xrightarrow[\text{weigh}]{L \text{ of these}}$ 2 grams
Benzene (liquid)	M_r =78	C_6H_6 $\xrightarrow[\text{weigh}]{L \text{ of these}}$ 78 grams
Calcium carbonate (solid)	M_r =100	$CaCO_3$ $\xrightarrow[\text{weigh}]{L \text{ of these}}$ 100 grams

The volume of a gas

We saw in Chapter 2 that gases do not have fixed shapes, nor do they occupy fixed volumes. They take up the space that their containers offer them. The volume of the gas in a container can be changed by changing either the pressure on it or its temperature.

Changing the pressure

If the pressure increases the volume decreases. This diagram shows the gas in a syringe being compressed when books are placed on top of the plunger. The more books, the greater the pressure. Look what happens to the volume.

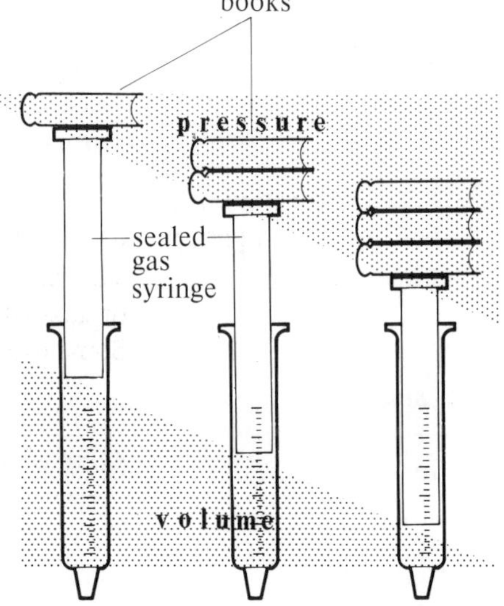

The first accurate results showing how the volume of a gas (V) changes with pressure (P) were produced by the famous scientist Robert Boyle. He found that for a fixed mass of gas, at a constant temperature:

$P \times V$ remains constant.

This became known as **Boyle's Law** and is often used in the form

$$P_1V_1 = P_2V_2$$

where V_1 is the volume at pressure P_1 and V_2 is the new volume at new pressure P_2.

Changing the temperature

If the temperature increases then the gas expands; that is, its volume increases. The diagram shows the gas in a syringe expanding, as the syringe is heated.

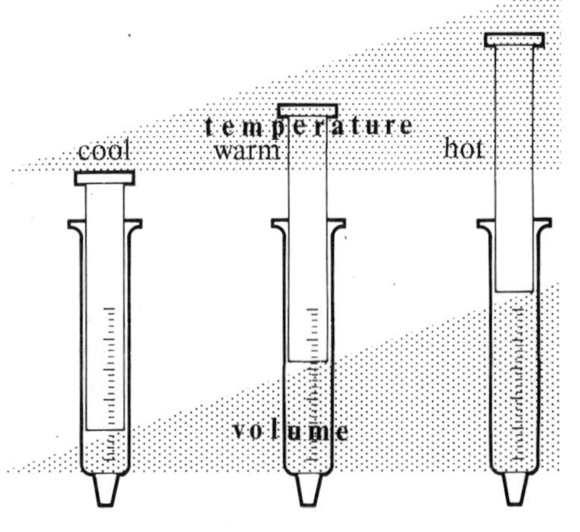

The behaviour of gases at different temperatures (T) was first examined properly by the Frenchman Jacques Charles. He found that for a given mass of gas, at constant pressure:

$\frac{V}{T}$ remains constant.

This became known as **Charles' Law** and is often used in the form:

$$\frac{V_1}{T_1} = \frac{V_2}{T_2}$$

where V_1 is the volume at temperature T_1 and V_2 the volume at temperature T_2. The temperature must be given in Kelvin (K). You can find the temperature in K by adding 273 to the temperature in °C. For example 10°C = (273 + 10)K = 283K.

These gas laws show how the volume of a gas changes, when either the temperature or the pressure changes. In both cases the gas is in a sealed syringe. No gas particles can get in or out, so the number of gas particles stays the same. It is only the space they take up which changes. Whenever we measure gas volumes therefore, we must (once again!) have a **standard temperature and pressure**. The standard temperature and pressure (shortened to S.T.P.) are 0°C and 760 mm of mercury.

Avogadro's Law

Table 13.2 shows the masses of 1000 cm^3 (or 1 dm^3) of two common gases measured at S.T.P.; from these, the numbers of particles present are calculated.

Table 23.2

Gas	*Mass of 1 dm³*	*Particle*	*Number of molecules present*
Hydrogen (H_2)	0·089 g	H_2 **M_r** = 2	2 g contains L particles ∴ 1 g contains $(L \times \frac{1}{2})$ ∴ 0·089 g contains $(L \times \frac{1}{2} \times 0{\cdot}089)$ = **0·045** × L
Carbon dioxide (CO_2)	1·96 g	CO_2 **M_r** = 44	44 g contains L particles ∴ 1 g contains $(L \times \frac{1}{44})$ ∴ 1·96 g contains $(L \times \frac{1}{44} \times 1{\cdot}96)$ = **0·045** × L

You can see from the table that both gases have the *same number of molecules in the same volume of gas*. They both have 0·045 × L molecules in 1 dm^3 of gas. The Italian scientist Avogadro was the first to discover this property of gases. He stated it in a law which became known as **Avogadro's Law**; it is sometimes called **Avogadro's Hypothesis**.

Avogadro's Law states that equal volumes of all molecular gases contain equal numbers of molecules under the same conditions of temperature and pressure.

If we apply our standards to Avogadro's Law—S.T.P. and L particles—we obtain a standard gas volume. This is the volume that one mole of any gas occupies at S.T.P. It is 22·4 dm^3.

One mole of any gas occupies a volume of 22·4 dm^3 at S.T.P.

By measuring the volume of a gas, the number of moles in it can be calculated, as shown in Table 23.3.

Table 23.3

Gas	*Volume at S.T.P.*	*Reasoning*	*Calculating the number of moles*
Air	$44{\cdot}8\ dm^3$	One mole occupies $22{\cdot}4\ dm^3$ at S.T.P.	$22{\cdot}4\ dm^3$ is 1 mole $\therefore$ $1\ dm^3$ is $(\frac{1}{22.4})$ mole $\therefore$ $44{\cdot}8\ dm^3$ is $(\frac{1}{22.4} \times 44{\cdot}8)$ moles = **2 moles**
Chlorine	$112\ cm^3$	$1000\ cm^3 = 1\ dm^3$ One mole occupies $22400\ cm^3$ at S.T.P.	$22400\ cm^3$ is 1 mole $\therefore$ $1\ cm^3$ is $(\frac{1}{22\,400})$ mole $\therefore$ $112\ cm^3$ is $(\frac{1}{22\,400} \times 112)$ moles = $\frac{1}{200}$ **moles**

The volume of a mole of gas at room temperature is often used, instead of its volume at 0°C. Room temperature is taken as 20°C. At this temperature and standard pressure, a mole of gas has a volume of $24\ dm^3$.

Volume and concentration of a solution

System	*Particle*	
Solution—mixture of matter in one phase: a solute dissolved in a solvent.		Solute and solvent particles are evenly mixed. • =solute =solvent
Dilute solution—small amount of solute dissolved in solvent.		Low number ratio solute particles : solvent particles
Concentrated solution—large amount of solute dissolved in solvent.		High number ratio solute particles : solvent particles

When we calculate amounts of substance present in a solution, both the volume and the concentration must be known.

The concentration is the number of moles of solute present in $1\ dm^3$ of solution.

The units for concentration are therefore **moles per dm^3**. This is shortened to **$mol\ dm^{-3}$**.

Remember that a mole of solute contains a fixed number of particles (L). So the concentration is a measure of the number of particles there are in a fixed volume of solution. The diagrams below should help to make this clear.

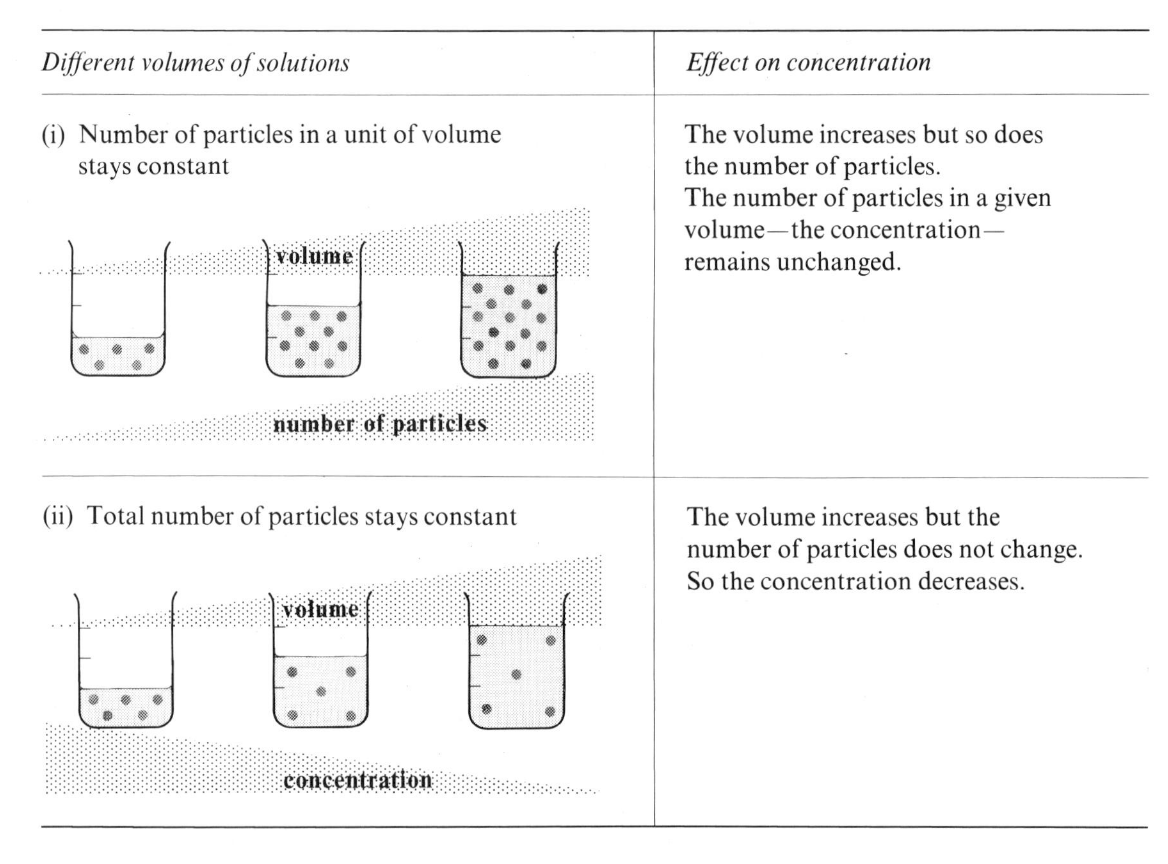

Different volumes of solutions	*Effect on concentration*
(i) Number of particles in a unit of volume stays constant	The volume increases but so does the number of particles. The number of particles in a given volume—the concentration—remains unchanged.
(ii) Total number of particles stays constant	The volume increases but the number of particles does not change. So the concentration decreases.

The correct units for concentration are moles per dm^3, or mol dm^{-3}. Concentrations are often given in other units; for example grams of solute instead of moles, and 100 cm^3 of solution instead of 1 dm^3. To change these into the correct units you must:

1 scale up the volume to dm^3
2 scale up the mass by the same factor
3 change the new mass into moles.

Example. A solution of sodium iodide (NaI) in water contains 3 g of solute in 100 cm^3 of solution. Calculate its concentration.
The calculation is shown below:

Volume of solution	*Mass of NaI*	*Moles of NaI*
There is more solute in 1000 cm^3 of solution than in 100 cm^3. 100 cm^3 — scale up — $10\times(100)=1000$ cm^3	3g in 100 cm^3 — scale up — 10×3g in 10×100 cm^3 There are 30 g of solute in 1000 cm^3 or 1 dm^3 of solution.	NaI — $M_r = (23+127) = 150$ ∴ 150 g is 1 mole ∴ 1 g is ($\frac{1}{150}$) mole ∴ 30 g is ($\frac{1}{150} \times 30$) moles = **0·2 moles**

The concentration of NaI is therefore **0·2 mol dm**$^{-3}$

The charge of an electric current

Electricity is the flow of electrons through the lattice of a conductor. (There is a lot more about it in Chapter 16.)

System	*Particle*
Electricity flows through metal wires.	Millions of electrons move through the metal lattice.
1 mole of electricity flows.	*L* electrons pass through.

Electrical charge is measured in **coulombs**.
The passage of *L* electrons (one mole of electrons) is the same as 96500 coulombs of charge. It is called a faraday of charge (F).

23.3 Worked examples

Here are some more calculations involving moles. They follow the order of section 23.2. The drawings in these examples are meant to give you an idea of the *mass* of each particle. They do not represent their *shapes*.

Mass

(i) Moles to masses

Calculate the mass of:
(a) 2 moles of water (b) 0·5 moles of iodine (c) 0·2 moles of salt

Working	*Reasoning*	
(a) 1 mole of water weighs 18 grams ∴ 2 moles weigh (2 × 18) = **36 grams**	H_2O	M_r = (2 + 16) = 18
(b) 1 mole of iodine weighs 254 grams ∴ 0·5 moles weigh (0·5 × 254) = **127 grams**	I_2	M_r = (2 × 127) = 254
(c) 1 mole of salt weighs 58·5 grams ∴ 0·2 moles weigh (0·2 × 58·5) = **11·7 grams**	NaCl	M_r = (23 + 35·5) = 58·5

So **mass** = **number of moles** × **mass of 1 mole**

(ii) Masses to moles

Calculate the number of moles in:
(a) 320 grams of magnesium oxide, MgO (b) 21·6 grams of silver, Ag

Working	*Reasoning*	
(a) There are 40 g of MgO in 1 mole ∴ 1 g of MgO is ($\frac{1}{40}$) moles ∴ 320 g of MgO is ($\frac{1}{40}$ × 320) moles = **8 moles**	MgO	M_r = (24 + 16) = 40
(b) There are 108 g of silver in 1 mole ∴ 1 g is ($\frac{1}{108}$) moles ∴ 21·6 g is ($\frac{1}{108}$ × 21·6) moles = **0·2 moles**	Ag	A_r = 108

So **number of moles** = $\dfrac{\textbf{mass}}{\textbf{mass of 1 mole}}$

Gas volumes
(i) Volumes to moles

Calculate the number of moles in:
(a) 56 dm^3 of xenon
(b) 2·24 dm^3 of hydrogen
(c) 448 cm^3 of chlorine
all measured at S.T.P.

Working	*Reasoning*
(a) 22·4 dm^3 of xenon is 1 mole ∴ 1 dm^3 is $(\frac{1}{22.4})$ moles ∴ 56 dm^3 is $(\frac{1}{22.4} \times 56)$ = **2·5 moles**	
(b) 22·4 dm^3 of hydrogen is 1 mole ∴ 1 dm^3 is $(\frac{1}{22.4})$ moles ∴ 2·24 dm^3 is $(\frac{1}{22.4} \times 2.24)$ = **0·1 moles**	1 mole of any gas occupies 22·4 dm^3 at S.T.P. (22·4 dm^3 = 22400 cm^3)
(c) 22400 cm^3 of chlorine is 1 mole ∴ 1 cm^3 is $(\frac{1}{22\,400})$ moles ∴ 448 cm^3 is $(\frac{1}{22\,400} \times 448)$ = **0·02 moles**	

So **number of moles** $= \dfrac{\textbf{volume in dm}^3\textbf{ at S.T.P}}{\textbf{22·4}}$

(ii) Masses to volumes (via moles)

Calculate the volume at S.T.P. occupied by:
(a) 10 grams of hydrogen, H_2
(b) 22 grams of carbon dioxide, CO_2

Working	*Reasoning*
(a) 2 grams of H_2 (1 mole) occupy 22·4 dm^3 ∴ 1 gram occupies $(\frac{22.4}{2})$ dm^3 ∴ 10 grams occupy $(\frac{22.4}{2} \times 10)$ = **112 dm^3**	1 mole of any gas occupies 22·4 dm^3 at S.T.P. H_2 $M_r = (2 \times 1) = 2$
(b) 44 grams of CO_2 (1 mole) occupy 22·4 dm^3 ∴ 1 gram occupies $(\frac{22.4}{44})$ dm^3 ∴ 22 grams occupy $(\frac{22.4}{44} \times 22)$ = **11·2 dm^3**	CO_2 $M_r = (12 + [2 \times 16]) = 44$

So **volume** = **number of moles** × **22·4** $= \dfrac{\textbf{mass}}{\textbf{mass of 1 mole}} \times$ **22·4**

Solutions
(i) Calculating concentrations

Calculate the concentration of:
(a) 11.7 g of NaCl in 500 cm^3 of solution in water
(b) 2·54 g of I_2 in 100 cm^3 of solution in tetrachloromethane

Working	*Reasoning*
(a) 11·7 g NaCl in 500 cm^3 ∴ (2 × 11·7) g in (2 × 500) cm^3 ∴ 23·4 g NaCl per dm^3 of solution. But 58·5 g NaCl is 1 mole ∴ 1 g is $(\frac{1}{58\cdot5})$ moles 23·4 g is $(\frac{1}{58\cdot5} \times 23\cdot4)$ moles = 0·4 moles ∴ concentration = **0·4 mol dm^{-3}**	Scale up to 1 dm^3 of solution (1000 cm^3). NaCl $M_r = 23 + 35\cdot5$ $= 58\cdot5$
(b) 2·54 g I_2 in 100 cm^3 ∴ (10 × 2·54) g in (10 × 100) cm^3 ∴ 25·4 g I_2 per dm^3 of solution But 254 g I_2 is 1 mole ∴ 25·4 g is 0·1 moles ∴ concentration = **0·1 mol dm^{-3}**	Scale up to 1 dm^3 of solution. I_2 $M_r = (2 \times 127)$ $= 254$

So **concentration** = $\dfrac{\textbf{mass of solute}}{\textbf{mass of 1 mole of solute} \times \textbf{volume of solution in dm}^3}$

(ii) Volumes from concentrations

What volume of a 0·4 mol dm^{-3} salt solution would contain:
(a) 0·1 mole of salt? (b) 2·34 g of salt?

Working	*Reasoning*
(a) 0·4 mol dm^{-3} ≡ 0·4 moles of salt in 1000 cm^3 of solution ∴ 1 mole of salt in $(\frac{1000}{0\cdot4})$ cm^3 of solution ∴ 0·1 moles of salt in $(\frac{1000}{0\cdot4} \times 0\cdot1)$ cm^3 of solution = **250 cm^3** of solution	scale down 0.4 mole in 1000 cm^3 0.1 mole in 250 cm^3
(b) 0·4 mol dm^{-3} ≡ (0·4 × 58·5) g of salt in 1000 cm^3 of solution or 23·4 g of salt in 1000 cm^3 of solution ∴ **2·34 g** of salt in 100 cm^3 of solution	M_r of NaCl = 58·5 scale down 23.4 g in 1000 cm^3 2.34 g in 100 cm^3

So **volume in dm^3** = $\dfrac{\textbf{mass of solute}}{\textbf{mass of 1 mole} \times \textbf{concentration in mol dm}^{-3}}$

Number of particles Masses, gas volumes, and electrical charge can all be related to the number of particles present by using the Avogadro Constant, L.

Calculate the number of:

(a) molecules in 36 g of water, H_2O
(b) atoms in 4 g of magnesium, Mg
(c) molecules in $11{\cdot}2\ dm^3$ of H_2 (at S.T.P.)
(d) electrons that pass when 197 coulombs of electricity flow.

For these calculations, let $L = 6 \times 10^{23}$.

Working	*Reasoning*
(a) 1 mole of water contains L molecules ∴ 18 g of water contain L molecules ∴ 1 g of water contains $(\frac{L}{18})$ molecules ∴ 36 g of water contain $(\frac{L}{18} \times 36) = 2 \times L$ molecules $=$ **12×10^{23} molecules**	1 mole contains L particles. H_2O $M_r = (2 \times 1) + 16$ $= 18$
(b) 1 mole of magnesium contains L atoms ∴ 24 g of magnesium contains L atoms ∴ 1 g of magnesium contains $(\frac{L}{24})$ atoms ∴ 4 g of magnesium contain $(\frac{L}{24} \times 4) = \frac{L}{6}$ atoms $=$ **1×10^{23} atoms.**	1 mole contains L particles. $A_r = 24$ Mg
(c) 1 mole of hydrogen contains L molecules ∴ $22{\cdot}4\ dm^3$ of hydrogen contain L molecules ∴ $1\ dm^3$ of hydrogen contains $(\frac{L}{22{\cdot}4})$ molecules ∴ $11{\cdot}2\ dm^3$ of hydrogen contain $(\frac{L}{22{\cdot}4} \times 11{\cdot}2)$ molecules $= \frac{L}{2}$ molecules $=$ **3×10^{23} molecules**	1 mole contains L particles. One mole of any gas occupies $22{\cdot}4\ dm^3$ at S.T.P.
(d) 1 mole of electricity contains L electrons. 1 mole of electricity ≡ 1 faraday of charge ∴ 96 500 coulombs of charge ≡ L electrons ∴ 1 coulomb of charge ≡ $(\frac{L}{96\,500})$ electrons ∴ 197 coulombs of charge ≡ $(197 \times \frac{L}{96\,500})$ $= \frac{L}{500}$ electrons $=$ **$1{\cdot}2 \times 10^{21}$ electrons.**	1 mole contains L particles 1 faraday of charge = 96 500 coulombs. 1 faraday of charge ≡ 1 mole of electricity.

So **number of particles** $=$ **$L \times$ number of moles**

Summary

When the number of grams of a substance is equal to its A_r or M_r, there are $6{\cdot}02 \times 10^{23}$ particles of the substance present.

This number $6{\cdot}02 \times 10^{23}$ is called the Avogadro Constant. The symbol L is used for it.

A mole of any system is the amount that contains L particles.

A mole of electricity contains L electrons. This quantity of electricity is called a faraday. The symbol F is used for it.

The faraday is equal to 96 500 coulombs of charge.

A mole of any gas, measured at standard temperature and pressure, occupies a volume of $22{\cdot}4\,dm^3$.

Avogadro's Law states that equal volumes of molecular gases, under the same conditions of temperature and pressure, contain equal numbers of molecules.

Boyle's Law and Charles' Law describe the way gases behave.

Boyle's Law states that pressure × volume is constant, if the temperature is kept constant:

$P \times V = \text{constant}, \text{ or } P_1V_1 = P_2V_2$

Charles' Law states that $\dfrac{\text{volume}}{\text{temperature (K)}}$ is constant, if the pressure is kept constant:

$\dfrac{V}{T} = \text{constant}, \text{ or } \dfrac{V_1}{T_1} = \dfrac{V_2}{T_2}$

The two laws can be combined together: $\dfrac{P_1V_1}{T_1} = \dfrac{P_2V_2}{T_2}$ where T is in Kelvin.

Questions

1 These are calculations of the numbers of particles present. Calculate the number of:
(a) sulphur atoms in 64 g of sulphur
(b) oxygen atoms in 22 g of carbon dioxide
(c) magnesium atoms in 0·01 moles of magnesium
(d) calcium atoms in 1 g of calcium
(e) H_2 molecules in 2·24 dm^3 of hydrogen
(f) CH_4 molecules in 112 cm^3 of methane
(g) electrons in 96·5 coulombs of charge
(h) atoms in a drop of water weighing 0·5 g

2 In these calculations, you must convert moles to masses. Calculate the mass in grams of:
(a) 2 moles of calcium metal, Ca
(b) 0·5 moles of magnesium metal, Mg
(c) 1 moles of iodine, I_2
(d) 0·1 moles of sulphur, S_8
(e) 0·25 moles of calcium carbonate, $CaCO_3$

3 In these calculations, you must convert masses to moles. Calculate the number of moles in:
(a) 14 g of silicon, Si
(b) 10·8 g of silver, Ag
(c) 1·2 g of carbon, C
(d) 31 g of phosphorus, P_4
(e) 0·56 g of calcium oxide, CaO

4 In these calculations, you must convert volumes to moles. How many moles are in the given volume of gas, measured at S.T.P.?
(a) 11·2 dm^3 of chlorine, Cl_2
(b) 1·12 dm^3 of sulphur dioxide, SO_2
(c) 0·224 dm^3 of methane, CH_4
(d) 2·24 dm^3 of nitrogen, N_2
(e) 22 400 cm^3 of oxygen, O_2
(f) 112 cm^3 of carbon dioxide, CO_2

5 In these calculations, you must convert moles to volumes. What volume at S.T.P. does each gas occupy?
(a) 2 moles of H_2 (b) 0·5 mole of O_2
(c) 0·25 mole of He (d) 3 moles of air

6 More calculations of volumes to moles. You need to know that
$10^{-1} = 1/10$ $10^{-2} = 1/10^2 = 1/100$ etc.
Calculate the following volumes at stp:
(a) 1·23 moles of nitrogen, N_2
(b) 2·05 moles of oxygen, O_2
(c) 10^{-3} moles of fluorine, F_2
(d) $1{\cdot}3 \times 10^{-2}$ moles of methane, CH_4
(e) $5{\cdot}4 \times 10^{-3}$ moles of helium, He
(f) $1{\cdot}36 \times 10^{-1}$ moles of air
(g) $1{\cdot}25 \times 10^{-3}$ moles of hydrogen, H_2

7 (a) Write out Boyle's Law in words, and then in the form of an equation.
(b) A fixed mass of gas occupies 50 cm^3 at a pressure of 3 atmospheres (atm); what is its volume at 1 atm and the same temperature?
(c) 120 cm^3 of gas at 1 atm is compressed at constant temperature to a pressure of 1·2 atm. What is its new volume?
(d) 1·5 dm^3 of gas expands to 2 dm^3 at constant temperature. If the starting pressure is 720 mm of mercury, what is the pressure after the expansion?
(e) 2 dm^3 of gas are at a pressure of 150 kN m^{-2}. If the pressure drops to 101·3 kN m^{-2} (normal pressure) at a constant temperature, what will the new volume be?

8 (a) Write out Charles's Law in words, and then in the form of an equation.
(b) At a pressure of 1 atm, 1 g of gas occupies 1 dm^3 at 10°C. If the pressure remains constant, calculate the volume of gas at:
(i) 0°C (ii) 100°C (iii) −15°C (iv) 25°C

9 (a) A fixed mass of gas occupies 150 cm^3 at 15°C and a pressure of 1·5 atm. What volume would it occupy at S.T.P.?
(b) 20 dm^3 of gas is at S.T.P.; what volume would it occupy at 25°C and 0·8 atm?
(c) A gas fills a 2 litre flask at 50°C and 5 atm. Calculate the volume it would occupy at S.T.P.
(d) How many moles of gas are present in the flask in part (c)?

10 Calculate the concentration of each solution in moles per dm^3. You are given the mass of solute dissolving in a certain volume of solvent.
(a) 3·65 g of HCl in 1 dm^3
(b) 0·98 g of H_2SO_4 in 1 dm^3
(c) 9·5 g of $MgCl_2$ in 0·5 dm^3
(d) 14·9 g of KCl in 2 dm^3
(e) 0·4 g of NaOH in 100 cm^3

11 What mass of each of the following substances must be dissolved in 250 cm^3 of water to give a concentration of 0·1 mol dm^{-3}?
(a) silver nitrate, $AgNO_3$
(b) hydrogen chloride, HCl
(c) sulphuric acid, H_2SO_4
(d) sodium hydroxide, NaOH
(e) potassium manganate (VII), $KMnO_4$
(f) potassium dichromate (VI), $K_2Cr_2O_7$

24 Chemical equations

Reacting masses

In Chapter 4 we saw that a chemical change can be represented by an equation. We are now going to see how the fixed mass relationships in chemical reactions can be explained by theory.

We shall use as our example the decomposition of calcium carbonate, which we used in Chapter 4.

Fact	*Theory*
(i) When calcium carbonate is heated, it decomposes to give calcium oxide and carbon dioxide.	Particles in calcium carbonate split up to give new particles.
(ii) The masses of calcium oxide and carbon dioxide produced are in a fixed ratio to the mass of calcium carbonate you start with. For every 1 g of $CaCO_3$, you always get 0·56 g of CaO and 0·44 g of CO_2.	?

The second of these points is a fact: masses can easily be measured. So far, however, we have made no effort to find a theory to explain this fixed ratio property of chemical change.

Let's think about the bonding in the particles involved in this reaction:

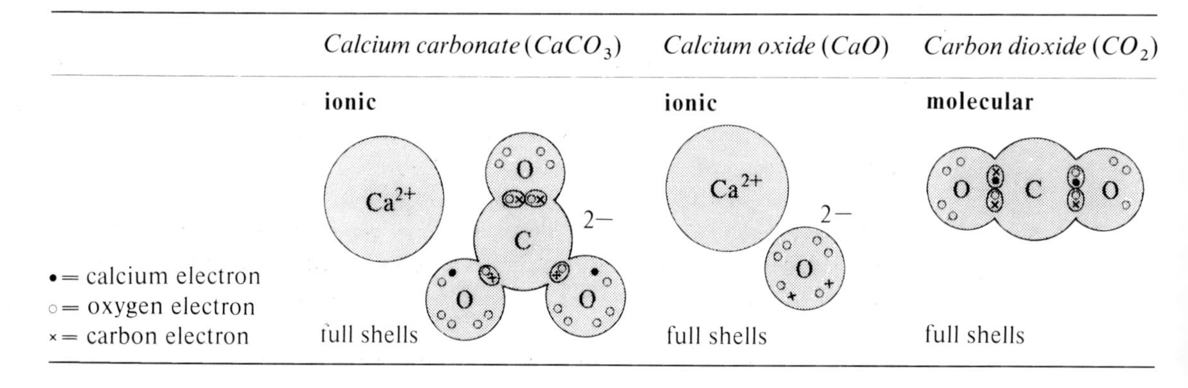

So we have a theory to explain the second fact. All we need do now is apply the ideas of Chapters 23 and 23:

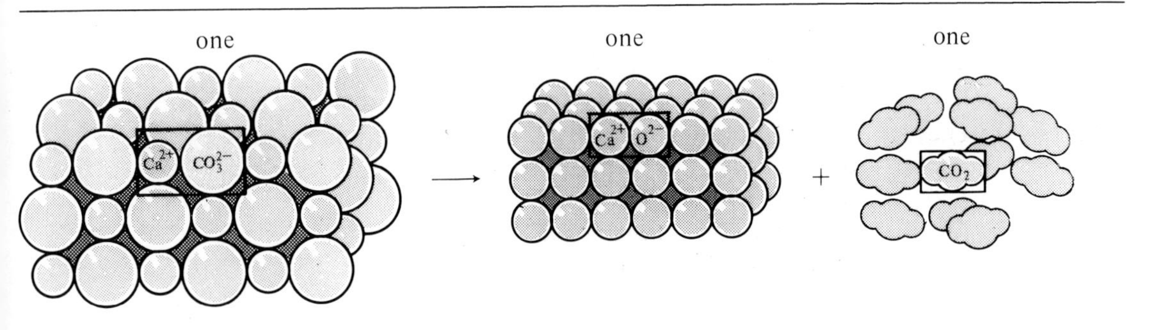

Repeat the individual step shown above Avogadro's Constant number of times:

$$L \times Ca^{2+}CO_3{}^{2-} \longrightarrow L \times Ca^{2+}O^{2-} \quad + \quad L \times CO_2$$

But L particles is the same as one mole of that system,
∴ one mole of $CaCO_3$ ⟶ one mole of CaO + one mole of CO_2

For $CaCO_3$, R.M.M. $= (40+12)+(3 \times 16) = 100$
For CaO, R.M.M. $= (40+16) = 56$
For CO_2, R.M.M. $= 44$

So 100 grams of $CaCO_3$ ⟶ 56 grams of CaO and 44 grams of CO_2
∴ 1 g of $CaCO_3$ ⟶ 0·56 g of CaO and 0·44 g of CO_2

Fixed mass ratio $CaCO_3 : CaO : CO_2 = 1 : 0{\cdot}56 : 0{\cdot}44$

We can now sum up the fixed ratio properties of a chemical change in terms of both fact and theory:

Fact	*Theory*
The mass ratio of reactants and products is fixed, for any particular chemical change.	The number ratio of reactant and product particles taking part in a particular chemical change is constant.

System equations and particle equations

Balanced equations can be used simply to replace sentences describing reactions. But this is not all that equations can do. The figures tell us more. They give the *actual number ratio* of the particles reacting. For example:

$2NH_3 \rightarrow N_2 + 3H_2$ means

2 molecules of ammonia	give	1 molecule of nitrogen	and	3 molecules of hydrogen

It is not possible to measure such tiny amounts as two or three molecules. If we want the equation to stand for real amounts that we can measure, the particle number ratio must be repeated Avogadro's Constant number of times. For the equation above:

$2 \times L$ molecules of ammonia	give	L molecules of nitrogen	and	$3 \times L$ molecules of hydrogen

This is the same as:

2 *moles* of ammonia	give	1 *mole* of nitrogen	and	3 *moles* of hydrogen

$2NH_3$ can therefore mean 2 *molecules* of ammonia or 2 *moles* of ammonia. Which meaning you use depends on whether you are thinking of the individual particles or of the whole system that can be seen and measured.

Equations can therefore be understood in two different ways:

as the number ratio of particles reacting—**a particle equation**
as the molar ratio of reactants and products—**a system equation.**

Equations often contain extra information that tells you which sort they are. Many of them have small state symbols in brackets after each formula: $_{(s)}$ for solid, $_{(l)}$ for liquid, $_{(g)}$ for gas and $_{(aq)}$ for aqueous solution. These letters tell the state in which the particles exist; the equations are particle equations.

System equations—particularly in older books—sometimes have arrows after a formula, ↓ or ↑. These mean:

↑—substance is a gas which escapes from the system
↓—substance is a solid which settles to the bottom of a solution.

Some examples that illustrate these symbols are shown below:

System equation	*Particle equation*
$CaCO_3 \rightarrow CaO + CO_2\uparrow$	$CaCO_{3\,(s)} \rightarrow CaO_{(s)} + CO_{2\,(g)}$
$2CuO + C \rightarrow 2Cu + CO_2\uparrow$	$2CuO_{(s)} + C_{(s)} \rightarrow 2Cu_{(s)} + CO_{2\,(g)}$

24.2 Calculations from equations : mass and volume

Mass Once you know the equation for a reaction you can calculate the exact amounts of reactants or products in it. You can do this by changing moles to grams as shown on page 250.

Example 1. Calculate the mass of carbon that reacts with 7·95 g of copper(II) oxide, in the reaction $2CuO + C \rightarrow 2Cu + CO_2\uparrow$.

Steps	*The working*
Equation	$2CuO + C \longrightarrow 2Cu + CO_2\uparrow$
Molar ratio	2 moles of copper(II) oxide react with 1 mole of carbon
Mass ratio using M_r's	CuO $M_r = (63.5 + 16) = 79.5$ C $A_r = 12$ $(2 \times 79{\cdot}5)$ g = 159 g of CuO react with 12 g of carbon
Scale down by 159	∴ 1 g of copper(II) oxide reacts with $(\frac{12}{159})$ g of carbon
Scale up by 7·95	∴ 7·95 g of copper(II) oxide reacts with $(\frac{12}{159} \times 7{\cdot}95)$ = **0·6 g of carbon**

Example 2. Calculate the mass of copper that is produced from the 7·95 g of copper oxide, in Example 1.

Steps	*The working*
Equation	$2CuO + C \longrightarrow 2Cu + CO_2\uparrow$
Molar ratio	2 moles of copper(II) oxide give 2 moles of copper
Mass ratio using M_r	CuO $M_r = (63.5 + 16) = 79.5$ Cu $A_r = 63.5$ $(2 \times 79{\cdot}5)$ = 159 g of CuO give $(2 \times 63{\cdot}5)$ = 127 g of copper
Scale down by 159	∴ 1 g of copper(II) oxide gives $(\frac{127}{159})$ g of copper
Scale up by 7·95	∴ 7·95 g of CuO gives $(\frac{127}{179} \times 7{\cdot}95)$ = **5·64 g of copper**

If you do not already know the equation for a reaction, you must first work it out as shown in the next example.

Example 3. Calcium metal reacts with water to give an insoluble suspension of calcium hydroxide. Hydrogen gas is also produced. Calculate the mass of hydrogen produced from 10 g of calcium.

Steps	*The working*
Reactants and products	atomic (Ca²⁺ Ca²⁺ Ca²⁺); molecular (H–O–H); ionic (OH⁻ Ca²⁺ OH⁻); H–H molecular
Unbalanced equation	$Ca + H_2O \longrightarrow Ca(OH)_2 + H_2$
Balance for hydrogen	Four on the right ∴ we need $2H_2O$ on the left to balance.
Balanced equation	$Ca + 2H_2O \longrightarrow Ca(OH)_2 + H_2$
Molar ratio	One mole of calcium gives one mole of hydrogen
Mass ratio	Ca $A_r = 40$ H_2 $M_r = 2$ ∴ 40 g of calcium give 2 g of hydrogen
Scaling	1 g of calcium gives $(\frac{2}{40})$ g of hydrogen ∴ 10 g of calcium give $(\frac{2}{40} \times 10)$ = **0·5 g of hydrogen**

Gas volumes Gases are so light that it is easier to measure their volumes than to weigh them. One mole of any gas occupies 22·4 dm^3 at S.T.P. Under room conditions—taken as 20°C and standard pressure—this volume is 24 dm^3. Moles of gases can therefore easily be converted to volumes.

Example 1. Calculate the volume of hydrogen (at S.T.P.) produced from 1 dm^3 of ammonia, in the reaction $2NH_3 \longrightarrow N_2 + 3H_2$.

Steps	*The working*
Equation	$2NH_3 \longrightarrow N_2 + 3H_2$
Molar ratio	2 moles of ammonia give 3 moles of hydrogen
Volume ratio	$(2 \times 22{\cdot}4)$ dm^3 of ammonia give $(3 \times 22{\cdot}4)$ dm^3 of hydrogen at S.T.P. so 44·8 dm^3 of ammonia give 67·2 dm^3 of hydrogen at S.T.P.
Scale down by 44·8	∴ 1 dm^3 of ammonia gives $(\frac{67.2}{44.8})$ = **1·5 dm^3 of hydrogen at S.T.P.**

Example 2. Calculate the volume of carbon dioxide (at S.T.P.) produced from 10 g of calcium carbonate, in the reaction
$CaCO_3 \longrightarrow CaO + CO_2\uparrow$.

Steps	*The working*
Equation	$CaCO_3 \longrightarrow CaO + CO_2\uparrow$
Molar ratio	One mole of $CaCO_3$ produces one mole of CO_2 gas
Mass:volume ratio	$CaCO_3$ $M_r = (40 + 12 + [3 \times 16]) = 100$ 100 g of $CaCO_3$ produces 22·4 dm^3 of CO_2 at S.T.P.
Scale down by 100	$\therefore$ 1 g of $CaCO_3$ produces $(\frac{22.4}{100})$ dm^3 of CO_2
Scale up by 10	$\therefore$ 10 g of $CaCO_3$ produces $(\frac{22.4}{100} \times 10)$ = **2·24 dm^3 of CO_2 at S.T.P.**

Example 3. In the manufacture of sulphuric acid, sulphur is burnt in oxygen to give sulphur(IV) oxide (sulphur dioxide). Sulphur(IV) oxide is then made to combine with more oxygen to give sulphur(VI) oxide (sulphur trioxide).
Calculate the volume of oxygen (at S.T.P.) needed to burn 25·6 kg of sulphur to sulphur dioxide.

Steps	*The working*		
Reactants and products	Solid sulphur exists as crown-shaped molecules, S_8. Oxidation number is zero S S S S S S S S (crown ring) molecular	O_2, a gas. Oxidation number is zero O=O molecular	SO_2, a gas. Oxidation number of S is IV S, O O molecular
Unbalanced equation	$S_8 + O_2 \longrightarrow SO_2$		
Balanced equation	$S_8 + 8O_2 \longrightarrow 8SO_2$		
Molar ratio	1 mole of sulphur burns in 8 moles of oxygen		
Mass:volume ratio	1 mole of oxygen occupies 22·4 dm^3 at S.T.P. $\therefore$ 8 moles occupy $8 \times 22{\cdot}4$ = 179·2 dm^3 $\therefore$ 256 g of sulphur burns in 179·2 dm^3 oxygen		
Scaling (1 kg = 1000 g)	$\therefore$ 1 g of sulphur burns in $(\frac{179.2}{256})$ dm^3 of oxygen $\therefore$ 25·6 kg of sulphur burns in $(\frac{179.2}{256} \times 25\,600)$ = **17 920 dm^3 of oxygen**		

Using Avogadro's Law

The law states that equal volumes of gases under the same conditions of temperature and pressure contain equal numbers of molecules.

Example 4. When North Sea gas burns, the equation is

$$CH_4 + 2O_2 \longrightarrow CO_2 + 2H_2O.$$

Calculate the volume of oxygen needed to burn 25 cm^3 of methane.

Steps	*Working*
Equation	$CH_4 + 2O_2 \longrightarrow CO_2 + 2H_2O$
Ratio of reacting molecules	One molecule of methane reacts with two of oxygen 1 molecule CH_4 : 2 molecules O_2
Apply Avogadro's Law	In 25 cm^3 there will be a very large number of methane molecules—call it M. M molecules of methane occupy 25 cm^3 $\therefore$ M molecules of oxygen occupy 25 cm^3.
Scaling	1 molecule of methane needs 2 molecules of oxygen $\therefore$ M molecules of methane need $(2 \times M)$ molecules of oxygen. $(2 \times M)$ molecules of oxygen will occupy (2×25) cm^3 = 50 cm^3 $\therefore$ **25 cm^3 of methane requires 50 cm^3 of oxygen** for complete reaction.

Example 5. Propane, C_3H_8, burns in oxygen to produce carbon dioxide and water. What volume of oxygen is needed to burn 40 cm^3 of propane? What volume of carbon dioxide is produced? (Volumes measured at S.T.P.)

Steps	*The working*
Reactants and products	Gases C_3H_8, O_2 and CO_2; liquid H_2O
Unbalanced equation	$C_3H_8 + O_2 \longrightarrow CO_2 + H_2O$
Balanced equation	$C_3H_8 + 5O_2 \longrightarrow 3CO_2 + 4H_2O$
Ratio of reacting molecules	1 molecule of propane reacts with 5 molecules of oxygen and produces 3 molecules of carbon dioxide.
Apply Avogadro's Law	1 volume of propane will need five times the volume of oxygen to burn it, and will produce 3 volumes of carbon dioxide.
Scaling	40 cm^3 of propane react with (5×40) cm^3 of oxygen and produce (3×40) cm^3 of carbon dioxide. So the volume of oxygen needed is **200 cm^3**. The volume of carbon dioxide produced is **120 cm^3**.

24.3 Calculations from equations: electrolysis

The electrolysis of molten lead(II) bromide is described on page 33. During **electrolysis** (see Chapter 16):

(i) There is a flow of electrons through the wire—a current.
(ii) Chemical changes occur at the electrodes.

1 Lead(II) bromide is ionic. In the solid phase the ions are trapped in a lattice. In the liquid phase they are free to move.

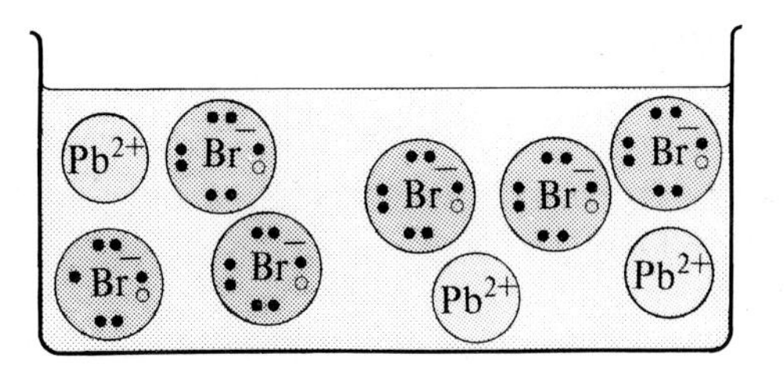

2 Pb^{2+} ions are attracted to the cathode. Br^- ions are attracted to the anode.

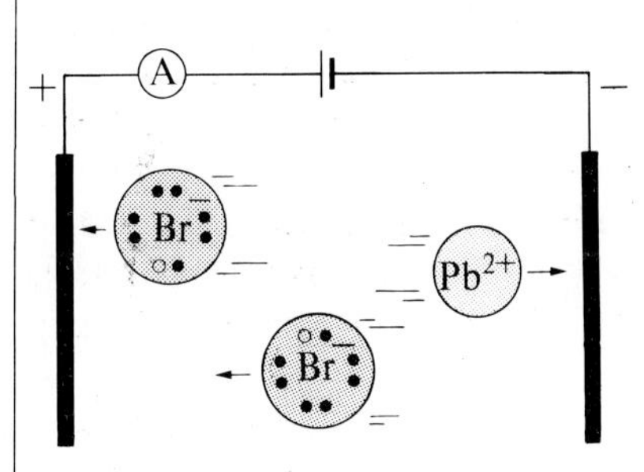

3 Br^- ions give up an electron each at the anode. These flow through the circuit. Pb^{2+} ions receive them at the cathode.

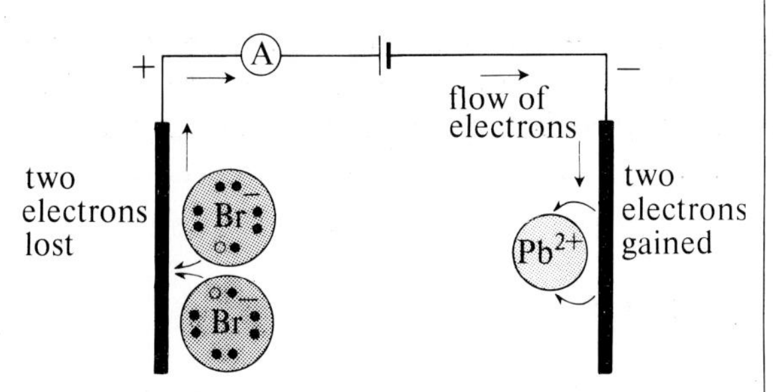

4 Bromine molecules and lead atoms form because of the electron transfer through the circuit.

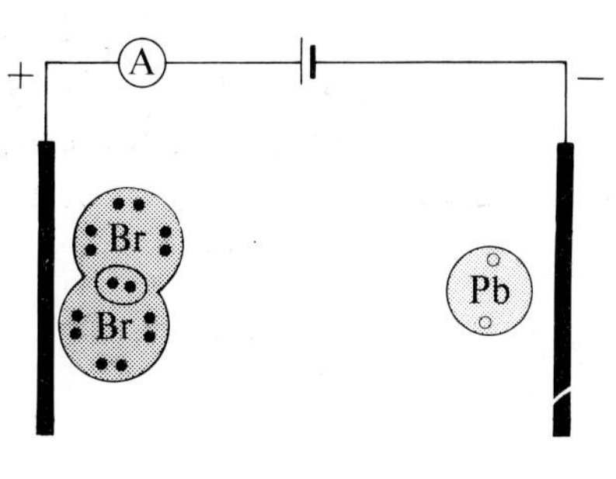

Fact	*Theory*
Liquid lead(II) bromide conducts electricity. During the passage of current, the compound is decomposed to its elements, lead and bromine. $PbBr_2 \xrightarrow{\text{electricity}} Pb\downarrow + Br_2\uparrow$	At the anode bromide ions give up their extra electrons. The bromine atoms produced bond covalently to give bromine molecules: $2Br^-_{(l)} \longrightarrow Br_{2\,(g)} + 2e^-$ At the cathode lead ions accept electrons that have been passed through the circuit from the anode. Lead atoms are formed: $Pb^{2+}_{(l)} + 2e^- \longrightarrow Pb_{(l)}$

From step 3 in the diagram opposite, you can see that the charge of electrons flowing through the circuit is produced by the bromide ions only and is accepted by the lead ions only. Since this charge can be measured by the ammeter, you can use it to calculate the amount of bromide ions, and lead ions, used up during the electrolysis.

The calculations

To do these calculations (and other electrolysis calculations) you need to know:

(i) 1 mole of electrons ≡ 96 500 coulombs of charge, or one faraday.

(ii) A current of 1 **ampere** passing for 1 second carries 1 coulomb of charge. 'Ampere' is usually shortened to **amp**.
So **amps** × **seconds** ≡ **coulombs**

(iii) The particle equation for the reaction at each electrode—that is, the equation showing the number of electrons involved. The particle equations for the electrolysis of lead bromide are given above.

Example 1. Calculate (a) the mass of lead deposited, and (b) the volume of bromide vapour produced (at S.T.P.), when a charge of 965 coulombs passes through liquid lead(II) bromide.

Steps	*The working*
(a) Equation	$Pb^{2+}_{(l)} + 2e^{-} \longrightarrow Pb_{(l)}$
Molar ratio	2 moles of electrons produce 1 mole of lead
Charge: mass ratio	Pb $A_r = 208$ (2 × 96 500) coulombs, i.e. 193000 coulombs, produce 208 g of lead
Scale down by 193000	1 coulomb of charge produces $(\frac{208}{193\,000})$ g of lead
Scale up by 965	∴ 965 coulombs produce $(\frac{208}{193\,000} \times 965)$ = **1·04 g of lead**

(b) Equation	$2Br^{-}_{(l)} \longrightarrow Br_{2(g)} + 2e^{-}$
Molar ratio	1 mole of bromine is produced with 2 moles of electrons
Charge: mass ratio	22.4 dm^3 of bromine (at S.T.P.) are produced with (2 × 96 500) coulombs
Scale down by 193 000	$\frac{22.4}{193\,000}$ dm^3 are produced with 1 coulomb
Scale up by 965	$\left(\frac{22.4}{193\,000} \times 965\right)$ = **0.112 dm^3 of bromine** is produced with 965 coulombs

Example 2. Aluminium is extracted by electrolysing aluminium oxide. Calculate the mass of aluminium produced when a current of 20 amps is passed through molten aluminium oxide for $1\frac{1}{2}$ hours.

Steps	*The reasoning and working*
1 Reactants and products	Aluminium oxide is ionic; Al^{3+} ions accept electrons at the cathode. So the reactants at the cathode are Al^{3+} ions and electrons; the product is Al
2 Equation	$Al^{3+} + 3e^- \longrightarrow Al$
3 Molar ratio	3 moles of electrons produce 1 mole of aluminium
4 Charge:mass ratio	$3 \times 96\,500$ coulombs produce 27 g of aluminium
5 Scale	$1\frac{1}{2}$ hours = 90 minutes = 5400 seconds 20 amps for 5400 seconds = 20×5400 coulombs = 108 000 coulombs $3 \times 96\,500$ coulombs produce 27 g of aluminium $\therefore$ 1 coulomb produces $\frac{27}{3 \times 96\,500}$ g $\therefore$ 108 000 coulombs produce $\frac{27 \times 108\,000}{3 \times 96\,500}$ g = **10·1 g of aluminium**

Summary

Chemical equations are shorthand ways of describing chemical reactions.

An equation can be understood in two different ways.
Take the equation $2H_2 + O_2 \longrightarrow 2H_2O$ for example.
The numbers tell you:

(i) the number ratio of the particles taking part in the reaction (two particles of hydrogen and one particle of oxygen give two particles of water), and

(ii) the molar ratio of the reactants and products (two moles of hydrogen and one mole of oxygen give two moles of water).

When we are thinking about the individual particles, the equation is called a particle equation. Particle equations include state symbols.

When we are thinking about the whole system and the moles of reactants and products, the equation is called a system equation.

Exact amounts of reactants and products can be calculated from chemical equations.

Questions

1 Write balanced equations for these reactions:
(a) Calcium metal and chlorine gas reacting to give calcium chloride.
(b) Copper metal and oxygen gas reacting to give copper(II) oxide.
(c) Magnesium metal and solid sulphur reacting to give magnesium sulphide.
(d) Carbon and carbon dioxide reacting to give carbon monoxide.
(e) Carbon and steam reacting to give hydrogen and carbon monoxide.

2 Write balanced equations for these reactions:
(a) Hydrogen sulphide and oxygen reacting to give sulphur(IV) oxide, and water.
(b) Sodium hydrogen carbonate decomposing to give sodium carbonate, carbon dioxide, and water.
(c) Ammonia and oxygen reacting to give nitrogen(II) oxide, and water.
(d) Potassium metal and carbon dioxide reacting to give potassium carbonate, and carbon.
(e) Lead(II) oxide and hydrogen reacting to give lead, and water.

3 Write balanced equations for these reactions:
(a) Copper(II) oxide and ammonia reacting to give copper, nitrogen(II) oxide, and water.
(b) Chromium(VI) oxide and sulphur(IV) oxide reacting to give chromium(III) sulphate.
(c) Nitrogen(II) oxide and hydrogen reacting to give ammonia, and water.
(d) Lead(II) nitrate(V) decomposing to give lead(II) oxide, nitrogen(IV) oxide, and oxygen.

4 Calculate the mass of magnesium oxide produced when 2.4 g of magnesium is burnt in air.

5 Calculate the mass of calcium oxide produced when 14.8 g of calcium hydroxide are heated according to the equation

$$Ca(OH)_2 \longrightarrow CaO + H_2O$$

6 Calculate the mass of calcium oxide produced when 5 g of calcium carbonate are decomposed.

7 Calculate the mass of iron produced from 100 g of iron(II) oxide by the reaction:

$$Fe_2O_3 + 3CO \longrightarrow 2Fe + 3CO_2$$

8 Calculate the mass of chromium(III) oxide produced when 25 g of ammonium dichromate(VI) are decomposed according to the equation:

$$(NH_4)_2Cr_2O_7 \rightarrow Cr_2O_3 + 4H_2O + N_2$$

In the following questions assume all volumes are measured at S.T.P. (standard temperature and pressure)

9 Calculate the volume of oxygen needed to burn 6 g of carbon in the reaction:

$$C + O_2 \longrightarrow CO_2$$

10 Calculate the volume of carbon monoxide needed to react with 100 g of iron(III) oxide in the reaction:

$$Fe_2O_3 \longrightarrow 2Fe + 3CO_2$$

11 Calculate the mass of copper produced if 10 dm^3 of carbon dioxide are made in the reaction:

$$2CuO + C \longrightarrow 2Cu + CO_2$$

12 Calculate the volume of oxygen needed to completely burn 20 cm^3 of the gas butane, C_4H_{10}.

13 Calculate the volume of oxygen needed to completely burn 50 cm^3 of the gas ethene, C_2H_4. What is the volume of carbon dioxide produced?

14 Calculate the volume of oxygen needed to completely burn 100 cm^3 of the gas pentane, C_5H_{12}. What volume of carbon dioxide is produced?

15 What is the maximum volume of propane, C_3H_8, that can be completely burned in 250 cm^3 of pure oxygen?

Section five

The chemistry of the elements

25 Hydrogen

25.1 The element

Hydrogen atom ${}^{1}_{1}H$

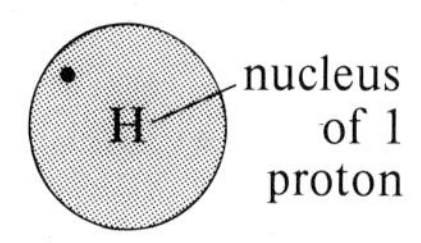

Hydrogen molecule H_2

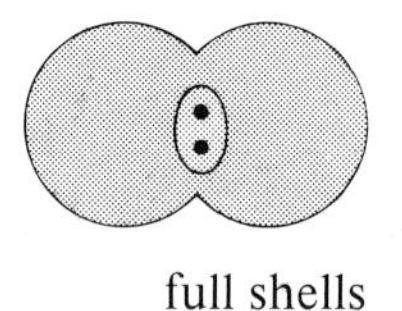

full shells

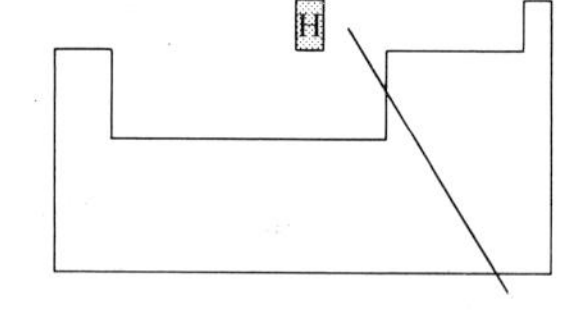

Appearance and preparation

Hydrogen is a gas at room temperature. It is the lightest known gas and has no smell or colour. It is usually made by adding dilute sulphuric acid to zinc metal:

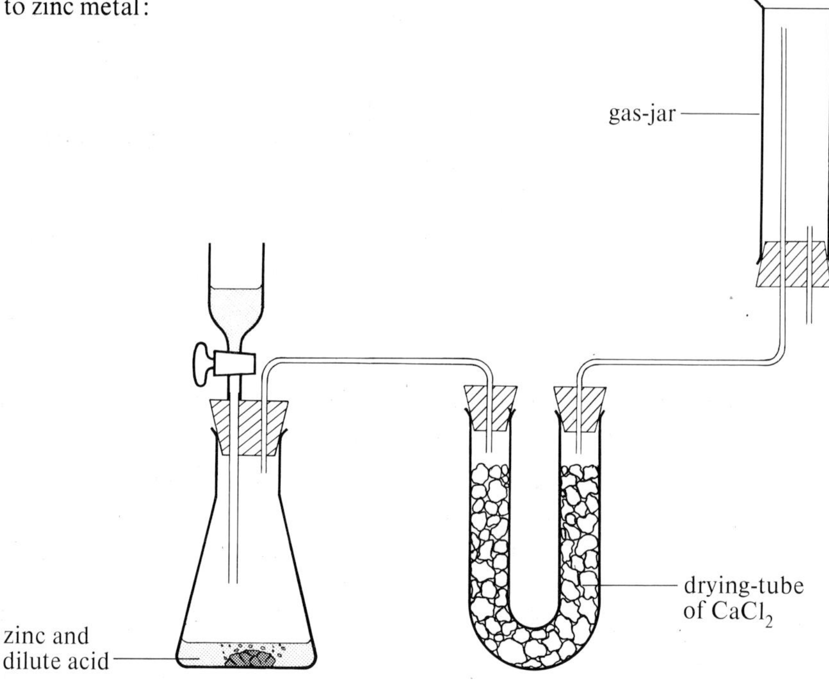

Fig. 1
Making hydrogen in the lab

The light gas pushes the air out of the gas-jar. If it is not important that the hydrogen is dry, we can collect the gas by bubbling it into an upside-down gas-jar full of water, instead of passing it through a drying tube.

$$Zn + H_2SO_4 \longrightarrow ZnSO_4 + H_2\uparrow$$

$$Zn_{(s)} + 2H_3O^+_{(aq)} \longrightarrow Zn^{2+}_{(aq)} + H_{2(g)} + 2H_2O_{(l)}$$

Physical properties The physical properties are shown in the table below. You don't need to learn the figures. Just get an idea of them.

Table 25.1

Substance	*m.p./°C*	*b.p./°C*	*Colour and smell*	*Density (compared with air)*	*Solubility in water at S.T.P.*
Hydrogen	−259	−253	None	0·07	$21{\cdot}4\,cm^3\,kg^{-1}$

Table 25.1 shows that hydrogen is a typical molecular substance (see p. 84).

25.2 Occurrence, extraction, and uses

Occurrence as the element Over 90 % of all matter in the universe contains hydrogen. It is combined in many common substances found on earth, but it occurs less frequently on earth than in other parts of the universe. Only 0·9 % by mass of the earth is combined hydrogen. There are very small amounts of it uncombined in the atmosphere. It is so light, however, that it tends to collect in the outer atmosphere, or escape from the atmosphere altogether.

Hydrogen is very light.

H_2 $M_r = 2$ O_2 $M_r = 32$ N_2 $M_r = 28$

Table 25.2 Composition of the atmosphere at sea level

Gas	Hydrogen	Nitrogen	Oxygen	Argon	Carbon dioxide
% volume in the air at sea level	0·0009 %	78·03 %	20·99 %	0·94 %	0·03 %

Occurrence in compounds The elements that hydrogen combines with most frequently are listed in Table 25.3.

Table 25.3 The common compounds of hydrogen

Combined with	*Compounds*	*Where found*
Oxygen	Water (H_2O)	Seas, rivers, lakes, and as water vapour in the air
Carbon, oxygen and nitrogen	Organic compounds	All living systems
Carbon	Hydrocarbons (fuels)	Deposits in the earth's crust: oil; coal; natural gas

water H_2O

methane – a fuel CH_4

Industrial extraction of hydrogen

Extracting hydrogen from its compounds is an important industrial process, because the element has many uses.
Water is the cheapest source of hydrogen. It is extracted from water in three main ways, shown in Table 25.4.

Table 25.4

Reduction of steam using red-hot coke	*Electrolysis of a brine solution*	*As a by-product of the oil industry*
carbon monoxide + hydrogen coke (carbon) 1000°C steam $C + H_2O \longrightarrow CO + H_2$ This is called the **Bosch process.**	Brine is a concentrated solution of common salt—sodium chloride. When it is electrolysed with a mercury cathode, sodium and chlorine are first produced. The sodium then reacts to give sodium hydroxide and hydrogen. The electrolysis is described in detail on page 283.	The hydrocarbons of oil can be reacted with excess steam in the presence of a heated nickel catalyst. Hydrogen and carbon monoxide form. Examples of the reactions are: $CH_4 + H_2O \longrightarrow CO + 3H_2$ and $C_7H_{16} + 7H_2O \longrightarrow 15H_2 + 7CO$

Uses

These are some of the important uses of hydrogen in industry:

(i) the manufacture of ammonia. $N_2 + 3H_2 \rightleftharpoons 2NH_3$

(ii) the 'hardening' of oils. Unsaturated hydrocarbons are changed to saturated ones by adding hydrogen. This reaction is used in the manufacture of margarine.

(iii) the manufacture of all types of organic compounds, for example the solvent methanol

$$\underset{\text{carbon monoxide}}{CO} + 2H_2 \longrightarrow \underset{\text{methanol}}{CH_3OH}$$

using a catalyst of zinc and chromium oxides at 300°C.

In the past, hydrogen was used for filling airships, like this World War One Zeppelin

25.3 Compound formation

Non-metal hydrides

Compounds of hydrogen with one other element are generally known as **hydrides**. However, some of the common ones have special names.
Hydrogen combines directly with many non-metals. The most reactive non-metals are those at the top right of the Periodic Table. Hydrogen's reactions with fluorine, chlorine, and oxygen are explosive.

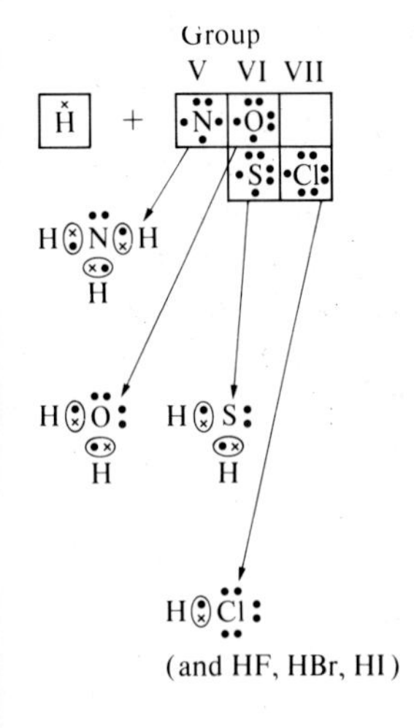

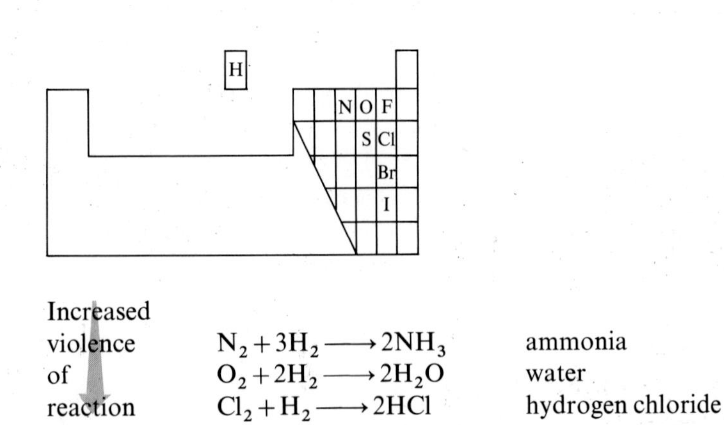

Chlorine's reaction with hydrogen is a good example of a photochemical reaction. A mixture of hydrogen and chlorine explodes violently if it is exposed to bright sunlight.
Sulphur, bromine, and iodine can also be made to combine with hydrogen, but the reactions are far less violent and often reversible.

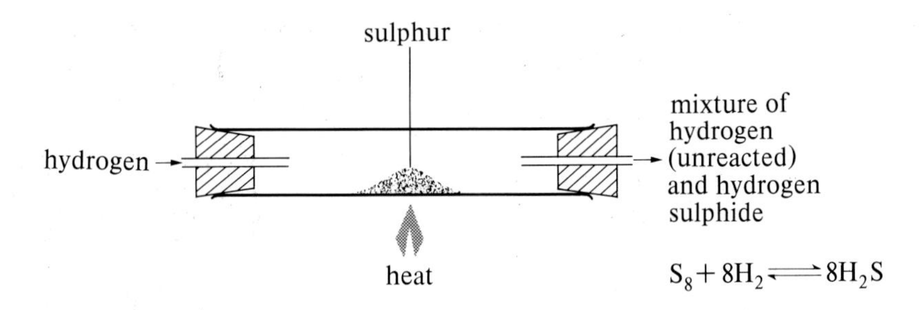

Carbon and phosphorus hydrides cannot be made by reacting the elements together.
All non-metal hydrides have typical *molecular* properties. They have low m.p.'s and b.p.'s; their solid phases are soft; they do not conduct electricity or heat well. The hydride of oxygen, water, is a very unusual compound, because of its hydrogen bonding. Its properties are discussed in Chapter 13.

Hydroxy compounds

Hydrogen reacts with oxygen to make water. However, water is not the only compound containing hydrogen and oxygen: hydrogen and oxygen combine with many other elements to make a whole range of compounds. All of these compounds have formulas that are *similar* to the formula of water. Half of the hydrogen part of water is replaced by a metal or another non-metal, as shown in the table below.

Metal hydroxides are ionic:

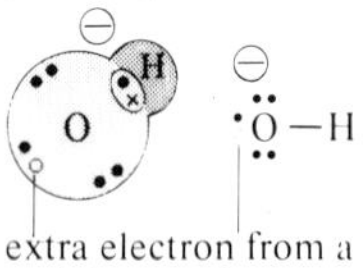

extra electron from a metal atom gives oxygen a full shell

Non-metal hydroxides are molecular:

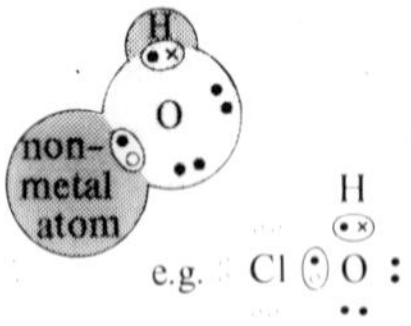

Table 25.5 Some common hydroxy compounds

	Elements combined	*Formula showing hydrogen and oxygen*	*Formula as usually written*
Metals	Sodium, hydrogen, oxygen	Na(OH)	NaOH
	Calcium, hydrogen, oxygen	$Ca(OH)_2$	$Ca(OH)_2$
	Iron, hydrogen, oxygen	$Fe(OH)_3$	$Fe(OH)_3$
Water	Hydrogen, oxygen	H(OH)	H_2O
Non-metals	Chlorine, hydrogen, oxygen	Cl(OH)	HClO
	Sulphur, hydrogen, oxygen	$SO_2(OH)_2$	H_2SO_4
	Nitrogen, hydrogen, oxygen	$NO_2(OH)$	HNO_3
	Carbon, hydrogen, oxygen	$CO(OH)_2$	H_2CO_3

These compounds are called **hydroxy compounds**. Their formulas all have the group **OH**: this is known as the **hydroxyl group**.

The formulas of *non-metal* hydroxy compounds are not usually written showing the OH groups. The hydrogen part of these formulas is usually written first. Because of this, the non-metal compounds are also named in a different way from the metal compounds.

Table 25.6 The differences in naming metal and non-metal hydroxy compounds

Type	*Usual way to write formula*	*Name*	
Metal	NaOH	Sodium hydroxide	
	$Ca(OH)_2$	Calcium hydroxide	
	$Fe(OH)_3$	Iron(III) hydroxide	
Non-metal	HClO	Hydrogen chlorate(I)	
	H_2SO_4	Hydrogen sulphate(VI)	(sulphuric acid)
	HNO_3	Hydrogen nitrate(V)	(nitric acid)
	H_2CO_3	Hydrogen carbonate(IV)	(carbonic acid)

Metal hydroxides:

are ionic
vary in solubility—NaOH soluble
$Ca(OH)_2$ slightly soluble
$Cu(OH)_2$ insoluble
give alkaline solutions when soluble.

Non-metal 'hydroxides':

are covalent
are soluble
give acidic solutions.

Making hydroxy compounds

(i) You can make the soluble hydroxy compounds, of both sorts, by reacting the element's oxide with water:

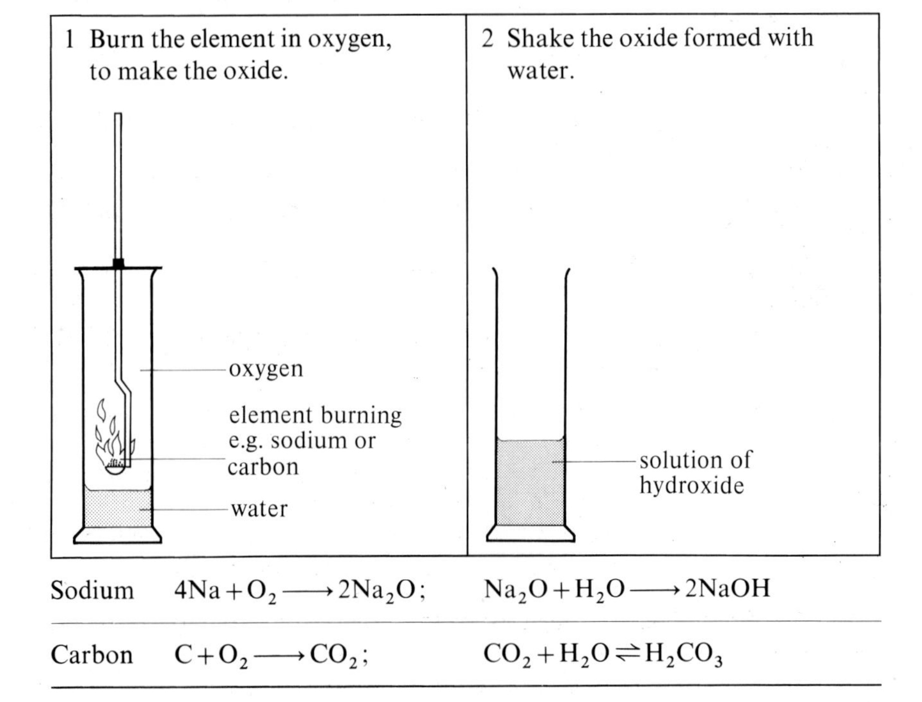
1 Burn the element in oxygen, to make the oxide.
oxygen
element burning e.g. sodium or carbon
water
2 Shake the oxide formed with water.
solution of hydroxide

Sodium	$4Na + O_2 \longrightarrow 2Na_2O$;	$Na_2O + H_2O \longrightarrow 2NaOH$
Carbon	$C + O_2 \longrightarrow CO_2$;	$CO_2 + H_2O \rightleftharpoons H_2CO_3$

(ii) Insoluble metal hydroxides are made by precipitation (see page 121). A solution of a metal compound is added to sodium hydroxide solution. The insoluble metal hydroxide is precipitated from the solution. It can be filtered from the reaction mixture.
For example, to make calcium hydroxide, $Ca(OH)_2$, you would begin with a soluble calcium compound, say calcium chloride:

$$CaCl_2 + 2NaOH \longrightarrow Ca(OH)_2\downarrow + 2NaCl$$
$$Ca^{2+}_{(aq)} + 2OH^-_{(aq)} \longrightarrow Ca(OH)_{2\,(s)}$$

25.4 Solubility and acid/base chemistry

Acids Solutions of non-metal hydroxides (like HNO_3, H_2SO_4) and hydrogen halides, (like HCl, HBr) all have acidic properties. The most common acids are:

Hydrogen compound	*Formula*	*Usual name for its solution in water*
Hydrogen chloride	HCl	Hydrochloric acid
Hydrogen sulphate(VI)	H_2SO_4	Sulphuric acid
Hydrogen nitrate(V)	HNO_3	Nitric acid
Hydrogen carbonate	H_2CO_3	Carbonic acid

Acids are *not* pure substances: they are solutions, normally in the solvent water. Acids have four main properties.

(i) They turn litmus and universal indicator red, so their pH is in the range 1–6.

(ii) They react with bases (like metal oxides) to produce salts and water.

Example: $CuO + H_2SO_4 \longrightarrow CuSO_4 + H_2O$

$$CuO_{(s)} + 2H_3O^+_{(aq)} \longrightarrow Cu^{2+}_{(aq)} + 3H_2O_{(l)}$$

(iii) They react with most metals to produce salts and hydrogen.

Example: $Mg + 2HCl \longrightarrow MgCl_2 + H_2\uparrow$

$$Mg_{(s)} + 2H_3O^+_{(aq)} \longrightarrow Mg^{2+}_{(aq)} + H_{2\,(g)} + 2H_2O_{(l)}$$

(iv) They react with metal carbonates to produce salts, water, and carbon dioxide.

Example: $CaCO_3 + 2HNO_3 \longrightarrow Ca(NO_3)_2 + CO_2\uparrow + H_2O$

$$CaCO_{3\,(s)} + 2H_3O^+_{(aq)} \longrightarrow Ca^{2+}_{(aq)} + CO_{2\,(g)} + 3H_2O$$

Bases and alkalis As you saw on pages 140–141, bases are substances that neutralise acids by turning them into salts. Alkalis are simply bases that dissolve in water.

All metal hydroxides are bases.

Group I metal hydroxides are soluble, so they are alkalis.

The hydride of nitrogen—ammonia, NH_3—is also able to neutralise acids. It is in fact very soluble in water, so it is an alkali. The salts made from ammonia are called ammonium salts.

Some acid/base reactions are given below:

Base	*+ acid*	$\longrightarrow$ *salt*	*+ water*
Copper(II) hydroxide	+ sulphuric acid	$\longrightarrow$ copper(II) sulphate(VI)	+ water
$Cu(OH)_2$	$+ H_2SO_4$	$\longrightarrow CuSO_4$	$+ H_2O$
$Cu(OH)_{2\,(s)}$	$+ 2H_3O^+_{(aq)}$	$\longrightarrow Cu^{2+}_{(aq)}$	$+ 4H_2O_{(l)}$
Calcium hydroxide	+ hydrochloric acid	$\longrightarrow$ calcium chloride	+ water
$Ca(OH)_2$	$+ HCl$	$\longrightarrow CaCl_2$	$+ H_2O$
$Ca(OH)_{2\,(s)}$	$+ 2H_3O^+_{(aq)}$	$\longrightarrow Ca^{2+}_{(aq)}$	$+ 4H_2O_{(l)}$

Alkali	*+acid*	*⟶ salt*	
Sodium hydroxide	+ nitric acid	⟶ sodium nitrate(V)	+ water
NaOH	$+ HNO_3$	$\longrightarrow NaNO_3$	$+ H_2O$
$OH^-_{(aq)}$	$+ H_3O^+_{(aq)}$	$\longrightarrow$	$2H_2O_{(l)}$
Ammonia	+ hydrochloric acid	⟶ ammonium chloride	
NH_3	+ HCl	$\longrightarrow NH_4Cl$	
$NH_{3\ (aq)}$	$+ H_3O^+_{(aq)}$	$\longrightarrow NH^+_{4\ (aq)}$	$+ H_2O_{(l)}$

25.5 Redox chemistry

Hydrogen as reducing agent

As the element, hydrogen has the oxidation number zero. In most of the reactions of hydrogen gas, a product is made in which hydrogen has oxidation number +I.

	Reactants	⟶	*products*	*Oxidation number changes*
(i)	Non-metal elements	⟶	hydrides	
	e.g. $Cl_2 + H_2$	⟶	2HCl	
(ii)	Oxides of unreactive metals	⟶	metals + water	
	e.g. $CuO + H_2$	⟶	$Cu + H_2O$	

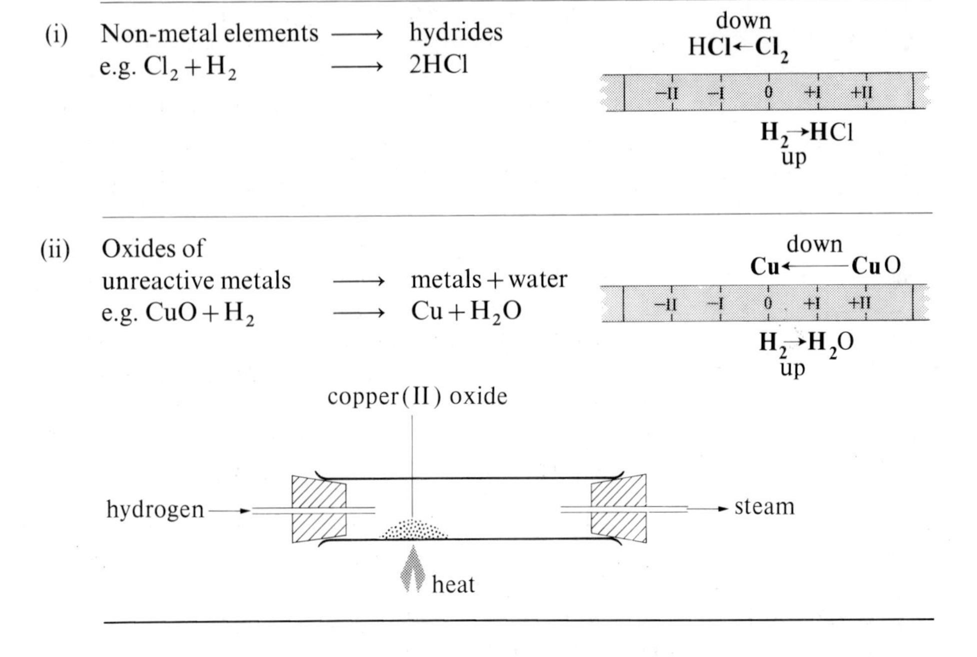

You can see that hydrogen's oxidation number goes up in these reactions, while another element's oxidation number comes down.
A substance is said to be reduced when it contains an element whose oxidation number comes down during a reaction. So hydrogen gas causes reduction: it acts as a **reducing agent**. It reduces chlorine to hydrogen chloride, and copper(II) oxide to copper.

Acids as oxidising agents

Acids contain hydrogen combined in oxidation number +I:

$\overset{+I}{H}_2SO_4$ $\overset{+I}{H}Cl$

Acids react with metals to produce salts and the element hydrogen, as a gas. So hydrogen's oxidation number is now zero.

Reactants		*products*		*Oxidation number changes*
Zinc + sulphuric acid Zn + H_2SO_4	⟶ ⟶	zinc sulphate(VI) $ZnSO_4$	+ hydrogen + H_2↑	down H_2 ← H_2SO_4 −II −I 0 +I +II Zn ⟶ $ZnSO_4$ up
Iron + hydrochloric acid Fe + 2HCl	⟶ ⟶	iron(II) chloride $FeCl_2$	+ hydrogen + H_2↑	down H_2 ← HCl −II −I 0 +I +II Fe ⟶ $FeCl_2$ up

The oxidation numbers of the metals have gone up, during these reactions—they are being oxidised. Therefore acids cause the oxidation of reactive metals. Acids act as **oxidising agents** in their reactions with metals. Zinc is oxidised to zinc sulphate(VI) and iron is oxidised to iron(II) chloride. (You should note that the acids themselves are *reduced* to hydrogen during the reactions: a substance that oxidises another substance is itself reduced.)

25.6 Special features

Hydrogen in the Periodic Table

In Chapter 7, you saw that elements with similar properties are often found in the same group in the Periodic Table. Hydrogen has properties that suggest it could be in *either* Group I *or* Group VII:

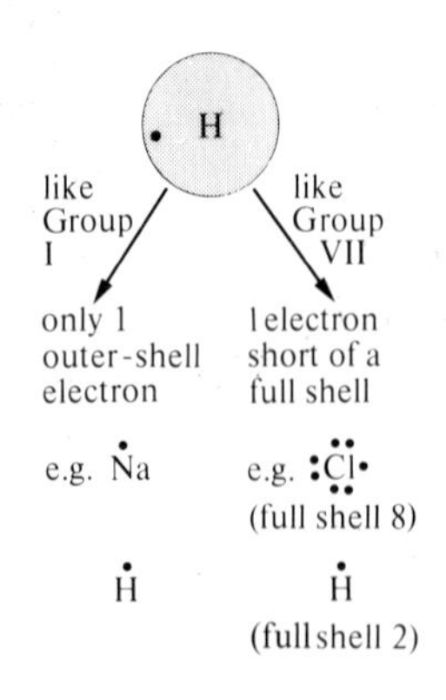

Like Group I

(i) Both hydrogen and sodium react violently with chlorine.

(ii) Both hydrogen and sodium always have a valency of one.

(iii) Electrolysis of most aqueous solutions produces hydrogen at the cathode, like a metal.

Like Group VII

(i) Both hydrogen and chlorine combine with sodium. The products are sodium hydride (NaH) and sodium chloride (NaCl).

(ii) Carbon forms molecular chlorides and hydrides. They are tetrachloromethane (CCl_4 carbon tetrachloride), and methane (CH_4).

(iii) Both hydrogen and chlorine have the physical properties typical of non-metals.

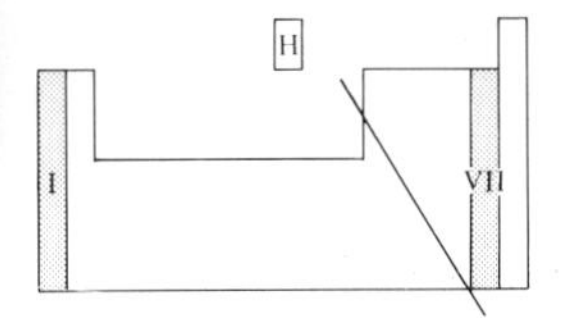

Because of this mixture of properties, hydrogen is not usually given any group number. Instead, it is placed in a separate box above the rest of the table. In some tables you will find it above the Group I elements; in other tables it is above the centre.

The unusual boiling-points of some hydrides

The boiling-points of some hydrides are shown on the graph below. The boiling-points of the Group VIII elements are shown with them, for comparison.

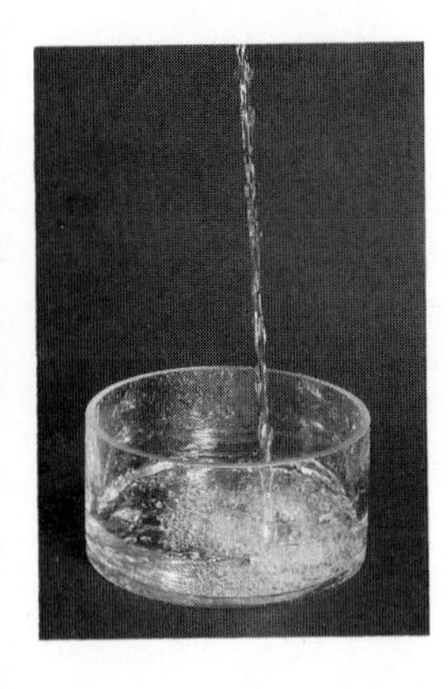

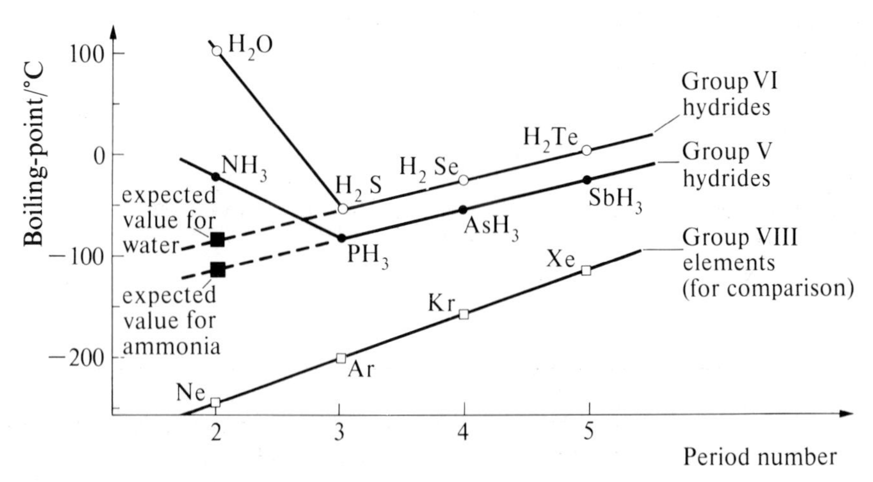

You can see that the b.p.'s of the Group VIII elements increase as you go down the group. This is to be expected, because the force of attraction between atoms increases as the number of electrons increases. The b.p.'s of the Group V and VI hydrides show a similar property, starting at period 3. But the period 2 hydrides *do not agree with the pattern.*

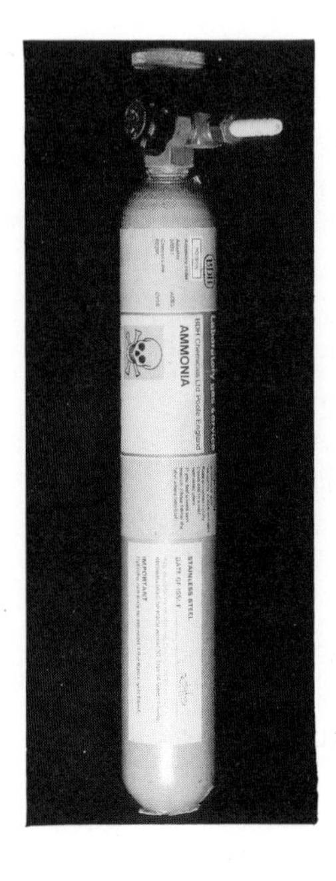

Hydride		*Expected b.p.*	*Actual b.p.*
Water	H_2O	About −90°C	+100°C
Ammonia	NH_3	About −120°C	−33°C

The reason is that there are some unusually strong forces within the liquid phases of water and ammonia, which hold the molecules together. These forces are called **hydrogen bonds**. A hydrogen bond is a strong attraction between a slightly positive hydrogen atom and a neighbouring molecule's lone pair.

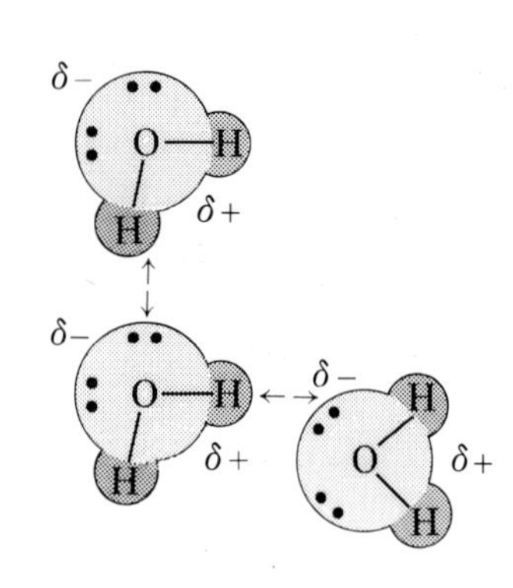

Reaction summary for hydrogen

Extraction In industry

(i) from coal: $C + H_2O \longrightarrow CO + H_2$

(ii) from oil: $C_7H_{16} + 7H_2O \longrightarrow 7CO + 15H_2$

(iii) electrolysis of brine: $2H_2O + 2e^- \longrightarrow 2OH^- + H_2$ at the cathode

In the laboratory
Reactive metal and dilute acid: $Zn + H_2SO_4 \longrightarrow ZnSO_4 + H_2\uparrow$

Reactions of the element

(i) It reduces other elements: $2Na + H_2 \longrightarrow 2NaH$
$N_2 + 3H_2 \longrightarrow 2NH_3$
$O_2 + 2H_2 \longrightarrow 2H_2O$
$Cl_2 + H_2 \longrightarrow 2HCl$

(ii) It reduces compounds: $CuO + H_2 \longrightarrow Cu + H_2O$

The oxides These are water and hydrogen peroxide.

(i) Water: see reactions of water in other summaries

(ii) Hydrogen peroxide: $2H_2O_2 \xrightarrow{MnO_2} 2H_2O + O_2\uparrow$ (see page 375)

The hydroxy compounds

(i) Metallic hydroxides are bases:
$NaOH + HCl \longrightarrow NaCl + H_2O$
$2NaOH + CuCl_2 \longrightarrow Cu(OH)_2\downarrow + 2NaCl$

(ii) Non-metallic hydroxy compounds are acids:
$SO_2(OH)_2$ or H_2SO_4 is sulphuric acid.
NO_2OH or HNO_3 is nitric acid.

The carbonate Hydrogen carbonate is also acidic: $H_2O + CO_2 \longrightarrow H_2CO_3$ (carbonic acid)

Questions

1 Complete the following equations, for reactions that produce hydrogen gas:
(a) $Zn + \ldots \rightarrow \ldots + H_2$
(b) $\ldots + H_2O \rightarrow \ldots OH + H_2$
(c) $\ldots + H_2SO_4 \rightarrow \ldots + H_2$
(d) $\ldots + H_2O \rightarrow \ldots O + H_2$
Now check that your equations balance.

2 (a) Sketch the apparatus you would use to make a sample of hydrogen in the laboratory. Label the diagram.
(b) Explain the purpose of each part of the apparatus.
(c) Give the equation for the reaction that makes the hydrogen.

3 (a) Describe three methods used in industry for making hydrogen.
(b) For each method give the equation for the reaction.

4 Give two major industrial uses of hydrogen gas. Try to explain why hydrogen is useful in each case.

5 (a) Write down the formulas for:
(i) two molecular hydrides
(ii) two ionic hydrides
(iii) two ionic hydroxy-compounds
(iv) two molecular hydroxy-compounds
(b) Beside each formula, write down whether it is *acidic, basic,* or *neutral.*

6 Draw diagrams (dot and cross) that show the electronic structures of the following compounds:
H_2S PH_3 BeH_2 BrOH KOH

7 (a) Draw a diagram of a hydrogen atom.
(b) Label it carefully.
(c) Using your diagram, explain why hydrogen could be thought to belong to Group I.
(d) Use your diagram to explain why hydrogen could be placed in group VII.

8 (a) Explain why hydrogen is placed above copper in the electrochemical series.
(b) Explain why hydrogen is placed below iron in the electrochemical series.

9 Pick out the oxidising and reducing *agents* in each of the following reactions of hydrogen:
(a) $3H_2 + N_2 \rightarrow 2NH_3$
(b) $Ag_2O + H_2 \rightarrow 2Ag + H_2O$
(c) $Mg + H_2O \rightarrow MgO + H_2$
(d) $2Li + H_2 \rightarrow 2LiH$
In each case, give reasons for your answer.

10 (a) What volume of hydrogen is made when 90 g of water reacts with excess calcium? (One mole of gas occupies 24 dm^3 at room conditions. The $\boldsymbol{A_r}$ of calcium is 40.)
(b) What mass of lithium will react completely with 10 g of hydrogen?
(The $\boldsymbol{A_r}$'s are: Li, 7; H, 1.)

11 For each of the following substances:
NaH H_2O HCl H_2S NaOH NH_3
(a) write down its name
(b) say what type of bonding it has
(c) say whether it is solid, liquid, or gas at room temperature.

12 Draw a table like this one and then fill in each section correctly.

	Metal hydroxide	*Non-metal hydroxide*
Method of making it		
Reaction		
Solubility		
Colour with indicator		

26 Group I: the alkali metals

26.1 The elements

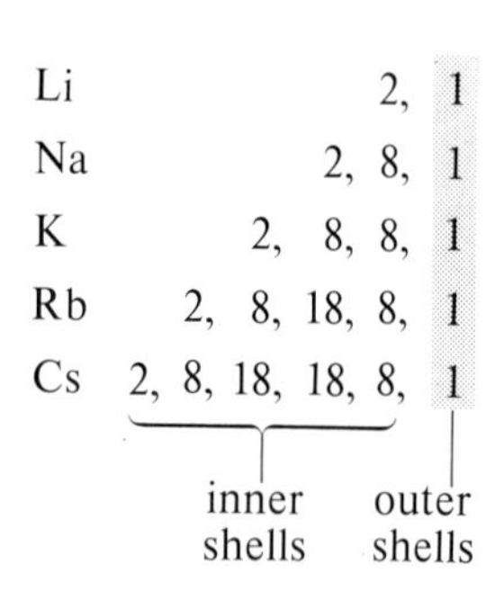

Li 2, 1
Na 2, 8, 1
K 2, 8, 8, 1
Rb 2, 8, 18, 8, 1
Cs 2, 8, 18, 18, 8, 1
inner shells
outer shells

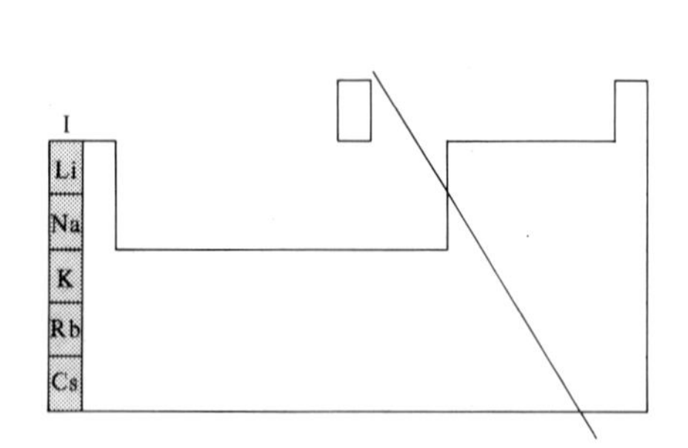

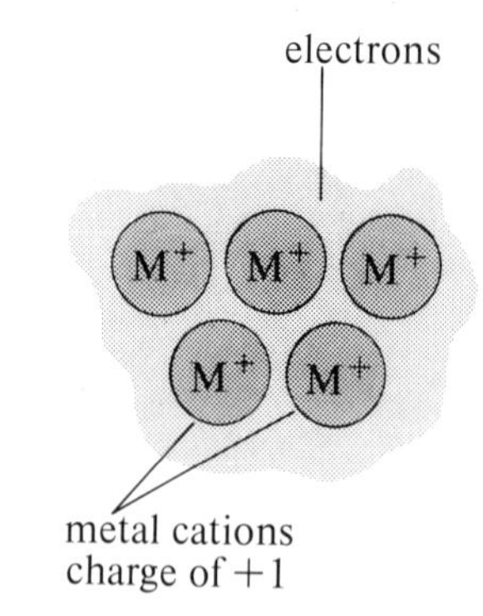

Appearance The elements are soft solids at room temperature, and are stored under oil. They are easily cut with a knife, and are shiny and silvery-grey when freshly cut.

Physical properties The physical properties of the Group I elements, excluding caesium, are listed in Table 26.1. You don't have to learn these values—just study their trends.

Table 26.1 Physical properties of the Group I elements

Element		*m.p./°C*	*b.p./°C*	*Density /g dm⁻³*	*Hardness /Moh*	*Conductivity /Ω⁻¹ cm⁻¹*
Lithium	Li	180	1336	530	0·6	11700
Sodium	Na	98	883	970	0·4	23800
Potassium	K	64	759	860	0·5	16400
Rubidium	Rb	39	700	1530	0·3	9100
Compare with a typical metal:						
Iron	Fe	1530	3000	7860	4–5	11300

The high conductivity values show that the Group 1 elements are metals. You can see that, compared with iron, these metals have unusually low fixed points and are unusually soft.

Sodium is used more than any of the other Group I metals, and is typical for the group; so we will concentrate on it in this chapter.

26.2 Occurrence, extraction, and uses

Occurrence How metals are found and how they are extracted depends on how *reactive* they are. The reactivity of the Group I metals has already been described on pages 62, 63, and 157. You should read these pages again.

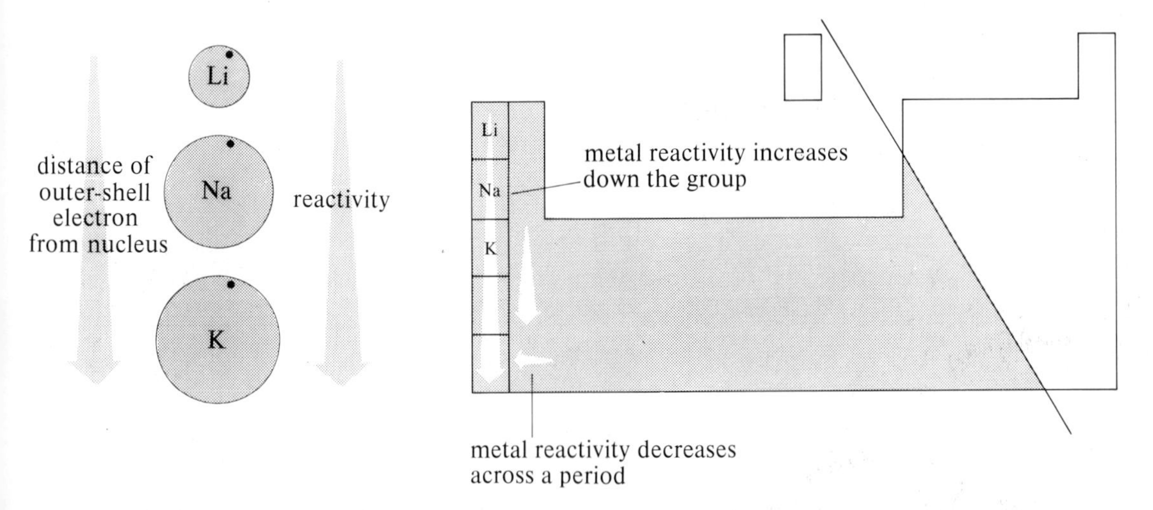

Group I is the most reactive group of metals. So it would be very surprising to find sodium in mines like you do gold! These metals are so reactive that they are always found combined with other elements.
Compounds of sodium occur naturally as:

the salt dissolved in the sea (NaCl)
deposits of rock-salt in the earth (NaCl)
deposits of saltpetre in the earth ($NaNO_3$)

Chemicals from salt It needs a lot of energy to make pure sodium from its compounds: since sodium is reactive, its compounds, once formed, are unreactive.
Sodium compounds are decomposed to sodium and non-metal products by electrolysis. Electrolysis is the best method for decomposing compounds of reactive metals. The cheapest compound to use is salt—sodium chloride. This is obtained either by evaporating sea-water or by mining rock-salt deposits.

Sea-water being evaporated to give salt, near Tripoli in Libya

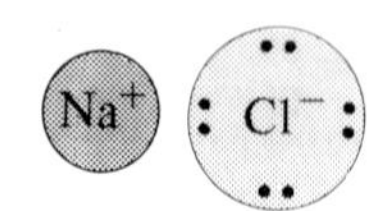

There are two different ways in which the salt can be used for electrolysis:

(i) as molten (liquid) sodium chloride, to produce sodium and chlorine.
(ii) as brine, a concentrated solution of sodium chloride in water.

The final products from this second process are sodium hydroxide, chlorine, and hydrogen. These are produced in two stages. In the first stage, sodium is produced by elecrolysis. In the second stage, the sodium is reacted to give sodium hydroxide and hydrogen—both important chemicals in industry.

Electrolysis of molten sodium chloride

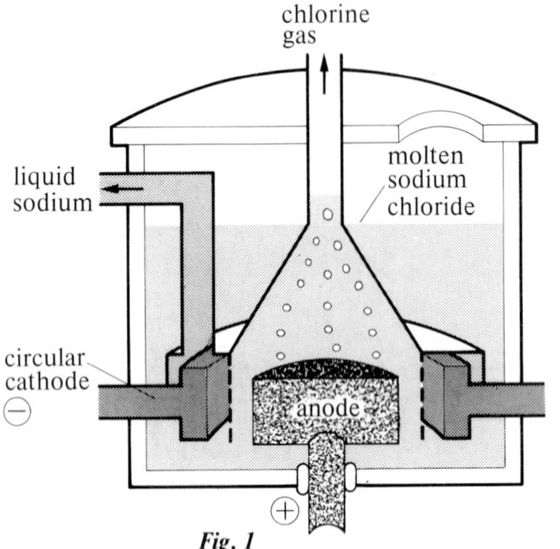

Fig. 1
The Down's Cell for electrolysing molten sodium chloride to produce molten sodium and chlorine gas

The process
Molten salt is decomposed by electrolysis in a cell. The current used is a very large one, and keeps the salt hot. The products of the electrolysis are liquid sodium and chlorine gas. Both these elements are very reactive, so the cell is made in such a way as to keep them apart. The anode is made of graphite and is at the centre of the cell. The hood over it channels the escaping chlorine. The cathode is a circular iron ring running around the anode. Notice how the liquid sodium gets led off from it.

Details
Raw material: 60 % NaCl, 40 % $CaCl_2$
Conditions: temperature, 600°C
power, 7 volts and 20 000 amps

Reactions
Anode: $2Cl^-_{(l)} \longrightarrow Cl_{2(g)} + 2e^-$ (oxidation)
Cathode: $2e^- + 2Na^+_{(l)} \longrightarrow 2Na_{(l)}$ (reduction)
Overall: $2NaCl \xrightarrow{\text{electricity}} 2Na + Cl_2$

The melting point of salt (801°C) is quite close to the boiling point of sodium (883°C). This could lead to the problem of sodium evaporating under the reaction conditions. To prevent that, enough calcium chloride is added to the salt, as an impurity, to lower its melting point to 600°C.

Electrolysis of dissolved sodium chloride (brine)

The process
A concentrated brine solution is electrolysed in a long narrow cell. The special feature of this cell is its liquid cathode, which is a thin layer of mercury flowing along the bottom of the cell. The anodes are graphite or a titanium alloy. The products of the electrolysis are sodium and chlorine. The chlorine bubbles off from the anode. The sodium dissolves in the mercury. In this way, it can flow out of the cell. It can then react with water in a lower tank, to give hydrogen and sodium hydroxide. The pure mercury is pumped around again.

Fig. 2
The moving mercury cell for electrolysing brine to produce sodium hydroxide, hydrogen, and chlorine

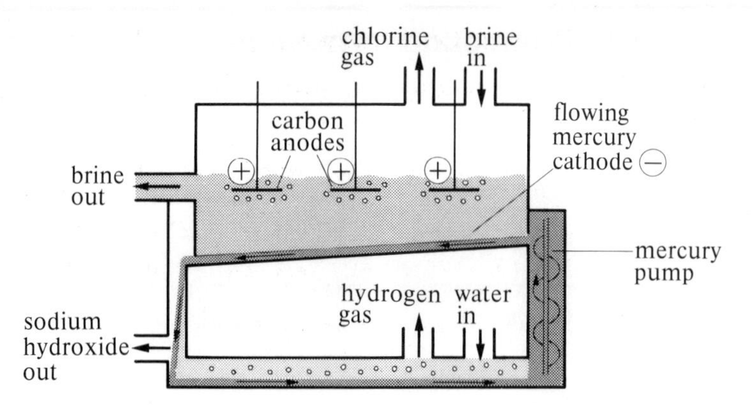

Details
Raw material: sodium chloride solution
Conditions: temperature, 20°C
power, 4·3 volts and 30 000 amps

Reactions
Cell anode: $2Cl^-_{(aq)} \longrightarrow Cl_{2(g)} + 2e^-$
cathode: $2e^- + 2Na^+_{(aq)} + Hg_{(l)} \longrightarrow 2Na/Hg_{(l)}$
overall: $2NaCl + Hg \xrightarrow{\text{electricity}} 2Na/Hg + Cl_2$
Lower tank: $2Na/Hg + 2H_2O \longrightarrow 2NaOH + H_2 + Hg$

Uses Sodium is used in industry, in the manufacture of:
(i) additives for petrol—'antiknock' or lead tetraethyl—to make engines run more smoothly.
(ii) sodium cyanide and sodium peroxide.
(iii) titanium metal from its ores.

Liquid sodium is used as a coolant in nuclear reactors. It is pumped in special pipes through the reactor. The liquid sodium extracts heat well from the reactor, because, like other metals, it is a good conductor of heat.

Salt in the diet One of the oldest uses of salt is as a preservative for food. When salt is added to food it keeps much longer.

Because a lot of food is highly salted, many people get a taste for salt and add it to nearly all their food.

Doctors now know that a highly salted diet is bad for your health. It leads to high blood pressure and heart disease. In America there has been a great campaign to cut down the amount of salt in everyone's diet, and as a result, the amount of heart disease in the population has fallen.

Heart disease in Britain is the highest in the world, and it is thought that a major cause of this is the large amount of salt and other sodium compounds, such as sodium glutamate, in popular foods like crisps and hamburgers.

26.3 Compound formation

Oxidation state

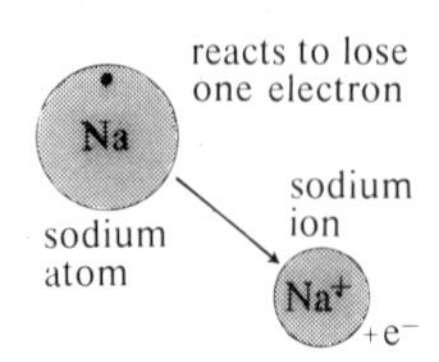

The Group I metals always form compounds in which the metals have the oxidation number +I. So their valency is always one. The compounds are all ionic and they are all soluble in water.

Some common sodium compounds are listed in Table 26.2.

Table 26.2

Compound	*Formula*	*Sodium's oxidation number*
Sodium hydroxide	$NaOH$	+I
Sodium chloride	$NaCl$	+I
Sodium carbonate	Na_2CO_3	+I
Sodium sulphate(VI)	Na_2SO_4	+I

These compounds can be made in the laboratory:

(i) sodium and water react to give sodium hydroxide (a vigorous reaction).
$2Na + 2HOH \longrightarrow 2NaOH + H_2\uparrow$

(ii) sodium hydroxide can be titrated with the appropriate acid to give the chloride, nitrate, and sulphate.

Compounds with non-metals

Like all metals, the Group I elements react with non-metals to produce compounds. We have already met some of these reactions in Chapter 7, pages 62 and 63; you should read these pages again.

Violent reactions occur with oxygen and all the Group VII elements, like chlorine. The Group I metals also combine with sulphur, phosphorus, and even nitrogen and hydrogen.

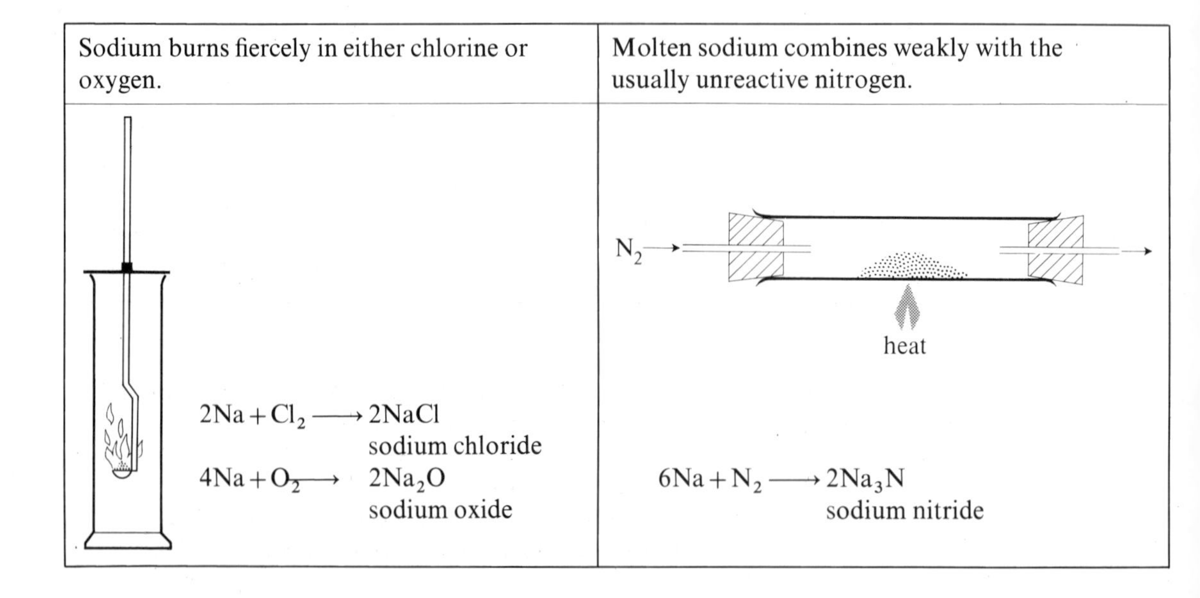

26.4 Solubility and acid/base chemistry

The metal oxides and hydroxides

Group I metals burn fiercely in oxygen, as shown on the opposite page. The white powder produced is a mixture of the metal oxide and a compound called a metal peroxide:

$2Na + O_2 \longrightarrow Na_2O_2$ sodium peroxide
$4Na + O_2 \longrightarrow 2Na_2O$ sodium oxide

Unlike other metal oxides, the Group I metal oxides and peroxides dissolve easily in water. Solutions of the metal hydroxides are produced:

$Na_2O + H_2O \longrightarrow 2NaOH$ sodium hydroxide
$Na_2O_2 + 2H_2O \longrightarrow 2NaOH + H_2O_2$ sodium hydroxide and hydrogen peroxide

If universal indicator is added to the hydroxide solution, it turns blue or deep purple. *The solutions are strongly alkaline.* They have a pH of 13 to 14. Sodium hydroxide solution is often called simply 'alkali'. When the right amount of it is added to an acidic solution, a neutral salt and water are produced:

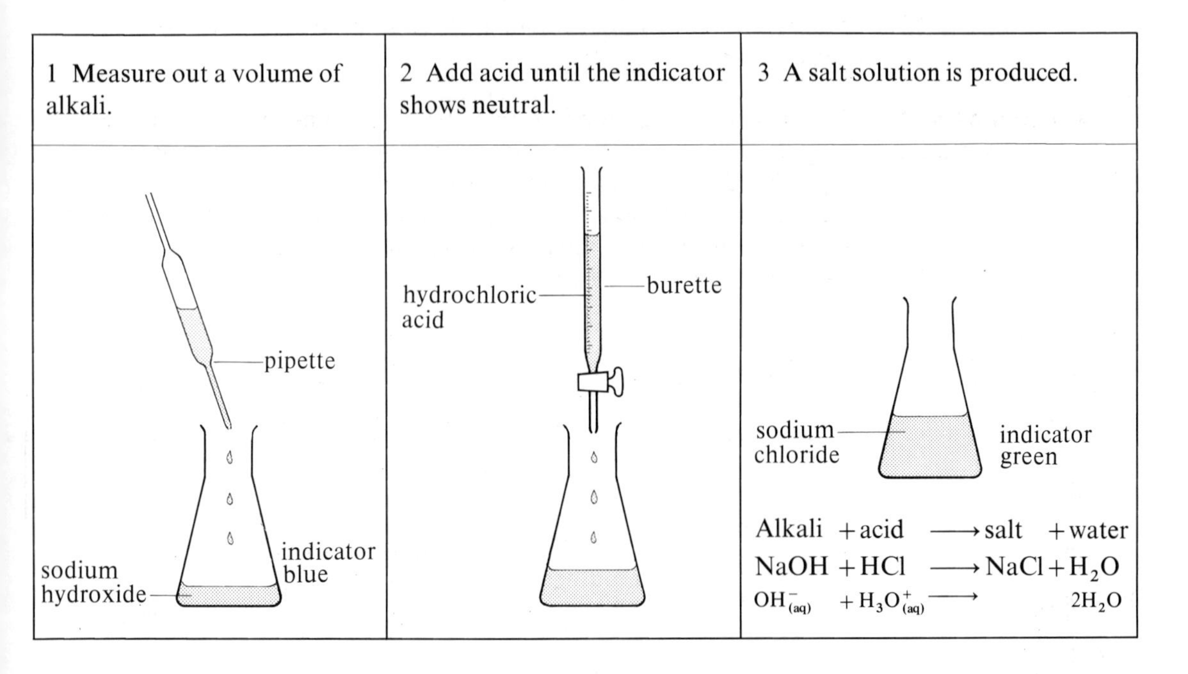

Solubility of the metal compounds

Most metal oxides, metal hydroxides, and metal carbonates are insoluble in water. However, *Group I metal oxides and carbonates are soluble.* In fact all the common compounds of the Group I metals dissolve well in water. This means that there can be no precipitation reactions that produce Group I compounds as products. Precipitation is a chemical reaction between two solutes in a solution, that produces an insoluble product.

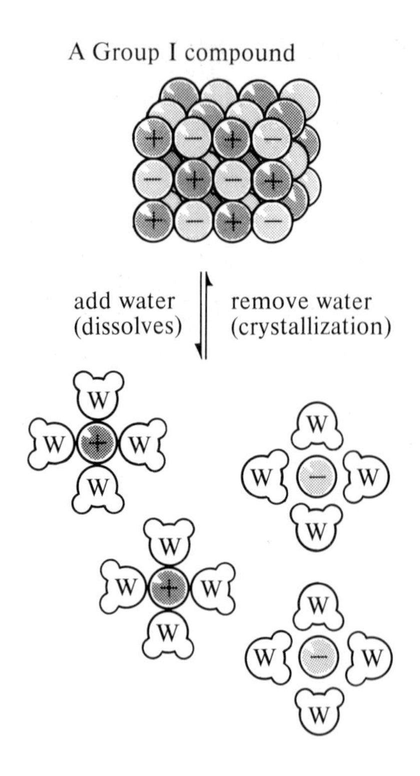

Since no Group I compounds are insoluble, none of them can be made by precipitation.

You can compare the solubilities of some sodium compounds in the table below.

Table 26.3

Compound	*Amount that dissolves in 100 g of water (0°C)*
Sodium choride	35·7 g
Sodium carbonate	21·5 g
Sodium hydroxide	42·0 g
Sodium nitrate(V)	73·0 g
Sodium sulphate(VI)	4·8 g

There is therefore only one way to remove a dissolved Group I compound from solution. You must evaporate the solvent, allowing the solid to crystallise. This is the process called crystallisation. When you evaporate off water from sodium chloride solution, for example, crystals of solid sodium chloride form.

26.5 Redox chemistry

The elements as reducing agents

You saw in section 26.2 that the Group I metals always have the oxidation number +I in their compounds. For example:

$$2Na_{(s)} + Cl_{2\,(g)} \longrightarrow 2NaCl_{(s)} \qquad 4Na_{(s)} + O_{2\,(g)} \longrightarrow 2Na_2O_{(s)}$$

Na**Cl** ← **Cl**$_2$ (down)
−I 0 +I
Na → **Na**Cl (up)

Na_2**O** ← **O**$_2$ (down)
−II −I 0 +I
Na → **Na**$_2$O (up)

In all the reactions of the metals:

(i) The Group I element goes up in oxidation number, from 0 to +I.
(ii) Another element goes down in oxidation number.

A substance is said to be reduced if it contains an element whose oxidation number goes down during the reaction. So Group I elements cause the reduction of chlorine and oxygen in the reactions shown above. *The Group I metals act as reducing agents.*

Since they are very reactive and form compounds very easily, they are powerful reducing agents.

An important industrial use for sodium's reducing power is in the

Right: Air Force One, the American presidential jet. Like all Boeing 707's, it contains some titanium. The production of titanium is one important use of sodium

production of titanium metal. Titanium is used in the manufacture of aircraft. It is extracted from its natural deposits as titanium(IV) chloride. This is reduced to titanium by sodium: titanium(IV) chloride vapour is passed into molten sodium, and a displacement reaction takes place.

$$TiCl_4 + 4Na \longrightarrow Ti + 4NaCl$$

down

Ti ← **$TiCl_4$**

0 +I +II +III +IV

Na → **NaCl**

up

Reaction with water

You may remember from Chapter 15 that the Group I metals are so reactive that they even react with cold water. These reactions are more examples of their behaviour as powerful reducing agents.

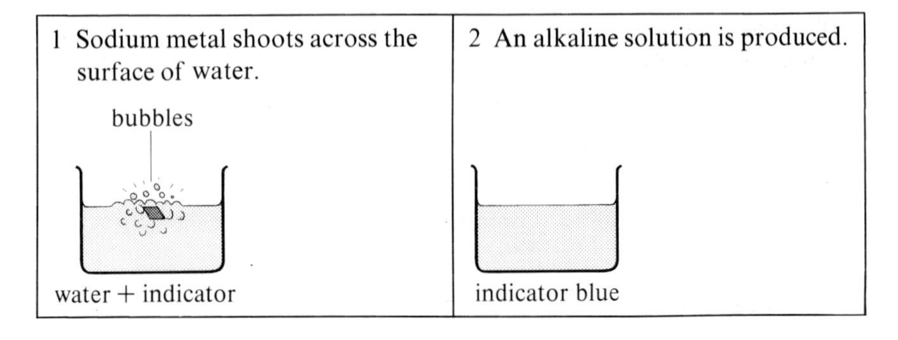

Hydrogen gas bubbles off as the metals dissolve exothermically. Sodium has reduced the water to hydrogen.

$$2Na + 2H_2O \longrightarrow 2NaOH + H_2\uparrow$$

$$2Na_{(s)} + 2H_2O_{(l)} \longrightarrow 2Na^+_{(aq)} + 2OH^-_{(aq)} + H_{2\,(g)}$$

down

H_2 ← **H_2O**

−II −I 0 +I +II

Na → **NaOH**

up

Because the reactions of Group I metals with water always produce an alkaline solution, they are often called **alkali metals.**

26.6 Special features

Solubility and stability to heat of Group I compounds

You have already learned, early in section 26.3, that all the common compounds of Group I metals are soluble in water. That is unusual. In Table 26.4 you can compare some Group I metal compounds with some others, both for solubility and for behaviour when heated.

Table 26.4 Comparison of Group I metal compounds with others

Metal group	*Metal compound*	*Amount dissolving in 100 g of water (0°C)*	*Behaviour when heated*
Group I	Sodium hydroxide NaOH	42·0	Melts at 318°C
	Sodium carbonate Na_2CO_3	21·5	Melts at 851°C
Group II	Calcium hydroxide $Ca(OH)_2$	0·18	Decomposes at 580°C $Ca(OH)_2 \longrightarrow CaO + H_2O$
	Calcium carbonate $CaCO_3$	~0	Decomposes at 825°C $CaCO_3 \longrightarrow CaO + CO_2$
Transition block	Copper hydroxide $Cu(OH)_2$	~0	Decomposes at 75°C $Cu(OH)_2 \longrightarrow CuO + H_2O$
	Copper carbonate $CuCO_3$	~0	Decomposes at 200°C $CuCO_3 \longrightarrow CuO + CO_2$

As a general rule we can say that most metal oxides, hydroxides, and carbonates are insoluble, and decompose on heating. However, the Group I hydroxides and carbonates are both soluble and stable to heat: they melt at high temperatures.

Flame tests

When a Group I compound is held in the edge of a bunsen flame, the flame becomes coloured. The compounds of each metal give a different colour. This simple test can be used to find out whether a Group I metal is present in a substance. For example:

sodium compounds $\longrightarrow$ yellow flame
potassium compounds $\longrightarrow$ lilac flame

Sodium is used as a coolant in nuclear reactors. Here are the pipes that carry liquid sodium from the heat exchangers, in the prototype fast reactor of Dounreay. Can you explain why sodium is used for this purpose? What problems might arise?

Reaction summary for sodium

This summary deals only with sodium and its compounds, because these are typical for Group I.

Extraction To give the pure metal: electrolysis of molten sodium chloride.
To give the hydroxide: electrolysis of brine with a mercury cathode.
In both cases, the reaction at the cathode is $Na^+ + e^- \longrightarrow Na_{(l)}$

Reactions of the element

(i) Reducing other elements:
$2Na + Cl_2 \longrightarrow 2NaCl$
$4Na + O_2 \longrightarrow 2Na_2O$
$2Na + O_2 \longrightarrow Na_2O_2$ (in excess oxygen)

(ii) Reducing compounds:
$2Na + 2H_2O \longrightarrow 2NaOH$ (alkaline solution) $+ H_2$

The oxides Soluble, basic oxides: $Na_2O + H_2O \longrightarrow 2NaOH$ (alkaline solution)
$Na_2O_2 + 2H_2O \longrightarrow 2NaOH + H_2O_2$

The hydroxide Soluble in water, giving an alkaline solution that conducts:

$NaOH_{(s)} + H_2O_{(l)} \longrightarrow Na^+_{(aq)} + OH^-_{(aq)}$

The solution shows the usual properties of bases.
The solid hydroxide is stable to heat: it melts but does not decompose.

The carbonate Soluble in water, and stable to heat: it melts but does not decompose.

Flame tests Sodium compounds give a yellow colour, when held in the edge of a bunsen flame.

Questions

1 (a) Draw diagrams showing the electronic structure of a lithium atom, a sodium atom, and a potassium atom.
(b) Explain why these elements are placed in Group I of the Periodic Table.
(c) Why do these elements have a valency of one?
(d) How do the atoms differ from each other?
(e) How does the reactivity of these elements change going down the group?
(f) When these elements react, do their atoms gain or lose electrons?
(g) Try to explain how the differences in reactivity going down the group are caused by the differences in electronic structure.

2 Sodium is extracted from molten sodium chloride by electrolysis.
(a) Draw a diagram of the type of cell used.
(b) Write down the cathode and anode reactions.
(c) Explain why the molten sodium chloride used in this process is deliberately made impure.
(d) What is to stop the two products, sodium and chlorine, which are very reactive, reacting together again, as soon as they are produced?

3 A concentrated solution of sodium chloride in water, called brine, is electrolysed in industry.
(a) Draw a diagram of the type of cell used in industry for this process.
(b) Explain why the cell is designed in the way you have shown.
(c) Write down the cathode and anode reactions for the process.
(d) Write down any other important reaction that takes place in this process.
(e) What other substance, apart from sodium chloride, does this process need?

4 (a) Write down the formulas of the following Group I compounds:
sodium oxide   sodium sulphate   potassium bromide   sodium carbonate   potassium hydroxide
(b) What is the charge on the Group I metal ions in each of these compounds?

5 All the Group I metals react with water.
(a) What would you *see* if a piece of sodium was put into a beaker of water?
(b) How would your observations differ if potassium was used instead of sodium?
(c) Write an equation for the reaction of one of these metals with water.
(d) Is the product of this reaction acidic, alkaline, or neutral?
(f) How could you test for the gas that came off?

6 Draw the electronic structures (dot and cross diagrams) of the following:
KBr   NaOH Li_2CO_3 $NaHCO_3$

7 Describe, with diagrams, how you would make a pure sample of sodium chloride in the laboratory from:
(a) some sea water
(b) rock salt
(c) hydrochloric acid.

8 Write equations for the reactions between:
(a) potassium and water
(b) sodium oxide and water
(c) sodium hydroxide and carbon dioxide
(d) sodium carbonate and hydrochloric acid.

9 Write down:
(a) two uses for metallic sodium
(b) uses for three sodium compounds.

10 (a) Explain what a flame test is.
(b) How would you distinguish between samples of sodium chloride and potassium chloride?

27 Group II: the alkaline earth metals

27.1 The elements

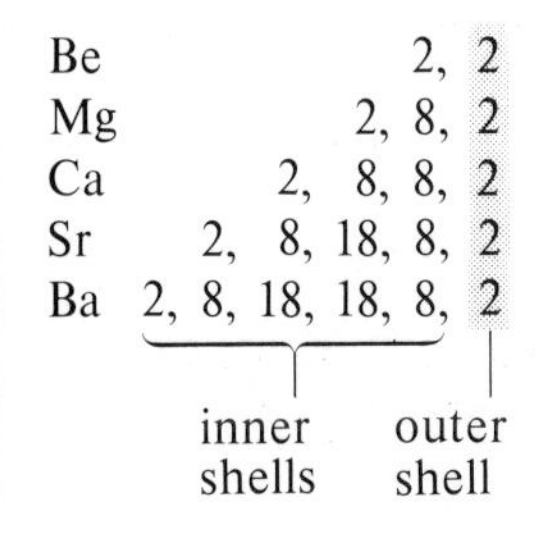

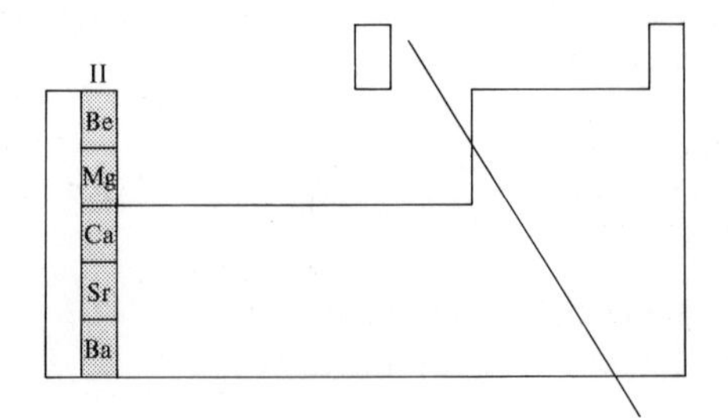

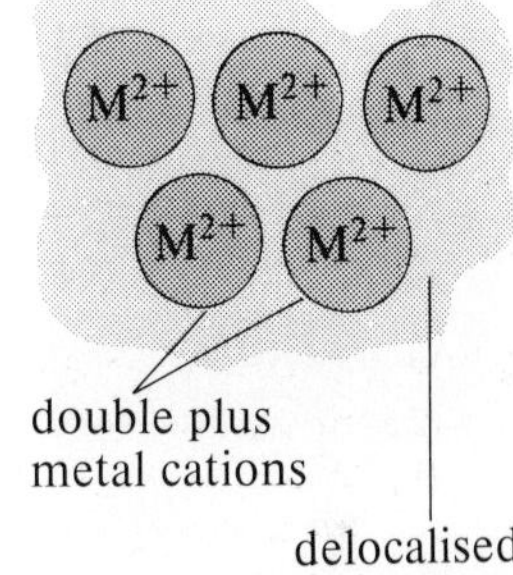

Appearance The more reactive elements, strontium and barium, must be stored under oil. The others are left in air, because a layer of oxide forms on them and protects them. They are all shiny and silvery-grey when freshly cut.

Physical properties The physical properties of the Group II elements are given in Table 27.1 below. You do not need to learn the values given here. Just get an idea of the trends.

Table 27.1

Element		*m.p./°C*	*b.p./°C*	*Density /g dm^{-3}*	*Hardness /Moh*	*Conductivity /Ω^{-1} cm^{-1}*
Beryllium	Be	1280	2500	1860	4·0	35700
Magnesium	Mg	650	1105	1750	2·0	25600
Calcium	Ca	850	1440	1550	1·5	29200
Strontium	Sr	770	1370	2600	1·8	4350
Barium	Ba	720	1600	3600	1·0	1700
Compare with a typical metal:						
Iron	Fe	1530	3000	7860	4–5	11300

Again the high conductivity values show that the Group II elements are metallic solids. But they are much lighter, and mostly softer, than our typical metal, iron.

27.2 Occurrence, extraction, and uses

Earth metals

Like the Group I metals, the Group II metals are very reactive. They are much too reactive to be found in an uncombined state in the Earth's crust. However, large amounts of their compounds are found in the earth. They are among the most common compounds making up the rocks of the Earth's crust. That is one of the reasons why the Group II elements are called **earth metals**.

Limestone cliffs

The map on the right shows the main areas of Great Britain where the rocks contain calcium and magnesium compounds. Do you live in one of these areas? If you do, you probably find that the water is **hard**—you need a lot of soap to make a lather when you wash. We will explain the reasons for this later on in the chapter.

The common calcium-containing rocks are **chalk** and **limestone**, which are both forms of calcium carbonate, $CaCO_3$. The most common magnesium-containing rock is **dolomite**, which is the unusual compound calcium magnesium carbonate, $CaMg(CO_3)_2$.

calcium-based rocks
magnesium-based rocks

Methods of extraction

Magnesium oxide and magnesium chloride are ionic. Magnesium metal is atomic. 'Extraction' means ions ⟶ atoms.

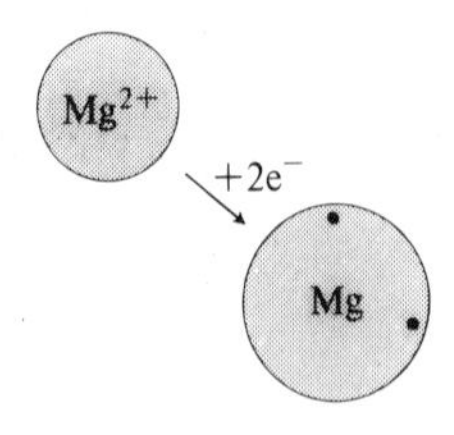

Of all the elements in Group II, magnesium is the most widely used as the metal. So we will look in detail at its extraction. You saw in Chapter 26, pages 281–283, how difficult it is to extract the reactive metal sodium from its compounds. Although magnesium is not as reactive as sodium, it still needs a lot of energy to extract it.
There are two common processes by which magnesium is produced:

magnesium chloride is decomposed using electrolysis;
magnesium oxide is reduced to magnesium by silicon, at very high temperatures.

These methods are described in Table 27.2.

Table 27.2 The two methods of extracting magnesium

Method	*(i) By electrolysis*	*(ii) By reduction with silicon*
The process	Magnesium ore is reacted with hydrochloric acid to make magnesium chloride. The solution is evaporated to give the solid salt. The salt is melted and electrolysed in a cell like the one shown on page 282. Magnesium and chlorine are produced.	Magnesium ore is roasted in a furnace to make magnesium oxide. The magnesium oxide is heated to high temperatures with silicon. Silica (silicon dioxide) and magnesium metal are produced.
Details	Raw material: magnesium hydroxide, hydrochloric acid Conditions: temperature 700°C, power, 6·6 volts and 40000 amps	Raw material: magnesium carbonate, silicon Conditions: temperature 700°C, very low pressure
Reactions	$Mg(OH)_2 + 2HCl \longrightarrow MgCl_2 + 2H_2O$ $MgCl_2 \xrightarrow{\text{electricity}} Mg + Cl_2$	$MgCO_3 \longrightarrow MgO + CO_2$ $2MgO + Si \longrightarrow 2Mg + SiO_2$

Calcium metal is extracted by electrolysing molten calcium chloride; calcium metal is produced at the cathode.

Uses Magnesium is a useful metal, because of its lightness. It is often alloyed with aluminium, for example, to make aircraft bodies.
Calcium metal is used to remove sulphur compounds from crude oil and to remove oxygen from molten metals. In both cases it is acting as a reducing agent. Several calcium compounds also have important uses.

(i) Calcium sulphate, found naturally as gypsum. It is the basis of the plaster used in the building industry. If it is heated gently, some of the water of crystallization is lost:

$$2CaSO_4{\cdot}2H_2O \xrightarrow{\text{heat}} (CaSO_4)_2{\cdot}H_2O + 3H_2O$$

The product, sometimes called plaster of Paris, reacts with water. It gets warm then sets hard.

(ii) Calcium hydroxide—lime—is used in agriculture to neutralize acid soils.

(iii) Calcium carbonate—limestone—is used to make cement by heating it with sand.

(iv) Calcium phosphate, when treated with sulphuric acid, forms the fertilizer calcium superphosphate, $Ca(H_2PO_4)_2$.

27.3 Group II chemistry is very like Group I chemistry

Both Groups I and II contain reactive metals. Their reactivity was first compared in Chapter 15 (see page 152) and again in Chapter 17 (see page 283). We found that:

> metal reactivity increases going down a group;
> metal reactivity decreases going across a period.

	I	II
2	Li	Be
3	Na	Mg
4	K	Ca

Down a group: Sodium is less reactive than potassium.

Across a period: Calcium is less reactive than potassium.

Conclusion: Sodium and calcium—diagonal to each other in the Periodic Table—have about equal reactivity.

Group II metals are slightly less reactive than the Group I metals of the same period. Those on a downward diagonal of the Periodic Table are, however, about equally reactive:

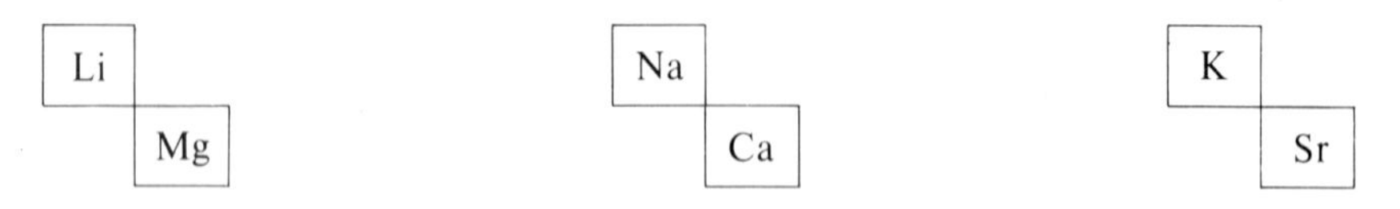

Group II properties are therefore much the same as those of Group I. We shall only outline the major similarities and differences in them.

Compound formation

The Group II metals combine with non-metals in much the same way as Group I metals do. Taking calcium as our example:

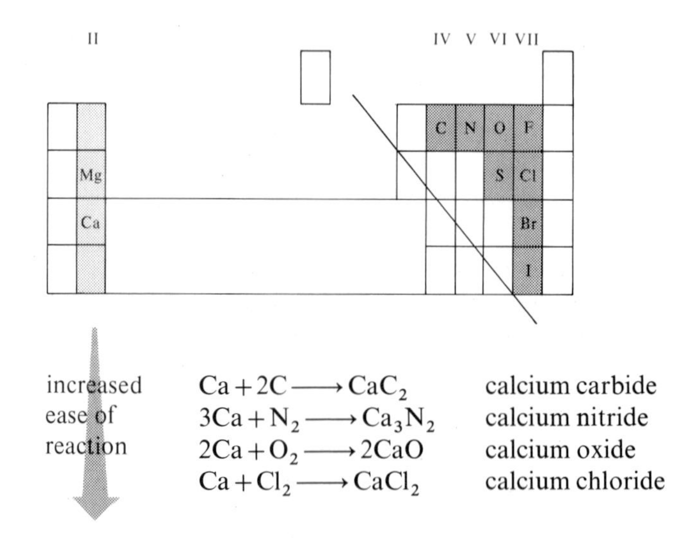

But there are three main differences between the compounds formed by the Group I and Group II metals.

Group II atoms become 2+ ions

Mg

Mg^{2+} $+2e^-$

(i) The oxidation number of the metal in the compounds
Group I metals always have the oxidation number +I in their compounds (Na^+ in NaCl).
Group II metals always have the oxidation number +II in their compounds (Ca^{2+} in $CaCl_2$).

(ii) Stability of the compounds to heat
Group I compounds are stable—they just melt—while some Group II compounds decompose. For example:

$$NaOH_{(s)} \xrightarrow{heat} NaOH_{(l)}$$

$$Na_2CO_{3\,(s)} \xrightarrow{heat} Na_2CO_{3\,(l)}$$

but $$Mg(OH)_{2\,(s)} \xrightarrow{heat} MgO_{(s)} + H_2O_{(l)}$$

$$MgCO_{3\,(s)} \xrightarrow{heat} MgO_{(s)} + CO_{2\,(g)}$$

(iii) Solubility of the compounds in water
Group I compounds are soluble, while some Group II compounds are only slightly soluble, or insoluble.

Solubility The solubilities of some different metal compounds are compared below.

Table 27.3 The solubilities of metal compounds from different groups (in water at 0°C)

Type of compound	*Group I e.g. sodium*	*Group II e.g. calcium*	*Other metals e.g. copper*
Hydroxide	Soluble NaOH: 42 g in 100 g	Slightly soluble $Ca(OH)_2$: 0·18 g in 100 g	Insoluble $Cu(OH)_2$: ~0
Carbonate	Soluble Na_2CO_3: 21·5 g in 100 g	Insoluble $CaCO_3$: ~0	Insoluble $CuCO_3$: ~0
Sulphate(VI)	Soluble Na_2SO_4: 4·8 g in 100 g	Insoluble $CaSO_4$: ~0	Soluble $CuSO_4$: 5·2 g in 100 g

In general we can say that the Group II metal compounds are less soluble than the Group I metal compounds but more soluble than most other metal compounds, with the exception of the sulphates.

We use the insolubility of Group II compounds in two common laboratory tests:

the lime water test for carbon dioxide gas;
the barium chloride test for metal sulphates.

Both the above tests are precipitation reactions (see page 124). They work because of the insolubility of calcium carbonate and of barium sulphate. They are described on page 296.

(i) Lime water test for carbon dioxide gas		*(ii) Barium chloride test for metal sulphates*	
1 Bubble the unknown gas into lime water.	2 A white suspension is produced if the gas is CO_2.	1 Add barium chloride, dissolved in hydrochloric acid, to the unknown solution.	2 A white suspension is produced if the solution contains a sulphate.
gas; calcium hydroxide solution (lime water)	white suspension given only if gas is CO_2		
$Ca(OH)_2 + CO_2 \longrightarrow CaCO_3\downarrow + H_2O$ base acid salt water		$BaCl_2 + MSO_4 \longrightarrow BaSO_4\downarrow + MCl_2$ $Ba^{2+}_{(aq)} + SO^{2-}_{4\,(aq)} \longrightarrow BaSO_{4\,(s)}$ M represents any metal.	

Redox chemistry The Group II metals act as reducing agents in just the same way as Group I metals do. For example, calcium reduces both chlorine and water:

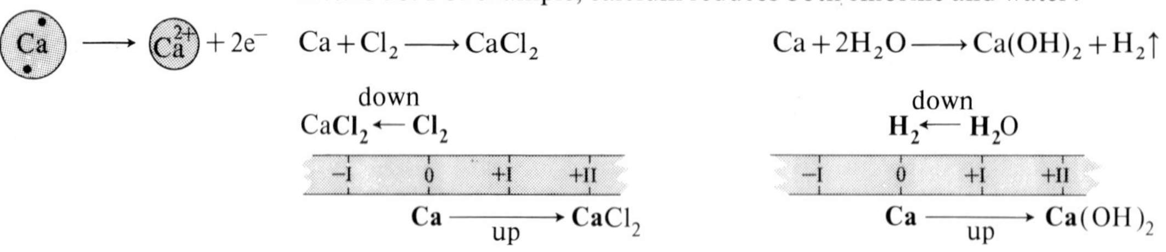

It is interesting to compare the reduction of water by sodium and by calcium:

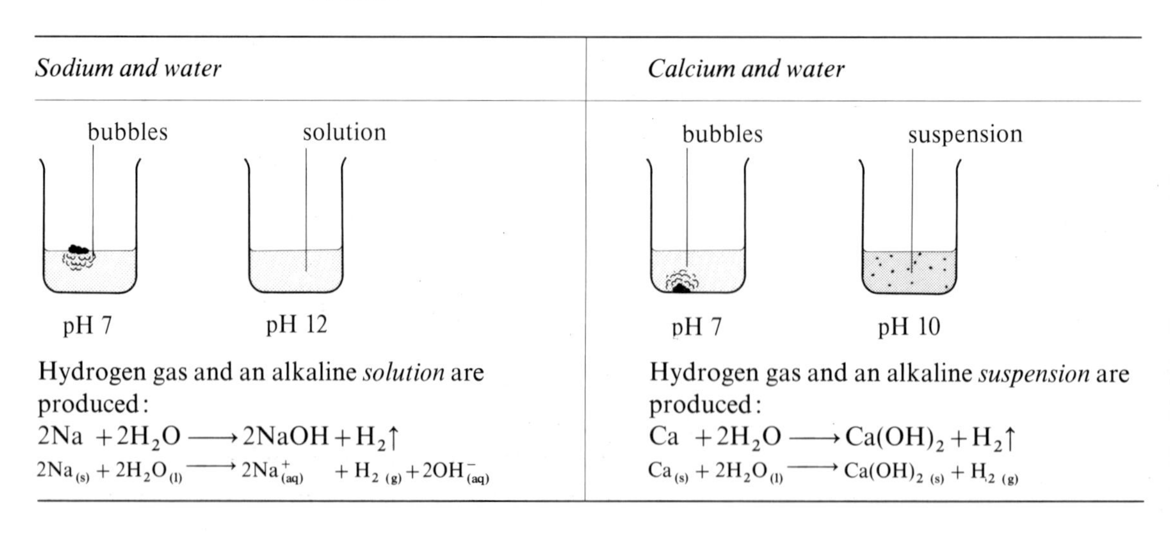

Hydrogen gas and an alkaline *solution* are produced:

$2Na + 2H_2O \longrightarrow 2NaOH + H_2\uparrow$

$2Na_{(s)} + 2H_2O_{(l)} \longrightarrow 2Na^+_{(aq)} + H_{2\,(g)} + 2OH^-_{(aq)}$

Hydrogen gas and an alkaline *suspension* are produced:

$Ca + 2H_2O \longrightarrow Ca(OH)_2 + H_2\uparrow$

$Ca_{(s)} + 2H_2O_{(l)} \longrightarrow Ca(OH)_{2\,(s)} + H_{2\,(g)}$

27.4 Special features

Hardness of water

In section 27.2 we saw that much of the rock in the Earth's crust is made of Group II compounds. You would expect rain-water to be able to dissolve the more soluble Group II compounds, like the chlorides and sulphates, from this rock.

Rain-water, however, is not just pure water. It is weakly acidic, because carbon dioxide from the air dissolves in it to produce a dilute solution of carbonic acid. We can demonstrate that in the laboratory by bubbling carbon dioxide into water in the presence of universal indicator:

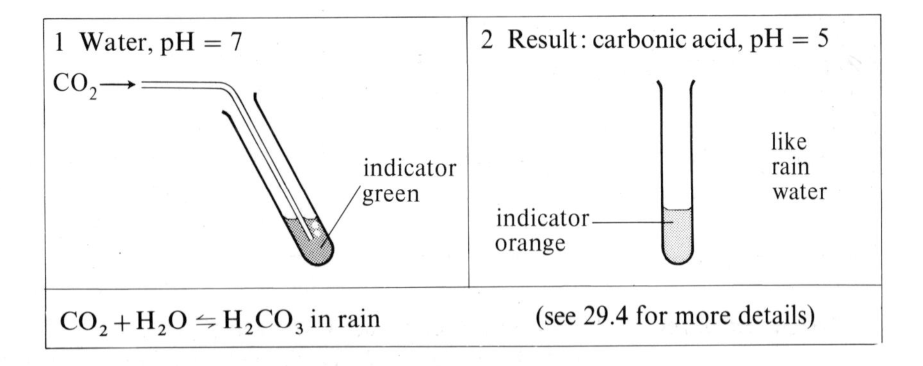

The acid nature of rain-water causes the Group II carbonates, which are normally almost insoluble, to dissolve from the rock as well. Soluble hydrogen carbonates are produced:

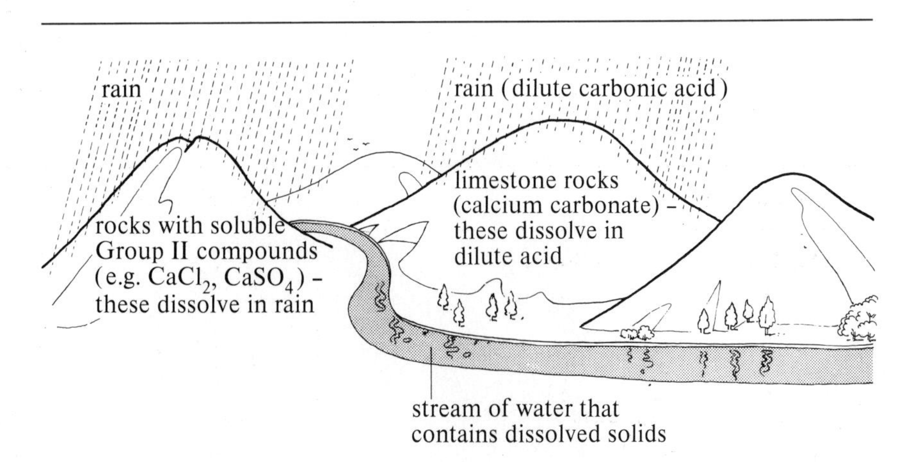

The reaction between insoluble calcium carbonate and carbonic acid is:

$$CaCO_3 + H_2CO_3 \longrightarrow Ca(HCO_3)_2$$

calcium hydrogen carbonate—soluble

River-water and water from reservoirs (see Chapter 12) may therefore contain a range of dissolved Group II compounds. The water is then said to be **hard**. This is because the solids reappear as hard, gritty deposits when the water is boiled. It is also hard to make soap lather in water of this sort. **Soft** water is water that does not contain dissolved solids. In soft water, it is easy to make soap lather.

Two types of hardness

There are two types of hardness in water, **temporary hardness** and **permanent hardness**. Temporary hardness can be cured simply by boiling the water. Permanent hardness cannot. Table 27.4 shows the cause of each type of hardness, and how to cure it.

Table 27.4

Temporary hardness	*Cause*	The soluble hydrogen carbonate salts, produced by the action of dilute carbonic acid on rock.
	Cure	Boil the water. Carbon dioxide is driven off. The hydrogen carbonates are decomposed to give deposits of insoluble carbonates. $Ca(HCO_3)_2 \xrightarrow{\text{boil}} CaCO_3\downarrow + CO_2\uparrow + H_2O$
Permanent hardness	*Cause*	Other dissolved solids in the water, like: $CaCl_2$, $MgCl_2$ $CaSO_4$, $MgSO_4$
	Cure	(i) Evaporate the water and then condense it back to a liquid again—distil it. (ii) Use a 'water softener', for example Zeolite, which removes the salts from the solution.

Hard water is not an entirely bad thing, but it does have some disadvantages.

Left: stalactites in an underground cave. These are formed when dripping water evaporates, leaving behind the dissolved calcium salts
Right: a domestic water softener

Disadvantages of hard water	*Advantages of hard water*
(i) Deposits of scale and 'fur' form in pipes, boilers, and kettles, as a result of heating hard water.	(i) It is healthier to drink hard water. There is more heart disease in soft water areas.
(ii) An insoluble scum is produced with soap.	(ii) Better beer is brewed in hard water areas!

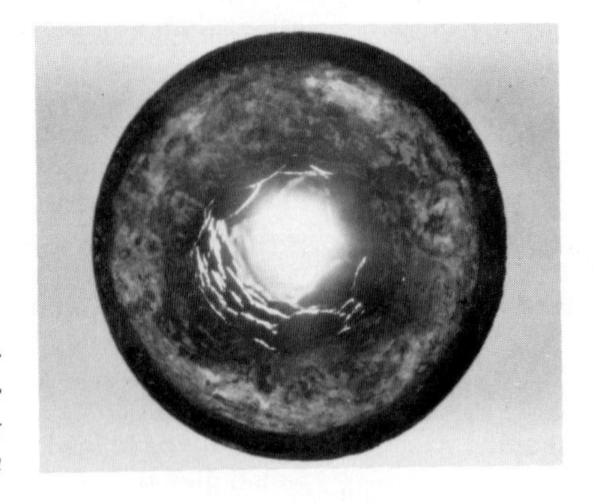

The 'fur' from hard water nearly blocking a pipe from a central heating system

You can tell how hard a sample of water is by adding drops of soap solution to it. Count the number of drops needed to be added to $100\,cm^3$ of water before a permanent lather is produced on shaking. The more drops you need, the harder the water.

Reaction summary for calcium

This summary deals only with calcium and its compounds, because calcium is the most common element in Group II and is typical for Group II.

Extraction Electrolysis of molten calcium chloride: $Ca^{2+} + 2e^- \longrightarrow Ca$ at the cathode

Reactions of the element

(i) Reducing other elements:

$$Ca + Cl_2 \longrightarrow CaCl_2$$
$$2Ca + O_2 \longrightarrow 2CaO$$
$$3Ca + N_2 \longrightarrow Ca_3N_2$$
$$Ca + 2C \longrightarrow CaC_2$$

(ii) Reducing compounds: $Ca + 2H_2O \longrightarrow Ca(OH)_2$ (alkaline suspension)

The hydroxide Slightly soluble in water, giving a solution called lime water:

$$Ca(OH)_{2\,(s)} + H_2O_{(l)} \longrightarrow Ca^{2+}_{(aq)} + 2OH^-_{(aq)}$$

The solid is unstable when heated: $Ca(OH)_{2\,(s)} \xrightarrow{heat} CaO_{(s)} + H_2O_{(g)}$

The carbonate Insoluble in water. The carbonate in rocks dissolves very slowly in rain:

$$CaCO_{3\,(s)} + H_2CO_3 \longrightarrow Ca(HCO_3)_2$$

It is unstable when heated: $CaCO_{3\,(s)} \longrightarrow CaO_{(s)} + CO_{2\,(g)}$

Flame test Calcium compounds give a brick-red colour, when held in the edge of a bunsen flame.

Questions

1 (a) Draw diagrams showing the electronic structures of the magnesium and calcium atoms.
(b) Explain why the elements magnesium and calcium are in Group II.
(c) What happens to their atoms when these two elements react, and what particles are formed?
(d) Explain why these two metals have a valency of two.

2 (a) Describe what you would see if a piece of calcium was dropped into a beaker of water containing universal indicator.
(b) Write an equation for the reactions taking place.
(c) In what two ways is this reaction very like the one between sodium and water?
(d) In what ways is this reaction different from the one between sodium and water?

3 Magnesium metal reacts only very slowly in cold water, but it does react well in steam.
(a) Draw the apparatus that you would use to react magnesium with steam.
(b) Describe what you would see during this reaction.
(c) Write an equation for this reaction.
(d) How would you test the gas which is produced in this reaction?
(e) If you part the solid product formed from this reaction into water, would it be acidic, neutral, or alkaline?

4 Group II compounds differ from Group I compounds in two important ways. Group II compounds are less soluble than the corresponding Group I compound, and the Group II compounds are less stable to heat.
(a) Illustrate the difference in solubility between the compounds of the two groups with some actual examples.
(b) Illustrate the difference in thermal stability between the two groups with some actual examples.

5 (a) What is meant by the term *hard water*?
(b) What are the advantages of hard water?
(c) What are the disadvantages of hard water?
(d) Explain the difference between temporary and permanent hardness in water.
(e) Explain how water becomes temporarily hard. Give equations.
(f) Explain how water becomes permanently hard.
(g) Explain how hardness might be removed.

6 (a) Calcium carbonate is very insoluble in pure water. Why is it more soluble in rain-water?
(b) The insolubility of calcium carbonate is used in a common laboratory test. Which test is it?
(c) Give the equation for the reaction that takes place if the test is positive.
(d) What happens if the test is carried on for too long?

7 Which of the following mixtures in solution would give precipitation reactions? Name the insoluble product in each case.
(a) $NaOH + KCl$
(b) $NaOH + CaCl_2$
(c) $Na_2CO_3 + CaCl_2$
(d) $NaCl + Mg(NO_3)_2$
(e) $Na_2SO_4 + BaCl_2$

8 You are asked to make a pure, hydrated sample of the salt magnesium chloride.
(a) What would you choose as reactants? Why?
(b) Write an equation for the reaction.
(c) Draw a series of diagrams to show clearly the practical steps involved.
(d) How would you dry the sample? Explain.

9 Ca CaO $CaCO_3$ $CaCl_2$ $MgSO_4$
Look at the substances listed above. Pick out the substances that are:
(a) bases (b) salts (c) reducing agents
(d) insoluble in water

28 Some industrially important metals

28.1 The elements

A typical metallic lattice:

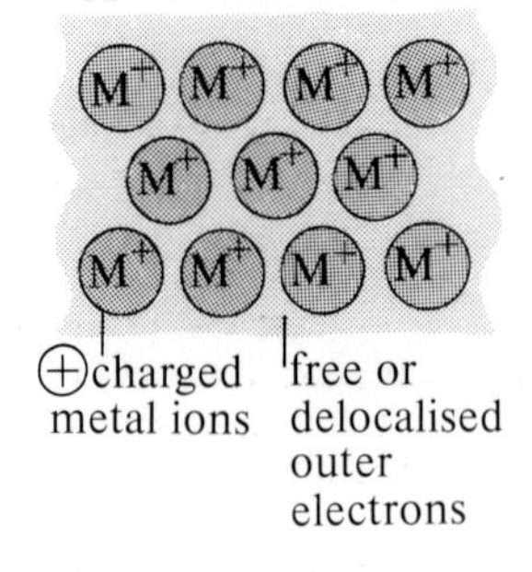

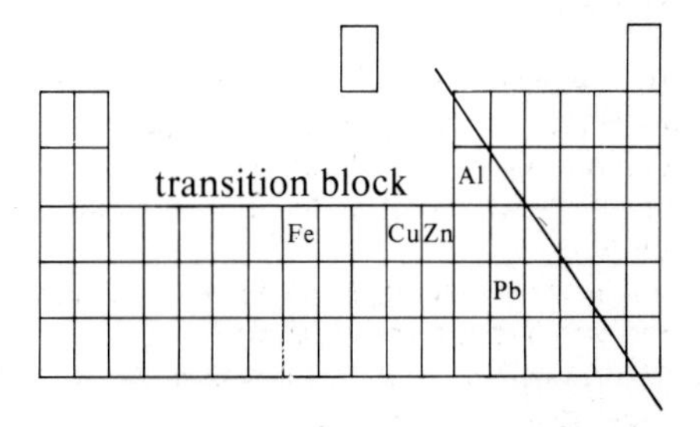

Appearance Aluminium, iron, copper, zinc, and lead are all metals that have a typically metallic appearance, and tend to tarnish in the air. They are used widely in industry.

Physical properties Their physical properties are shown in Table 28.1. You do not need to learn the values given here, but you should get an idea of the trends in their properties.

Table 28.1 The physical properties of some industrially important metals

Metal	*m.p./°C*	*b.p./°C*	*Density /g dm^{-3}*	*Conductivity /Ω^{-1} cm^{-1}*	*Hardness /Moh*	*Colour*
Aluminium	660	2057	2702	40800	2–3	Silver-white
Iron	1530	3000	7860	11200	4–5	Silver-grey
Copper	1083	2336	8920	64100	2·5–3	Red-brown
Zinc	419	907	7140	18200	2·5	Silver-grey
Lead	327	1620	11340	5300	1·5	Silver-grey

All these metals are **ductile**—this means they can be drawn into wires—and **malleable**—this means they can be hammered into shape without breaking. These qualities partly account for their usefulness.

28.2 Occurrence and extraction

Occurrence

These metals are examples of finite or non-renewable resources. There are only certain amounts of the ores containing these metals in the earth's crust, and only some of the ore can be easily mined.

Calculations have been done to work out how long the sources of these metals will last if we go on using them at the present rates. The results are quite worrying:

Metal	*No. of years supply left*
aluminium	100
iron	240
copper	36
zinc	23
lead	26

In fact the situation might be even worse than this because the demand for all these metals is increasing.

There are two main ways to tackle this problem. The first is to recycle more of the scrap metal, which is now thrown away. The second is to find other substances to use instead of the metals.

Apart from their industrial uses, these metals all have biological importance in different ways. Iron, zinc, and copper are certainly essential for human life, and aluminium may be, because they are found in the body. Lead should not be found in the body because it is highly toxic. The most common cause of lead poisoning used to be from drinking water that had run through lead pipes. Pipes are now made of copper. The main cause of lead poisoning now comes from breathing in air polluted with lead. This lead gets into the air from petrol, which has lead compounds in it to help it burn smoothly. Leaded petrol has been outlawed in America and some European countries.

Metal	*Amount in man (parts per million)*	*Function*
iron	50	in haemoglobin
copper	4	in enzymes controlling oxidation
zinc	25	in enzymes controlling healing
aluminium	0·5	not known
lead	toxic at 0·8	damages brain and nervous system

Extraction

You saw on page 157 that the metals aluminium, iron, copper, zinc, and lead are not as reactive as the Groups I and II metals. You may remember that they came lower down in the electrochemical series.

K

size of atoms increases up the list

reactivity of metals increases up the list

Cu

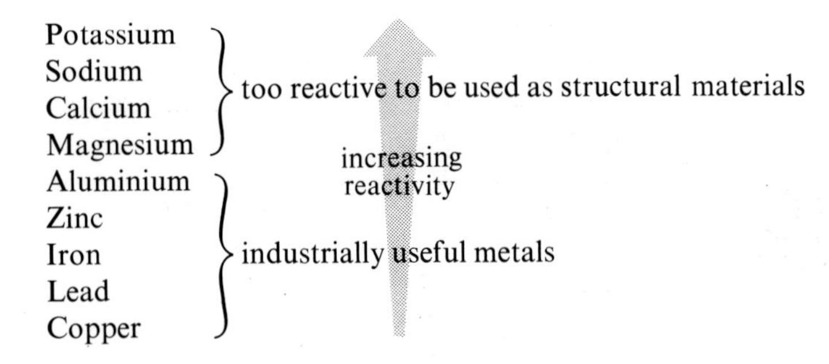

Even though the industrially useful metals are less reactive, we still do not find them in a pure, uncombined state in the earth's crust. Only very small amounts of the least reactive metal, copper, are found pure. The rest are always found as ores. Some common ores of the metals are shown in Table 28.2

Table 28.2

Metal	*Common ore*	*Chemical name*	*Formula*
Aluminium	Bauxite	Aluminium oxide	Al_2O_3
Iron	Haematite	Iron(III) oxide	Fe_2O_3
Copper	Copper glance	Copper(I) sulphide	Cu_2S
Zinc	Zinc blende	Zinc(II) sulphide	ZnS
Lead	Galena	Lead(II) sulphide	PbS

Methods of extraction

A metal ore contains a metal combined with non-metal. The oxidation number of the metal in its ore is always positive:

$$\overset{+III}{Al}{}_2O_3 \qquad \overset{+II}{Zn}S$$

Extraction means making the pure metal from its ore. Pure metals always have oxidation number zero, so extraction is a process of *reduction*:

Metal ores are ionic. Pure metals are atomic. 'Extraction' means ions ⟶ atoms

$$\overset{0}{Al} \qquad \overset{0}{Zn}$$

0 +I +II +III

Al ⟵ down ⟵ Al_2O_3

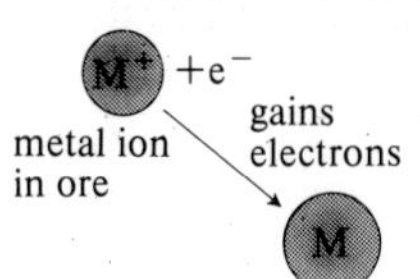

The method of extracting a metal depends on its reactivity, as shown in Table 28.3.

Electrolysis is the most powerful form of reduction. But it is also the most expensive. It is not necessary to use electrolysis to extract the less reactive metals from their ores. The reducing agent, carbon, can be used. And copper ores can be decomposed to copper just by heating.

Table 28.3 The method of extraction depends on reactivity

Electrochemical series	*Method of extraction from ores*
Potassium Sodium Calcium Magnesium Aluminium	Reduction using electrolysis
Zinc Iron Lead	Reduction by heating with carbon
Copper	Decomposition by heating alone

K
size
Cu
difficulty in extracting

Extraction of aluminium

The process:

The aluminium ore, bauxite, is first purified. This is done by dissolving it in sodium hydroxide solution and filtering out the solid impurities. The dissolved bauxite is then precipitated back out of solution. Next, a special impurity called cryolite (Na_3AlF_6) is added to the 'pure' bauxite. This lowers the melting point of the ore, but does not affect the extraction. The molten ore is electrolysed in a carbon-lined cell, called the Hall cell. Aluminium is produced as a liquid at the cathode. It is 99·5–99·8 % pure.

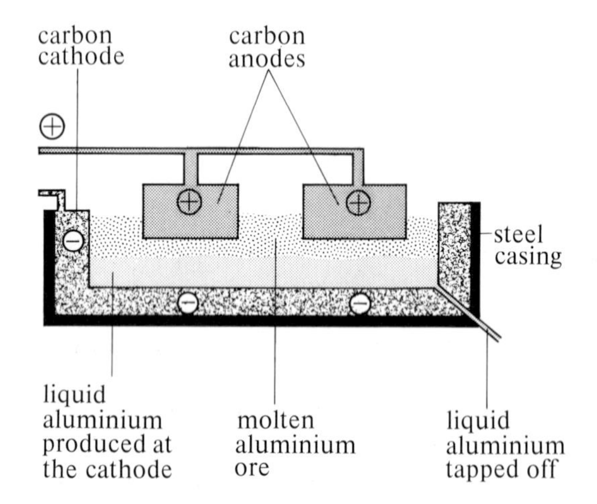

Details

Raw materials: bauxite (Al_2O_3)
cryolite (Na_3AlF_6)
Conditions: temperature, 950°C
power, 4–5 volts and 100000 amps

Reactions

Cathode: $Al^{3+} + 3e^- \longrightarrow Al$

Anode: $2O^{2-} \longrightarrow O_2 + 4e^-$

Overall: $2Al_2O_{3\,(l)} \xrightarrow{\text{electricity}} 4Al_{(l)} + 3O_{2\,(g)}$

Important points

1 Pure aluminium oxide melts at a temperature of 2045°C. It would be impossible to work the cell at this temperature. However, adding the impurity lowers the m.p. to 950°C.
2 Oxygen is produced at the carbon anode. At the high temperature, it combines with the anode which burns away.

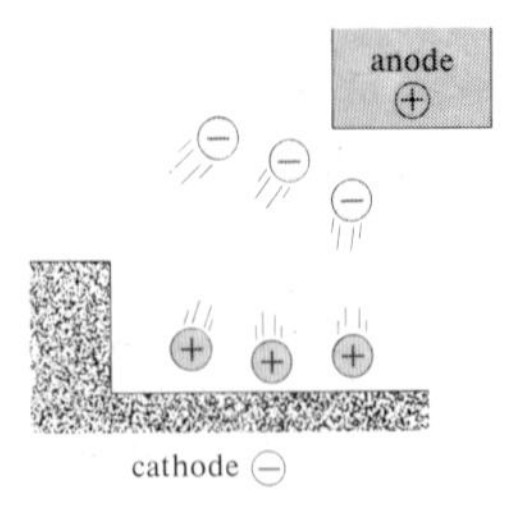

See page 226 for some details of the Lynemouth plant.

Extraction of iron: the blast furnace

The process:

Iron ore—mostly haematite—is crushed and mixed with coke and limestone. This mixture is called the **charge**.

Ore	Fe_2O_3	make up the charge
Coke	C	
Limestone	$CaCO_3$	

The charge is loaded into the top of a tall furnace, called a **blast furnace**. Hot air—the blast—is pumped into the lower part of the furnace. The ore is reduced to iron as the charge falls through the furnace. A waste material called slag is formed at the same time. The slag floats on the surface of the liquid iron produced. Each layer can be tapped off separately.

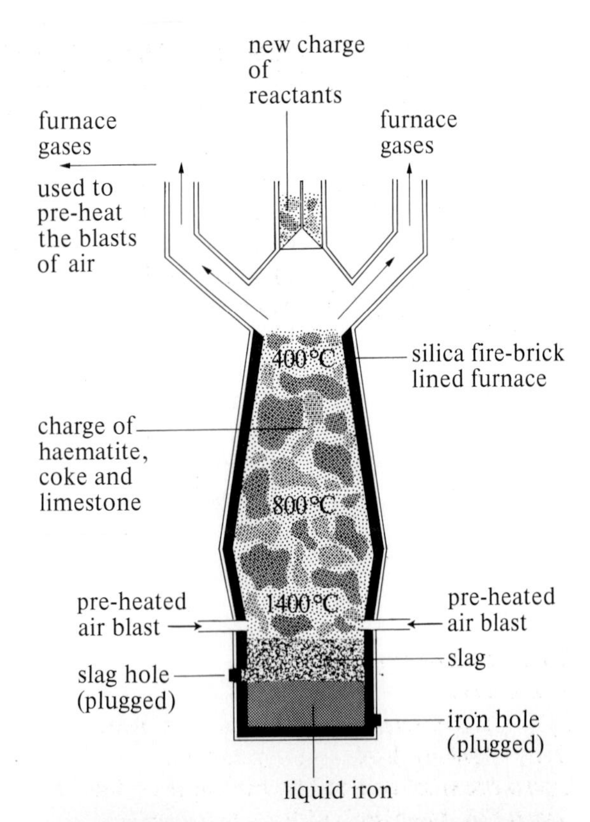

Details

Raw materials: iron ore, coke, limestone, air

Conditions: temperature at bottom of furnace, 1400°C
at top of furnace, 400°C

Reactions

1 Coke burns in the blast: $C + O_2 \longrightarrow CO_2$
2 Limestone decomposes:
$CaCO_3 \longrightarrow CaO + CO_2$
3 The carbon dioxide produced reacts with more coke: $CO_2 + C \longrightarrow 2CO$
4 The iron ore is *reduced* by carbon monoxide gas: $Fe_2O_3 + 3CO \longrightarrow 2Fe + 3CO_2$
5 Earthy impurities in the ore (mainly silica) react with calcium oxide to form a slag:
$CaO + SiO_2 \longrightarrow \underset{\text{slag}}{CaSiO_3}$

Important points

1 The iron produced is impure. It is called pig-iron or cast-iron. The major impurities in it are non-metals: carbon, phosphorus, and sulphur.
2 Slag is used to make breeze blocks and road foundations.

The top of a large blast furnace at Llanwern in Wales. The main part of the furnace is in the shed below. The pipes remove the furnace gases.

Recycling scrap iron

On page 302 we saw that iron ore is a finite resource. Even though the reserves of iron ore, 240 years at present rates of use, are larger than for many other metals, it is important to think of ways of making them last as long as possible for our descendants. In the case of iron, the main way of doing this is to collect and recycle much of the scrap iron lying around. There are three main types of scrap. The first kind is called **capital scrap**, and this is all the rusty old iron and steel objects that have finished their useful lives and are cluttering up the countryside. The second type of scrap is called **circulating scrap** and this is made in the iron and steel works itself. Up to 27% of the iron made in the plant is scrap, and is recycled. The last type of scrap is called **process scrap** and is produced in industry. The car industry buys in steel to make cars, and 30% of this metal becomes scrap. The last two kinds of scrap are best for steel making, because they do not contain such a wide range of impurities as the capital scrap that comes from all sorts of places.

Greater and greater efforts are being made to collect and recycle scrap metal of all kinds. Recycling scrap metal has other advantages apart from just reducing the amount of ore that has to be mined.

Recycling means that less energy is needed to make new metal. The ore and coke do not have to be transported, and less coke is needed because the scrap is richer in metal than ore, having already been reduced.

Less pollution is produced because less coke is burnt, and less fuel is used in transport. This means less sulphur and nitrogen oxides, and dust in the atmosphere.

Finally, collecting and recycling scrap means that there is less metal rubbish lying around the landscape.

So as much scrap iron as possible is added to the iron from the blast furnace, and the mixture is converted into steel.

Alloys A pure metal often does not have the exact properties that are needed for a particular job. For example, it is necessary sometimes to use a metal that is both hard *and* a good conductor of electricity. Look at Table 28.1 again. Here you can see that:

copper conducts well but is soft,
zinc is harder but does not conduct very well.

By melting the right mixture of copper and zinc, a liquid mixture of the two metals is obtained. When this cools to a solid, it has both the properties that are needed. Such a mixture is an **alloy**. This one is called brass.
Alloys are solid solutions of metals.

Steels 'Steels' are the most widely used alloys. They are based on iron. The iron is converted to steel in a **converter**, which is like a huge bucket.

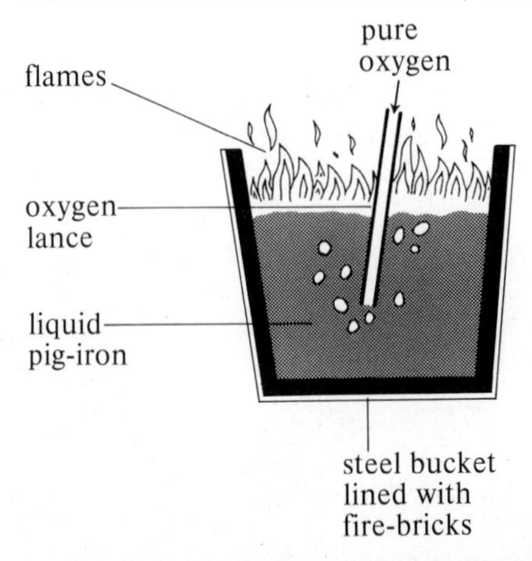

Oxygen is pumped through a pipe into the molten pig-iron. The non-metal impurities all react more readily with the oxygen than iron does. They burn fiercely at the surface of the liquid. You can tell when they have all burnt up by the colour of the flame.
When burning is finished, the oxygen pipe is taken out. Then small amounts of 'alloying' elements are added to make the steel.
See Table 28.4 for three different steels.

A 300–tonne converter at the British Steel works at Scunthorpe being loaded with iron for making into steel

Table 28.4 Composition of some alloys

Alloy	*Particular properties*	*Contains*	%
Stainless steel	Resists corrosion	Iron	74
		Chromium	18
		Nickel	8
High speed steel	Very strong, even at high temperatures	Iron	75·7
		Tungsten	18
		Chromium	6
		Vanadium	0·3
Silicon steel	Better conducting properties	Iron	97·6
		Silicon	2
		Carbon	0·4
Brass	Easily worked but resistant to corrosion	Copper	60–70
		Zinc	30–40
Bronze	Harder and more resistant to wear than brass	Copper	70–90
		Tin	10–30
Magnalium	Light but strong	Aluminium	90
		Magnesium	10
Solder	Low m.p. but forming a fairly strong solid	Lead	50
		Tin	50

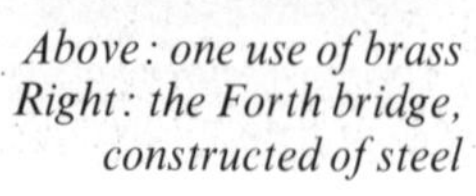

Above: one use of brass
Right: the Forth bridge, constructed of steel

28.3 Uses of the metals

Metal	*Use*	*Reason*
Aluminium	Structural material for ships, planes, cars, saucepans Electric cables	Strong but light; oxide layer prevents corrosion. Light, but good conductor.
Zinc	Coating (galvanising) steel Alloys: brass (Zn/Cu) bronze (Zn/Sn/Cu)	Reactive—gives sacrificial protection to iron; does not corrode easily. Modifies the properties of the other elements.
Iron	Structural material for all industries (in the form of steel).	Strong and cheap; properties can be made suitable by alloying.
Lead	Roofing Car batteries Solder (Pb/Sn alloy)	Very malleable and does not corrode. Design of battery makes recharging possible. Low m.p.
Copper	Electric cables Pipes Alloys (see above) Coins (alloyed with nickel)	Very good conductor. Very ductile, does not corrode easily. A traditional metal for coins.

Concorde: one use of aluminium

Left: zinc-coated (galvanised) dustbins
Right: a copper hot-water tank

28.4 Compound formation

Variable valency

You saw in Chapters 26 and 27 that the metals in Groups I and II always have just one valency.

Group I: the valency is 1 ($NaCl$, Na_2O)
Group II: the valency is 2 ($MgCl_2$, MgO)

With the exception of aluminium and zinc, the industrially important metals each have *two* different valencies.

Table 28.5 The valencies of the industrially important metals

Metal	*Some typical compounds*	*Metal valency*
Aluminium	$AlCl_3$, $Al(OH)_3$	Always 3
Zinc	$ZnCl_2$, $Zn(OH)_2$	Always 2
Iron	$FeCl_2$, $Fe(OH)_2$	2
	$FeCl_3$, $Fe(OH)_3$	3
Lead	$PbCl_2$, PbO	2
	$PbCl_4$, PbO_2	4
Copper	$CuCl$, Cu_2O	1
	$CuCl_2$, CuO	2

Non-metal compounds

In several earlier chapters, you became used to the idea that metals react with non-metals. Oxygen and chlorine are the most common non-metals. So we shall discuss the properties of the more important metal oxides and chlorides.

Oxides

Aluminium oxide

Aluminium is quite high in the electrochemical series. Despite this, it is not as reactive towards non-metals, acids or water as might be expected. This is because the metal surface is usually protected by a very thin layer of aluminium oxide. A new aluminium surface reacts very rapidly with the oxygen in the air:

$$4Al + 3O_2 \longrightarrow 2Al_2O_3$$

The oxide layer prevents other substances from reaching the metal. If the layer is scraped off, a new layer forms straight away to protect the metal again. Sometimes the layer is deliberately made thicker. This is done by making the metal the anode in an electrolytic cell. The electrolyte (e.g. dilute sulphuric acid) reacts to produce oxygen at the aluminium anode. The oxygen immediately reacts with the aluminium. This process is called **anodising** the aluminium. Anodised aluminium resists corrosion very well and is used particularly in the fittings on yachts.

Rust flakes, × 9400

Iron and rust

Iron oxide forms if iron is heated in air. A much more important reaction takes place if iron is left in moist air at room temperature:

$$4Fe + \underbrace{3O_2 + 6H_2O}_{\text{moist air}} \longrightarrow \underbrace{4Fe(OH)_3}_{\text{rust}}$$

The product, iron(III) hydroxide, is a red-brown substance which is the basis of **rust**. Some simple experiments using iron nails can show the best conditions for rusting.

Experiment 1: Test tubes are set up, containing nails in dry air, air-free water, and ordinary water, as shown in the diagrams below.

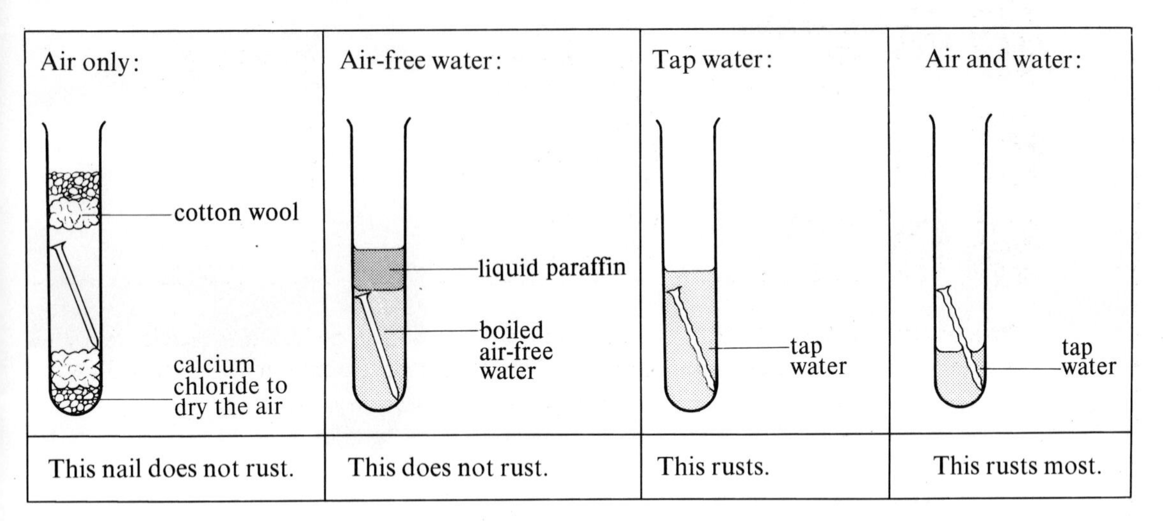

Experiment 2: This time, only tap water in small beakers is used.

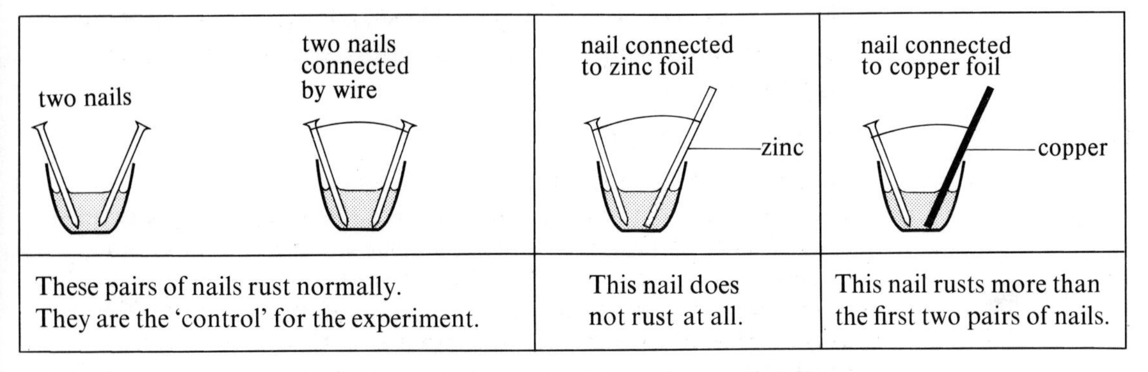

Using this experiment you can show that iron will not rust if it is connected to a more reactive metal like zinc or magnesium. But it will rust very badly if connected to a less reactive metal like copper or chromium.

The rust reaction wastes millions of pounds every year. Preventing rusting is therefore a major problem. Here are some of the ways to tackle it:

(i) Coat the metal with grease or paint, so that no air can reach the surface.

(ii) Coat the metal with an unreactive metal. Tin is used on cans, chromium on car parts.

(iii) Coat the metal with a more reactive metal. In **galvanising**, iron is coated with zinc. For the same reasons, zinc blocks are attached to ships. The more reactive zinc protects the iron by reacting with the air instead. This is sometimes called **sacrificial protection**.

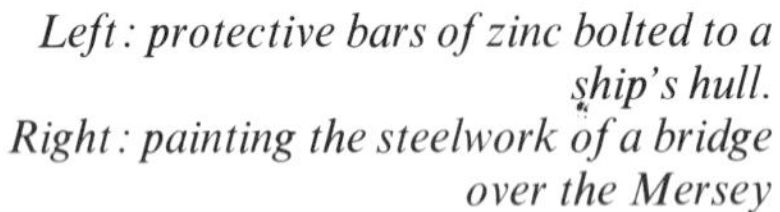

Left: protective bars of zinc bolted to a ship's hull.
Right: painting the steelwork of a bridge over the Mersey

Chlorides All the metals form chlorides if they are heated in a stream of chlorine gas.

Iron can form two chlorides, $\overset{II}{Fe}Cl_2$ and $\overset{III}{Fe}Cl_3$. It is iron(III) chloride that is produced in the reaction of iron with chlorine:

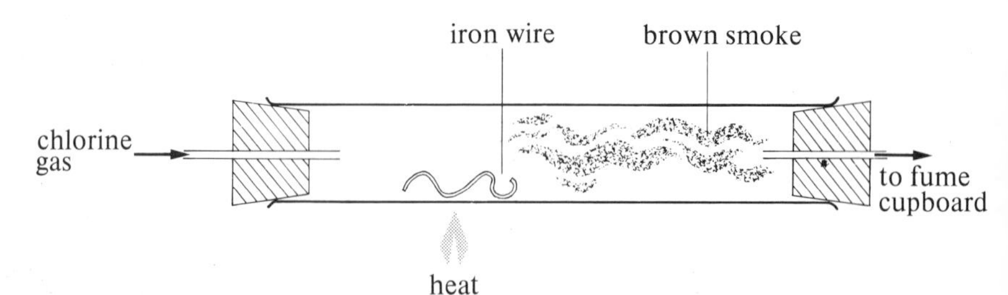

$2Fe_{(s)} + 3Cl_{2\,(g)} \longrightarrow 2FeCl_{3\,(s)}$

We can make iron(II) chloride by using hydrogen chloride gas instead of chlorine gas.

$Fe_{(s)} + 2HCl_{(g)} \longrightarrow FeCl_{2\,(s)} + H_{2\,(g)}$

28.5 Solubility and acid/base chemistry

Metal salts Solutions of the salts of these metals can be made easily. Simply react any of the following substances with dilute acid:

the metals (except copper—too unreactive)
the metal oxides
the metal carbonates

For example

(i) Iron reacts with sulphuric acid to produce the salt iron(II) sulphate:

metal acid salt hydrogen

$$Fe + H_2SO_4 \longrightarrow FeSO_4 + H_2\uparrow$$

$$Fe_{(s)} + 2H_3O^+_{(aq)} \longrightarrow Fe^{2+}_{(aq)} + H_{2\,(g)} + 2H_2O_{(l)}$$

(ii) Copper(II) oxide neutralises sulphuric acid to make copper(II) sulphate:

base acid salt water

$$CuO + H_2SO_4 \longrightarrow CuSO_4 + H_2O$$

$$CuO_{(s)} + 2H_3O^+_{(aq)} \longrightarrow Cu^{2+}_{(aq)} + 3H_2O_{(l)}$$

(iii) Zinc carbonate neutralises sulphuric acid to make zinc sulphate:

carbonate acid salt water carbon dioxide

$$ZnCO_3 + H_2SO_4 \longrightarrow ZnSO_4 + H_2O + CO_2\uparrow$$

$$ZnCO_{3\,(s)} + 2H_3O^+_{(aq)} \longrightarrow Zn^{2+}_{(aq)} + 3H_2O_{(l)} + CO_{2\,(g)}$$

All these salts are formed in the solution. We can evaporate off the water and obtain crystals. To make chloride or nitrate salts, you would use hydrochloric or nitric acid instead of sulphuric acid, in the above reactions.

Precipitation of insoluble compounds Some of the compounds of these metals are insoluble in water. This means that we can make them by precipitation reactions. (See page 124.) Important insoluble compounds include

the metal hydroxides
the metal carbonates
lead halides and lead sulphate.

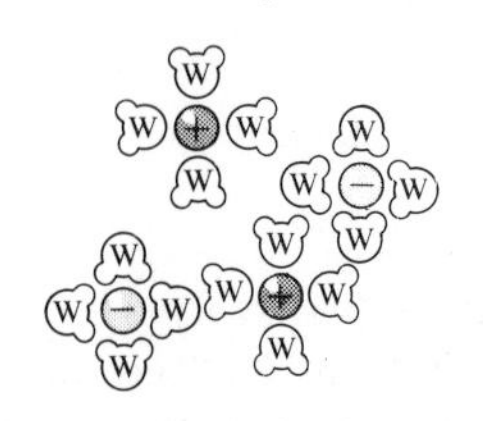

ions in solution move together

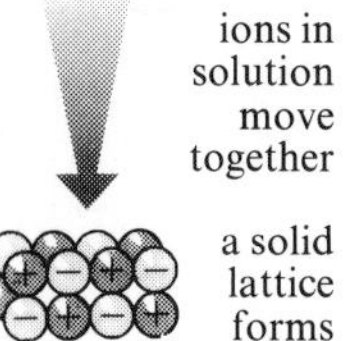

a solid lattice forms

Precipitation is the opposite of dissolving. For example:

$$Al^{3+}_{(aq)} + 3OH^-_{(aq)} \longrightarrow Al(OH)_{3(s)}$$

$$Cu^{2+}_{(aq)} + 2OH^-_{(aq)} \longrightarrow Cu(OH)_{2(s)}$$

(i) The metal hydroxides

The hydroxides of these metals can be made by precipitation. Add a few drops of sodium hydroxide solution to *any* of their soluble salts in solution.

Table 28.6 Making the metal hydroxides by precipitation

Adding sodium hydroxide to metal nitrates in solution

1 Add a few drops of NaOH.	2 A jelly-like solid forms.
metal salt solution	insoluble metal hydroxide

Reaction	Precipitate colour
$Al(NO_3)_3 + 3NaOH \longrightarrow 3NaNO_3 + Al(OH)_3\downarrow$	white
$Zn(NO_3)_2 + 2NaOH \longrightarrow 2NaNO_3 + Zn(OH)_2\downarrow$	white
$Pb(NO_3)_2 + 2NaOH \longrightarrow 2NaNO_3 + Pb(OH)_2\downarrow$	white
$Fe(NO_3)_2 + 2NaOH \longrightarrow 2NaNO_3 + Fe(OH)_2\downarrow$	green
$Fe(NO_3)_3 + 3NaOH \longrightarrow 3NaNO_3 + Fe(OH)_3\downarrow$	rust-brown
$Cu(NO_3)_2 + 2NaOH \longrightarrow 2NaNO_3 + Cu(OH)_2\downarrow$	royal blue

The colour of the hydroxide precipitate can sometimes tell us the type of metal salt present in an unknown solution. For example, if a green precipitate forms on adding sodium hydroxide, the metal salt present is probably an iron(II) salt. If a rust-brown precipitate forms, it is probably an iron(III) salt.

All metal hydroxides are bases; they therefore dissolve in acids, e.g.

base	acid		salt	water	
$Zn(OH)_2$	$+ 2HCl$	$\longrightarrow$	$ZnCl_2$	$+ 2H_2O$	(1)

$$Zn(OH)_{2\,(s)} + 2H_3O^+_{(aq)} \longrightarrow Zn^{2+}_{(aq)} + 4H_2O_{(l)}$$

But the white hydroxides of aluminium, zinc, and lead also dissolve in *alkali*. In other words, after being made, as shown in Table 28.6, the precipitates dissolve again if extra sodium hydroxide is added.

acid	base		salt	water	
$Zn(OH)_2$	$+ 2NaOH$	$\longrightarrow$	Na_2ZnO_2	$+ 2H_2O$	(2)

$$Zn(OH)_{2\,(s)} + 2OH^-_{(aq)} \longrightarrow Zn(OH)_4^{2-}{}_{(aq)}$$

Notice that zinc hydroxide is acting as a *base* in reaction (1); but it is acting as an *acid* in reaction (2)! A substance that behaves in this way is called an **ampholyte**. Its behaviour is said to be **amphoteric**. We can understand ampholytes by comparing their acid-base properties with a temperature scale.

Temperature scale	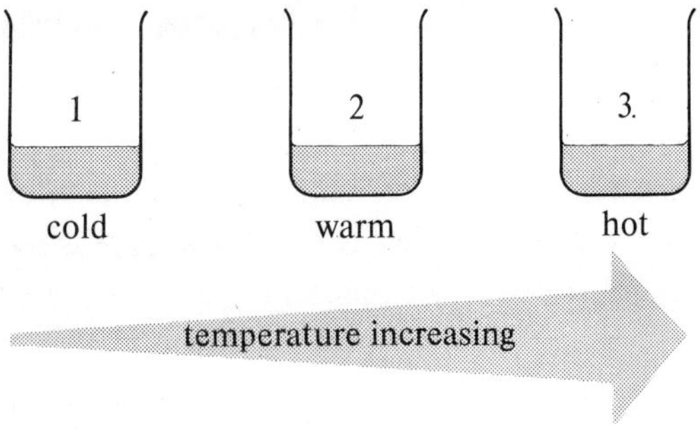 1 cold, 2 warm, 3. hot — temperature increasing → Beaker 2 is hot compared with beaker 1. Beaker 2 is cold compared with beaker 3.
Acid-base scale	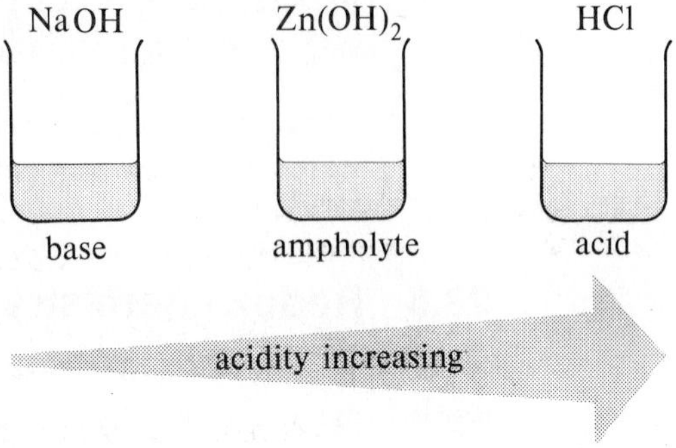 NaOH base, $Zn(OH)_2$ ampholyte, HCl acid — acidity increasing → $Zn(OH)_2$ is acidic compared with NaOH. $Zn(OH)_2$ is basic compared with HCl.

To test for amphoteric behaviour:

1. Dissolve a metal salt in water.
2. Add a few drops of NaOH. A precipitate forms.
3. Add more NaOH. If the precipitate redissolves, the hydroxide is amphoteric.

Examples of ampholytes are: Al_2O_3 ZnO PbO
$Al(OH)_3$ $Zn(OH)_2$ $Pb(OH)_2$

(ii) Metal carbonates

The carbonates of these metals are insoluble: all metal carbonates except those of Group I are insoluble. So if you add sodium carbonate solution to the solution of a soluble metal salt, an immediate precipitate forms:

$$ZnCl_2 + Na_2CO_3 \longrightarrow 2NaCl + ZnCO_3\downarrow \qquad Zn^{2+}_{(aq)} + CO_3^{2-}{}_{(aq)} \longrightarrow ZnCO_{3\,(s)}$$
$$MgSO_4 + Na_2CO_3 \longrightarrow Na_2SO_4 + MgCO_3\downarrow \qquad Mg^{2+}_{(aq)} + CO_3^{2-}{}_{(aq)} \longrightarrow MgCO_{3\,(s)}$$

The last reaction is interesting for another reason. It is an *endothermic* precipitation. The system takes in energy from the surroundings, so the surroundings cool down.

(iii) Lead halides and lead sulphate

The halides and sulphate of lead can be made by precipitation:

$$Pb(NO_3)_2 + Na_2SO_4 \longrightarrow \underset{\text{white}}{PbSO_4}\downarrow + 2NaNO_3$$
$$Pb^{2+}_{(aq)} + SO_{4\,(aq)}^{2-} \longrightarrow PbSO_{4\,(s)}$$

$$Pb(NO_3)_2 + 2NaCl \longrightarrow \underset{\text{white}}{PbCl_2}\downarrow + 2NaNO_3$$
$$Pb^{2+}_{(aq)} + 2Cl^-_{(aq)} \longrightarrow PbCl_{2\,(s)}$$

$$Pb(NO_3)_2 + 2NaBr \longrightarrow \underset{\text{cream}}{PbBr_2}\downarrow + 2NaNO_3$$
$$Pb^{2+}_{(aq)} + 2Br^-_{(aq)} \longrightarrow PbBr_{2\,(s)}$$

$$Pb(NO_3)_2 + 2NaI \longrightarrow \underset{\text{yellow}}{PbI_2}\downarrow + 2NaNO_3$$
$$Pb^{2+}_{(aq)} + 2I^-_{(aq)} \longrightarrow PbI_{2\,(s)}$$

Although the halides are insoluble in cold water, they are quite soluble in hot water. They have steep solubility curves (see page 19).

28.6 Redox chemistry

The pure metals

By now you should be familiar with the fact that metals combine with non-metals. In all these reactions:

the oxidation number of the metal goes up;
the oxidation number of the non-metal comes down.

For example, iron and chlorine react to give anhydrous iron(III) chloride:

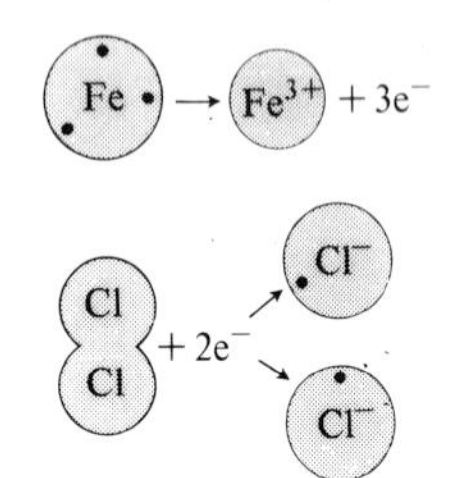

$$2Fe + 3Cl_2 \longrightarrow 2FeCl_3$$

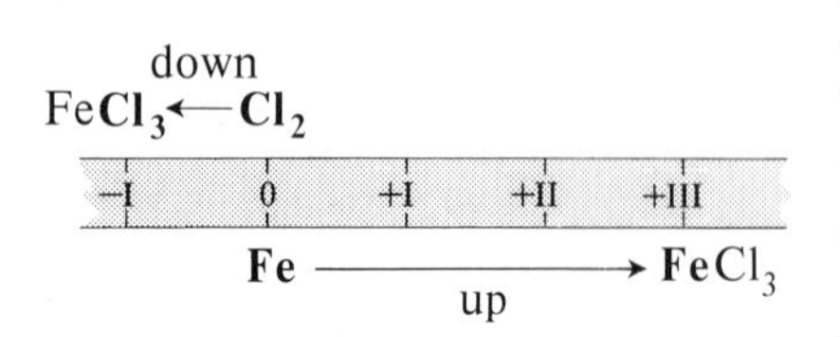

Iron has caused the reduction of chlorine; iron is therefore acting as a reductant. In the same way all these metals act as reductants. The more reactive the metal, the stronger a reductant it is.

Compounds of the metals

Iron, copper, and lead can exist in more than one oxidation state in their compounds:

$\overset{+III}{FeCl_3}$ or $\overset{+II}{FeCl_2}$ $\qquad$ $\overset{+II}{CuI_2}$ or $\overset{+I}{CuI}$ $\qquad$ $\overset{+IV}{PbCl_4}$ or $\overset{+II}{PbCl_2}$

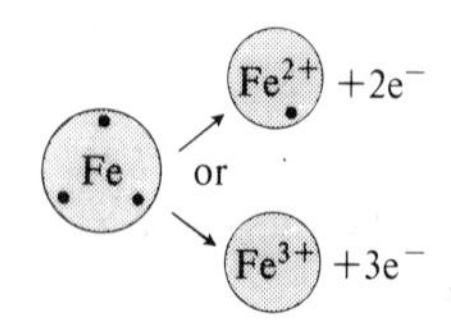

In many reactions, the metal can go from one state to the other during reaction. Since its oxidation number changes, these reactions are redox reactions. They happen most often in aqueous solution. Examples are:

(i) Iron(II) to iron(III)

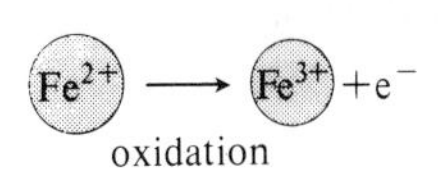

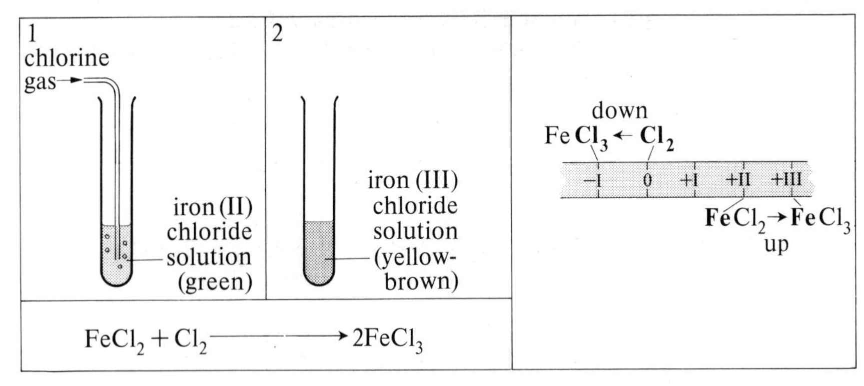

Iron(II) chloride has been oxidised by chlorine; chlorine has been reduced. Iron(II) salts can therefore act as reductants.

(ii) Iron(III) to iron(II)

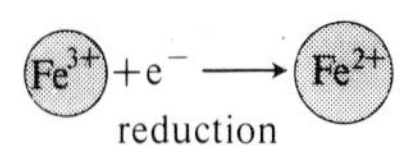

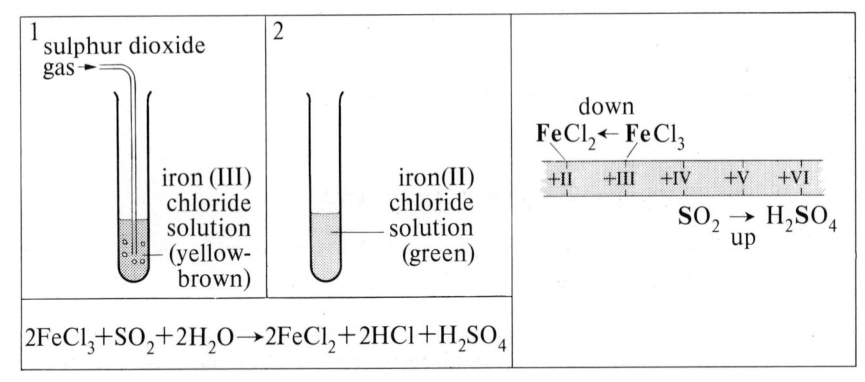

Iron(III) chloride has been reduced by sulphur dioxide; sulphur dioxide has been oxidised. Iron(III) salts can therefore act as oxidants.

(iii) Copper(II) to copper(I)

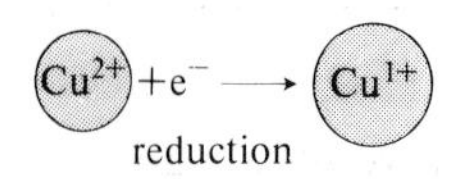

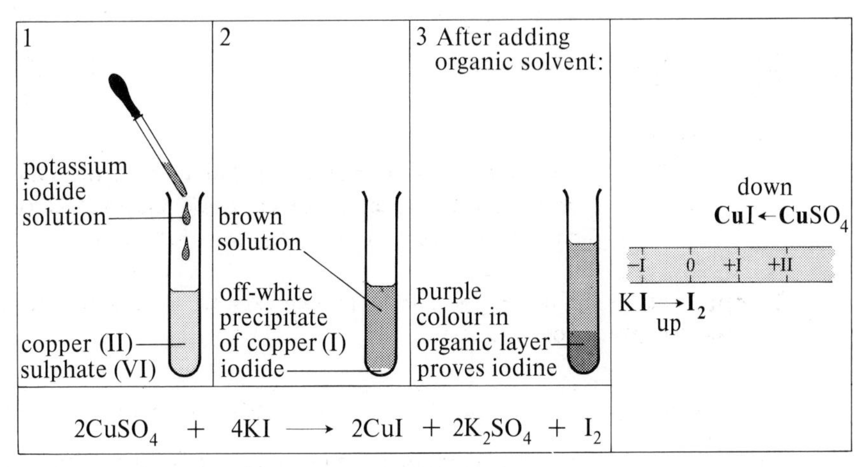

Copper(II) sulphate has been reduced by potassium iodide; potassium iodide has been oxidised. Copper(II) salts can act as oxidants.

Reaction summary for aluminium, iron, and copper

Aluminium	*Iron*	*Copper*
Preparation Electrolysis of aluminium oxide. $Al^{3+} + 3e^- \longrightarrow Al$ at cathode	Chemical reduction in blast furnace. $Fe_2O_3 + 3CO \longrightarrow 2Fe + 3CO_2$	Thermal decomposition in furnace. $Cu_2S \xrightarrow{air} 2Cu + SO_2$
Reactions of element: $4Al + 3O_2 \longrightarrow 2Al_2O_3$ oxide layer formed $2Al + 3Cl_2 \longrightarrow Al_2Cl_6$ or $2AlCl_3$ $2Al + 3H_2SO_4 \longrightarrow Al_2(SO_4)_3 + 3H_2$	$2Fe + 3H_2O + O_2 \longrightarrow 2Fe(OH)_3$ rust $Fe + 2HCl \longrightarrow FeCl_2 + H_2$ $2Fe + 3Cl_2 \longrightarrow 2FeCl_3$ $Fe + S \longrightarrow FeS$ $Fe + H_2SO_4 \longrightarrow FeSO_4 + H_2$	$2Cu + O_2 \longrightarrow 2CuO$ $Cu + Cl_2 \longrightarrow CuCl_2$ $2Cu + S \longrightarrow Cu_2S$ $Cu + H_2SO_4 \longrightarrow$ no reaction with dilute acid $Cu + 2H_2SO_4 \longrightarrow CuSO_4 + SO_2 + 2H_2O$ with conc. acid
Oxide: $Al_2O_3 + 3H_2SO_4 \longrightarrow Al_2(SO_4)_3 + 3H_2O$ $Al_2O_3 + 2NaOH \longrightarrow 2NaAlO_2 + H_2O$	$FeO + H_2SO_4 \longrightarrow FeSO_4 + H_2O$ $Fe_2O_3 + 3H_2SO_4 \longrightarrow Fe_2(SO_4)_3 + 3H_2O$	$CuO + H_2SO_4 \longrightarrow CuSO_4 + H_2O$
Hydroxide: $AlCl_3 + 3NaOH \longrightarrow Al(OH)_3 + 3NaCl$ $Al(OH)_3 + 3HCl \longrightarrow AlCl_3 + 3H_2O$ $Al(OH)_3 + NaOH \longrightarrow NaAlO_2 + 2H_2O$ (amphoteric)	$FeCl_2 + 2NaOH \longrightarrow Fe(OH)_2 + 2NaCl$ $FeCl_3 + 3NaOH \longrightarrow Fe(OH)_3 + 3NaCl$ $Fe(OH)_2 + 2HCl \longrightarrow FeCl_2 + 2H_2O$ $Fe(OH)_3 + 3HCl \longrightarrow FeCl_3 + 3H_2O$	$CuCl_2 + 2NaOH \longrightarrow Cu(OH)_2 + 2NaCl$ $Cu(OH)_2 + 2HCl \longrightarrow CuCl_2 + 2H_2O$
Carbonate: Not formed.	Unstable to heat. $FeCO_3 \longrightarrow FeO + CO_2$ $FeCO_3 + H_2SO_4 \longrightarrow FeSO_4 + CO_2 + H_2O$	Unstable to heat. $CuCO_3 \longrightarrow CuO + CO_2$ $CuCO_3 + H_2SO_4 \longrightarrow CuSO_4 + CO_2 + H_2O$
Change of valency: Only one oxidation state	$2Fe^{2+}_{(aq)} + Cl_{2\,(g)} \longrightarrow 2Fe^{3+}_{(aq)} + 2Cl^-_{(aq)}$	$2Cu^{2+}_{(aq)} + 2I^-_{(aq)} \longrightarrow 2Cu^+_{(aq)} + I_{2\,(s)}$ then $Cu^+_{(aq)} + I^-_{(aq)} \longrightarrow CuI_{(s)}$
Flame test: None	None	Green

Questions

1 Look at the list of metals given below.
Sodium, aluminium, iron, copper, magnesium, zinc.
Write an equation which shows:
(a) one of the metals reacting with cold water
(b) one of the metals reacting with steam
(c) one of the metals, which does not react with water, reacting with dilute acid
(d) one of the metals reacting with the solution of the salt of one of the other metals

2 Describe what you would see if:
(a) an iron nail is put into a solution of copper sulphate
(b) chlorine is passed into a solution of iron(II) sulphate
(c) sulphur dioxide gas is passed into a solution of iron(III) chloride
(d) potassium iodide solution is added to copper(II) sulphate solution

3 Write equations for the reactions described in question 2 above.

4 Describe what you would see, and write the equation for the reaction that takes place, when sodium hydroxide solution is added, first drop by drop, and then in excess to:
(a) dilute copper sulphate solution
(b) dilute aluminium chloride solution
(c) dilute iron(II) sulphate solution
(d) dilute iron(III) chloride solution

5 (a) Name the ore from which aluminium is extracted.
(b) Draw a labelled diagram of the apparatus used in the extraction of aluminium.
(c) Write down the electrode reactions.
(d) List the conditions under which the process takes place.
(e) Why do the anodes need to be replaced regularly?
(f) Explain why cryolite is used in this process.

6 (a) Draw and label a diagram of the blast furnace.
(b) List the substances that are loaded into the top of the furnace and explain why each one is added.
(c) What is pumped into the bottom of the furnace?
(d) Write down the important reactions that take place in the blast furnace.
(e) What impurities does the iron that comes out of the furnace contain?

7 (a) What is steel?
(b) Describe with a diagram how iron is converted into steel.
(c) List three elements that are alloyed with steel, and explain the effect of each alloying element on the properties of the steel.

8 (a) Explain what rust is, and how it forms.
(b) Zinc paint is often used to prevent rusting. Explain how this works.
(c) Give two other methods of preventing rusting. Explain how each method works.

9 Starting from the pure metal in each case, describe with diagrams and the appropriate equations how you would prepare pure dry samples of:
(a) zinc sulphate
(b) iron(III) chloride
(c) copper carbonate

10 Neutralization, combination, precipitation, and redox are all different types of reaction. What type of reaction occurs when the following pairs of solution are added together?
(a) sodium iodide and lead nitrate
(b) sodium iodide and copper(II) sulphate
(c) sodium hydroxide and iron(II) sulphate
(d) chlorine water and iron(II) sulphate
(e) sodium carbonate and zinc chloride
(f) sodium carbonate and hydrogen chloride

11 (a) Explain the meaning of the word *amphoteric*.
(b) Write two equations that illustrate amphoteric behaviour.

29 Group IV: carbon

29.1 The element

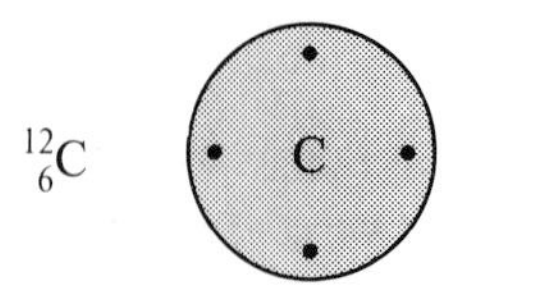

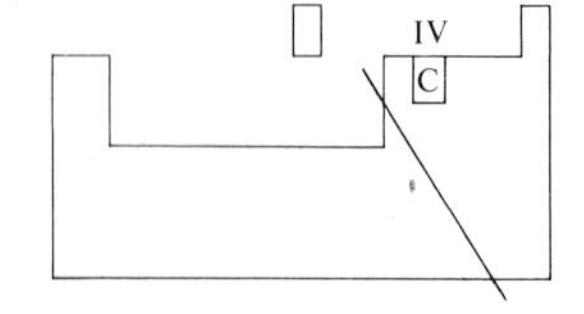

Appearance Diamonds are very hard colourless crystals which sparkle in the light. Graphite is a soft, flaky grey solid. *Both these substances are pure carbon, in different forms.*

We can show that they are both carbon by burning a known mass of each. The same amount of carbon dioxide is produced in each case.

$$C + O_2 \longrightarrow CO_2$$

Diamond and graphite are **allotropes** of carbon. When an element can exist in more than one form in the same state, we say it shows **allotropy**. Carbon has two forms in the solid state.

Physical properties Compare diamond and graphite in the table below:

Table 29.1 The physical properties of carbon's allotropes

Allotrope	*m.p./°C*	*b.p./°C*	*Density /g dm^{-3}*	*Conductivity /Ω^{-1} cm^{-1}*	*Hardness /Moh*	*Solubility in water*
Diamond	3700	4200	3510	Almost zero	10	Insoluble
Graphite	Sublimes at 4200		2220	2857	0·5–1	Insoluble

The properties of diamond are typical of macromolecular solids. So are the properties of graphite except that:

it conducts very well and so is used for electrodes;
it is soft and flaky—it feels rather slippery.

We can explain these differences in their properties by looking at their structures, drawn in Fig. 1. In diamond, each carbon atom makes four covalent bonds, producing a giant lattice in three dimensions. In graphite, each carbon atom makes only three covalent bonds, producing a giant lattice

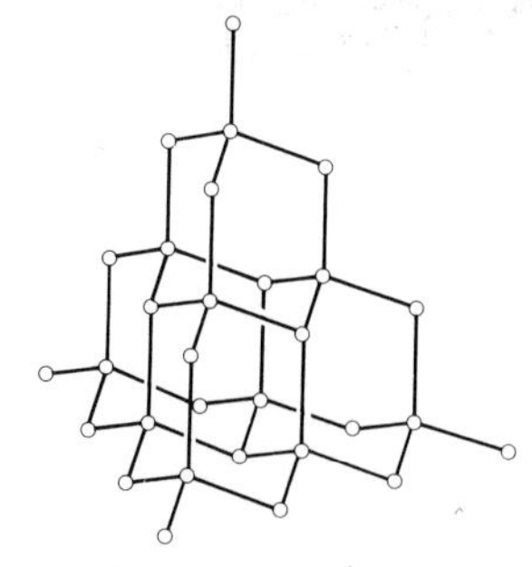

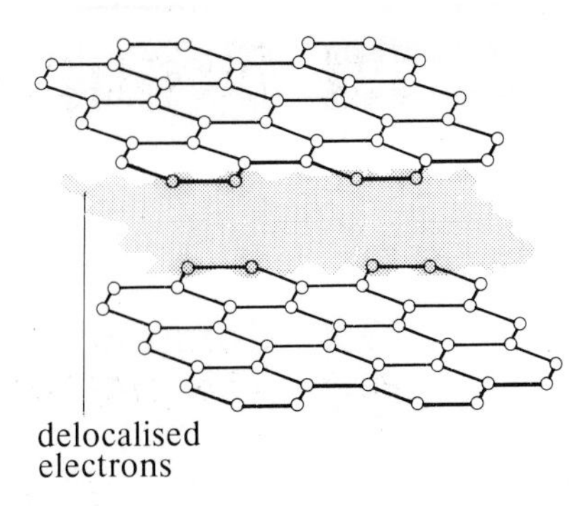

Fig. 1
The structures of diamond (left) and graphite (right)

in only two dimensions. The remaining electron from each atom is delocalised between the layers. It is these free electrons which allow graphite to conduct electricity.

Since the layers in graphite are not bonded covalently to each other, they can slide past each other; this gives graphite its slippery feel.

29.2 Occurrence and extraction

As the element

Both diamond and graphite are found naturally. They are both mined because of their economic importance. Here are some of their more important uses.

Table 29.2

Allotrope	*Use*	*Reason*
Diamond	Gemstones	Rarity and beauty
	Tips for drills, saws etc. Styluses in record players	Hardness
Graphite	Furnace lining	High m.p.
	Electrodes	Conductor
	Lubricants	Slippery, high m.p.
	Pencils	Slippery, grey and soft
(Impure graphite)	Steel-making	Reductant
	Carbon black for ink	Colour

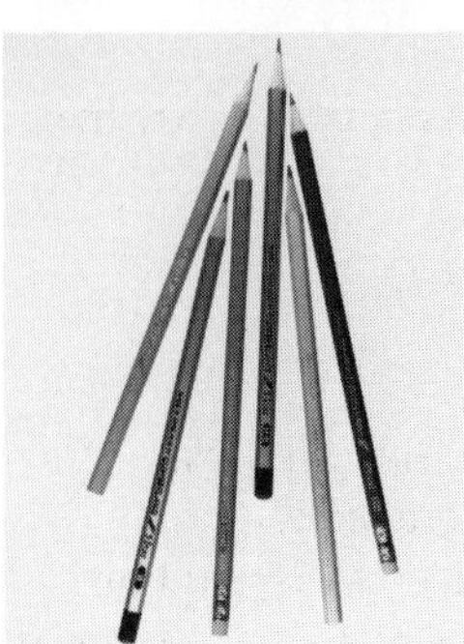

The demand for these allotropes cannot be met by mining alone. Both are now manufactured in large amounts. 30 % of diamonds are made from graphite by using high temperature and pressure. 70 % of graphite is made by purifying coal.

Above left: the Koh-I-Noor diamond, one of the largest in the world. It is a single molecule
Left: graphite is mixed with clay to make pencil leads

As carbon compounds

Literally millions of carbon compounds occur naturally. Three major sources are coal, oil, and natural gas. So many uses have been found for these carbon compounds that a whole industry has grown up to deal with them. This is the **petrochemical industry**. Its purpose and business are discussed in the next chapter.

29.3 Compound formation

Compounds with metals

Carbon is a non-metallic element. In Chapters 26, 27 and 28 we saw that non-metals usually combine well with metals. Carbon does *not* follow this pattern: it forms metal carbides with difficulty. However it can be made to combine with the more reactive metals. For example if carbon and calcium are heated together, calcium carbide is produced:

ionic

$$Ca + 2C \xrightarrow[2\,000^\circ C]{heat} CaC_2 \quad \text{calcium carbide}$$

Compounds with non-metals

Carbon forms a very large number of compounds with other non-metals. These compounds are all molecular substances. Many of them are classed as **organic compounds**. They are a very important part of carbon's chemistry, and are discussed in detail in the next chapter.

The properties of the simpler compounds of carbon and non-metals are outlined here: oxides, hydrides, and chlorides.

The oxides

Carbon forms two different oxides: carbon dioxide and carbon monoxide. Both oxides can be made by burning carbon. Take the burning of coal, which is nearly pure carbon, for example:

O = C = O

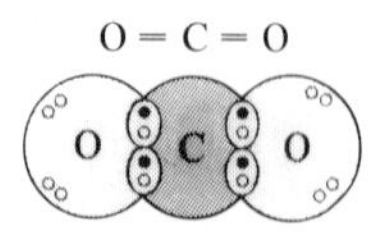

C ≡ O

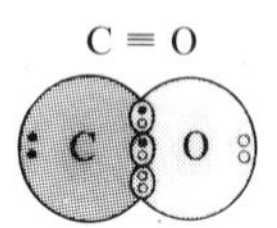

here any carbon monoxide is converted to carbon dioxide:
$2CO + O_2 \longrightarrow 2CO_2$
carbon dioxide reacts with more carbon to give carbon monoxide:
$CO_2 + C \longrightarrow 2CO$
carbon burns to give carbon dioxide:
$C + O_2 \longrightarrow CO_2$

You can see that carbon therefore has more than one possible oxidation state in its compounds:

$\overset{\text{IV}}{CO_2}$ $\quad$ $\overset{\text{II}}{CO}$

Table 29.3 Comparing the two oxides of carbon

Oxide	*m.p./°C*	*b.p./°C*	*Density**	*Colour and smell*	*Effect on damp litmus*
Carbon dioxide	Sublimes at −78		1·52	None	Blue ⟶ pink (acid)
Carbon monoxide	−205	−190	0·97	None	None

* Compared with air.

Respiration

Blood contains a substance called haemoglobin. Oxygen that you breathe in from the air reacts with the haemoglobin, at the surface of the lungs, to form a compound called oxyhaemoglobin. This is then carried in the blood to the cell tissues. In the cells the oxyhaemoglobin decomposes:

$$\text{oxygen} + \text{haemoglobin} \underset{\text{in cells}}{\overset{\text{in lungs}}{\rightleftharpoons}} \text{oxyhaemoglobin}$$

This reversible reaction keeps the cells supplied with oxygen for oxidising carbon compounds obtained from food. The oxidation of these compounds is called **respiration** and produces the energy that keeps us alive:

$$\underset{\text{glucose}}{C_6H_{12}O_6} + 6O_2 \longrightarrow 6H_2O + 6CO_2 + \text{energy}$$

The water and carbon dioxide are carried away by the blood. We breathe out the carbon dioxide through our lungs.

Unlike carbon dioxide, carbon monoxide is poisonous. This is because it reacts with haemoglobin in a reaction that is *not* reversible:

$$\text{carbon monoxide} + \text{haemoglobin} \longrightarrow \text{carboxyhaemoglobin}$$

Gradually the haemoglobin gets used up. Its concentration in the blood falls, so less and less oxygen gets taken to the cells. The victim dies of oxygen starvation.

This is one reason why burning hydrocarbon fuels in the engines of cars and trucks causes pollution. In the cylinders of an engine, there is only a limited amount of air. The fuel reacts to produce carbon monoxide which then passes out in the exhaust and into the atmosphere.

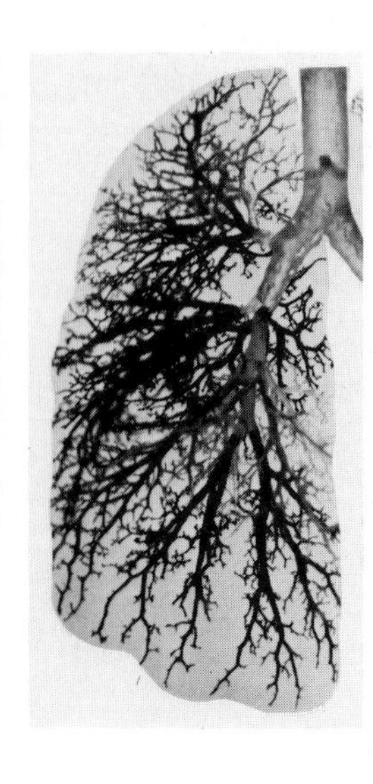

Left: the bronchioles of the left lung. The bronchioles are the passages which carry air into your lungs. At the end of each branch are tiny air sacs. It is at the surface of these sacs that oxygen combines with haemoglobin, and the blood gives up carbon dioxide.

Preparation of the oxides

The oxides of carbon are not made in the laboratory by burning carbon. It is easier and safer to collect gases that are given off as a result of a reaction between:

a solid reactant (put in the bottom of a flask) and
a liquid reactant (poured onto the solid through a tap-funnel).

Apparatus		*Solid reactant*	+ *Liquid reactant*	$\longrightarrow$ *products*
liquid reactant; gas product; solid reactant	For CO_2:	calcium carbonate $CaCO_3$	+ hydrochloric acid $2HCl$	$\longrightarrow$ $CaCl_2 + H_2O + CO_2\uparrow$ (salt, water, carbon dioxide)
	For CO:	sodium methanoate $HCOONa$	+ conc. sulphuric acid H_2SO_4	$\longrightarrow$ $NaHSO_4 + H_2O + CO\uparrow$ (salt, water, carbon monoxide)

Laboratory tests for the oxides

For carbon dioxide: the gas turns lime water cloudy—see opposite page.
For carbon monoxide: it burns with a blue flame

$$2CO + O_2 \longrightarrow 2CO_2$$

Hydrides and chlorides

The simplest hydride of carbon is methane, CH_4. (You will meet many more hydrides in the next chapter.) The simplest chloride is tetrachloromethane, CCl_4.

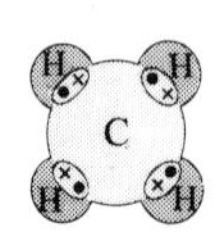

Methane is found in large quantities as 'natural gas'. It is used as a fuel because it burns well and gives out energy to the surroundings as it burns. Carbon dioxide and water are produced:

$$CH_4 + 2O_2 \longrightarrow CO_2 + 2H_2O$$

Tetrachloromethane is made by reacting methane with chlorine. A jet of methane will burn in chlorine to produce tetrachloromethane. A mixture of the two gases explodes, however, if sunlight shines on it. You may remember that a mixture of hydrogen and chlorine also does this.

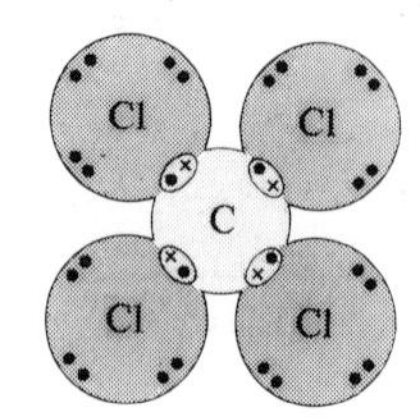

$$CH_4 + 4Cl_2 \longrightarrow CCl_4 + 4HCl$$

Tetrachloromethane is a liquid at room temperature. It is used as a solvent for dry-cleaning and in the laboratory (see page 326).

29.4 Solubility and acid/base chemistry

Solutions of carbon dioxide in water

Carbon dioxide dissolves in water. A reaction occurs to produce carbonic acid, H_2CO_3. It is a weak acid.

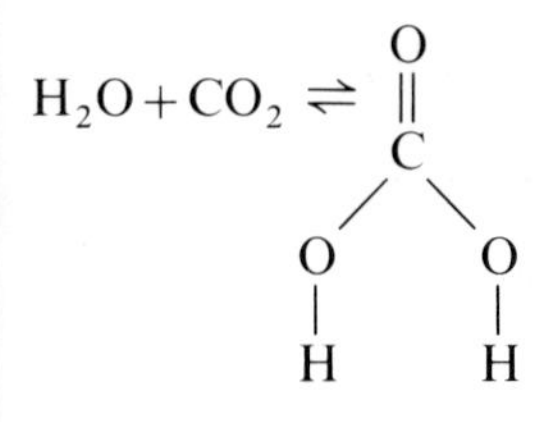

$$H_2O + CO_2 \rightleftharpoons H_2CO_3$$

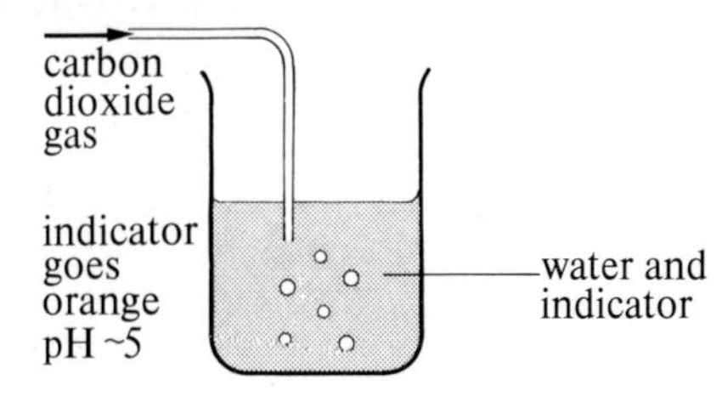

Under standard conditions, the solubility of CO_2 in water is 1 dm^3 in 1 dm^3 of water. The solution produced has a pH of 5.

Carbonic acid decomposes easily again to give off carbon dioxide gas. The whole process is an example of a *reversible reaction.*

Soda-water and other fizzy drinks contain dissolved carbon dioxide. The sharp taste of soda-water is due to the acidity of the solution. When the can or bottle is opened, the pressure is released. The carbonic acid in the solution starts to decompose and bubbles of carbon dioxide are produced. Rainwater is another example of a carbon dioxide solution, but it is much more dilute.

Saturated solutions of carbon dioxide—carbonic acid—show typical acid reactions with bases. Look at its reactions with calcium hydroxide and calcium carbonate:

(i) base + acid ⟶ insoluble salt + water — lime water test for CO_2

$$Ca(OH)_2 + H_2CO_3 \longrightarrow CaCO_3\downarrow + 2H_2O$$
$$2OH^-_{(aq)} + H_2CO_{3\,(aq)} \longrightarrow CO^{2-}_{3\,(aq)} + 2H_2O_{(l)};$$
$$CO^{2-}_{3\,(aq)} + Ca^{2+}_{(aq)} \longrightarrow CaCO_{3\,(s)}$$

(ii) base + acid ⟶ soluble salt + water — this reaction causes 'hard' water

$$CaCO_3 + H_2CO_3 \longrightarrow Ca(HCO_3)_2 + H_2O$$
$$CaCO_{3\,(s)} + H_2CO_{3\,(aq)} \longrightarrow Ca^{2+}_{(aq)} + 2HCO^-_{3\,(aq)}$$

Escaping carbon dioxide produces the fizz in drinks

The first of these two reactions is the lime water test for carbon dioxide. When carbon dioxide is bubbled into a solution of calcium hydroxide (lime water) a precipitate of calcium carbonate forms. This appears as cloudiness in the lime water. But if the supply of carbon dioxide continues, the cloudiness gradually disappears again. This is because the concentration of carbonic acid is increasing: the second reaction above starts to happen. The calcium carbonate dissolves and soluble calcium hydrogen carbonate is produced.

The second reaction above also occurs when rain-water falls on limestone rocks. Rain-water contains dissolved carbon dioxide and limestone is calcium carbonate. The soluble calcium hydrogen carbonate produced causes water in limestone areas to be 'hard'. Hard water is discussed in more detail on pages 297–299.

Tetrachloromethane as a non-aqueous solvent

Tetrachloromethane is a useful solvent because it is immiscible with water. Some solids dissolve better in tetrachloromethane than in water. When such a solid is present in aqueous solution, we can extract it by shaking with tetrachloromethane. Iodine is a good example:

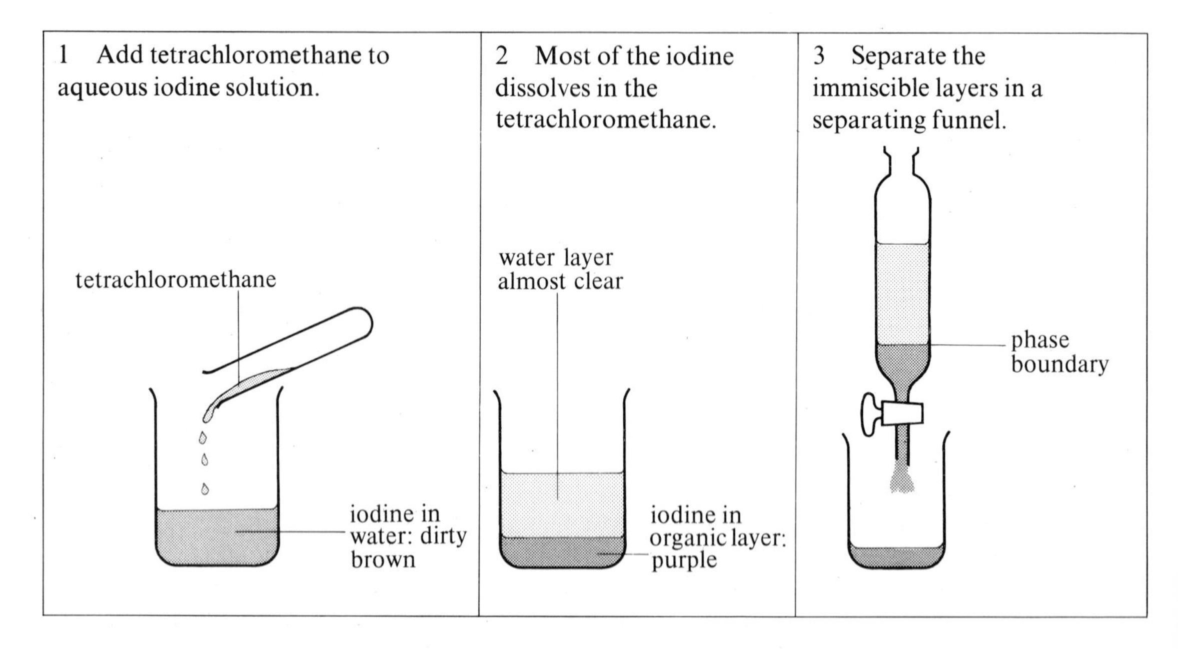

29.5 Redox chemistry

Carbon as reducing agent

Pure carbon has the oxidation number zero, like all elements. Carbon forms its oxides quite easily. In them, carbon has the oxidation number +II (in CO) or +IV (in CO_2).

Carbon's oxidation number therefore goes up when it is converted to one of its oxides. Another element's oxidation number must come down during the reaction. A substance is said to be reduced when it contains an element whose oxidation number comes down. So carbon acts as a reducing agent. This is clear in these examples:

(i) carbon reduces oxygen

$$C + O_2 \longrightarrow CO_2$$

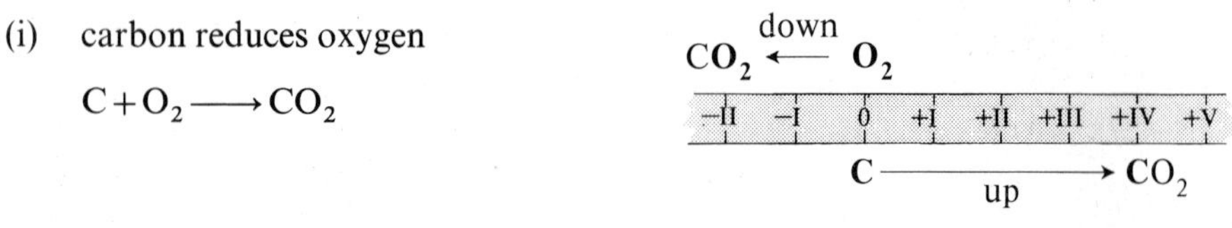

(ii) carbon reduces copper(II) oxide

$$C + 2CuO \xrightarrow{\text{heat}} 2Cu + CO_2$$

down
Cu ← CuO
0 +I +II +III +IV
C → CO_2
up

Carbon monoxide as reducing agent

Carbon monoxide is quite easily converted to carbon dioxide. Here again, carbon's oxidation number goes up, this time from +II to +IV. The carbon monoxide is oxidised. Therefore it acts as a reducing agent, like carbon. For example:

(iii) carbon monoxide also reduces oxygen

$$2CO + O_2 \longrightarrow 2CO_2$$

down: $CO_2 \longleftarrow O_2$

−III −II −I 0 +I +II +III +IV

up: $CO \longrightarrow CO_2$

(iv) carbon monoxide is the reducing agent in the blast furnace (see page 305).

$$Fe_2O_3 + 3CO \longrightarrow 2Fe + 3CO_2$$

down: $Fe \longleftarrow Fe_2O_3$

0 +I +II +III +IV

up: $CO \longrightarrow CO_2$

Carbon dioxide as oxidising agent

When carbon dioxide reacts to form carbon or carbon monoxide, the reverse of reactions (i) and (iii) above is happening. Carbon's oxidation number comes down, therefore another element's oxidation number goes up.

So carbon dioxide is acting as an oxidising agent. For example:

(i) carbon dioxide oxidises burning magnesium

$$2Mg + CO_2 \longrightarrow 2MgO + C$$

down: $C \longleftarrow CO_2$

0 +I +II +III +IV

up: $Mg \longrightarrow MgO$

(ii) in the blast furnace carbon dioxide oxidises carbon (see page 306).

$$C + CO_2 \rightleftharpoons 2CO$$

down: $CO \longleftarrow CO_2$

0 +I +II +III +IV

up: $C \longrightarrow CO$

But carbon dioxide does not react easily to produce carbon monoxide or carbon. (The opposite reaction *does* happen easily). For this reason, carbon dioxide is used in fire extinguishers. The heavy gas blankets the fire and prevents oxygen from reaching the flames, without itself reacting.

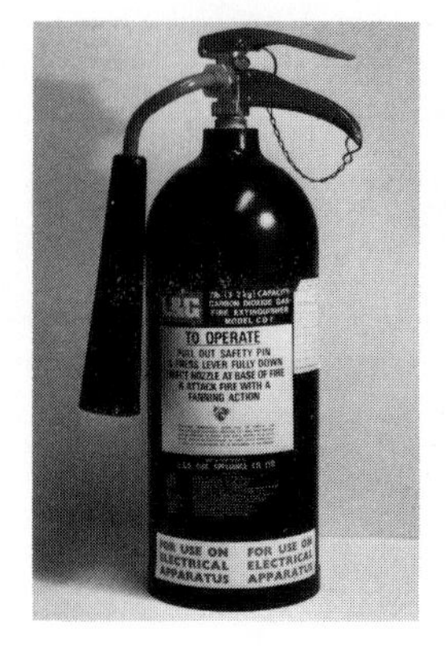

Left: a carbon dioxide fire extinguisher. It contains liquid carbon dioxide under pressure. When the safety pin is pulled out and the lever pressed, the carbon dioxide sprays out as a gas.

29.6 Special features: organic chemistry

Carbon forms so many special compounds that the whole of the next chapter is devoted to them. Most of these compounds were originally extracted from living organisms so this part of carbon chemistry is called **organic chemistry**.

Reaction summary for carbon

Extraction Found as coal, graphite, and diamonds.

Reactions of the element

(i) It reduces other elements: $C + O_2 \longrightarrow CO_2$
$2C + O_2 \longrightarrow 2CO$

(ii) It reduces compounds: $C + CO_2 \rightleftharpoons 2CO$
$C + 2CuO \longrightarrow 2Cu + CO_2$

(iii) It oxidises calcium: $Ca + 2C \longrightarrow CaC_2$

The oxides These are carbon dioxide and carbon monoxide.

Carbon dioxide

(i) It is a typical non-metal acidic oxide:
$H_2O + CO_2 \rightleftharpoons H_2CO_3$
carbonic acid

$H_2CO_3 + Ca(OH)_2 \longrightarrow CaCO_3\downarrow + 2H_2O$
The reaction above is the lime water test.

$H_2CO_3 + CaCO_3 \longrightarrow Ca(HCO_3)_2$
The reaction above causes 'hard water' in limestone areas.

(ii) It can act as an oxidising agent: $CO_2 + C \rightleftharpoons 2CO$
$CO_2 + 2Mg \longrightarrow 2MgO + C$
It will oxidise only the reactive metals.

Carbon monoxide

(i) It is a neutral oxide.

(ii) It can act as a reducing agent:
$CO + CuO \longrightarrow Cu + CO_2$
$3CO + Fe_2O_3 \longrightarrow 2Fe + 3CO_2$
$2CO + O_2 \longrightarrow 2CO_2$

(iii) It is poisonous: it reacts irreversibly with haemoglobin in the blood.

The hydroxy compound This is non-metallic and therefore acidic: $CO(OH)_2$ or H_2CO_3 is carbonic acid, a weak acid.

Questions

1 (a) Write down one physical property and one chemical property that show carbon as a typical non-metal.
(b) Which property of graphite is typical of a metal? Explain why graphite has this property.
(c) Diamond and graphite are allotropes. What does the term *allotrope* mean?
(d) How could you prove that diamond and graphite are both made of carbon?

2 (a) Draw part of a diamond lattice.
(b) How many covalent bonds does each carbon atom in the middle of a diamond lattice form.
(c) How many electrons does each atom use to form these bonds?
(d) How many outer electrons does each carbon atom have?
(e) Explain why diamond is a good insulator.

3 (a) Draw part of a graphite lattice.
(b) How many covalent bonds does each carbon atom in the middle of a graphite lattice form.
(c) How many electrons does each atom use to form these bonds?
(d) How many outer electrons does each carbon atom have?
(e) Explain why graphite is a good conductor.
(f) Give two examples of it conducting.

4 (a) Show the bonding in carbon dioxide.
(b) Does it smell?
(c) Is it heavier or lighter than air?
(d) Is its solution in water acid, neutral, or alkaline.
(e) How do you make it in the laboratory? Give an equation.
(f) How do you test for it?

5 (a) Show the bonding in carbon monoxide.
(b) Does it smell?
(c) Is it heavier or lighter than air?
(d) Is its solution in water acid, neutral, or alkaline.
(e) How do you make it in the laboratory? Give an equation.
(f) How do you test for it?

6 (a) How much carbon dioxide is there in the atmosphere?
(b) Name two processes that release carbon dioxide into the atmosphere.
(c) Name a process that removes carbon dioxide from the atmosphere.
(d) What would you observe if carbon dioxide is bubbled into limewater for a long time?
(e) Give the equations for the reactions that have occurred in (d) above.

7 (a) The apparatus shown below was set up. The gas in the syringe was passed backwards and forwards over the heated carbon until no more change was detected. What was the final volume of the gas? Explain your answer.

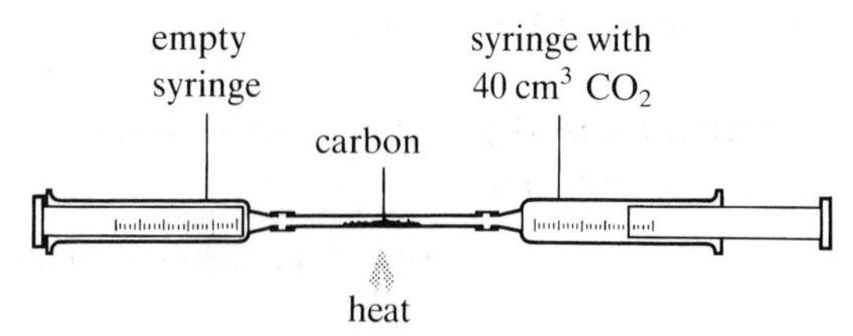

$C_{(s)} + CO_{2(g)} \rightarrow 2CO_{(g)}$

8 (a) What would you expect to see if the test-tube on the right was heated? Write the equation for any change that might occur.
(b) Would you expect your answer to be different if magnesium oxide was used instead of silver oxide? Explain.

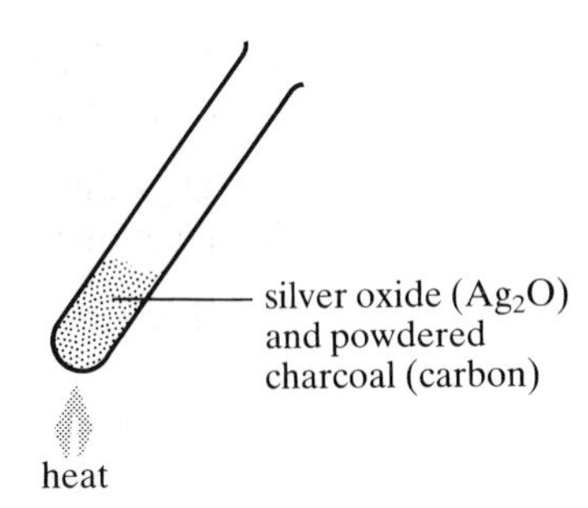

9 Limestone and chalk are two natural forms of calcium carbonate, which is insoluble in water. In view of this fact, how does rain-water dissolve out great gorges and potholes in limestone districts?

10 From the list below, pick out the terms that can be *correctly* used to describe the properties of (a) carbon, and (b) carbon dioxide:
oxidizing agent   element   metal   reducing agent   acid   can be oxidized   compound   can be reduced   allotropic

11 (a) Explain how the blood carries oxygen.
(b) What is the oxygen in the blood used for?
(c) Why is carbon monoxide toxic?

30 Organic chemistry

30.1 Carbon the chain-former

While the atoms of most non-metallic elements bond together to form small molecules, carbon atoms can also bond together to form large molecules with many atoms in them.

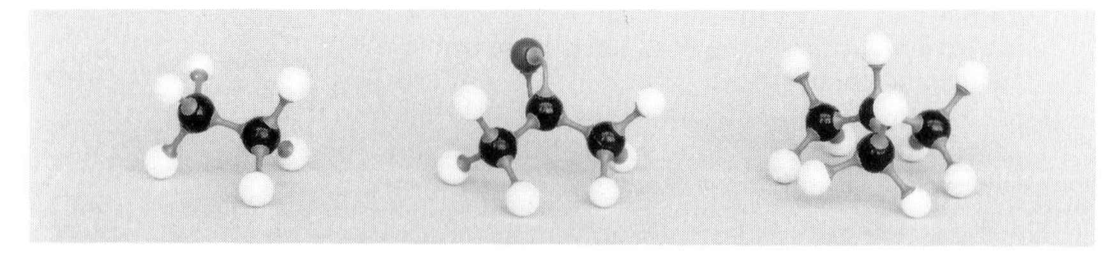

Typical small molecules but carbon can form long chains too

We have already seen carbon form such giant molecules as diamond and graphite. In this chapter we are going to see that carbon atoms can bond together in so many combinations that there are more compounds formed by carbon than by all the other elements put together.
The reasons for this are that carbon atoms can:

form four bonds;
form strong bonds with other carbon atoms;
form chains of atoms—this property is called **catenation**

Hydrogen, oxygen, and nitrogen atoms often join to carbon chains. These chains can be almost any length. A different number of atoms in the chain means a different compound.
The number of carbon atoms in a chain is called the **chain length**.

Example	H H | | H—C—C—H | | H H	H H H | | | H—C—C—C—H | | | H H H	H H H H H H H H | | | | | | | | H—C—C—C—C—C—C—C—C—H | | | | | | | | H H H H H H H H
Chain length	2	3	8

30.2 The importance of carbon chain compounds

Long-chain carbon compounds are found in all living organisms, both animals and plants.
Some common examples are shown on the next page.

Table 30.1

Compound	*Found in*	*Chain length*
Haemoglobin	Blood	5000
Starches	Animals and plants	300–10000
Cellulose	Plants	20000

Because so many carbon chain compounds are found in living systems—organisms—the chemistry of carbon chain compounds is called **organic chemistry**.

Not only are we built from carbon chains, we also depend on them every day of our lives. They are found in:

- foods —proteins, fats, starches, vitamins
- clothing —wool, cotton, silk, nylon, rayon, terylene
- packaging—wood, cardboard, paper, cellophane, polythene
- fuels —petrol, diesel oil, paraffin
- plastics —melamine, polystyrene, PVC, non-stick teflon

So the study of organic chemistry is obviously very important.

How many things made from carbon chains can you see in this picture?

30.3 Chain length and chain type

This chapter is mostly about extracting substances that contain carbon chains, and making use of them. Because there are so many of these substances, the study of them can be very confusing unless you have two important ideas clear in your mind:

1 **The length of the carbon chain in a compound controls most of its physical properties (such as m.p., b.p. and solubility).**
Compounds with short chains are gases or liquids that boil easily.
Compounds with medium chains are liquids.
Compounds with long chains are solids.
For example:

Chain length	3	6	16
Formula	H H H | | | H—C—C—C—H | | | H H H	H H H H H H | | | | | | H—C—C—C—C—C—C—H | | | | | | H H H H H H	$C_{16}H_{34}$
m.p./°*C*	−187	−94	18
b.p./°*C*	−41	69	287

2 **The types of link in the chain or at its ends control the compound's chemical properties (i.e. what it reacts with).**
Carbon chains can have different types of chain links and chain ends. For example:

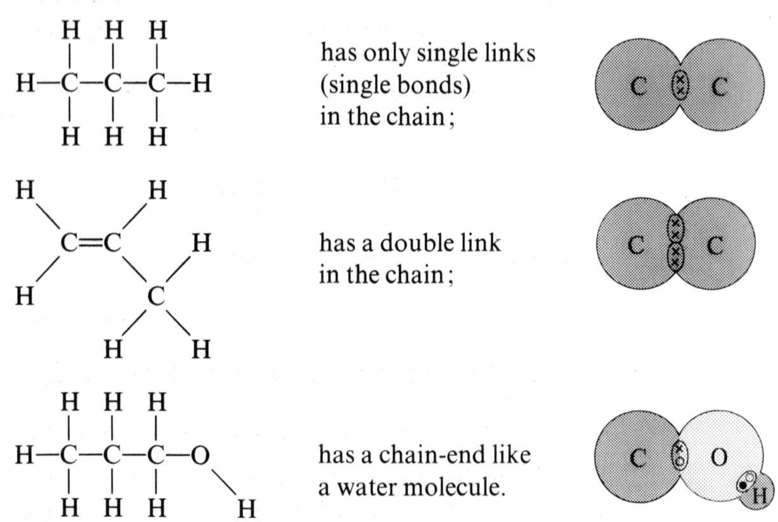

Functional groups

An important chain-link or chain-end is called a **functional group**.
A functional group is an atom or group of atoms in the chain whose reactions determine the properties of the whole chain.

The most important functional groups we will study are:

a double chain-link

```
\     /
 C=C
/     \
```

e.g.

```
H     H
 \   /
   C           H
 /   \        /
H      C=C
      /    \
     H      H
```

a chain-end like water

```
—O
   \
    H
```

e.g.

```
     H  H
     |  |
  H—C—C—O
     |  |   \
     H  H    H
```

a chain-end like carbonic acid

```
   O
   ||
   C
  / \
     O—H
```

e.g.

```
          O
          ||
   H      C
    \   /   \
      C      O—H
    /   \
   H     H
```

Homologous series

A whole group of compounds may have the same functional group but different chain lengths. This means that these compounds have:

different physical properties (due to different chain lengths)
the same chemical reactions (due to the same functional group)

A group of compounds like this is called a **homologous series**.
The simplest homologous series is the **alkanes**. Below are the first four members of the series. Note how their boiling-points increase with chain-length.

Name	*Formula*	*State at 20°C*	*b.p./°C*	*General formula*
Methane	CH_4	Gas	−164	
Ethane	C_2H_6	Gas	−87	C_nH_{2n+2}
Propane	C_3H_8	Gas	−42	
Butane	C_4H_{10}	Gas	−0·5	

30.4 Naming carbon chain compounds

The name of a carbon compound is divided into two parts:

the first part of the name tells you the chain length (Table 30.2);
the second part of the name tells you the functional group (Table 30.3).

Table 30.2

Chain length	*First part of name*
C_1	meth-
C_2	eth-
C_3	prop-
C_4	but-
C_5	pent-
C_6	hex-

Table 30.3

Functional group	*Second part of name*
H | —C—H | H	-ane
C=C	-ene
C—O | H	-ol
O ‖ —C \\ O—H	-oic acid

Table 30.4 Examples of naming compounds

Molecule	*Chain length*	*Functional group*	*Name*
H | H—C—H | H	1 ⇒ meth-	H | —C—H ⇒ -ane | H	methane
H H H | | | H—C—C—C—H | | | H H H	3 ⇒ prop-	H | —C—H ⇒ -ane | H	propane
H | H C—H | // H—C—C | | H H	3 ⇒ prop-	C=C ⇒ -ene	propene
H H \\ / C=C / \\ H H	2 ⇒ eth-	C=C ⇒ -ene	ethene

30.5 Sources of carbon chains

Photosynthesis

A major source of carbon for chains is the carbon dioxide in the air. Plants are able to combine carbon dioxide with water. Light energy from the sun is needed for this endothermic reaction. The process is called **photosynthesis**. The products are long-chain carbon compounds. These compounds contain carbon, hydrogen, and oxygen and are known as **carbohydrates**:

$$6CO_2 + 6H_2O \xrightarrow{\text{sunlight}} C_6H_{12}O_6 \quad (+6O_2\uparrow)$$

The $C_6H_{12}O_6$ units can join together to make long-chain carbohydrates. Cellulose and starch are two examples of these.

Cellulose and starch make up the structure and cells of plants. Animals eat plants; the carbohydrates in the plants are changed into different carbon chains in the animals' bodies.

The effect of photosynthesis on the atmosphere is the opposite of respiration.

Respiration in animals

$$C_6H_{12}O_6 + 6O_2 \rightarrow 6H_2O + 6CO_2$$

Respiration takes oxygen out of the atmosphere and replaces it with carbon dioxide, while photosynthesis cancels this out by doing the opposite. The result is that for millions of years, the amounts of carbon dioxide and oxygen in the atmosphere have been constant.

However, measurements taken regularly since 1858 at Mauna Loa in Hawaii show that the amount of carbon dioxide in the atmosphere has been rising rapidly in the last hundred years.

The reasons for this are:

larger amounts of coal and oil are burnt every year, making ever larger amounts of carbon dioxide

the forests of the world get smaller every year, making less oxygen by photosynthesis

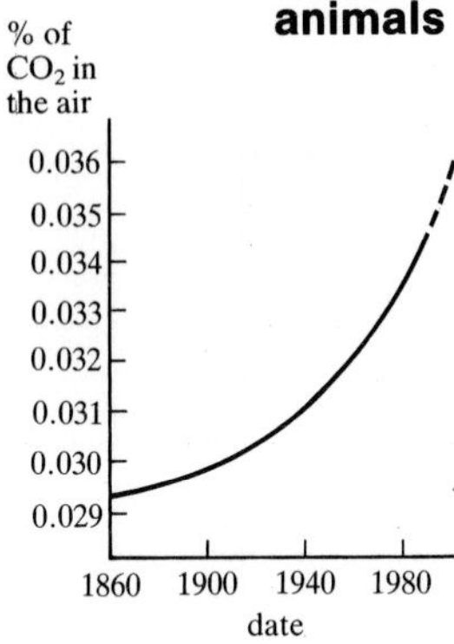

Amounts of carbon dioxide in the atmosphere, measured in Hawaii.

Scientists are worried that this extra carbon dioxide will trap more heat from the sun and cause the atmosphere to heat up. This is called the 'greenhouse effect'. If this happened the ice caps would melt, upsetting the climate of the whole world.

Oil and coal

Time of decay	*Source of chains*
	Wood
	Peat
	Brown coal
	Black coal
	Anthracite

From the beginning of life on earth, plants and animals have been dying and decaying. Their remains, containing carbon chains, are deposited in the earth. The deposits from prehistoric times are now our main sources of carbon chains for industry. They are **oil** and **coal**. Oil is the decayed remains of billions of tiny prehistoric sea animals. Coal is the decayed remains of prehistoric forests.

It takes a very long time for plants to decay to give coal. Decay involves the gradual loss of water from the carbohydrates that make up the plants. The chains are therefore losing hydrogen and oxygen and getting closer and closer to pure carbon. The chains in wood, for example, contain 32% C, 47%H, and 19%O, by mass. After many millions of years decay, wood ends up as anthracite. The chains in anthracite contain 78%C, 19%H, and 2%O by mass.

The effects of photosynthesis and decay are summarized in the **carbon cycle**.

The carbon cycle

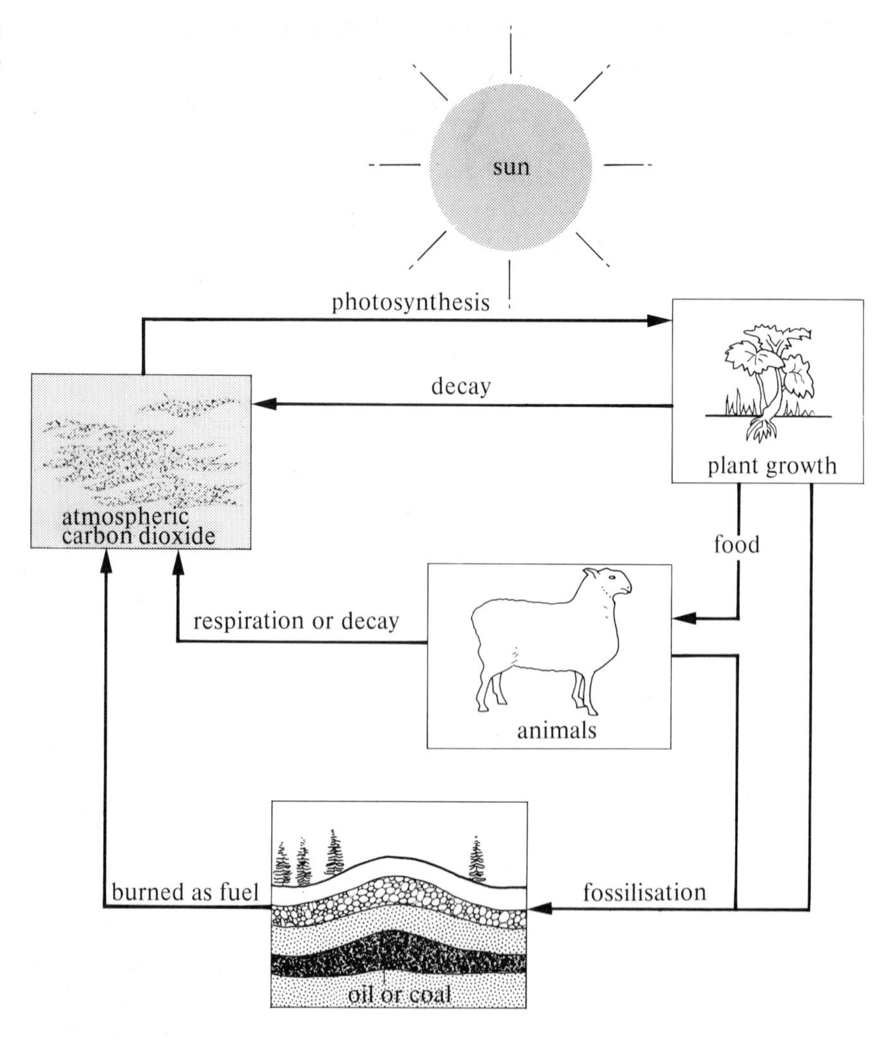

Which steps of the carbon cycle do these show?

30.6 Separating hydrocarbons

Oil Both oil and coal are in fact mixtures of compounds. Most of the compounds in them contain only hydrogen and carbon. Such compounds are generally called **hydrocarbons**.

There are many different hydrocarbons in oil, all with different chain lengths. You have already seen that chain length is mainly responsible for the physical properties of a compound. The physical properties of a compound often determine how we can use it; therefore it makes sense to separate the components of oil according to their chain lengths. This is done by **fractional distillation** of the oil. The photograph and diagram below show the column used in industry. The column is colder at the top than the bottom. The compounds are separated as they go up the column, because of their different boiling-point ranges. Each group of compounds, or **fraction**, that comes off during the distillation has a name—kerosene, gasoline, and so on. The uses for each fraction are shown in Table 30.5.

To obtain hydrogen, the hydrocarbons from oil are sometimes reacted with steam in the presence of a heated nickel catalyst (see the table on page 270).

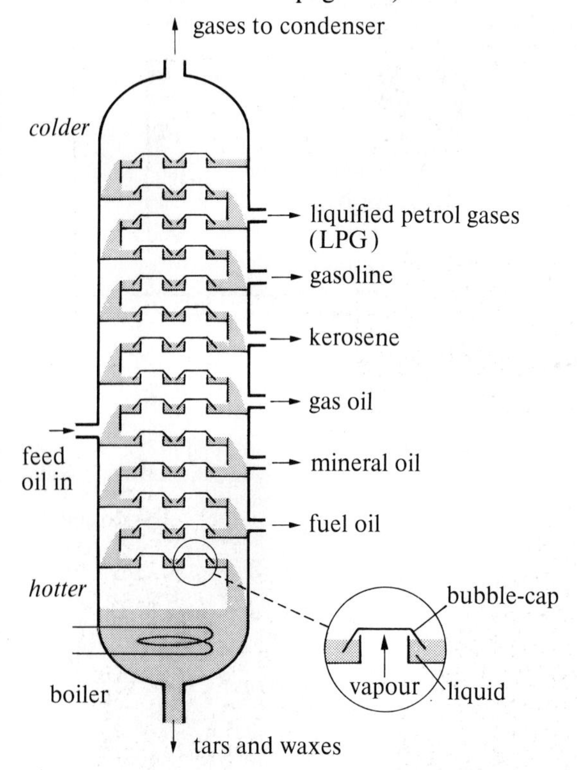

Fig. 1
Crude oil is a mixture which is separated by fractional distillation. The pipes that take off the different fractions can be clearly seen on the photograph, which shows the industrial fractionating column at the Shell refinery at Stanlow

Table 30.5 Fractions from oil and their uses

Name	*Chain length*	*Range of b.p.'s in fraction*	*Uses*
Liquified petrol gases (LPG)	C_1–C_4	up to 25°C	Calor gas, Gaz
Gasoline (petrol)	C_4–C_{12}	20–200°C	Petrol for cars
Kerosene	C_{11}–C_{15}	174–275°C	Paraffin, jet fuel, petrochemicals
Gas oil (diesel)	C_{15}–C_{19}	200–400°C	Fuel for central-heating boilers, petrochemicals
Mineral oil (lubricating oil)	C_{20}–C_{30}	over 350°C	Lubrication, petrochemicals
Fuel oil	C_{30}–C_{40}	over 400°C	Fuels for power stations and ships, petrochemicals
Wax, grease	C_{41}–C_{50}	Solid	Candles, wax paper, grease for bearings
Bitumen	C_{50} upwards	Solid	Roofing, roads

Coal Coal, unlike oil, is solid; so it cannot be fractionally distilled to give useful products. Instead, coal can be **destructively distilled** as shown in Fig. 2. At the high furnace temperatures, its carbon chains break up, and hydrogen leaves them. Impure carbon, coke, is left behind. Steam is blasted in to react with some of the coke, giving hydrogen and carbon monoxide. The main products are coke, hydrogen, carbon monoxide and methane.
The coke is used as a solid fuel. Certain mixtures of the gases are used as fuel in industry. They are

water gas: CO and H_2
coal gas: CO and CH_4

Because of the increasing cost of oil, greater efforts are now being made to produce liquid fuels from coal.

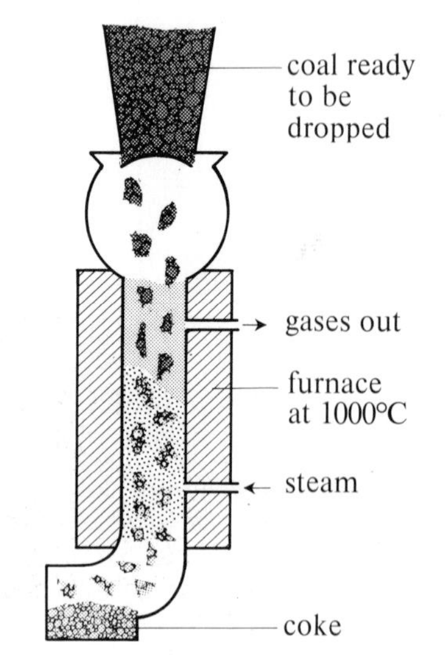

Fig. 2 *The destructive distillation of coal*

30.7 Using hydrocarbons

Table 30.5 shows that hydrocarbons have two main uses. The first is as **fuels**, and the second is to make **petrochemicals**.

Fuels

A fuel is any substance that gives out energy when it reacts with the oxygen in the air. The products of burning the fuel should be harmless, or using the fuel will cause pollution. Hydrogen is an ideal fuel because it burns, giving out a lot of energy, to make water.

We burn hydrocarbons to drive engines, to heat buildings, and to power industry, but unfortunately, the products of burning these hydrocarbons are *not* harmless. The diagram below shows the main pollution products put into our atmosphere each year.

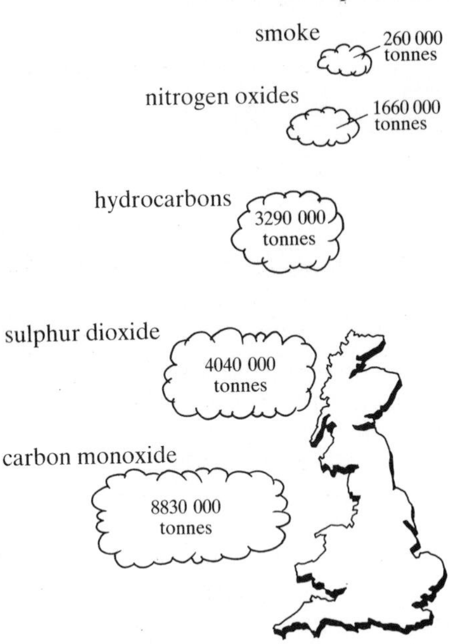

These figures are produced by the Department of the Environment, and are the figures for 1982. Each pollutant brings its own problems: the level of carbon monoxide is not thought to be harmful to the general public, although people with heart disease may be affected; nitrogen dioxide is known to be poisonous, but levels in the atmosphere are still very low; sulphur dioxide and smoke are bad for the lungs. The oxides of nitrogen and sulphur dissolve in rain causing 'acid rain'. While most people are quite clear what makes rain acidic, there is no clear agreement about the effects of the acid rain. What cannot be denied is that by 1984, half the trees in some of the forests of Europe are damaged and may die, and that many lakes no longer have fish in them. It has also been suggested that Britain is responsible for much of this pollution. This may also be true, but it is certainly not all one-way traffic. These maps show that when an Easterly wind blows from Europe, the acidity of our rain is worse than when the wind is from the West. The effect of the clean rain from the Atlantic is shown on the left-hand map. Whatever the arguments, acid rain is becoming a matter for public concern.

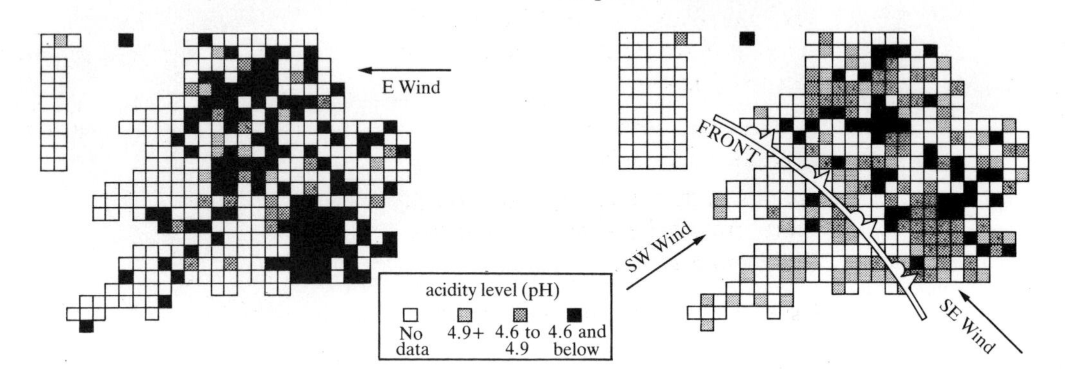

Sources of atmospheric pollution

This pollution of the atmosphere comes from a number of different sources, and each source produces a slightly different mixture of pollutants as the diagrams show.

Source	*Railways*	*Farming*	*Diesels*	*Houses*	*Factories*	*Power*	*Cars*
Total tonnes	67 000	270 000	475 000	999 000	2190 000	3481 000	8471 000
% smoke	—	—	—	16	2	—	—
$\%SO_2$	15	—	7	21	51	78	3
$\%NO_2$	54	6	36	6	15	19	4
Hydrocarbons	11	14	7	7	28	1.5	6
CO	20	80	50	50	4	1.5	87

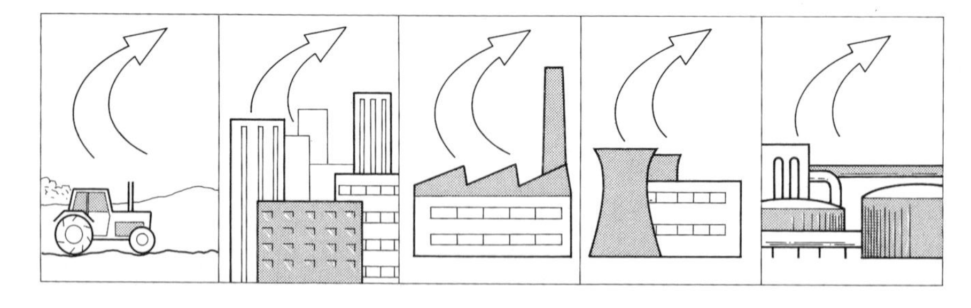

Petrochemicals

An enormous number of uses are found for carbon compounds. The hydrocarbon chains in oil are the starting material for the manufacture of many other carbon chains. The *petrochemical industry* produces new compounds by:

changing the length of the chains
changing the atoms joined to the chains.

The new substances made from oil in the petrochemical industry include:

all the common plastics like polythene, PVC, polystyrene;
all the synthetic fibres like nylon, terylene, and acetate;
glues, paints, dyes, drugs, cosmetics.

Paint and polystyrene: two common substances made from petrochemicals

30.8 Two important types of hydrocarbon

Saturated hydrocarbons

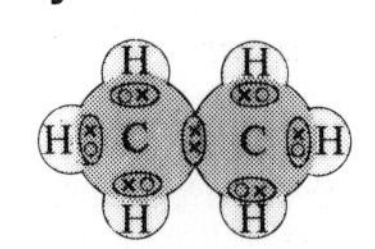

High quality crude oil—like that from the North Sea—is almost entirely a mixture of alkanes. Alkanes are very good fuels, reacting well with oxygen. They also react with chlorine. But otherwise they are very unreactive.
One reason for this is that there are no spare bonds in an alkane chain.
Each carbon in the chain has as many hydrogens bonded onto it as possible.
The compound is said to be **saturated**.

Unsaturated hydrocarbons

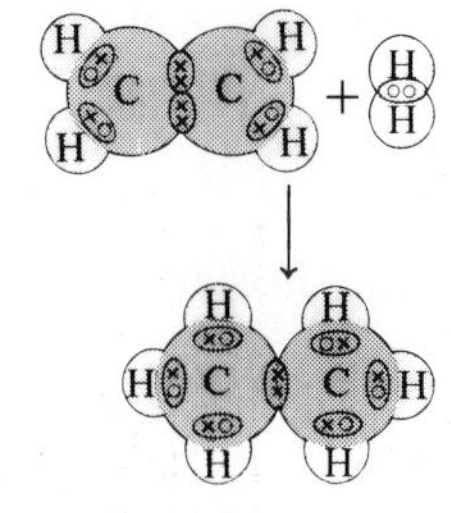

Alkenes are much more reactive because they contain a carbon–carbon double bond. One of these bonds can be broken, allowing more atoms to add onto the chain:

$$\begin{array}{ccc} H & & H \\ & \diagdown\;\diagup & \\ & C{=}C & \\ & \diagup\;\diagdown & \\ H & & H \end{array} + H_2 \longrightarrow \begin{array}{ccccccc} & & H & & H & & \\ & & | & & | & & \\ H & - & C & - & C & - & H \\ & & | & & | & & \\ & & H & & H & & \end{array}$$

Because more atoms can be added to the chain without breaking it the chain is said to be **unsaturated**.

Alkynes are also unsaturated. They contain a carbon–carbon triple bond.

$$H{-}C{\equiv}C{-}H + 2H_2 \longrightarrow \begin{array}{ccccccc} & & H & & H & & \\ & & | & & | & & \\ H & - & C & - & C & - & H \\ & & | & & | & & \\ & & H & & H & & \end{array}$$

Comparing the two types

Compare the reactions of an alkane and an alkene with chlorine.
For the alkane to react, a hydrogen atom must be broken from the chain to make room for a chlorine atom:

$$\begin{array}{ccccccc} & & H & & H & & \\ & & | & & | & & \\ H & - & C & - & C & - & H \\ & & | & & | & & \\ & & H & & H & & \end{array} + Cl_2 \longrightarrow \begin{array}{ccccccc} & & H & & H & & \\ & & | & & | & & \\ H & - & C & - & C & - & Cl \\ & & | & & | & & \\ & & H & & H & & \end{array} + HCl$$

This is called a **substitution** reaction, because the chlorine atom has taken the place of a hydrogen atom.
But the alkene can add on the chlorine without losing any hydrogen:

$$\begin{array}{ccc} H & & H \\ & \diagdown\;\diagup & \\ & C{=}C & \\ & \diagup\;\diagdown & \\ H & & H \end{array} + Cl_2 \longrightarrow \begin{array}{ccccccc} & & H & & H & & \\ & & | & & | & & \\ H & - & C & - & C & - & H \\ & & | & & | & & \\ & & Cl & & Cl & & \end{array}$$

This is called an **addition** reaction.

To summarise:

(i) saturated compounds have only single bonds, and can only undergo substitution reactions.

(ii) unsaturated compounds contain double (or triple) bonds and can add on new atoms.

From saturated to unsaturated: cracking

The alkanes are fairly unreactive, because the chains are saturated. Alkanes burn as fuels and also burn in chlorine but have few other reactions. However, they can be decomposed at about 400°C with the help of an aluminium oxide catalyst. This process is called **cracking**.

In industry, alkanes of medium chain-length from oil are cracked. The chains break up to give smaller pieces. The products are shorter chain-length alk*anes* together with very short chain alk*enes*.

Cracking is important for two reasons:

(i) It turns the less useful fraction of oil into more widely-used shorter-chain compounds, like petrol and kerosene.

(ii) It produces *unsaturated* hydrocarbons, like ethene. These—as you have seen—are more reactive than saturated hydrocarbons. Ethene is the starting material for many important manufactured substances, such as plastics and methylated spirits.

Fig. 3
Cracking medium-chain hydrocarbons to produce more useful compounds

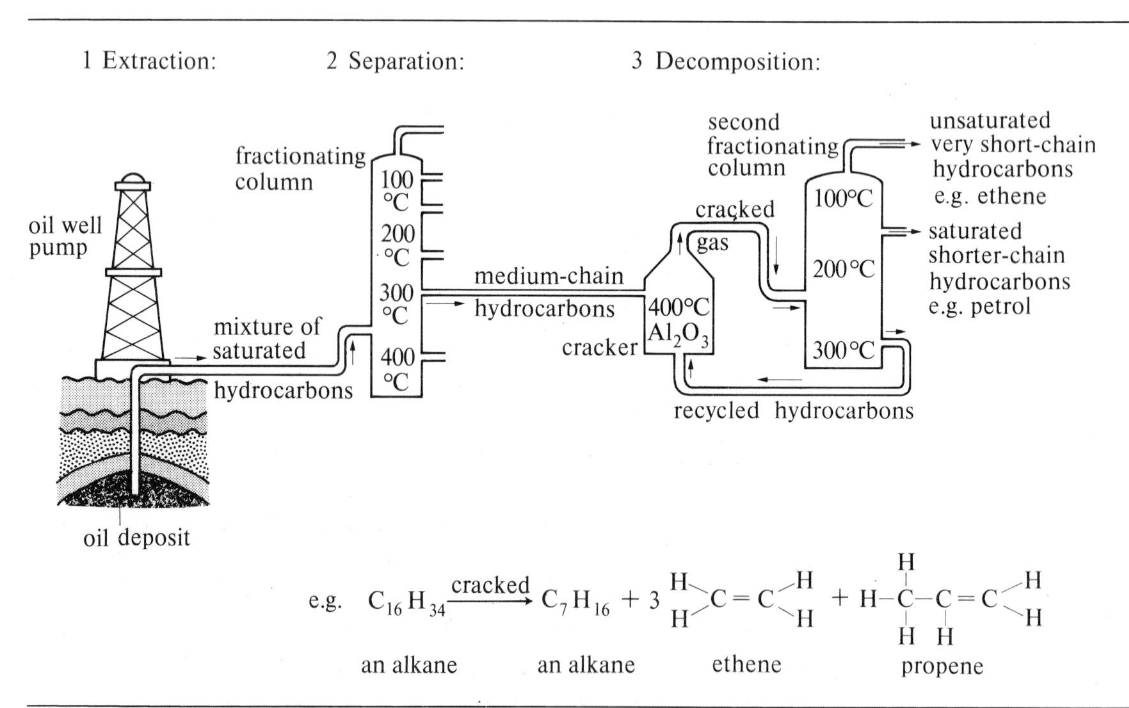

A catalytic cracker. The left-hand column is the catalytic cracker. The right-hand column is a fractionating column for separating the products of cracking. During the reaction, carbon is deposited on the catalyst; this is burnt off in the central column

From unsaturated to saturated: addition

Unsaturated hydrocarbons are more reactive than saturated ones. This is because other atoms can easily be added to the unsaturated chains. We can make ethene do addition reactions in the laboratory with:

(i) bromine dissolved in water
(ii) potassium manganate(VII) in alkaline solution.

Ethene (double links)	*reacts with (adds on)*	*to produce (single links)*	*Observations*
$\mathrm{H_2C{=}CH_2}$ (H, H on each C, C=C)	$+\mathrm{Br{-}Br}$ brown	$\longrightarrow \mathrm{H{-}\underset{Br}{\overset{H}{C}}{-}\underset{Br}{\overset{H}{C}}{-}H}$ colourless	The brown colour of bromine water disappears. Any unsaturated hydrocarbon will cause this colour change.
$\mathrm{H_2C{=}CH_2}$ (H, H on each C, C=C)	$[\mathrm{O}]$ from $\mathrm{KMnO_4}$ + $\mathrm{H{-}OH}$ purple	$\longrightarrow \mathrm{H{-}\underset{OH}{\overset{H}{C}}{-}\underset{OH}{\overset{H}{C}}{-}H} + \mathrm{MnO_2}$ brown	The purple colour of the manganate(VII) turns to a dirty brown. This is also a test for unsaturation.

The two addition reactions shown in the table above are used for testing hydrocarbons in the laboratory. If a hydrocarbon causes the colour changes, it is unsaturated. If it does not, it is saturated.

There are two other very important addition reactions of ethene. Both these are carried out on a large scale in industry:

(i) adding water to make ethanol (alcohol)
(ii) adding ethene to itself to make polythene (polymerisation)

(i) Adding water

$$\mathrm{H_2C{=}CH_2} + \mathrm{H{-}OH} \xrightarrow[\text{temperature } 300^\circ\text{C, pressure 70 atmospheres}]{\text{catalyst: orthophosphoric acid}} \mathrm{H{-}CH_2{-}CH_2{-}OH}$$

We can *reverse* this reaction in the laboratory, to make ethene. Concentrated sulphuric acid is added to ethanol and the mixture heated to 170°C. The concentrated sulphuric acid acts as a **dehydrating agent**.

$$\mathrm{H{-}CH_2{-}CH_2{-}OH} \xrightarrow[170^\circ\text{C}]{\text{conc. } H_2SO_4} \mathrm{H_2C{=}CH_2} \;(-H_2O)$$

conc. sulphuric acid

ethanol

heat

wash bottle of alkali to remove acidic impurities

ethene

water

Fig. 4
Dehydrating ethanol to ethene in the lab

(ii) Polymerisation

$$\mathrm{H_2C{=}CH_2} + \mathrm{H_2C{=}CH_2} + \mathrm{H_2C{=}CH_2} \xrightarrow[\text{temp. } 100-150^\circ\text{C, pressure: up to 200 atmos.}]{\text{'Ziegler catalysts'}} \text{and on} \;{-}\mathrm{CH_2{-}CH_2{-}CH_2{-}CH_2{-}CH_2{-}CH_2}{-}\; \text{and on}$$

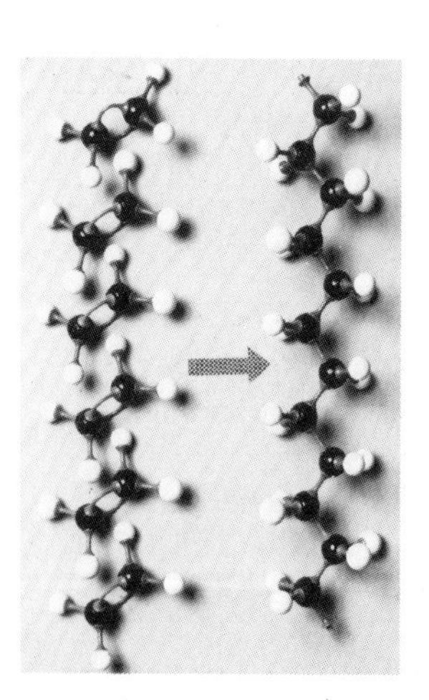

The Ziegler catalysts are titanium and aluminium compounds. During the reaction, ethene molecules add to each other. The process is called **polymerisation**. In Greek, 'poly' means 'many'. The product is a **plastic**. It is named **polythene** because *ethene* has been *poly*merised. The process is the opposite of cracking, because this time short-chain unsaturated hydrocarbons are being turned into long-chain saturated compounds. Some other common plastics are listed on page 345.

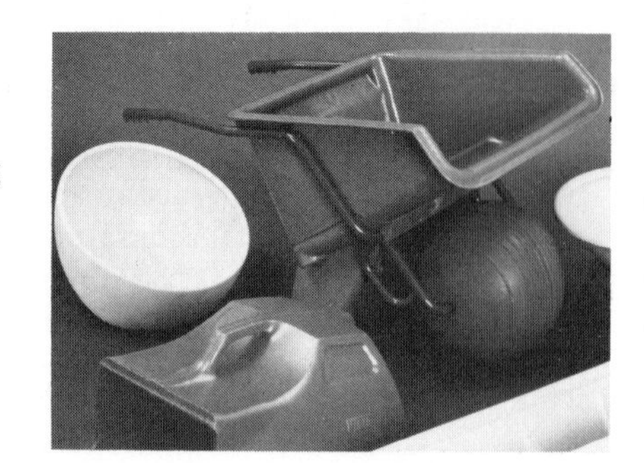

Left: polymerisation of ethene
Right: polythene for gardeners!

Polypropene	made by polymerising	$CH_3(H)C{=}CH_2$	propene
PVC (polyvinylchloride)	made by polymerising	$Cl(H)C{=}CH_2$	vinyl chloride
Polystyrene	made by polymerising	$C_6H_5(H)C{=}CH_2$	styrene

30.9 Alcohols

Structure and properties

If we write a water molecule as a chain, it looks like this: H—O—H
Some carbon chains end in the same way as a water molecule:

$$H_3C{-}O{-}H \qquad H_3C{-}CH_2{-}O{-}H$$

methanol ethanol

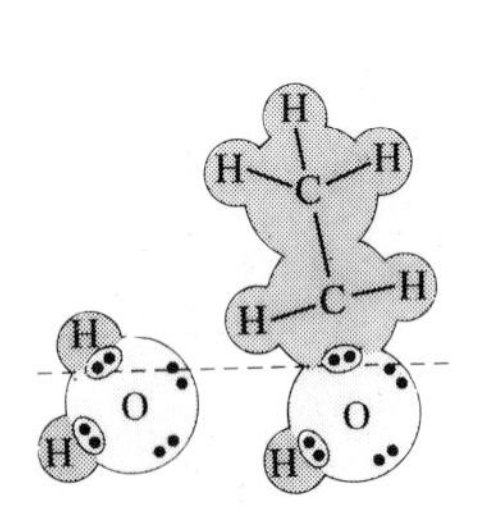

These compounds are called **alcohols**. Methanol and ethanol are the two simplest alcohols.

The structures of ethanol and water are similar. The two liquids also have properties that are similar, as shown in Table 30.6.

Table 30.6

Compound	*Structure*	*b.p./°C*	*With sodium chloride*	*With sodium*
Water	H—O—H	100	Salt dissolves in both. Both solutions conduct electricity.	Sodium reacts with both. Hydrogen gas is produced in both cases.
Ethanol	$H_3C{-}CH_2{-}O{-}H$	78		

Preparation of ethanol

Ethanol is made in very large amounts. Two different methods are used—each method makes the alcohol for a different purpose.

Type	*Made by*	*Use*
Methylated spirits (mainly ethanol)	Reacting ethene + water (see page 344)	Solvent, fuel, starting material for other chemicals.
Fermented alcohol (mainly ethanol)	Fermenting sugars with yeast	Beers, wines, spirits for drinking.

Fermentation

Fermentation is a process of decomposition. The carbon chains of sugar, found in certain types of grain and fruit, are broken down into simpler chains. Ethanol and carbon dioxide are the main products. The process is catalyzed by yeast. Yeast contains special substances called **enzymes** which cause the decomposition to occur. There are two stages:

(i) $C_{12}H_{22}O_{11} + H_2O \xrightarrow{\text{maltase}} 2C_6H_{12}O_6$
sugar

(ii) $C_6H_{12}O_6 \xrightarrow{\text{zymase}} 2C_2H_5OH + 2CO_2$
ethanol carbon dioxide

Maltase and zymase are enzymes in yeast.
Fermentation produces carbon dioxide gas. When the process is carried out in a closed container, the gas dissolves in the solution. On opening the container the gas escapes, causing the 'fizz' in cider, beer, and sparkling wines.

Oxidation of ethanol

The ethanol made by reacting ethene and water is used mostly as a solvent or a fuel, methylated spirits. Ethanol burns easily.

$$C_2H_5OH + 3O_2 \longrightarrow 2CO_2 + 3H_2O$$

Burning is an oxidation reaction; but ethanol can also be oxidised in another way. The product is a sour-tasting substance present in wine that has been left open to the air. The substance turns blue litmus red. It is ethanoic acid, an example of an **organic acid**.

Right: beer fermenting. Yeast is reacting with sugar in a liquor made by treating malted barley and hops with hot water. The fermentation is complete after about three days.

30.10 Organic acids

Structure and properties

All organic acid chains end in —COOH like carbonic acid. For example:

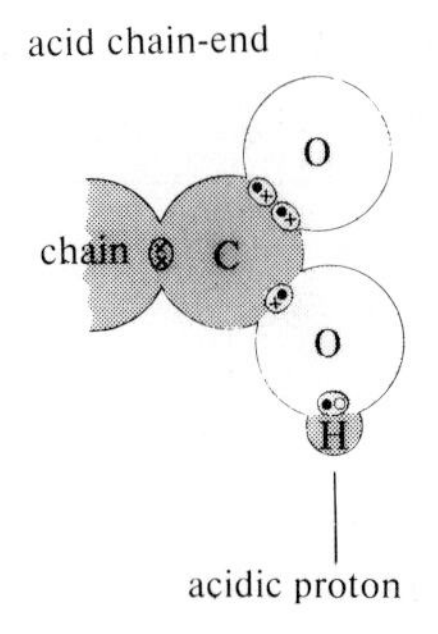

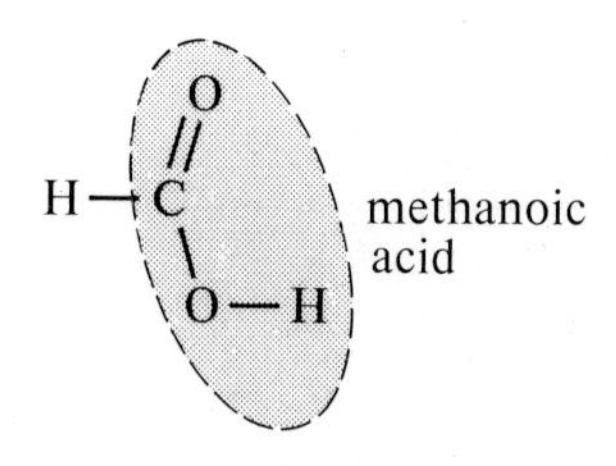

H H O
H—C—C—C
H H O—H
propanoic acid

Organic acids also have properties similar to carbonic acid. Take ethanoic acid, CH_3COOH, for example:

Acid	*Chain*	*pH*	*Taste*	*Reaction with alkali (sodium hydroxide) produces:*
Carbonic	H—O—C(=O)—O—H	5	Sharp (soda-water)	a salt H—O—C(=O)—O^-Na^+ sodium hydrogen carbonate
Ethanoic	H—C(H)(H)—C(=O)—O—H	3	Sour (vinegar)	a salt H—C(H)(H)—C(=O)—O^-Na^+ sodium ethanoate

Preparation of ethanoic acid

Organic acids are made by oxidising alcohols. The oxidation is usually done in solution, as shown for ethanoic acid in Fig. 5.

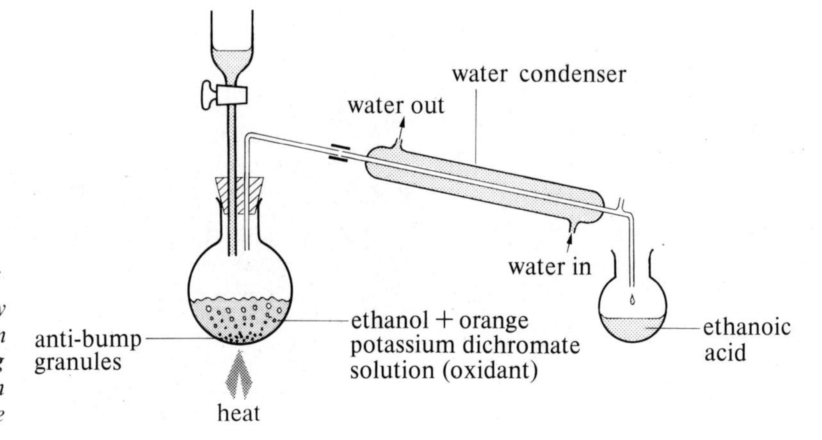

Fig. 5
Ethanoic acid made by oxidising ethanol in solution. The oxidising agent is potassium dichromate

The oxidation is done in solution to prevent it from going too far—we have already seen that alcohols, not in solution, are easily oxidised by burning to carbon dioxide and water. A good oxidising agent in solution is potassium dichromate(VI) dissolved in dilute acid.
Compare the products of the two methods of oxidising ethanol:

Ethanol	*oxidised*	*to produce*

<pre>   H H   H
 	
		/
		H—C—C—O
		
			|
			   H H</pre>
				by burning (total oxidation)	<pre>O H O
					|| / //
					C O cool O—C
					|| \ together / \
					O H H O—H
					                       carbonic acid</pre>

<pre>   H H   H
 	
		/
		H—C—C—O
		
			|
			   H H</pre>
				in solution (partial oxidation)	<pre> H O
					
						//
						H—C—C
						
							\
							   H O—H
							ethanoic acid</pre> (carbon chain ending like carbonic acid)

Notice that only the —OH end of the ethanol chain is oxidised when ethanoic acid is made.

Properties of ethanoic acid

Ethanoic acid occurs naturally in 'sour' wine—vinegar. The oxygen dissolved in the wine oxidises ethanol in the solution to ethanoic acid. A typical bottle of vinegar contains about 12% ethanoic acid. Ethanoic acid does all the reactions typical of acids:

(i) Acid	base (oxide)	salt	water	
$2CH_3COOH +$	$CaO \longrightarrow$	$Ca(CH_3COO)_2 +$	H_2O	
ethanoic acid	calcium oxide	calcium ethanoate		
(ii) Acid	base (carbonate)	a salt	carbon dioxide	water
$2CH_3COOH +$	$CuCO_3 \longrightarrow$	$Cu(CH_3COO)_2$	$+CO_2\uparrow$	$+H_2O$
ethanoic acid	copper(II) carbonate	copper(II) ethanoate		
(iii) Acid	reactive metal	salt	hydrogen	
$2CH_3COOH +$	$Mg \longrightarrow$	$Mg(CH_3COO)_2$	$+H_2\uparrow$	
ethanoic acid	magnesium	magnesium ethanoate		

30.11 Esters

If we write a carbonic acid molecule to look like a chain, its structure is:

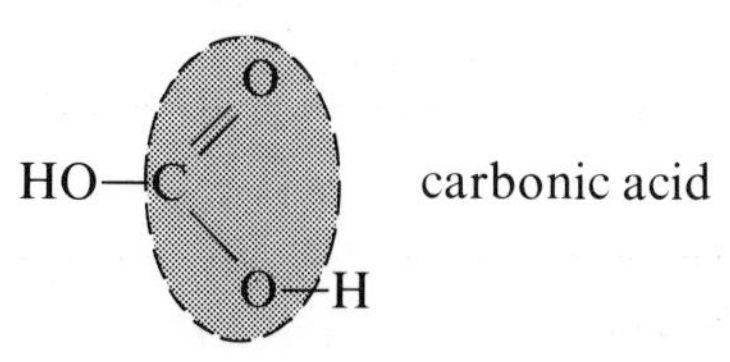

It is the two ends of a water molecule, HO— and —H, linked by 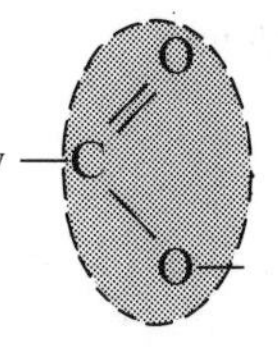

This special link is found in the chains of many naturally-occurring compounds. These compounds are called **esters**. A simple example of an ester is ethyl ethanoate:

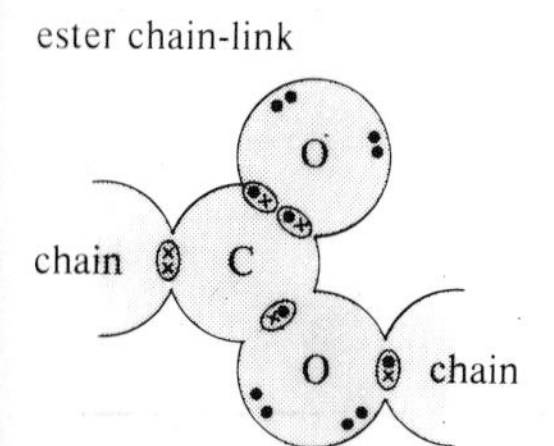

$$\mathrm{H{-}\overset{H}{\underset{H}{C}}{-}\overset{O}{\overset{\|}{C}}{-}O{-}\overset{H}{\underset{H}{C}}{-}\overset{H}{\underset{H}{C}}{-}H}$$

ethyl ethanoate

It is two ends of a carbon chain, $\mathrm{H{-}\overset{H}{\underset{H}{C}}{-}}$ and $\mathrm{{-}\overset{H}{\underset{H}{C}}{-}\overset{H}{\underset{H}{C}}{-}H}$, linked by $\mathrm{{-}\overset{O}{\overset{\|}{C}}{-}O{-}}$

Two types of esters are found naturally:

Chain length	*Found in*	*Properties*	*Used for*
Short*	Ripe fruit	Liquids with fruity smells	Flavouring, perfumes, and solvents
Long	Vegetable oils, animal fats	Oily liquids or fatty solids	Foods, soap-making

* These esters can also be made artificially.

Long-chain esters from animal and vegetable sources (fish, soya beans, palm kernels) are used to make Stork margarine. Short-chain esters from lemons are used to make lemon flavouring

Preparation of esters In the laboratory we make esters by reacting organic acids with alcohols. For example:

water removed

$$CH_3-C(=O)-O-H + H-O-CH_2-CH_3 \longrightarrow CH_3-C(=O)-O-CH_2-CH_3 + H-OH$$

ethanoic acid + ethanol ⟶ ethyl ethanoate + water

This reaction is rather like the reaction of an acid and an alkali. Compare the two reactions of ethanoic acid:

(i) $CH_3COO-H + H-O-C_2H_5 \longrightarrow CH_3COOC_2H_5 + H-OH$

acid + alcohol ⟶ ester + water

(ii) $CH_3COO-H + HO-Na \longrightarrow CH_3COONa + H-OH$

acid + alkali ⟶ salt + water

Although the reactions are alike, there are two important differences.

The nature of the reactions is different:

Reaction	*How fast*	*How far*
Acid and alkali	Very quick	Completely reacted
Acid and alcohol	Very slow	Does not react to completion

The acid and alcohol reaction is reversible. It is so slow at room temperature that it would take about a year to reach equilibrium. Equilibrium, you may remember, means that the amount of product reaches a constant level.
We can speed up the rate of reaction by:

heating
adding a catalyst—concentrated sulphuric acid.

Adding concentrated sulphuric acid helps the reaction because it is a **dehydrating agent**. It removes water well.

The products of the two reactions are different:

Product	*Appearance*	*m.p./°C*	*Solubility in 100 g of water at 20°C*	*Conductivity of solution in water*
The ester, $CH_3COOC_2H_5$ (ethyl ethanoate)	Liquid	−83·6	Less than 1 g	Zero
The salt, CH_3COONa (sodium ethanoate)	White solid	324	146 g	High

Compare the properties of the two *ethanoates* with the properties of tetrachloromethane and sodium chloride:

CCl_4 (tetrachloromethane)	Liquid	−23	Less than 1 g	Zero
NaCl (sodium chloride)	White solid	801	36 g	High

From the comparisons above, you should see that:

ethyl ethanoate is a typical **molecular** substance (see page 84)

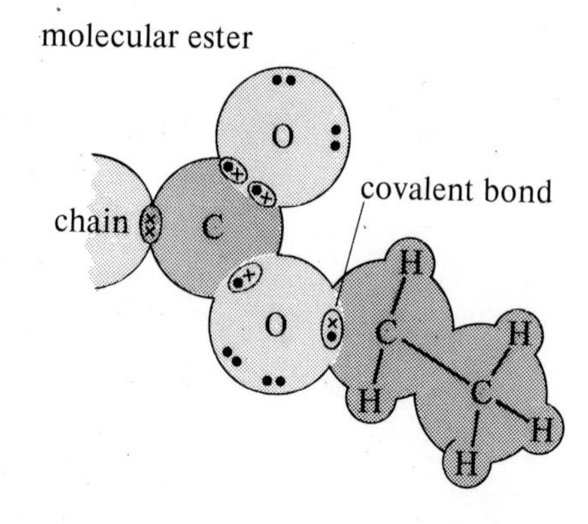

sodium ethanoate is a typical **ionic** substance (see page 85)

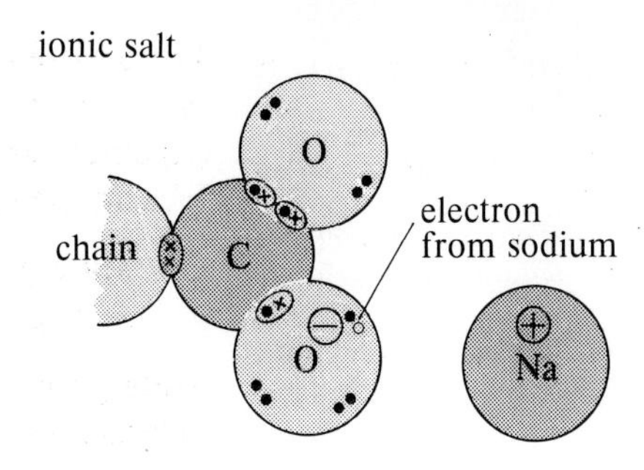

Hydrolysis of esters

The ester link is made by removing water in *acid conditions.*
The ester link can be broken by adding water in *alkaline conditions.*
In other words, you can reverse the reaction that makes an ester. You just take an ester and boil it in a solution of sodium hydroxide:

$$CH_3COOC_2H_5 + NaOH \xrightarrow{\text{boil}} CH_3COONa + C_2H_5OH$$

ester + alkali ⟶ salt + alcohol

This reaction is often called the **hydrolysis** of an ester.

Fats, oils, and soaps

Long-chain esters are found in all animal fats and vegetable oils. Each of these chains contains three ester links:

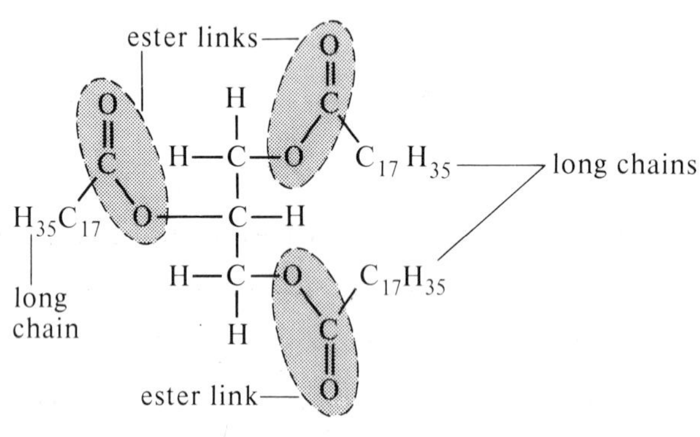

In fats, the long chains are saturated; in oils, they are unsaturated.

These esters can be hydrolysed just like ethyl ethanoate. If you boil a piece of fat or some vegetable oil in concentrated alkali, you get a sodium salt and an alcohol. The sodium salt is the salt of a long-chain acid:

$$C_{17}H_{35}-COO^-Na^+$$

Top: olive oil contains unsaturated long-chain esters
Bottom: hydrolysis of a long-chain ester produces soap

It is the main ingredient of soap.

Soaps are very useful because they allow grease to dissolve in water. Greases are molecular substances. Water is a good solvent for ionic substances. A soap works because it has both molecular and ionic properties:

A typical grease	H H H H H H H H H H H H H—C—C—C—C—C—C—C—C—C—C—C—C—H H H H H H H H H H H H H	Molecular, so mixes well with other oils and fats
A salt	H O H—C—C (=O) H O^-Na^+ sodium ethanoate	Ionic, so mixes well with water
A soap	H H H H H H H H H H H O H—C—C—C—C—C—C—C—C—C—C—C—C (=O) H H H H H H H H H H H O^-Na^+	One end is molecular, the other end is ionic. So mixes well with oils and fats *and* with water

30.12 Chain length and chain type: a summary

(i) The length of its carbon chain controls most of a compound's *physical properties*.

Alkanes	H H H—C—C—H H H	H H H H H H—C—C—C—C—C—H H H H H H	H H H H H—C—C—$(CH_2)_{30}$—C—C—H H H H H
	C_2	C_5 chain-length	C_{34}
Alkenes	H H C=C H H	H H H H H—C—C=C—C—C—H H H H H	H H H C=C—$(CH_2)_{30}$—C=C H H H
State at 15°C	gas ⟶	liquid ⟶	solid

(ii) Its functional group controls most of a compound's chemical properties.

Class of compound	*Example*	*Functional group*	*Typical chemical property*
Alkene	ethene H H C=C H H	>C=C<	Decolourizes bromine water >C=C< + Br_2 ⟶ —C(Br)—C(Br)— brown colourless
Alcohol	ethanol H H H H—C—C—O H H	—C—O—H	Can be dehydrated to an alkene H H H—C—C—OH $\xrightarrow[H_2SO_4]{conc.}$ C=C (with H on each carbon) H H
Acid	ethanoic acid H O H—C—C H O—H	—C(=O)—O—H	Reacts with sodium carbonate solution acid carbonate salt $2CH_3COOH + Na_2CO_3 \longrightarrow 2CH_3COO^-Na^+$ $CO_2\uparrow + H_2O$
Ester	methyl methanoate O H—C H O—C—H H	—C(=O)—O—C—	Can be hydrolyzed by alkali ester alkali salt $HCOOCH_3 + NaOH \longrightarrow HCOO^-Na^+$ $+ CH_3OH$ alcohol

(iii) The position of the links in the chain can affect a compound's *physical and chemical* properties. Look at the acid and ester in the table above. Each has the same molecular formula, $C_2H_4O_2$, but each has different properties. This is because the atoms are linked in a different way in each. Compounds of this sort are called **isomers**.

Isomers are compounds with the same molecular formula but different structural formulas.

The simplest examples are the two isomers with molecular formula C_4H_{10}:

butane: H—C—C—C—C—H (each C carrying two H above and below)

isobutane: $H_3C—C(CH_3)(H)—CH_3$

Reaction summary for organic compounds

Alkanes Unreactive, saturated:

$C_3H_8 + 5O_2 \longrightarrow 3CO_2 + 4H_2O$

$CH_4 + Cl_2 \longrightarrow CH_3Cl + HCl$ substitution—it continues until CCl_4

$C_8H_{18} \longrightarrow H_2C{=}CH_2 + C_6H_{14}$ cracking

Alkenes More reactive, because they are unsaturated:

$H_2C{=}CH_2 + H_2 \longrightarrow H_3C{-}CH_3$

$H_2C{=}CH_2 + Br_2 \longrightarrow H_2C(Br){-}CH_2(Br)$

$H_2C{=}CH_2 + HOH \xrightarrow{H_2SO_4} CH_3CH_2OH$

$nCH_2{=}CH_2 \longrightarrow {-}(CH_2{-}CH_2)_n{-}$

(addition reactions)

Alcohols Preparation:

(i) in industry $CH_2{=}CH_2 + H_2O \xrightarrow[70\ \text{atmos.}]{H_3PO_4,\ 300\ C} CH_3CH_2OH$

(ii) by fermentation $C_{12}H_{22}O_{11} + H_2O \xrightarrow[\text{catalyst}]{\text{enzyme}} 2C_6H_{12}O_6 \xrightarrow[\text{catalyst}]{\text{enzyme}} 4CH_3CH_2OH + 4CO_2$

(iii) in the laboratory $CH_2{=}CH_2 + H_2O \xrightarrow[\text{heat}]{\text{conc. } H_2SO_4} CH_3CH_2OH$

Oxidation:

(i) by burning $CH_3CH_2OH + 3O_2 \longrightarrow 2CO_2 + 3H_2O$

(ii) by an oxidising agent $CH_3CH_2OH \xrightarrow[\text{dil. } H_2SO_4]{K_2Cr_2O_7} CH_3CO_2H$

Reaction with acid produces an ester:

$CH_3CH_2OH + CH_3CO_2H \xrightarrow[H_2SO_4]{\text{conc.}} H_2O + CH_3CO_2CH_2CH_3$

Acids Preparation: oxidation of alcohols—see oxidation (ii) above.
They react with metals, oxides, hydroxides and carbonates to give salts; and they react with alcohols to give esters. Compare the reactions:

$CH_3CO_2H + NaOH \longrightarrow CH_3CO_2Na + H_2O$

complete reaction giving ionic salt

$CH_3CO_2H + CH_3CH_2OH \longrightarrow CH_3CO_2CH_2CH_3 + H_2O$

partial reaction giving covalent ester

Esters Preparation: reactions of acids with alcohols—see last reaction above.
They can be hydrolysed by NaOH to give a salt and an alcohol:

$CH_3CO_2CH_2CH_3 + NaOH \longrightarrow CH_3CO_2Na + CH_3CH_2OH$

hydrolysis—gives a soap when a large ester is used

Questions

1 (a) Describe the fractional distillation of crude oil.
(b) Draw the fractionating column, and beside it list, in the appropriate places, the names and uses of the main fractions that are produced.

2 Explain the meaning of the following words:
hydrocarbon fraction saturated cracking polymerisation functional group

3 Draw the structural formulas of the following molecules and then say which are unsaturated:
methane propanol butene hexane ethyne pentene

4 (a) Give examples of saturated and unsaturated hydrocarbons.
(b) What test would you carry out in the laboratory to tell them apart?
(c) Compare and contrast their reactions with halogens.

5 Explain carefully the difference between fractional distillation and cracking of oil. What are the products of each process?

6 (a) Describe how you would make ethene from ethanol in the laboratory.
(b) Give an equation for the reaction.
(c) Explain why the *polymerization* of ethene is important.
(d) Describe how ethanol is made from ethene in industry.

7 (a) Draw the apparatus that you would use to convert ethanol to ethanoic acid.
(b) Write an equation for this reaction.
(c) Describe what you would see in the reaction flask during the reaction.
(d) What type of reaction is this?

8 Draw the structural formulas of the following molecules:
ethane ethene ethanol ethanoic acid sodium ethanoate ethyl ethanoate

9 (a) Write the equation for the reaction between ethanoic acid and sodium hydroxide solution.
(b) What is this type of reaction called?
(c) Write the equation for the reaction between ethanoic acid and ethanol.
(d) What is this type of reaction called?
(e) What are the conditions for the second reaction?
(f) Compare the rates of these two reactions.
(g) Compare the extent of the two reactions.
(h) Compare the bonding in the products of the two reactions.

10 (a) Describe the properties of a typical short-chain ester.
(b) What is it used for?
(c) Describe the properties of a typical long-chain triester.
(d) Where is it found?
(e) What is its main use?

11 (a) What is a soap?
(b) How does a soap work?
(c) Describe how a soap is made, giving an equation.
(d) Explain the meaning of the word *saponification*.

12 Alcohol for drinking is made by fermenting sugars and starches.
(a) Write an equation for the fermentation of cane sugar, $C_{12}H_{22}O_{11}$.
(b) What is the purpose of an enzyme in this process?
(c) What other product is made during fermentation?
(d) How many moles of alcohol are made from each mole of cane sugar?

13 (a) What are *isomers*?
(b) Draw out the structural formulas of all the isomers with molecular formula C_6H_{14}.
(c) Would you expect these isomers to have similar chemical properties? Explain your answer.
(d) Would they have similar physical properties?

31 Group V: nitrogen

31.1 The element

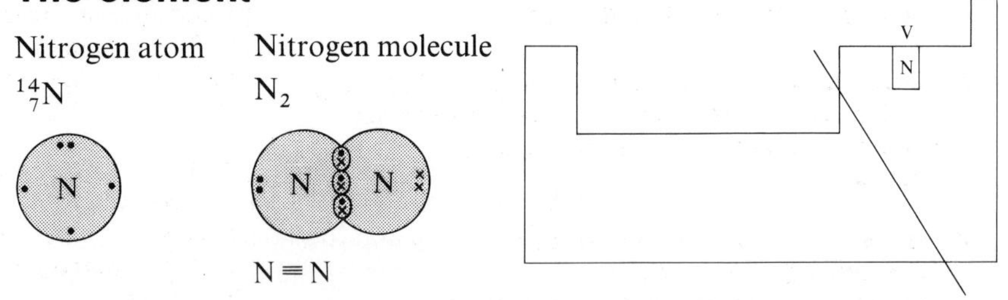

Appearance and preparation

Nitrogen is a colourless unreactive gas. It is found in large amounts free in the air. Its formula is N_2.

To obtain a sample of nitrogen gas in the laboratory, a solution of ammonium nitrate(III) is boiled. Ammonium nitrate(III), NH_4NO_2, decomposes to nitrogen and water:

$$NH_4NO_2 \xrightarrow{\text{heat}} 2H_2O + N_2 \uparrow$$

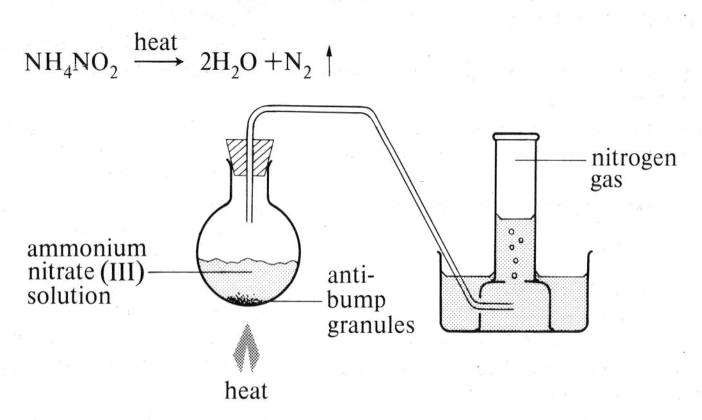

Fig. 1
Making nitrogen in the lab

Physical properties

These are listed in the table below.

Element	*m.p./°C*	*b.p./°C*	*Density (c.f. air)*	*Colour*	*Smell*	*Solubility in water at S.T.P.*
Nitrogen	−210	−196	0·97	None	None	Less than 1 cm^3 of gas in 100 cm^3 water.

These figures show that nitrogen is a typical *molecular* substance (see page 84). It has no colour or smell and is neutral to litmus. It is a difficult gas to test for.

31.2 Occurrence and extraction

As the element Free nitrogen makes up most of the atmosphere around us:

Gas	*% by volume in the air*
Nitrogen	78·0
Oxygen	21·0
Noble gases	0·9
Carbon dioxide	0·03

Uses Nitrogen has several important uses in industry. It is used as the starting material for the manufacture of ammonia and nitric acid. It is also used as an 'inert' gas in welding and food packaging.

The nitrogen required by industry is taken from the air. Fractional distillation of liquid air is the best method for this:

1. The air is cooled until it becomes a liquid mixture, temperature −200°C.
2. Water and carbon dioxide solidify at that temperature and are removed.
3. The liquid air is allowed to boil gently, at −192°C, in a fractionating column.
4. Nitrogen has the lowest b.p. and can pass up the column as a vapour. The other components condense and run back down as liquids.
5. 99·9% pure nitrogen is collected at the top of the column.

metal column

N_2

−196 °C

−192 °C

vapour from liquid air

Fig. 2 *Separating liquid air by fractional distillation. The tower in the photo contains the column, which is insulated to keep it cold*

As nitrogen compounds Apart from the nitrogen in the air, there are other sources of the element in nature:

(i) Plants and animals contain **proteins**. Proteins are organic compounds of nitrogen.

(ii) The soil and river-water contain metal nitrates and ammonium salts.

All these nitrogen compounds are formed using the nitrogen in the air.

The nitrogen cycle

Sometimes very complicated reactions take place before they are produced. Some of these compounds slowly decompose to make nitrogen again. The whole process is called the **nitrogen cycle**. Compare it with the carbon cycle on page 336.

The nitrogen cycle

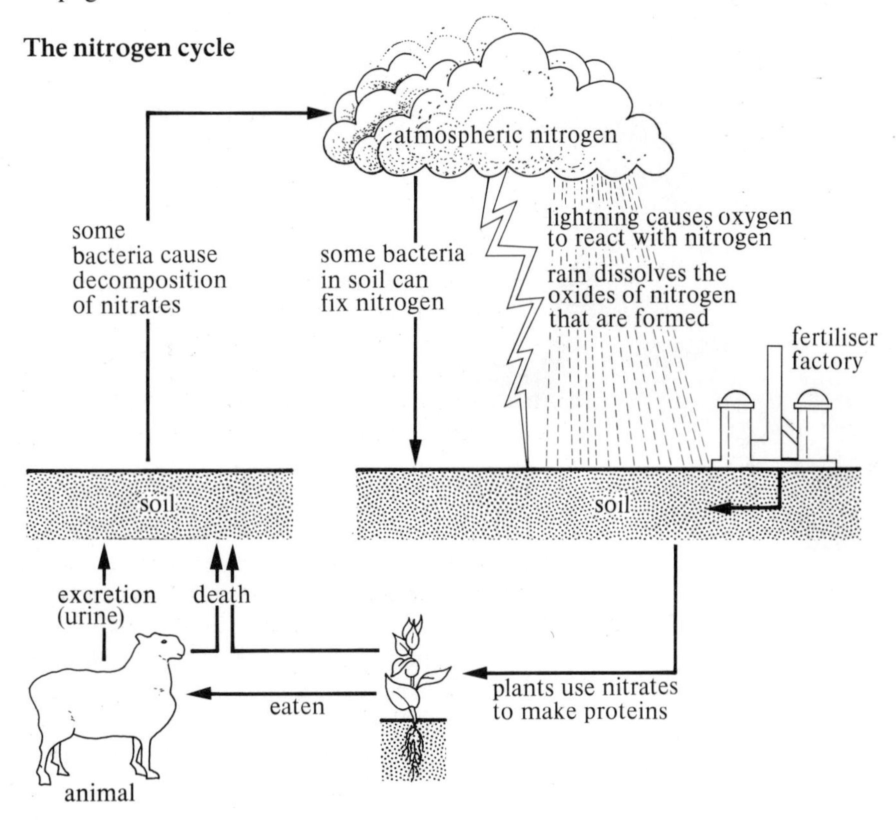

Plants and animals need nitrogen to grow. Any process in which the nitrogen from the air is turned into nitrogen compounds in the soil is called **fixing nitrogen**. Nitrogen is fixed naturally in thunderstorms, and by bacteria in the soil. Fertilisers contain artificially fixed nitrogen.

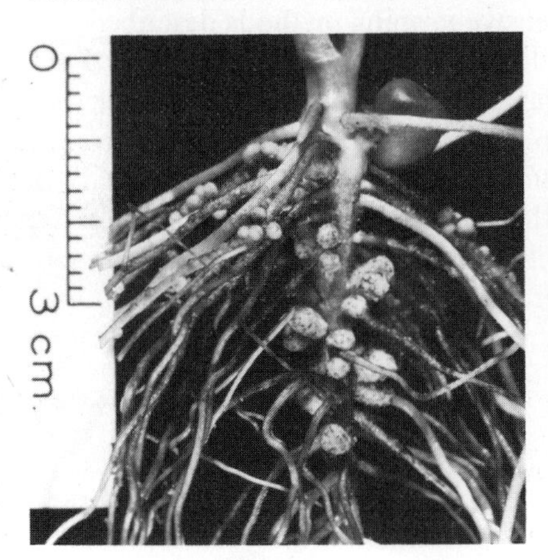

Right: root nodules of a bean plant. They contain bacteria which can fix nitrogen, converting it to nitrates. Plants with such nodules (e.g. beans, peas, clover) are described as leguminous

Far right: loading a tractor with ammonium nitrate fertiliser (Nitram)

Fertilizers

An experiment to show what elements plants need in order to grow.

Laboratory experiments like the one above show that, in addition to nitrogen, plants need other elements in the soil if they are to grow well. While nitrogen is needed for making proteins, phosphorus is vital for plant enzymes, magnesium is the centre of the chlorophyl molecule, calcium is needed in the cell walls, and potassium is essential for controlling the rate of photosynthesis.

Without these elements in fair quantities, and others like iron, copper, and zinc in trace amounts, the plant will not grow well.

A balanced fertilizer (see page 364) supplies the needs of the plant that are lacking in the soil. The knowledge of exactly what plants require has led to the intensive farming methods described in Chapter 21.

Measured doses of the right fertilizer applied at the right time have produced huge increases in the yields of arable crops. The result has been the grain surpluses of America, Canada, Australia, and more recently the European Community.

However, the very heavy cropping has begun to destroy the soil balance, possibly by killing the bacteria that lives in the soil. The result is that large parts of the grain belt of America in particular are becoming barren dust bowls. This is not a new development. The northern Sahara desert was once the granary of the Roman Empire.

31.3 Compound formation

Compounds with metals

Nitrogen is an unreactive non-metal. Only under rather violent conditions will it react with other elements.

When reactive metals are strongly heated in a stream of nitrogen, a slow reaction occurs. The products are metal nitrides:

ionic nitrides

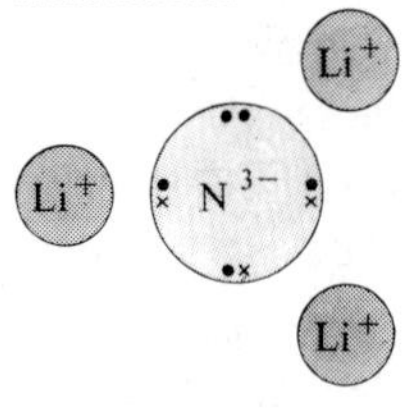

Magnesium ⟶ magnesium nitride

$3Mg + N_2 \longrightarrow Mg_3N_2$

or

Lithium ⟶ lithium nitride

$6Li + N_2 \longrightarrow 2Li_3N$

Compounds with non-metals

(i) With hydrogen

Nitrogen forms the important hydride ammonia, NH_3. Ammonia is a gas at room temperature (b.p. $-33°C$), with a strong pungent smell. It dissolves in water to give an alkaline solution.

molecular ammonia

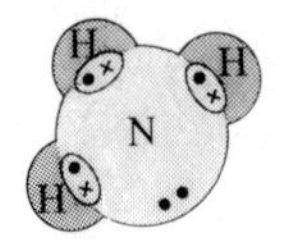

Ammonia is produced when water is added to a metal nitride:

$Li_3N + 3H_2O \longrightarrow 3LiOH + NH_3\uparrow$

However, it is best prepared in the laboratory using the standard apparatus for gas preparation. For ammonia:

the solid reactant is ammonium chloride;

the liquid reactant is sodium hydroxide solution.

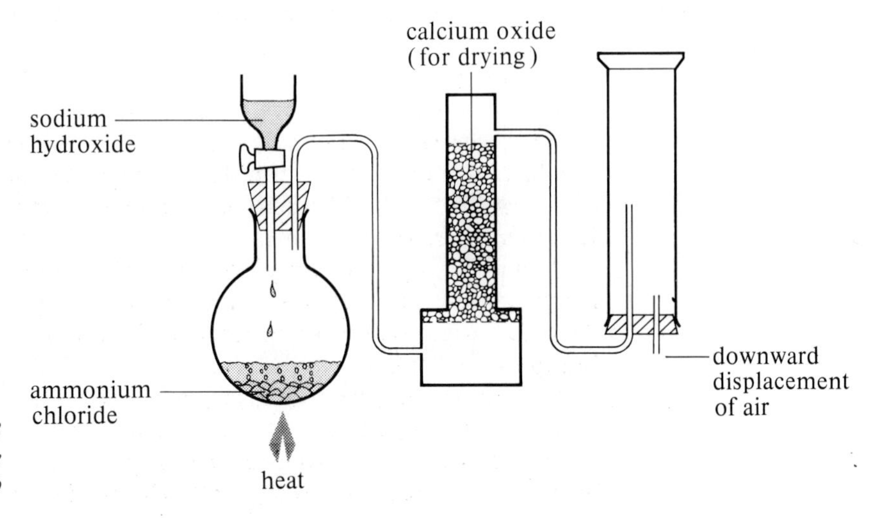

Fig. 3
Making ammonia in the lab

$NH_4Cl + NaOH \xrightarrow{boil} NaCl + H_2O + NH_3\uparrow$

Ammonia is lighter than air and is collected by downward displacement. Ammonia cannot be collected over water: it is very soluble.

Manufacture of ammonia

In industry, ammonia is made in huge quantities by reacting together nitrogen and hydrogen. The process is called the **Haber process**.

The process

Nitrogen is extracted from the air and hydrogen is produced by reducing water. The gases are mixed and passed into a reaction chamber. High pressures and temperatures and an iron catalyst are used. Only about 10 % of the reactants react to produce ammonia. The ammonia is liquified by sudden cooling and the unreacted nitrogen and hydrogen are recycled into the reaction chamber again.

Details

Raw materials: Air and water ⟶ nitrogen + hydrogen

Conditions: Temperature, up to 500°C
Pressure, up to 1000 atmospheres
Catalyst, iron

Reaction

$$N_2 + 3H_2 \rightleftharpoons 2NH_3$$

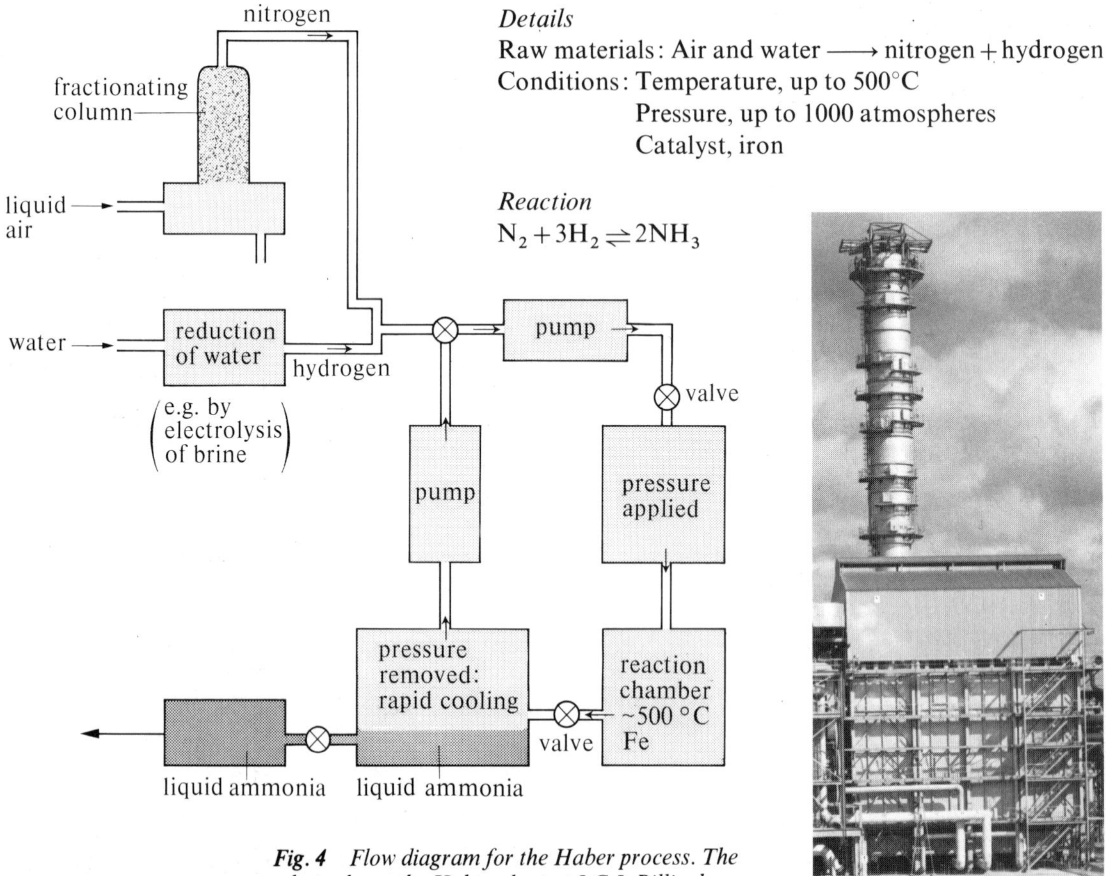

Fig. 4 *Flow diagram for the Haber process. The photo shows the Haber plant at I.C.I. Billingham*

Important points

Ammonia decomposes at temperatures as high as 500°C. So while *some* nitrogen and hydrogen is combining to make ammonia, *some* of the ammonia is decomposing back to the elements. In other words the reaction is reversible. The reaction mixture is passed quickly over the catalyst: only a 10 % conversion occurs. Ammonia liquifies much more easily than nitrogen and hydrogen. Unreacted nitrogen and hydrogen gases are recycled.

A catalytic converter for a car's exhaust system

(ii) With oxygen

Nitrogen combines with oxygen only when a large amount of energy is supplied to the mixture.

In thunderstorms, lightning flashes supply enough energy to make nitrogen and oxygen combine.

In car engines, the spark and subsequent explosion of the petrol-air mixtures causes nitrogen to combine with oxygen in the cylinders. Nitrogen monoxide (nitrogen(II) oxide), NO, is the main product. It is poisonous. In America, the exhaust systems of new cars are now fitted with special catalysts, to convert the poisonous gases into safer products:

$$2CO + 2NO \xrightarrow[\text{in exhaust pipe}]{\text{catalyst}} 2CO_2 + N_2$$

poisonous gases　　　safe gases

Manufacture of nitric acid

The process:

A mixture of nitrogen oxide and nitrogen dioxide is made by reacting ammonia with oxygen. A platinum catalyst is used for this exothermic reaction. The mixture of nitrogen oxides and air is then reacted with water to make hydrogen nitrate(V), nitric acid.

Details

Raw materials: ammonia, air, water

Conditions: temperature, 900°C
pressure, 4 atmospheres
catalyst, platinum

Reactions:

$$\left.\begin{array}{l} 4NH_3 + 5O_2 \xrightarrow{\text{platinum}} 4NO + 6H_2O \\ 2NO + O_2 \xrightarrow{\text{platinum}} 2NO_2 \end{array}\right\}$$

$$4NO_2 + O_2 + 2H_2O \longrightarrow 4HNO_3$$

Important points

The platinum catalyst is heated electrically to 900°C. Once the reaction starts, however, no more heating is needed. The oxidation of ammonia is exothermic. Unreacted gases are recycled to prevent waste. The nitric acid produced is often combined with ammonia to make ammonium nitrate, NH_4NO_3, for fertilizers.

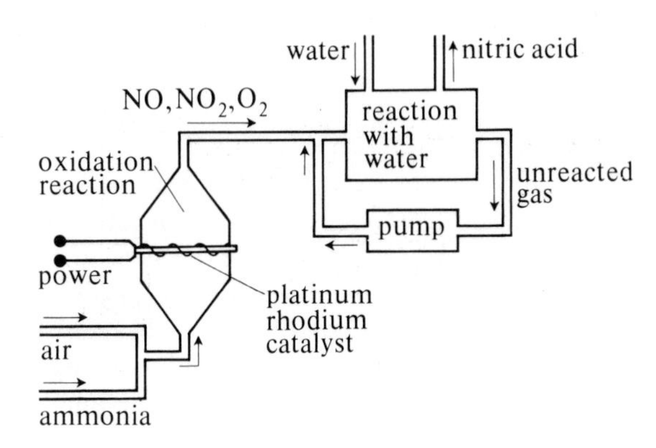

The nitric acid plant at I.C.I. Billingham. Because it uses ammonia, it is built next to the Haber plant (p. 362). Above is the flow diagram for the process.

I.C.I. Billingham—the siting of a plant

As we saw on page 360, plants need a range of elements to grow healthily. Fertilizers are added to the soil to make up for shortages of some of these elements.

The I.C.I. plant at Billingham contains an ammonia plant, a nitric acid plant, a sulphuric acid plant, and a phosphoric acid plant. These are all integrated together to produce two types of fertilizer. 'Straight N' fertilizer to stimulate leaf growth, and 'Compound NPK' fertilizer to improve crop yields. The raw materials and the relationship of the different plants is shown below.

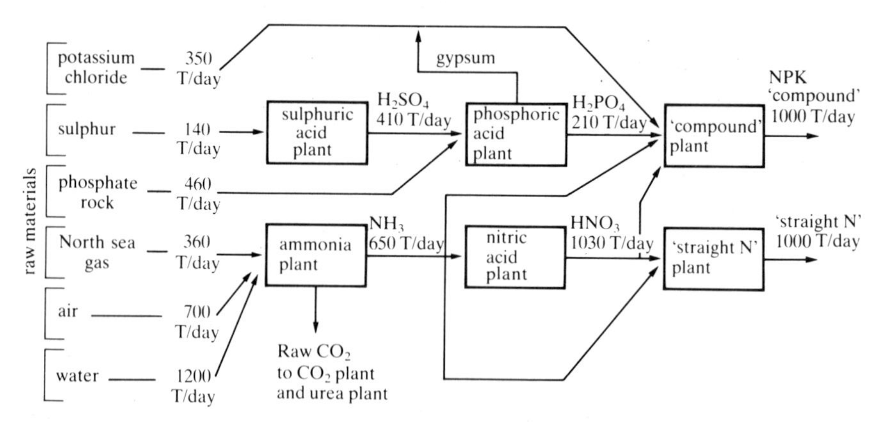

The plant was originally built at Billingham, because it was on top of a large deposit of calcium sulphate, from which sulphuric acid could be made, and because it was near the Yorkshire coalfield. The potassium chloride comes from Boulby, in Cleveland.

Now it is cheaper to import liquid sulphur and phosphate rock from abroad, and to get hydrogen from North Sea gas. Can you see why the position of the plant makes this so?

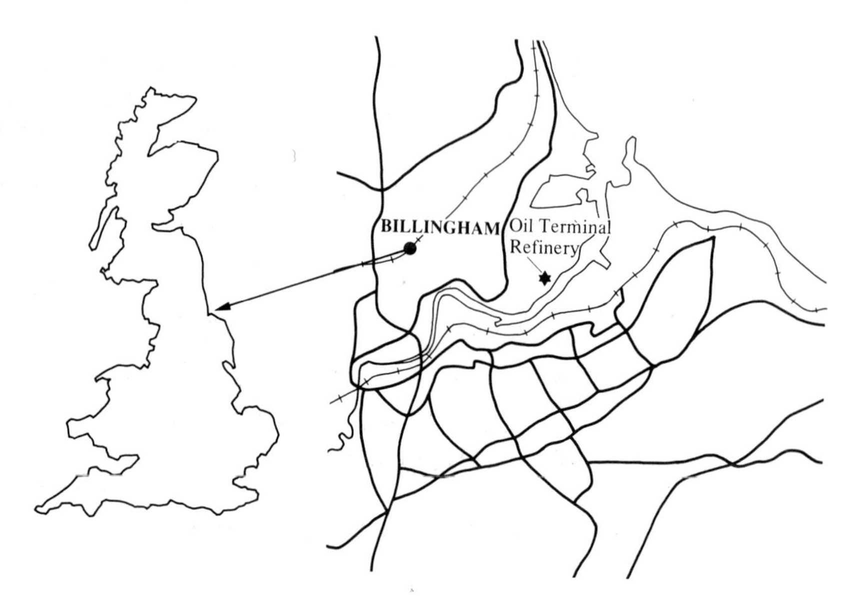

31.4 Solubility and acid/base chemistry

Ammonia as an alkali

Ammonia is very soluble in water: you can do the fountain experiment with it (see page 135).
A solution of ammonia in water is alkaline:

$$\underbrace{NH_{3(g)} + H_2O_{(l)}}_{99\%} \rightleftharpoons \underbrace{NH_{4(aq)}^{+} + OH_{(aq)}^{-}}_{1\%}$$

weak alkali

it is a weak alkali (only 1 % dissociation)
it turns red litmus blue;
it has a pH value greater than 7 (usually about 10–12)
it neutralises acid solutions to produce ammonium salts. For example we can make ammonium sulphate, $(NH_4)_2SO_4$, by neutralising sulphuric acid:

ammonium salts have the molecular cation NH_4^+

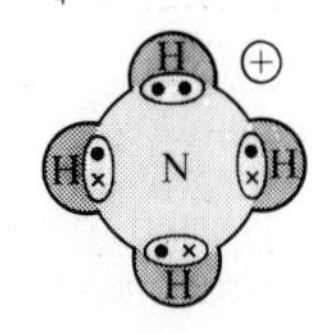

alkali	acid	salt
$2NH_3$	$+H_2SO_4 \longrightarrow$	$(NH_4)_2SO_4$

$$NH_{3\,(aq)} + H_3O^{+}_{(aq)} \longrightarrow NH^{+}_{4\,(aq)} + H_2O_{(l)}$$

Because ammonia in water reacts to produce the hydroxide ion, solutions of ammonia are able to cause precipitation of insoluble hydroxides. When a couple of drops of ammonia are added to a solution of a metal compound, an insoluble metal hydroxide precipitate often forms. For example:

soluble	ammonia solution		insoluble	soluble
$MgCl_2$	$+2NH_3+2H_2O$	$\longrightarrow$	$Mg(OH)_2\downarrow$	$+2NH_4Cl$
$Mg^{2+}_{(aq)}$	$+2OH^{-}_{(aq)}$	$\longrightarrow$	$Mg(OH)_{2\,(s)}$	

Nitric acid and nitrates

You saw that nitrogen dioxide (nitrogen(IV) oxide) combines with water and oxygen to produce hydrogen nitrate(V), HNO_3. Solutions of this compound:

turn blue litmus red;
have pH values less than 7 (usually 1 or 2).

These solutions are usually called nitric acid. Nitric acid shows all the properties of a typical acid (see page 141). The salts made by neutralizing nitric acid are called metal nitrates(V). Examples of the neutralization are:

	acid	metal oxide		salt	water
(i)	$2HNO_3$	$+MgO$	$\longrightarrow$	$Mg(NO_3)_2$ magnesium nitrate(V)	$+H_2O$
	$2H_3O^{+}_{(aq)}$	$+MgO_{(s)}$	$\longrightarrow$	$Mg^{2+}_{(aq)}$	$+3H_2O_{(l)}$

	acid	metal hydroxide		salt	water
(ii)	$2HNO_3$	$+Cu(OH)_2$	$\longrightarrow$	$Cu(NO_3)_2$ copper(II) nitrate(V)	$+2H_2O$
	$2H_3O^{+}_{(aq)}$	$+Cu(OH)_{2\,(s)}$	$\longrightarrow$	$Cu^{2+}_{(aq)}$	$+4H_2O_{(l)}$

	acid	metal carbonate	salt	carbon dioxide	water
(iii)	$2HNO_3 + CaCO_3$		$\longrightarrow Ca(NO_3)_2$ calcium nitrate(V)	$+CO_2\uparrow$	$+H_2O$
	$2H_3O^+_{(aq)} + CaCO_{3\,(s)}$		$\longrightarrow Ca^{2+}_{(aq)}$	$+CO_{2\,(g)}$	$+3H_2O_{(l)}$

31.5 Redox chemistry

Ammonia as a reductant

Nitrogen has an oxidation number −III in ammonia: $\overset{-III}{N}\overset{I}{H}_3$. In some of ammonia's reactions, the products are either:

pure nitrogen, N_2—oxidation number zero, or
nitrogen oxides—like NO, where nitrogen's oxidation number is +II.

Nitrogen's oxidation number goes up when ammonia is converted to either of the above products. So ammonia is being oxidised. Therefore another substance is being reduced, since both oxidation and reduction take place together. This means that ammonia is acting as a reducing agent.

(i) Ammonia reduces oxygen in the manufacture of nitrogen(II) oxide for making nitric acid.

$$4NH_3 + 5O_2 \longrightarrow 4NO + 6H_2O$$

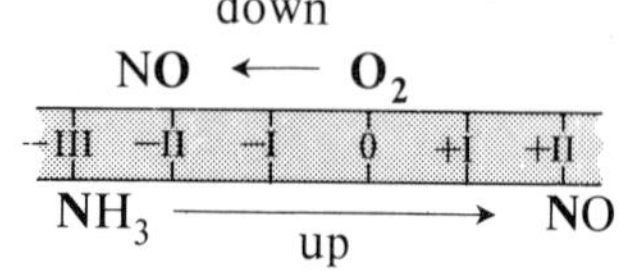

(ii) Ammonia can be used as a reducing agent to reduce copper(II) oxide to copper.

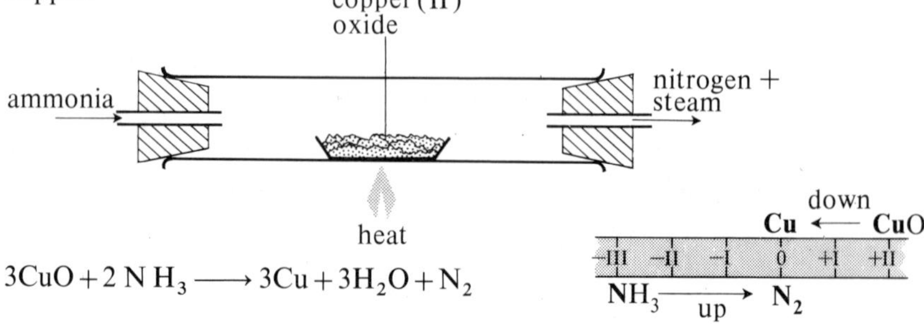

$$3CuO + 2NH_3 \longrightarrow 3Cu + 3H_2O + N_2$$

Nitric acid as an oxidant

Nitrogen has an oxidation number +V in nitric acid, $H\overset{+V}{N}O_3$. In many of nitric acid's reactions, nitrogen oxides are produced:

NO —nitrogen's oxidation number is +II.
NO_2—nitrogen's oxidation number is +IV.

So in reactions of nitric acid, nitrogen's oxidation number comes down. It is being reduced. Therefore another substance is being oxidised, and nitric acid is acting as an oxidising agent. It is in fact a very powerful oxidising agent. For example:

(i) When reactive metals react with nitric acid, it is often not possible to detect hydrogen coming off; this is because the acid oxidises the hydrogen to water.

(ii) Concentrated nitric acid oxidises carbon to carbon dioxide.

$$C + 2HNO_3 \longrightarrow CO_2 + 2NO_2 + H_2O$$

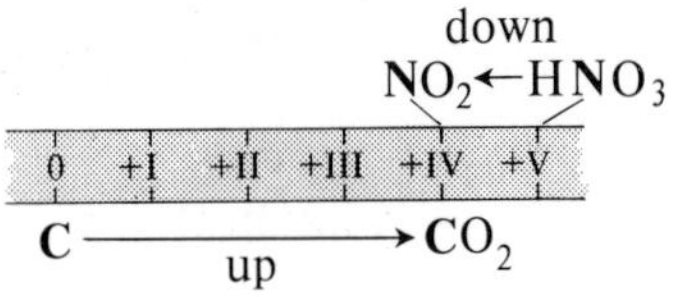

down
$NO_2 \leftarrow HNO_3$
0 +I +II +III +IV +V
$C \longrightarrow CO_2$
up

(iii) Nitric acid oxidises copper and unreactive metals low in the electrochemical series. Nitrogen oxide is produced when the acid is diluted a little, nitrogen dioxide is produced when concentrated acid is used.

$$3Cu + 8HNO_3 \longrightarrow 3Cu(NO_3)_2 + 4H_2O + 2NO\uparrow$$
cold, dilute

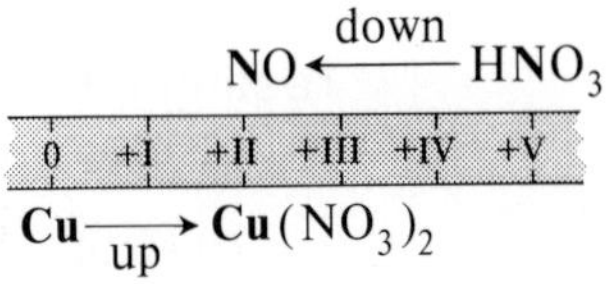

down
$NO \longleftarrow HNO_3$
0 +I +II +III +IV +V
$Cu \longrightarrow Cu(NO_3)_2$
up

$$Cu + 4NO_3 \longrightarrow Cu(NO_3)_2 + 2H_2O + NO_2\uparrow$$
concentrated

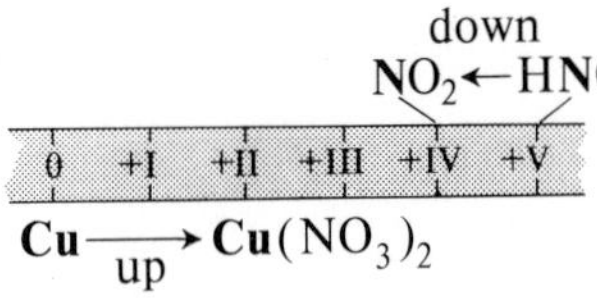

down
$NO_2 \leftarrow HNO_3$
0 +I +II +III +IV +V
$Cu \longrightarrow Cu(NO_3)_2$
up

31.6 Special features

Complex salts of ammonia

On page 365, you saw that ammonia is an alkali. Like sodium hydroxide, it gives precipitates of insoluble metal hydroxides, when added to solutions of most metal compounds:

insoluble solution	a few drops of ammonia solution		pale blue	soluble
$CuSO_4$	$+2NH_3 + 2H_2O$	$\longrightarrow$	$Cu(OH)_2\downarrow$	$+(NH_4)_2SO_4$
$Cu^{2+}_{(aq)}$	$+2OH^-_{(aq)}$	$\longrightarrow$	$Cu(OH)_{2\,(s)}$	

An unusual effect is seen when excess ammonia solution is added to certain metal hydroxide precipitates, like $Cu(OH)_2$, AgOH and $Zn(OH)_2$. *The precipitates dissolve again.* Often the solution also changes colour. The reason is that ammonia reacts further with the precipitates to produce a **complex salt**. All complex salts are soluble, so the precipitates redissolve:

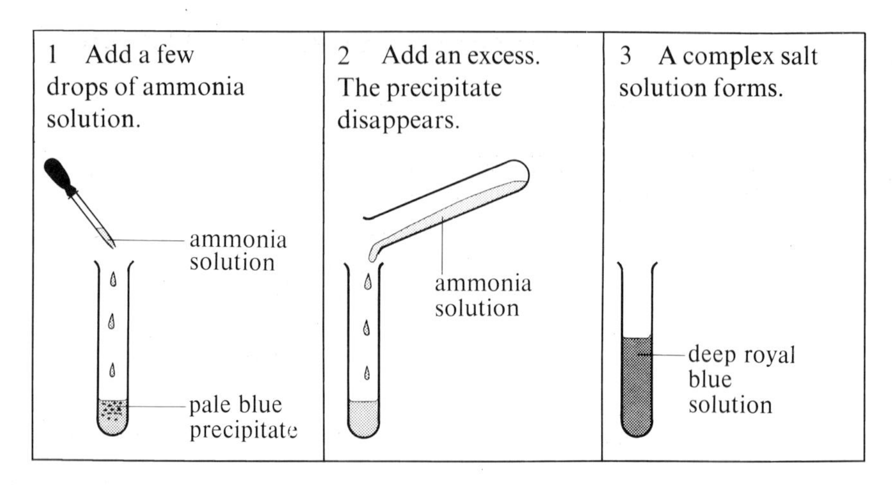

$$\underset{\text{a few drops}}{CuSO_4 + 2NH_3} + 2H_2O \longrightarrow \underset{\text{pale blue}}{Cu(OH)_2\downarrow} + (NH_4)_2SO_4 \xrightarrow[\text{extra ammonia}]{+2NH_3} \underset{\text{deep blue solution; a complex salt}}{[Cu(NH_3)_4]SO_4} + 2H_2O$$

The complex salt above is tetra–ammino copper(II) sulphate(VI).

Testing for nitrates

There are two ways in which you can test a solid to see if it is a metal nitrate(V):

(i) Heat

Heat the solid strongly. Nitrates are not very stable when heated and decompose in a particular way.
The nitrates of less reactive metals decompose easily with gentle heating. Clouds of brown nitrogen dioxide can be seen.

$$2Cu(NO_3)_2 \xrightarrow{\text{heat}} 2CuO + \underset{\text{brown, acidic}}{4NO_2\uparrow} + O_2\uparrow$$

The nitrates of more reactive metals need much stronger heating. They decompose in a different way:

$$\underset{\text{sodium nitrate(V)}}{2NaNO_3} \xrightarrow[\text{heat}]{\text{strong}} \underset{\text{sodium nitrate(III)}}{2NaNO_2} + O_2\uparrow$$

(ii) The brown ring test

The steps are:

1 Dissolve the unknown solid.
2 Add some iron(II) sulphate to the solution.
3 Add concentrated sulphuric acid slowly to the mixture. Hold the tube at an angle. The oily liquid sinks to the bottom.
4 A brown ring forms between the two liquid layers if the solid is a nitrate(V).

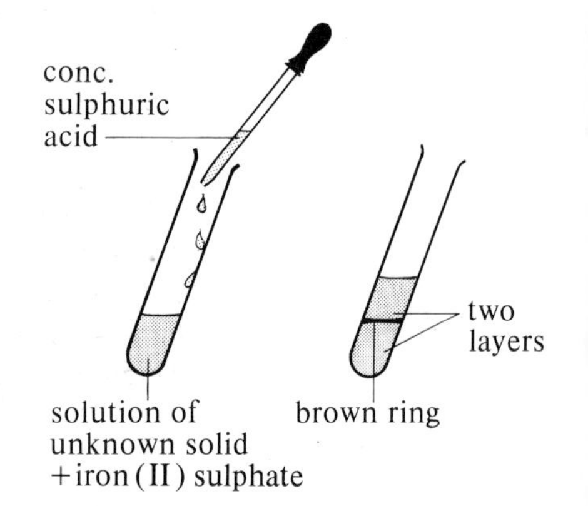

Reaction summary for nitrogen

Extraction Fractional distillation of air.

Reactions of the element Nitrogen is unreactive. It only reacts with very reactive metals or under violent conditions.
Oxidising other elements:
$3Mg + N_2 \longrightarrow Mg_3N_2$
$6Na + N_2 \longrightarrow 2Na_3N$; $Na_3N + 3H_2O \longrightarrow 3NaOH + NH_3$

$N_2 + 3H_2 \underset{\text{catalyst}}{\overset{\text{Fe}}{\rightleftharpoons}} 2NH_3$ (Haber process, 500°C, 500 atmospheres)

The hydride The hydride is ammonia. It has the following properties.

(i) Soluble alkaline gas: $NH_3 + H_2O \rightleftharpoons NH_4^+ + OH^-$
$NH_3 + HCl \rightleftharpoons NH_4Cl$

(ii) Reducing agent: $2NH_3 + 3CuO \longrightarrow 3Cu + 3H_2O + N_2$

$4NH_3 + 5O_2 \longrightarrow 4NO + 6H_2O$
then $2NO + O_2 \longrightarrow 2NO_2$
and $4NO_2 + O_2 + 2H_2O \longrightarrow 4HNO_3$
} oxidation of ammonia to make nitric acid

(iii) Complexing agent: $Cu(OH)_{2\,(s)} + 4NH_{3\,(aq)} \rightleftharpoons Cu(NH_3)_4{}^{2+}{}_{(aq)} + 2OH^-_{(aq)}$

The oxides There are three oxides: dinitrogen oxide (N_2O), nitrogen oxide (NO) and nitrogen dioxide (NO_2). Here is how they are made:

$NH_4NO_3 \longrightarrow 2H_2O + N_2O$
$3Cu + 8HNO_3 \longrightarrow 3Cu(NO_3)_2 + 4H_2O + 2NO$
cold, less concentrated acid
$Cu + 4HNO_3 \longrightarrow Cu(NO_3)_2 + 2H_2O + NO_2$
hot, concentrated acid

The hydroxy compound The hydroxy compound is nitric acid. It has these properties:

(i) strong acid: $HNO_{3\,(l)} + H_2O_{(l)} \longrightarrow H_3O^+_{(aq)} + NO^-_{3(aq)}$
It reacts with metals and bases.

(ii) oxidising agent: $C + 2HNO_3 \longrightarrow CO_2 + 2NO_2 + H_2O$
$Cu + 4HNO_3 \longrightarrow Cu(NO_3)_2 + 2H_2O + NO_2$

The nitrates All nitrates are soluble in water.
All decompose on heating.
Group I: $2NaNO_3 \longrightarrow 2NaNO_2 + O_2$
Others: $2Pb(NO_3)_2 \longrightarrow 2PbO + 4NO_2 + O_2$
Ammonium nitrate: $NH_4NO_3 \longrightarrow 2H_2O + N_2O$

Questions

1 Draw diagrams to show the electronic structure of:
(a) a nitrogen atom (b) a nitrogen molecule

2 Give an example of:
(a) a molecular nitride (b) an ionic nitride.
(c) Draw the electronic structure of each.

3 (a) What is the main source of nitrogen?
(b) How is nitrogen extracted?
(c) What is pure nitrogen used for?
(d) Explain why plants need nitrogen.
(e) What does the *fixing* of nitrogen mean?
(f) Describe two natural ways in which nitrogen is fixed.
(g) Describe an artificial way in which nitrogen is fixed.
(h) Why is it necessary to fix nitrogen artificially?

4 (a) Sketch the apparatus you would use to prepare and collect a gas in the laboratory. What reactants would you use, and what is the equation for the reaction to make?
(b) ammonia
(c) nitrogen dioxide
(d) dinitrogen oxide

5 (a) Name the industrial process for making ammonia.
(b) Write the equation for the overall reaction.
(c) What are the conditions used in industry for this process?
(d) Explain why in spite of these conditions, only about 10% of the reactants are made into ammonia.
(e) What are the main uses for ammonia?

6 (a) How does ammonia react with water?
(b) What will be the pH of the solution?
(c) Describe an experiment that shows how soluble ammonia is in water.
(d) How can you test for ammonia in the laboratory?

7 Write an equation for the reaction of ammonia with:
(a) hydrogen chloride gas
(b) nitric acid
(c) copper(II) oxide

8 (a) Write down the equations for the reactions in which nitric acid is manufactured.
(b) What are the conditions used in industry?
(c) Where does the heat for the reaction come from?
(d) What are the main uses for nitric acid?

9 (a) Nitric acid reacts with copper, sodium hydroxide, calcium carbonate, carbon, and copper oxide.
Give an equation for the reaction in each case.
(b) For each reaction, say whether nitric acid is reacting as an acid, a base, an oxidizing agent or a reducing agent.

10 An unknown solution is thought to be ammonium nitrate. You would need to perform two tests to prove this. Describe each test, and say what you would see if the test were positive.

11 (a) What would you see if a little ammonia gas were bubbled through copper(II) sulphate solution?
(b) Give an equation for this reaction.
(c) What would happen if much more ammonia were bubbled through the solution?
(d) Give an equation for this reaction.

12 The oxides of nitrogen are some of the pollutants produced by the internal combustion engine.
(a) How does the nitrogen get into the engine in the first place?
(b) What makes the unreactive nitrogen react in the engine?
(c) How can these nitrogen oxides be removed from car exhausts?

13 (a) Draw a diagram illustrating the nitrogen cycle, and label each step in the cycle.

32 Group VI: oxygen and sulphur

32.1 The elements

Oxygen atom, ${}^{16}_{8}O$

Oxygen molecule, O_2

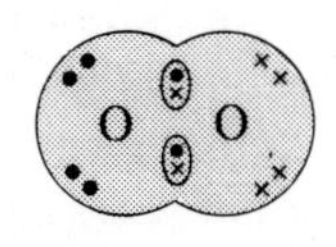

VI
O
S

Sulphur atom, ${}^{32}_{16}S$

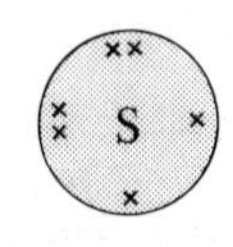

Sulphur molecule, S_8, crown-shaped

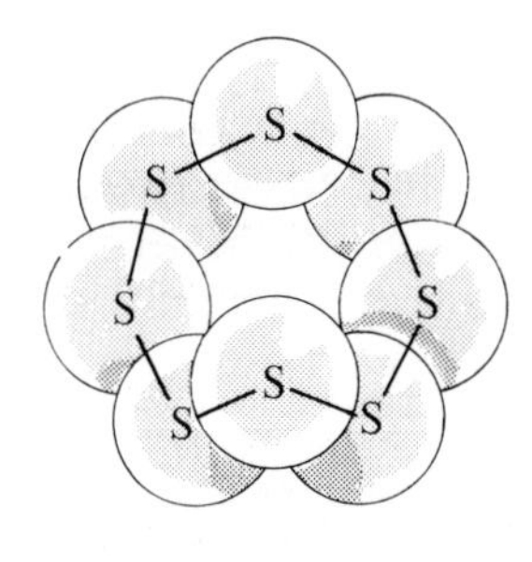

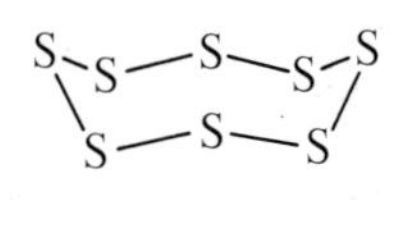

Appearance The Group VI elements, oxygen and sulphur, are common non-metals. Oxygen is a colourless gas at room temperature; sulphur is a yellow solid. Both elements have allotropes which are described later in this chapter. Oxygen's formula is O_2. Sulphur's formula is S_8.

Physical properties Their physical properties are shown in Table 32.1. Compare the values, but don't try to learn them! Notice how different the densities are.

Table 32.1

Element	*m.p./°C*	*b.p./°C*	*Density /g dm^{-3}*	*Colour*	*Smell*	*Conductivity when solid /Ω^{-1} cm^{-1}*	*Solubility (at S.T.P. in 100 g of water)*
Oxygen	−218	−183	1·31	None	None	Almost zero	0·007 g
Sulphur	119	444	2070	Yellow	Slight	Almost zero	Almost insoluble

Table 32.1 shows that oxygen and sulphur are molecular substances (see page 84). They melt fairly easily, are electrical insulators and do not dissolve well in the polar solvent water.

32.2 Occurrence and extraction

As the element Both oxygen and sulphur occur in large quantities as the pure elements. Oxygen makes up 20·95 % of the air around us. Sulphur is found trapped between rock layers in the earth in volcanic regions.

Oxygen It is rather surprising that we should find oxygen uncombined in the atmosphere: *oxygen is the second most reactive non-metal known.* It is thought that the oxygen in the air resulted from violent changes that took place on the Earth millions of years ago. When the Earth broke free as a fragment from the sun, its temperature was very high, and remained high for a long time. Under these conditions, the water present on it decomposed to produce hydrogen and oxygen.

Preparation

Oxygen has many important uses in industry. It is extracted from the air by fractional distillation of liquid air (see page 359).

It is prepared in the laboratory by the decomposition of hydrogen peroxide, using manganese(IV) oxide as a catalyst. The standard apparatus for preparing a gas is used:

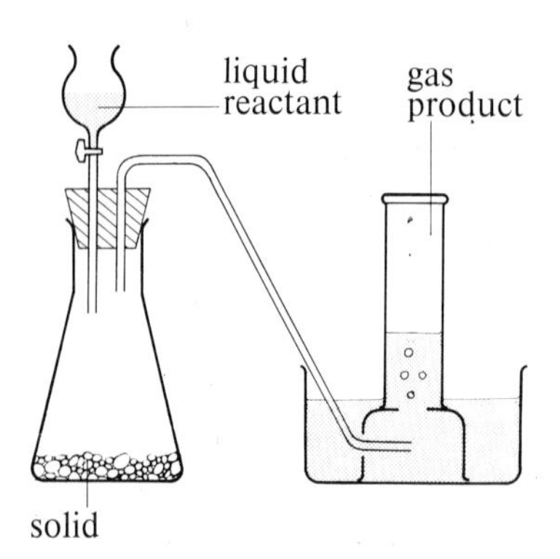

liquid reactant: hydrogen peroxide solution

solid (catalyst): manganese(IV) oxide

reaction: $2H_2O_{2\,(l)} \longrightarrow 2H_2O_{(l)} + O_{2\,(g)}$

Sulphur The sulphur deposits are thought to be the result of volcanic eruptions. The heat from volcanic eruptions, millions of years ago, caused the decomposition of rocks containing sulphur compounds. Liquid sulphur was produced. As the Earth cooled again, the sulphur solidified, trapped between layers of rock. The largest deposits in the world are found in the U.S.A., in Texas and Louisiana.

Sulphur is extracted from the earth using a special type of pump called a **Frasch pump**.The process is called the Frasch process and is shown in Fig. 1.

Fig. 1 The Frasch process

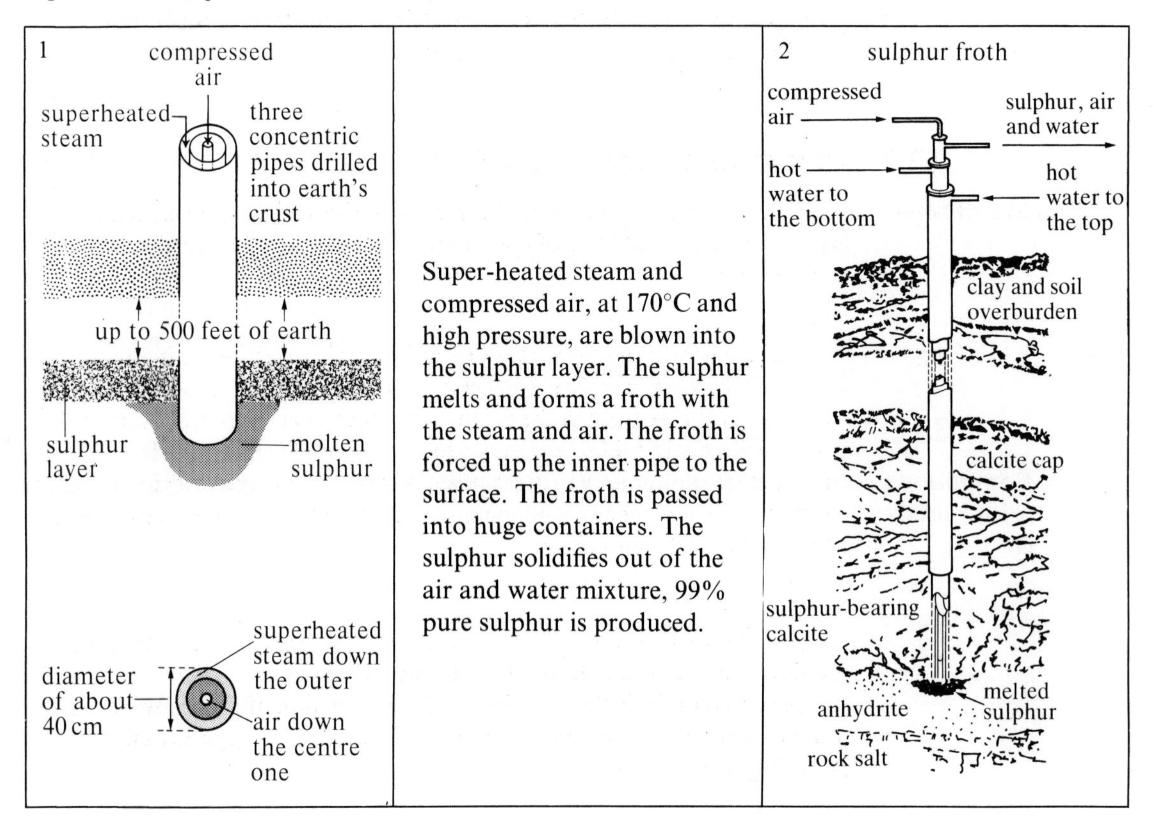

As compounds Table 32.2 shows the percentage of the main elements in the top 25 km of the earth's crust. These occur mostly as compounds. You can see that oxygen is the most common. Sulphur is lowest on this list, but is more common than many other non-metals.

Table 32.2

Element		*% found in earth's crust*
Oxygen	O	46·6
Silicon	Si	27·7
Aluminium	Al	8·1
Iron	Fe	5·0
Calcium	Ca	3·6
Hydrogen	H	0·22
Carbon	C	0·19
Sulphur	S	0·12

The main oxygen compounds in the earth's crust are:

metal oxides (e.g. Fe_2O_3, Fe_3O_4, Al_2O_3),
metal silicates (in clay and sand),
metal carbonates ($CaCO_3$, $MgCO_3$).

The main sulphur compounds are:

metal sulphides (e.g. PbS, FeS, ZnS, CuS),
metal sulphates (e.g. $CaSO_4.2H_2O$ (gypsum)),
hydrogen sulphide gas (small amounts occur in natural gas).

These sulphur compounds are usually air-oxidised in industry to make sulphur(IV) oxide. Sulphur(IV) oxide is the raw material for the manufacture of sulphuric acid (see page 378).

32.3 Compound formation

Compounds with metals

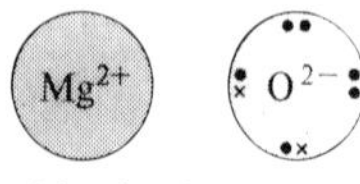

Metal oxides and sulphides are ionic.

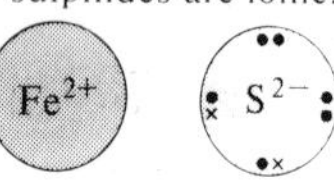

Oxygen and sulphur are fairly reactive non-metals. You have already seen that they combine readily with metals. The products of their reaction with metals are:

oxides (like MgO, ZnO)
sulphides (like MgS, ZnS).

The valency of oxygen and sulphur in these compounds is *always* two. That means the oxidation number of oxygen and sulphur in them is always −II.

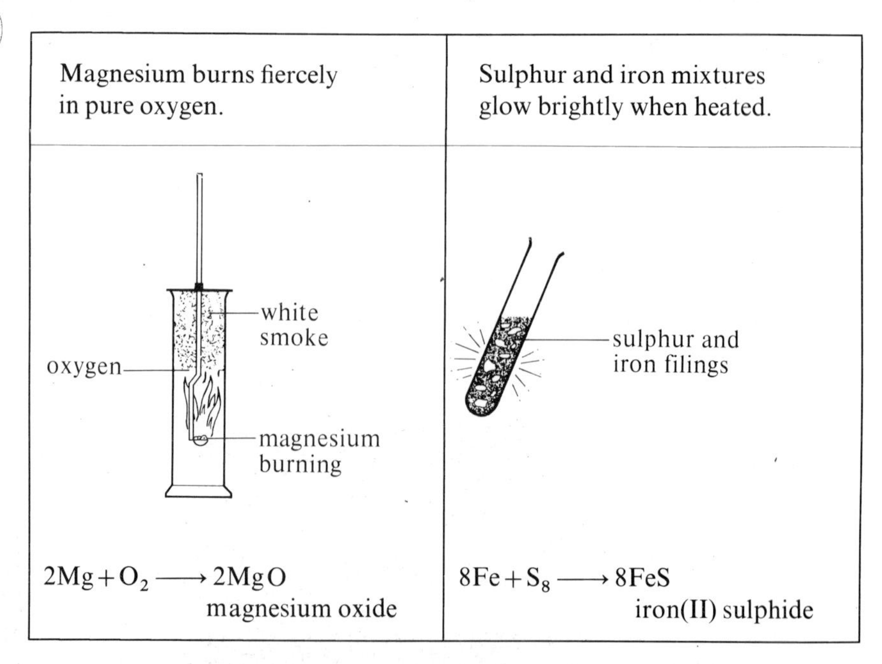

Oxygen reacts more violently with metals than sulphur does. It is a more reactive non-metal than sulphur. Remember that non-metal reactivity increases across a period, and decreases down a group.

a peroxide ion

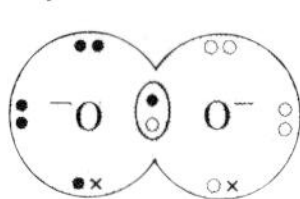

When a reactive metal—sodium, for example—is burned in oxygen, a metal **peroxide** may be formed. Peroxides contain oxygen with an oxidation number of −I. (Oxygen is usually −II.)

A little oxygen: $4Na + O_2 \longrightarrow 2Na_2O$ sodium oxide
Excess oxygen: $2Na + O_2 \longrightarrow Na_2O_2$ sodium peroxide

Compounds with non-metals

We know that oxygen and sulphur combine well with metals. They also combine with almost every other non-metal except the noble gases. The compounds can usually be formed by direct combination between the elements. The reaction between phosphorus and oxygen, for example, is used to show the amount of oxygen in the air: about 21 %.

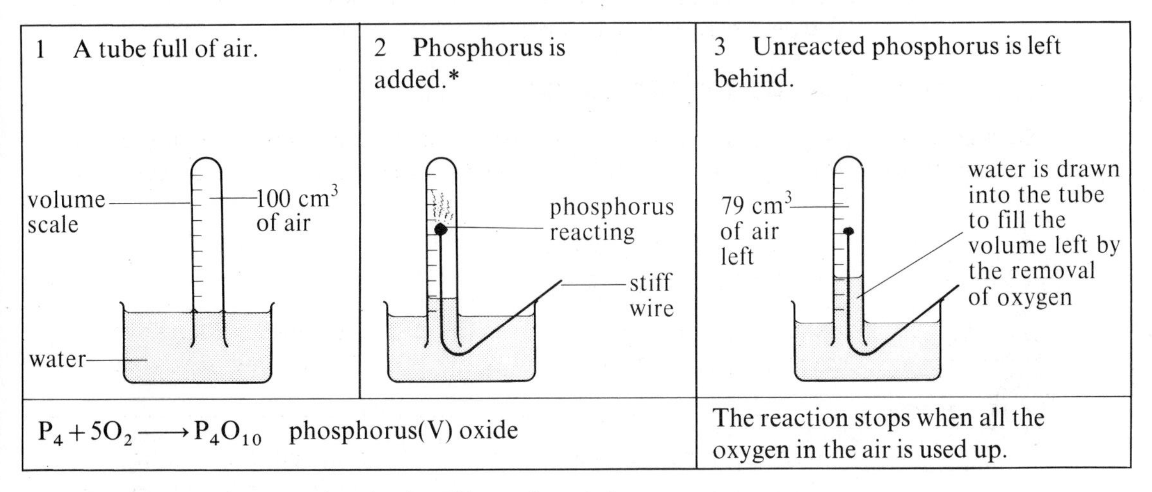
1 A tube full of air.
volume scale
100 cm³ of air
water
2 Phosphorus is added.*
phosphorus reacting
stiff wire
3 Unreacted phosphorus is left behind.
79 cm³ of air left
water is drawn into the tube to fill the volume left by the removal of oxygen
P₄ + 5O₂ ⟶ P₄O₁₀ phosphorus(V) oxide
The reaction stops when all the oxygen in the air is used up.

* Phosphorus is very poisonous and catches fire. This experiment is dangerous.

The most important non-metal compounds of oxygen and sulphur are:

the hydrogen compounds
the sulphur oxides and their related compounds.

(i) The hydrogen compounds

Oxygen forms two compounds with hydrogen:
an oxide—water, H_2O
a peroxide—hydrogen peroxide, H_2O_2

Aqua Pura

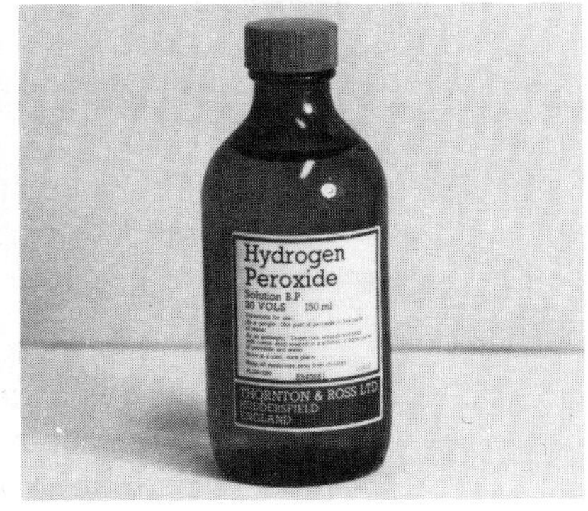
Hydrogen Peroxide

The oxides of hydrogen. Right: Aqua Pura, which is bottled spring drinking water Far right: hydrogen peroxide, which is used as an antiseptic and a bleaching agent. It is stored in dark bottles, because light catalyses its decomposition to water and oxygen

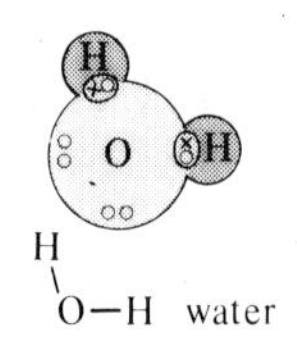

Water is made when hydrogen and oxygen are mixed and exploded together by a spark. It is a very important compound. The whole of Section two—Chapters 12 to 16—deals with its many properties.

Hydrogen peroxide is made by reacting a metal peroxide with acid. Usually barium peroxide is added to ice-cold sulphuric acid:

$$BaO_2 + H_2SO_4 \longrightarrow BaSO_4\downarrow + H_2O_2$$

$$BaO_{2\,(s)} + 2H_3O^+_{(aq)} \longrightarrow Ba^{2+}_{(aq)} + H_2O_{2\,(aq)}; \quad Ba^{2+}_{(aq)} + SO^{2-}_{4\,(aq)} \longrightarrow BaSO_{4\,(s)}$$

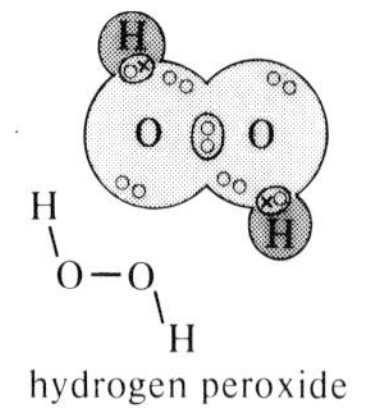

A precipitate of barium sulphate is produced. This is filtered off to leave a solution of hydrogen peroxide.
Hydrogen peroxide decomposes slowly to water and oxygen. The process is speeded up by light—hydrogen peroxide is always stored in dark bottles—and by catalysts; these include blood and manganese(IV) oxide, MnO_2.

$$2H_2O_2 \xrightarrow[\text{light}]{\text{catalysts or}} 2H_2O + O_2$$

Hydrogen sulphide, H_2S, is the main hydrogen compound of sulphur. It is an unpleasant-smelling gas at room temperature. It is made by reaction between a metal sulphide and dilute acid.

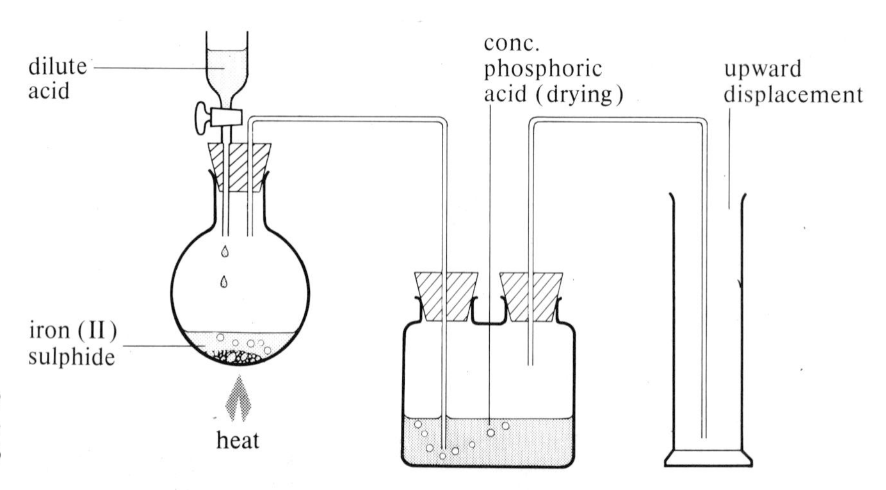

Fig. 2
Making hydrogen sulphide in the lab

$$FeS + 2HCl \longrightarrow FeCl_2 + H_2S\uparrow$$

$$FeS_{(s)} + 2H_3O^+_{(aq)} \longrightarrow Fe^{2+}_{(aq)} + H_2S_{(g)}$$

Although oxygen and sulphur are in the same group, their hydrides are quite different: water and hydrogen peroxide are odourless liquids at room temperature, and do not affect litmus; hydrogen sulphide is a gas that smells of bad eggs and turns blue litmus red—it is acidic.

The sulphur oxides

Sulphur forms two oxides: SO_2—Sulphur dioxide (sulphur(IV) oxide)
SO_3—sulphur trioxide (sulphur(VI) oxide)

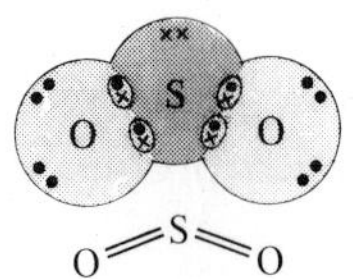

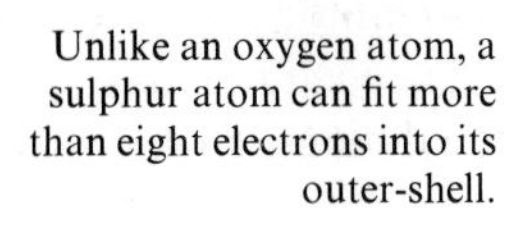

Unlike an oxygen atom, a sulphur atom can fit more than eight electrons into its outer-shell.

Sulphur dioxide is formed when sulphur burns in oxygen:

$$S_8 + 8O_2 \xrightarrow{\text{blue flame}} 8SO_2$$

In the laboratory, sulphur dioxide is made by adding dilute acid to a metal sulphite (IV), for example zinc sulphite, $ZnSO_3$.

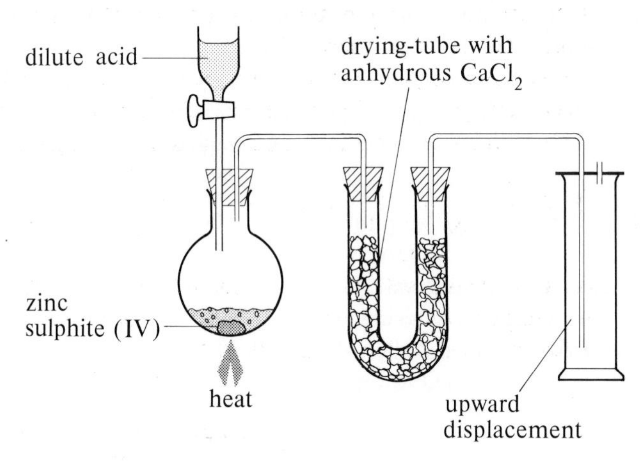

Fig. 3
Making sulphur dioxide in the lab: compare this with the way to make CO_2 in the lab. You add dilute acid to a metal carbonate.

$$ZnSO_3 + 2HCl \longrightarrow ZnCl_2 + H_2O + SO_2\uparrow$$
$$ZnSO_{3\,(s)} + 2H_3O^+_{(aq)} \longrightarrow Zn^{2+}_{(aq)} + 3H_2O_{(l)} + SO_{2\,(g)}$$

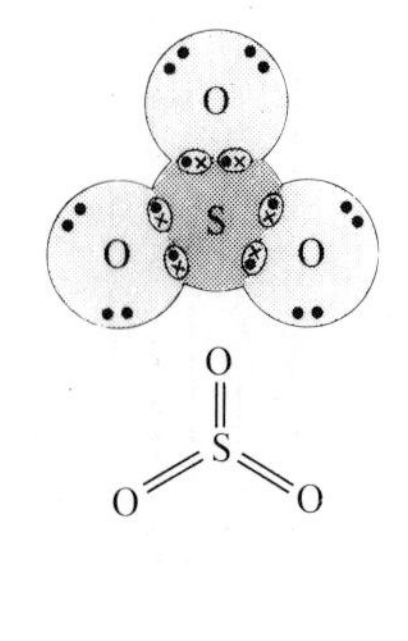

Sulphur trioxide is formed when sulphur dioxide and oxygen are passed together over a bed of heated catalyst. The catalyst can be either platinized asbestos or vanadium(V) oxide. The gases must be dry before they can react well:

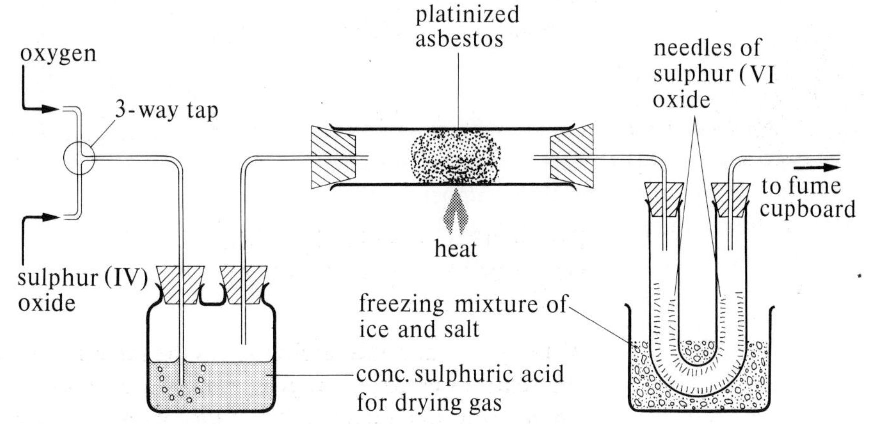

Fig. 4
Making sulphur trioxide in the lab

$$2SO_2 + O_2 \rightleftharpoons 2SO_3$$

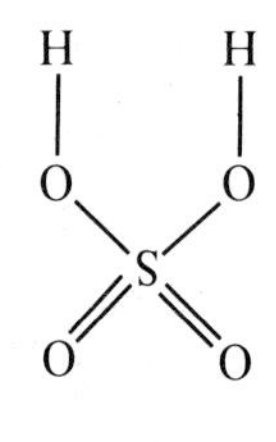

Only about 5% of the gases reacts to make sulphur trioxide. The product solidifies in the U-tube, cooled by the freezing mixture. The gas left over is led away to the fume cupboard. There must be *no water present* because both the oxides of sulphur react with water.

The reaction of sulphur trioxide with water is a particularly important one. Hydrogen sulphate(VI), or **concentrated sulphuric acid**, is produced:

$SO_3 + H_2O \longrightarrow H_2SO_4$ highly exothermic

The main industrial use of sulphur and its oxides is in the manufacture of sulphuric acid. Sulphuric acid is an important compound needed to make fertilisers (ammonium sulphate), and in the manufacture of man-made fibres, paints, detergents and soaps.

The manufacture of sulphuric acid: the Contact process

The process:
Sulphur is melted and burned in a large 'sulphur burner'. The sulphur dioxide produced is mixed with air. The gases are dried and purified and passed over heated beds of vanadium(V) oxide catalyst. The sulphur trioxide produced is then reacted with water in a specially designed tower.

Details:
Raw materials: sulphur, air and water.
Conditions for the reaction of sulphur dioxide with oxygen (from the air):

temperature, 450°C
pressure, atmospheric
catalyst, vanadium(V) oxide

Reactions:

$S_8 + 8O_2 \longrightarrow 8SO_2$
$2SO_2 + O_2 \rightleftharpoons 2SO_3$
$SO_3 + H_2O \longrightarrow H_2SO_4$

Important points

(i) All the reactions are exothermic. Coolers are used throughout the plant. If the temperature rises too much the catalyst decomposes.

(ii) The reaction of sulphur trioxide and water, the last one above, is done in 98% sulphuric acid. This stops the water boiling and prevents the formation of an acid 'mist'.

(iii) Any unreacted gases—'waste gases'—are passed back through the purifier and drier to be reacted again.

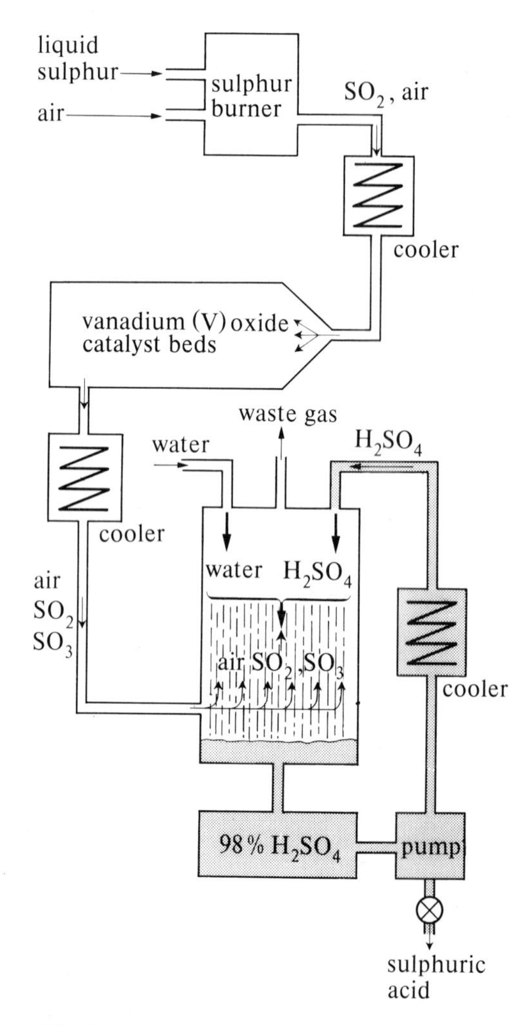

Fig. 5
Flow diagram of the Contact process for making sulphuric acid

The I.C.I. Contact plant. The catalyst beds are in the large tank on the right. The pipes take the gases over the beds several times until most of the sulphur dioxide is oxidised. The patterned tank on the left is a cooler

Sulphur oxides as pollutants

The oxides of sulphur are the main cause of acid rain. This map shows the countries producing the most sulphur dioxide in Europe and the main areas of damage to forests and lakes.

It also shows that Britain is one of the countries that has not agreed to reduce the amount of sulphur dioxide it produces.

Key

- → prevailing wind
- (cloud symbol) country that emitted over 1 million tonnes SO_2 in 1982
- (tree symbol) significant forest damage
- (fish symbol) significant damage to lakes and streams
- ▲ country that has announced at least 30% desulphurization programme

32.4 Solubility and acid/base chemistry

Metal oxides and sulphides

Most metal oxides and sulphides are insoluble in water. Only the Group I metal oxides and sulphides are really soluble, producing alkaline solutions. Group II metal oxides and sulphides are slightly soluble, also giving alkaline solutions:

$$Na_2O_{(s)} + H_2O_{(l)} \longrightarrow 2Na^+_{(aq)} + 2OH^-_{(aq)}$$

$$CaO_{(s)} + H_2O_{(l)} \longrightarrow Ca^{2+}_{(aq)} + 2OH^-_{(aq)} \longrightarrow Ca(OH)_{2\,(s)}$$

Most metal sulphides can therefore be precipitated out of solution, because they are insoluble.

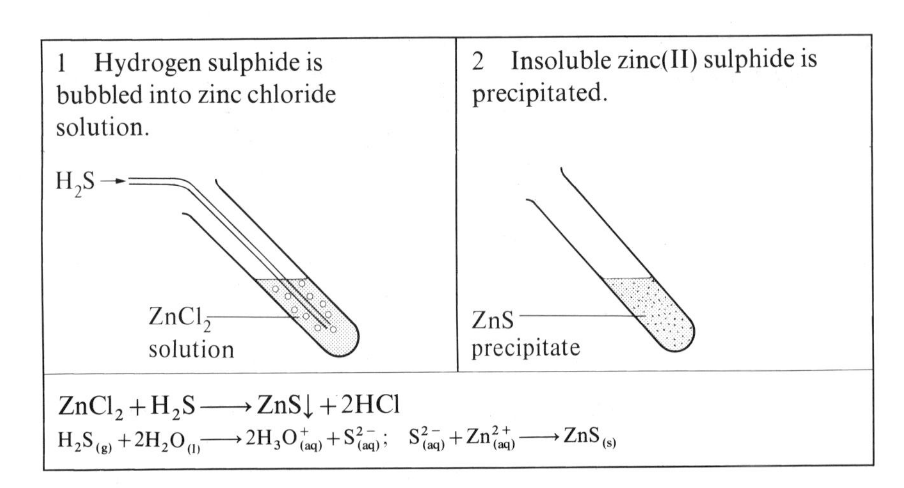

1 Hydrogen sulphide is bubbled into zinc chloride solution.
H_2S
$ZnCl_2$ solution
2 Insoluble zinc(II) sulphide is precipitated.
ZnS precipitate
$ZnCl_2 + H_2S \longrightarrow ZnS\downarrow + 2HCl$
$H_2S_{(g)} + 2H_2O_{(l)} \longrightarrow 2H_3O^+_{(aq)} + S^{2-}_{(aq)}$; $S^{2-}_{(aq)} + Zn^{2+}_{(aq)} \longrightarrow ZnS_{(s)}$

A precipitation reaction of this sort is used to test for hydrogen sulphide gas. A strip of filter paper is dipped in a solution of a lead salt, then held in the gas. If it goes black the gas is H_2S—black lead sulphide has been precipitated:

$$H_2S + Pb(NO_3)_2 \longrightarrow \underset{\text{black}}{PbS} + 2HNO_3$$

$$H_2S_{(g)} + 2H_2O_{(l)} \longrightarrow 2H_3O^+_{(aq)} + S^{2-}_{(aq)};\quad S^{2-}_{(aq)} + Pb^{2+}_{(aq)} \longrightarrow PbS_{(s)}$$

Metal oxides are **bases**. They have the property of neutralizing acids to produce salt solutions. Metal sulphides react in a very similar way:

$$\underset{\text{base}}{CuO} + \underset{\text{acid}}{H_2SO_4} \longrightarrow \underset{\text{salt}}{CuSO_4} + \underset{\text{water}}{H_2O}$$

$$CuO_{(s)} + 2H_3O^+_{(aq)} \longrightarrow Cu^{2+}_{(aq)} + 3H_2O_{(l)}$$

$$\underset{\text{metal sulphide}}{CuS} + \underset{\text{acid}}{H_2SO_4} \longrightarrow \underset{\text{salt}}{CuSO_4} + \underset{\text{hydrogen sulphide}}{H_2S\uparrow}$$

$$CuS_{(s)} + 2H_3O^+_{(aq)} \longrightarrow Cu^{2+}_{(aq)} + H_2S_{(aq)} + 2H_2O_{(l)}$$

Non-metal oxides and sulphides

Almost all non-metal oxides react with water, dissolving to produce acid solutions. Important exceptions to this general rule are:

water itself, H_2O	carbon monoxide, CO
nitrogen(I) oxide, N_2O	carbon disulphide, CS_2

Sulphur dioxide's reaction with water is reversible. It is just like the reaction of carbon dioxide with water.

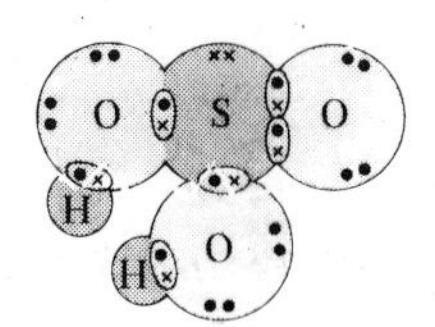

$$SO_2 + H_2O \underset{\text{decomposition}}{\overset{\text{combination}}{\rightleftharpoons}} H_2SO_3$$ sulphuric(IV) acid, sulphurous acid

$$CO_2 + H_2O \underset{\text{decomposition}}{\overset{\text{combination}}{\rightleftharpoons}} H_2CO_3$$ carbonic acid

Sulphuric acid and sulphates

You saw on page 378 that sulphur trioxide or sulphur(VI) oxide, SO_3, combines violently with water. Hydrogen sulphate(VI) is produced. This is better known as concentrated sulphuric acid, H_2SO_4. The concentrated acid has some unusual properties which are described in section 32.6. Besides these properties, solutions of the compound in water show all the usual properties of acids. The salts made by neutralising sulphuric acid are **metal sulphates(VI)**:

Acid	metal oxide		salt	water
H_2SO_4 +	MgO	$\longrightarrow$	$MgSO_4$ +	H_2O
$2H_3O^+_{(aq)}$ +	$MgO_{(s)}$	$\longrightarrow$	$Mg^{2+}_{(aq)}$	+ $3H_2O_{(l)}$

Acid	metal hydroxide		salt	water
H_2SO_4 +	$Cu(OH)_2$	$\longrightarrow$	$CuSO_4$ +	$2H_2O$
$2H_3O^+_{(aq)}$ +	$Cu(OH)_{2\,(s)}$	$\longrightarrow$	$Cu^{2+}_{(aq)}$	+ $4H_2O_{(l)}$

Acid	metal carbonate		salt	carbon dioxide	water
H_2SO_4 +	$ZnCO_3$	$\longrightarrow$	$ZnSO_4$	+ $CO_2\uparrow$	+ H_2O
$2H_3O^+_{(aq)}$ +	$ZnCO_{3\,(s)}$	$\longrightarrow$	$Zn^{2+}_{(aq)}$	+ $CO_{2\,(g)}$	+ $2H_2O_{(l)}$

Sulphate test

The sulphates of two of the Group II metals, $CaSO_4$ and $BaSO_4$, are insoluble in water. We can use this property to test whether an unknown substance is a sulphate:

1. Dissolve the unknown substance.
2. Add some dilute hydrochloric acid to the solution.
3. Add a few drops of barium chloride to the solution.
4. An immediate white precipitate of barium sulphate will appear if the unknown substance is a sulphate.

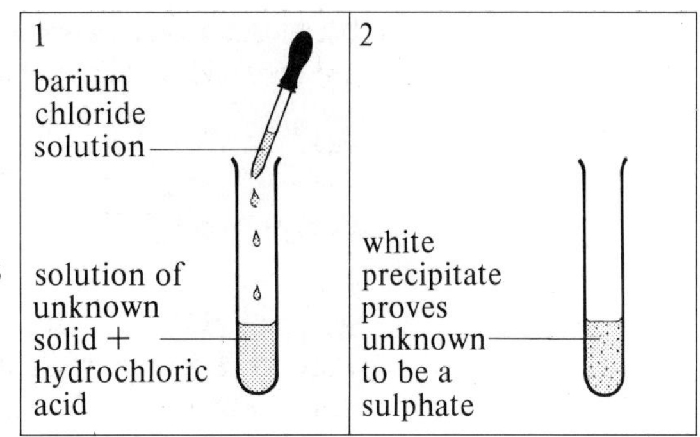

For example:

soluble	soluble		insoluble	soluble
$BaCl_2$ +	Na_2SO_4	$\longrightarrow$	$BaSO_4\downarrow$	+ 2NaCl
$Ba^{2+}_{(aq)}$ +	SO_4^{2-}	$\longrightarrow$	$BaSO_{4\,(s)}$	

32.5 Redox chemistry

Oxidation numbers You met many compounds of oxygen and sulphur in the earlier sections of this chapter. The oxidation numbers of oxygen and sulphur in these compounds were not always the same. Sulphur has more possible oxidation numbers than oxygen:

Table 32.3

Element	*Oxidation numbers*
Oxygen	−II in oxides of metals and hydrogen (except H_2O_2), e.g. ZnO, H_2O −I in hydrogen peroxide, H_2O_2 0 in the element, O_2
Sulphur	−II in sulphides of metals and hydrogen, e.g. CuS, H_2S 0 in the element, S_8 +I and +II in the chlorides of sulphur, S_2Cl_2 and SCl_2 +IV in sulphur dioxide (SO_2), and all the sulphite (IV) salts, e.g. $ZnSO_3$ +VI in sulphur trioxide (SO_3), sulphuric acid (H_2SO_4), and the sulphate (VI) salts, e.g. $ZnSO_4$

If the oxidation number of oxygen or sulphur in a substance increases during a reaction, we say the substance is being oxidised. For example, in the conversion of SO_2 to SO_3, the SO_2 is oxidised. In the same way, H_2O_2 is oxidised when it is converted to oxygen, O_2.

If the oxidation number of oxygen or sulphur in a substance decreases during a reaction, the substance is being reduced. Examples of reductions are the conversions of O_2 to H_2O and SO_2 to S_8.

Reducing agents Reducing agents are substances that cause the reduction of other substances. Any substance that is being oxidised is itself acting as a reducing agent. The most important reducing agents from Table 32.3 are:

hydrogen sulphide, which is easily oxidized to sulphur; S(−II) to S(0)
hydrogen peroxide, which is easily oxidized to oxygen; O(−I) to O(0)
sulphur dioxide, which is easily oxidized to sulphate(VI); S(IV) to S(VI)

Bleaching is an important industrial redox reaction. The bleach acts as an oxidizing agent. Cloth is fed into a bleaching unit that uses hydrogen peroxide. The hydrogen peroxide is itself reduced to water. Hydrogen peroxide can act either as an oxidizing or a reducing agent.

In the following reactions the sulphur and oxygen compounds act as reducing agents.

(i) Hydrogen sulphide reduces chlorine gas.

$$8H_2S + 8Cl_2 \longrightarrow S_8 + 16HCl$$

down
$HCl \leftarrow Cl_2$
−II −I 0 +I
$H_2S \longrightarrow S_8$
up

(ii) Hydrogen peroxide reduces potassium manganate(VII) solution.

$$H_2O_2 + 2KMnO_4 \longrightarrow 2KOH + 2MnO_2\downarrow + 2O_2\uparrow$$

purple brown

down
$MnO_2 \leftarrow KMnO_4$
−I 0 +I +II +III +IV +V +VI +VII
$H_2O_2 \rightarrow O_2$
up

(iii) Sulphur(IV) oxide reduces iron(III) chloride solution.

$$SO_2 + 2FeCl_3 + 2H_2O \longrightarrow 2FeCl_2 + 2HCl + H_2SO_4$$

yellow-brown green

down
$FeCl_2 \leftarrow FeCl_3$
+II +III +IV +V +VI
$SO_2 \rightarrow H_2SO_4$
up

Oxidizing agents Oxidizing agents are substances that cause the oxidation of other substances. Any substance that is being reduced is itself acting as an oxidizing agent. The most important oxidizing agents from Table 32.3 are:

oxygen gas, which is easily reduced to an oxide; O(0) to O(−II)

conc. sulphuric acid, which is easily reduced to sulphur dioxide; S(VI) to S(IV).

For example:

(i) Oxygen oxidises sulphur(IV) oxide.

$$O_2 + 2SO_2 \longrightarrow 2SO_3$$

up
$SO_2 \longrightarrow SO_3$
−II −I 0 +I +II +III +IV +V +VI
$SO_3 \longleftarrow O_2$
down

(ii) Concentrated sulphuric acid oxidises copper.

$$Cu + 2H_2SO_4 \longrightarrow CuSO_4 + 2H_2O + SO_2\uparrow$$

down
$SO_2 \leftarrow H_2SO_4$
0 +I +II +III +IV +V +VI
$Cu \longrightarrow CuSO_4$
up

32.6 Special features

Sulphuric acid's wide range of properties

Dilute sulphuric acid solutions have all the properties that we have come to expect of acids. The pure, concentrated acid has two other important properties apart from its acid property:

as an oxidant,
as a dehydrating agent.

Page 383 dealt with sulphuric acid's action as an oxidising agent. As a dehydrating agent, sulphuric acid is able to remove water from compounds. Three important examples are shown below:

(i) hydrated copper(II) sulphate to anhydrous copper(II) sulphate

$$\underset{\text{blue}}{CuSO_4.5H_2O} \xrightarrow[-5H_2O]{\text{conc. } H_2SO_4} \underset{\text{white}}{CuSO_4}$$

(ii) Ethanol to ethene (see page 344)

```
   H  OH     conc. H2SO4      H         H
   |  |        170°C           \       /
H—C—C—H   ------------->       C=C
   |  |         -H2O           /       \
   H  H                       H         H
```

(iii) Cane sugar to carbon: 'charring'

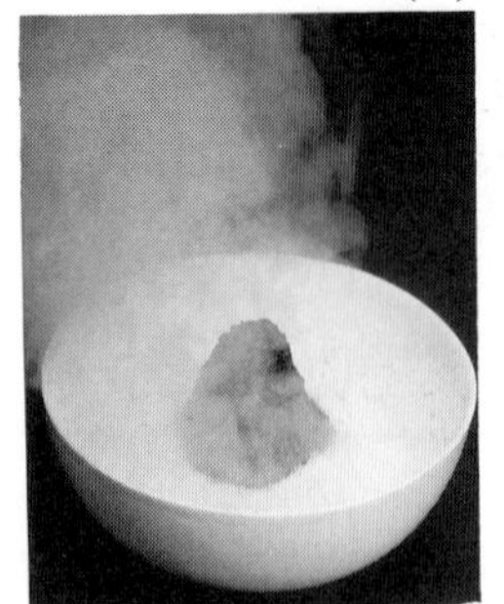

$$C_{12}H_{22}O_{11} \xrightarrow[-11H_2O]{\text{conc. } H_2SO_4} 12C$$

The carbon is then oxidised by the sulphuric acid:

$$C + 2H_2SO_4 \longrightarrow CO_2\uparrow + 2SO_2\uparrow + 2H_2O$$

down: $SO_2 \leftarrow H_2SO_4$

0 | +I | +II | +III | +IV | +V | +VI

$C \xrightarrow{\text{up}} CO_2$

Allotropy of the elements

Both oxygen and sulphur can exist in more than one physical form in the same state. They both exhibit allotropy, like carbon (page 320).
Oxygen, O_2 and **ozone**, O_3, are gaseous allotropes of oxygen.
Monoclinic sulphur and **rhombic sulphur** are solid allotropes of sulphur.

Ozone

ozone molecule

There is very little ozone in the oxygen of the air at ground level. At very high altitudes, however, normal oxygen can be converted into ozone. This process is caused by the high energy rays from the sun.

$$3O_2 \xrightarrow[\text{rays}]{\text{high energy}} 2O_3$$

An **ozone layer** has formed in the outer atmosphere. It has the useful

property of filtering out some of the high energy radiation from the sun. Radiation of this sort can cause skin cancer and damage to the eyes, therefore any man-made process that disturbs the ozone layer is likely to be harmful. Recently two processes have caused worry:
The gases used to force the contents of an aerosol can into the air react with ozone, using it up.
High flying jet aircraft release exhaust gases into the ozone layer. These also react with the ozone.

The exhaust gases from high-flying jets react with the ozone layer

Monoclinic and rhombic sulphur

These two allotropes of sulphur look very much the same, and both contain S_8 molecules. But the molecules are packed together differently in each. Only one allotrope, rhombic sulphur, occurs naturally—unlike carbon's allotropes, diamond and graphite, which are both mined.

At room temperature, sulphur is most stable as rhombic sulphur. Above 96°C, sulphur is most stable as monoclinic sulphur. We can change one allotrope into the other by melting the sulphur to a liquid, or by making a sulphur solution in the solvent toluene. These methods are shown in the diagram below:

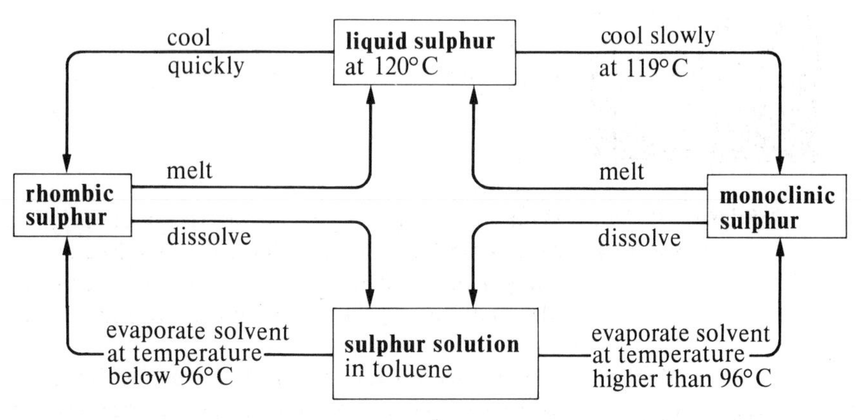

A model of the chains in plastic sulphur

Liquid sulphur is an unusual liquid. It is pale golden and runny at about 120°C, but it gets darker and thicker as it is heated. The sulphur molecules start to break up and form chains which become tangled with one another. If you pour the dark, thick liquid sulphur into cold water, another sort of sulphur forms. It is called **plastic sulphur** and is like yellow chewing-gum.

Reaction summary for oxygen

Extraction

In industry: Fractional distillation of air.
In the laboratory: Catalytic decomposition of hydrogen peroxide

$$2H_2O_2 \xrightarrow{MnO_2} 2H_2O + O_2\uparrow$$

Oxygen is also evolved by heating nitrates.

Reactions of the element

(i) Oxidises other elements:
$2H_2 + O_2 \longrightarrow 2H_2O$
$4Na + O_2 \longrightarrow 2Na_2O$ (limited supply of oxygen)
$2Na + O_2 \longrightarrow Na_2O_2$ (excess oxygen)
$S_8 + 8O_2 \longrightarrow 8SO_2$
$P_4 + 5O_2 \longrightarrow P_4O_{10}$
$2Cu + O_2 \longrightarrow 2CuO$ etc.

(ii) Oxidises compounds:
$2CO + O_2 \longrightarrow 2CO_2$
$2SO_2 + O_2 \rightleftharpoons 2SO_3$
$4NH_3 + 5O_2 \longrightarrow 4NO + 6H_2O$

The hydrides

These are water and hydrogen peroxide.
(i) Water
oxidises reactive metals: $2Na + H_2O \longrightarrow 2NaOH + H_2\uparrow$
accepts protons from acids: $H_2O + HCl \longrightarrow H_3O^+ + Cl^-$
gives protons to bases: $H_2O + NH_3 \rightleftharpoons NH_4^+ + OH^-$
reacts with non-metal oxides: $H_2O + CO_2 \rightleftharpoons H_2CO_3$
$H_2O + SO_2 \rightleftharpoons H_2SO_3$

(ii) Hydrogen peroxide
Preparation: $BaO_2 + H_2SO_4 \longrightarrow H_2O_2 + BaSO_4\downarrow$
is an oxidising agent: $H_2O_2 + 2FeCl_2 + 2HCl \longrightarrow 2FeCl_3 + 2H_2O$
is a reducing agent: $H_2O_2 + 2KMnO_4 \longrightarrow 2KOH + 2MnO_2 + 2O_2$

The oxides

(i) Metal oxides
basic: $CuO + H_2SO_4 \longrightarrow CuSO_4 + H_2O$
amphoteric: $ZnO + H_2SO_4 \longrightarrow ZnSO_4 + H_2O$
$ZnO + 2NaOH \longrightarrow Na_2ZnO_2 + H_2O$

(ii) Non-metal oxides
acidic: SO_2, CO_2—see above.
neutral: H_2O, CO.

Reaction summary for sulphur

Extraction Found as the element, and extracted by the Frasch process.

Reactions of the element
(i) Oxidises some elements:
$Fe + S \longrightarrow FeS$
$2Cu + S \longrightarrow Cu_2S$
$H_2 + S \longrightarrow H_2S$
(ii) Reduces other elements:
$S + O_2 \longrightarrow SO_2$
$2S + Cl_2 \longrightarrow S_2Cl_2$
(iii) Reduces some compounds: $S + 2H_2SO_4 \longrightarrow 3SO_2\uparrow + 2H_2O$

The hydride H_2S
(i) is a sparingly soluble acidic gas: $H_2S + H_2O \rightleftharpoons H_3O^+ + HS^-$
(ii) is a reducing agent: $2H_2S + SO_2 \longrightarrow 2H_2O + 3S\downarrow$
(iii) causes precipitation of insoluble metal sulphides:
$CuSO_4 + H_2S \longrightarrow CuS\downarrow + H_2SO_4$

The oxides Sulphur(IV) oxide, SO_2 (sulphur dioxide) and sulphur(IV) oxide, SO_3 (sulphur trioxide). Sulphur dioxide:
(i) is a sparingly soluble acidic gas: $SO_2 + H_2O \rightleftharpoons H_2SO_3$
(ii) is a reducing agent:
$2SO_2 + O_2 \rightleftharpoons 2SO_3$
$SO_2 + 2H_2S \longrightarrow 3S\downarrow + 2H_2O$
SO_3 is very acidic: $SO_3 + H_2O \longrightarrow H_2SO_4$

The hydroxy compound Sulphuric acid is made in the Contact process:
$S + O_2 \longrightarrow SO_2$
$2SO_2 + O_2 \rightleftharpoons 2SO_3$ (V_2O_5 catalyst)
$SO_3 + H_2O \longrightarrow H_2SO_4$

Sulphuric acid behaves—
(i) as a strong acid: $H_2SO_4 + H_2O \longrightarrow H_3O^+ + HSO_4^-$
The dilute acid reacts with metals and bases.
(ii) as an oxidising agent:
$2H_2SO_4 + C \longrightarrow CO_2 + 2SO_2\uparrow + 2H_2O$
$Cu + 2H_2SO_4 \longrightarrow CuSO_4 + SO_2\uparrow + 2H_2O$
(iii) as a dehydrating agent:

$CH_3CH_2OH \xrightarrow[170°C]{H_2SO_4} CH_2 = CH_2$

$C_{12}H_{22}O_{11} \xrightarrow[-11H_2O]{H_2SO_4} 12C$

$CuSO_4.5H_2O \xrightarrow[-5H_2O]{H_2SO_4} CuSO_4$

} use the concentrated acid

The sulphates All are soluble in water except $PbSO_4$, $BaSO_4$ and $CaSO_4$.
All are stable to heat.

Questions

1 (a) Draw a fully labelled diagram of an oxygen atom.
(b) Use the diagram to explain oxygen's valency of two.
(c) Draw a diagram to show the electronic structure of an oxygen molecule.

2 Oxygen reacts with nearly all the elements. Non-metal oxides and metal oxides have different properties however.
(a) List three properties typical of metal oxides.
(b) What type of bonding would you expect in a metal oxide?
(c) Give an example.
(d) List three properties typical of non-metal oxides.
(e) What type of bonding would you expect in a non-metal oxide?
(f) Give two examples, one which is molecular, and one which is macromolecular.

3 (a) How is pure oxygen obtained from the atmosphere? Draw a diagram of the apparatus used, and explain the idea behind it.
(b) Give two large scale uses for oxygen.

4 Draw diagrams that show the electronic structure of the following:
H_2O_2 CO_2 Na_2O Al_2O_3 Na_2O_2

5 Oxygen is made in the laboratory by decomposing hydrogen peroxide with a catalyst.
(a) Draw a diagram of the apparatus you would use to make and collect the gas.
(b) Write an equation for the reaction.
(c) What catalyst is used?
(d) Hydrogen peroxide oxidizes lead sulphide to lead sulphate. Write an equation for this reaction. How does the equation show that lead sulphide has been oxidized?

6 (a) Write an equation for the reaction that takes place when zinc and sulphur are warmed together.
(b) Large beds of unreacted sulphur are found deep in the earth between layers of rock in Texas. How were these beds probably formed?
(c) How is the sulphur removed from the beds? Draw a diagram and explain the process carefully.

7 Sulphuric acid is made from the raw materials sulphur, air, and water, in the Contact Process.
(a) Give equations for each step of the process.
(b) What catalyst is used in the reaction of sulphur dioxide and oxygen?
(c) Why is sulphuric acid used as a solvent in the final step?

8 Sulphur dioxide can be made by burning sulphur in air.
(a) Why is this not a suitable method for making the gas in the laboratory?
(b) How would you make the gas in the laboratory?
(c) Draw the apparatus that you would use.
(d) Give the equation for the reaction.

9 Give an example of a reaction in which sulphuric acid acts as:
(a) an acid
(b) an oxidizing agent
(c) a dehydrating agent

10 Write equations for the reactions of sulphuric acid with:
(a) copper
(b) sodium carbonate
(c) hydrated copper sulphate
(d) carbon

11 Describe what you think you would see in each of the following:
(a) concentrated sulphuric acid is poured onto sugar
(b) sulphur dioxide gas is bubbled into an acidic solution of potassium manganate(VII)
(c) hydrogen peroxide is added to an acidic solution of potassium manganate(VII)

33 Group VII: the halogens

33.1 The elements

Halogen atom

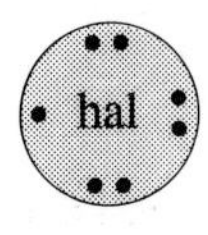

F	2, 7
Cl	2, 8, 7
Br	2, 8, 18, 7
I	2,8,18,18, 7

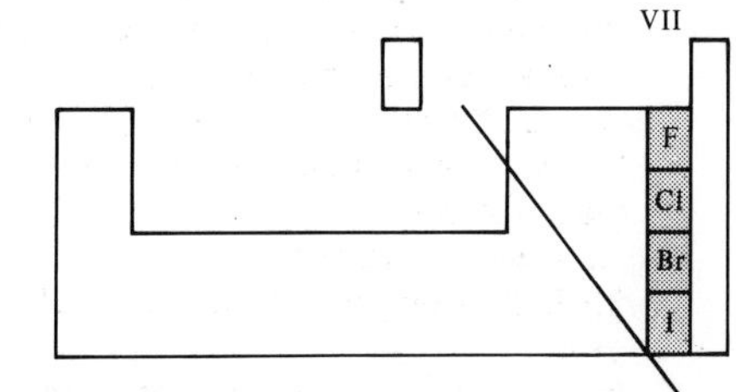

Halogen molecule

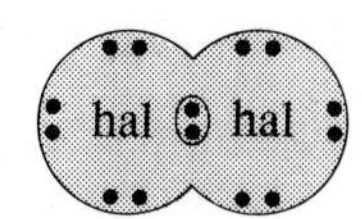

F—F
Cl—Cl
Br—Br
I—I

Appearance and preparation

The Group VII elements are given a special name. They are called the **halogens**. They are coloured, non-metallic elements. At room temperature:

fluorine is a very pale yellow gas
chlorine is a yellow-green gas
bromine is a red-brown, fuming liquid
iodine is a black-purple solid.

We usually obtain a halogen in the laboratory from a solution of its hydrogen compound. For example chlorine is made from a solution of hydrogen chloride, HCl. The hydrogen halide is oxidised by manganese(IV) oxide.

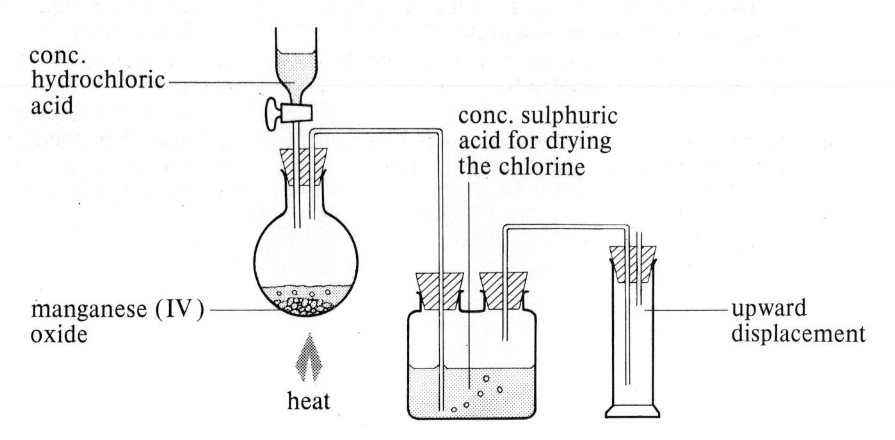

Fig. 1
Making chlorine in the lab

$$MnO_2 + 4HCl \longrightarrow MnCl_2 + Cl_2\uparrow + 2H_2O$$

Physical properties

The physical properties are listed in Table 33.1. Compare the values, but do not try to learn them!

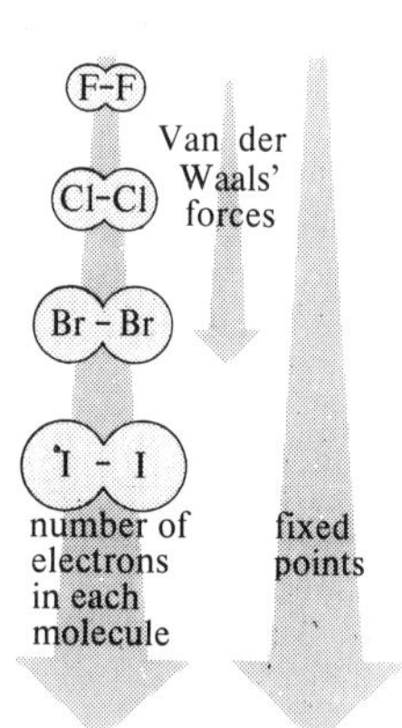

Table 33.1

Element	*m.p. /°C*	*b.p. /°C*	*Density at 10°C /g dm^{-3}*	*Colour*	*Solubility in water (at S.T.P.)*
Fluorine	−220	−188	1·69	Pale yellow	Reacts with water
Chlorine	−101	−35	3·21	Yellow-green	1·46 g in 100 g water
Bromine	−7	59	2930	Red-brown	4·17 g in 100 g water
Iodine	114	184	4930	Purple-black	0·02 g in 100 g water

The figures in Table 33.1 show that the halogens are typical molecular substances: they all have low m.p. and b.p. values and all are non-conducting. (See page 84.) They all have a pungent and irritating smell. Notice the differences in the densities of the halogens at room temperature. Iodine and bromine are much more dense than the other two. This is because iodine is a solid and bromine a liquid at room temperature, while chlorine and fluorine are gases.

Above: the solubility of chlorine in water allows it to be used as a disinfectant in swimming pools

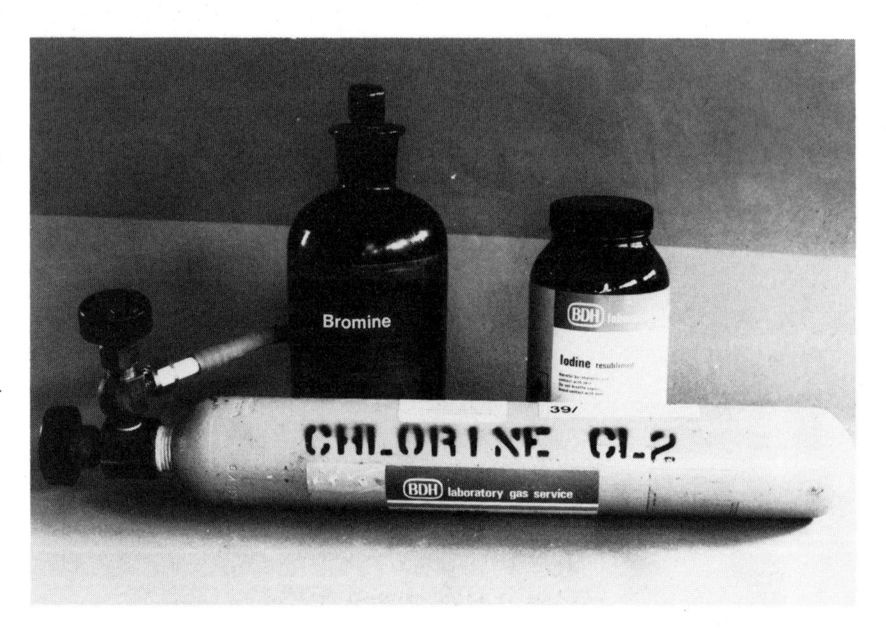

Right: chlorine is a gas, bromine a liquid and iodine a solid

33.2 Occurrence and extraction

Reactivity of the halogens

Non-metal atoms gain electrons when the non-metal reacts. The halogens need only gain one electron to achieve a full shell:

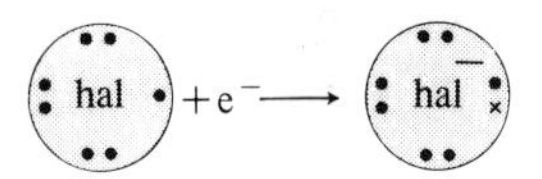

Small atoms gain electrons most easily. The nucleus is closer to the outer electrons in smaller atoms.

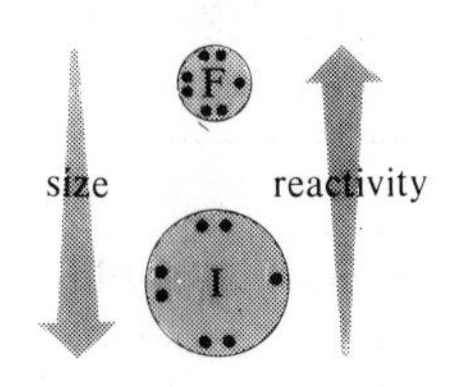

We first met the reactivity of the halogens in Chapter 7. Their reactions with aluminium and phosphorus show that they are reactive non-metals. A fuller account of their reactivity is given in Chapter 17, on pages 181–185. You should read these pages again.

The diagram below shows the reactivity trend for non-metals:

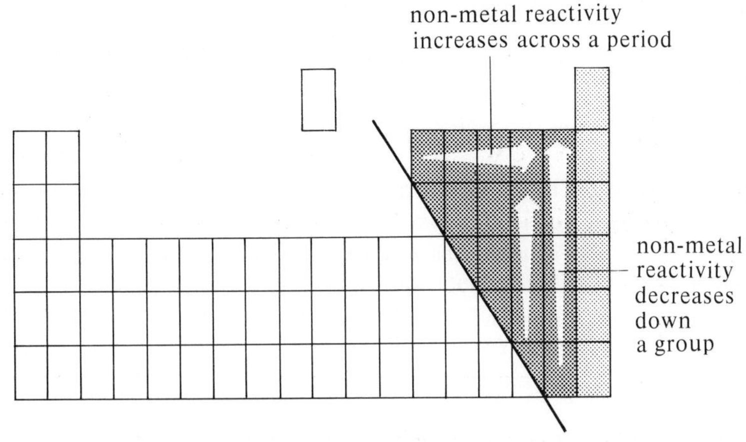

You can see that Group VII is the most reactive group of non-metals. For this reason, it would be surprising to find the elements in their pure, uncombined states. In fact the halogens are so reactive that they are always found combined with other elements.

Halogen compounds

The main source of halogens is the salts found in the sea. The salts found in the sea around the British Isles are listed in Table 33.2.

Table 33.2

Compound	*Mass/g present in 100 g of sea-water*
Sodium chloride, $NaCl$	2·6
Magnesium chloride, $MgCl_2$	0·3
Magnesium sulphate, $MgSO_4$	0·2
Calcium sulphate, $CaSO_4$	0·1
Potassium chloride, KCl	0·1
Magnesium bromide, $MgBr_2$	0·01
Magnesium iodide, MgI_2	0·0003

You can see from Table 33.2 that the sea can provide large quantities of sodium chloride. The other main sources of chlorine are deposits of **rocksalt** ($NaCl$) and **carnallite** ($KCl.MgCl_2.6H_2O$).

Fluorine is found in the earth as **fluorite**, CaF_2, and **cryolite**, Na_3AlF_6.
Iodine, as the compound sodium iodate, $NaIO_3$, occurs as a 2% impurity in saltpetre, $NaNO_3$.
Chlorine is extracted in larger quantities than other halogens. Its extraction has already been dealt with in the chapter on the Group I elements, page 282. It is obtained by electrolysing molten sodium chloride or brine.

$2NaCl \longrightarrow 2Na + Cl_2\uparrow$

Uses of chlorine: in the manufacture of hydrochloric acid, plastics, bleaches and solvents (e.g. CCl_4)
as a germ-killer in swimming-pools.

33.3 Compound formation

Compounds with metals

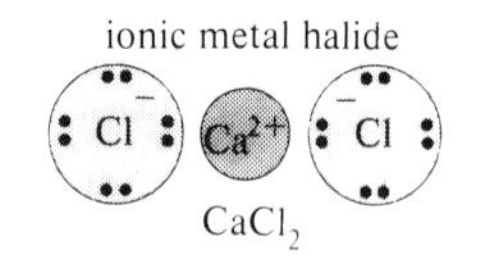

The halogens are the most reactive non-metals. They combine well with nearly all metals to produce compounds.
The reactions with reactive metals, like sodium and calcium, are violent. These metals burn brightly in chlorine gas.
Less reactive metals require heating in a stream of chlorine for reaction to begin. Once the reaction starts, the metal glows brightly because the combination is exothermic.

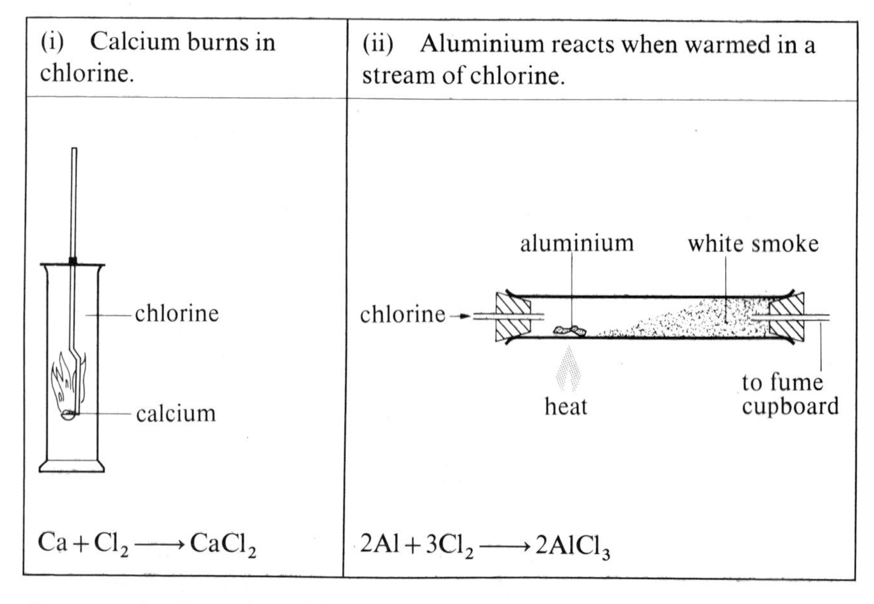

Compounds of metals and halogens are often called **metal halides**. The metal halides are mostly typical ionic substances (see page 85):

they have high m.p. and b.p. values;
they are conductors of electricity when molten;
they are hard but brittle when solid;
they dissolve in water to give solutions that conduct electricity.

Compounds with hydrogen

The halogens all combine directly with hydrogen. You can tell the order of reactivity of the halogens by their reactions with hydrogen:

Table 33.3

Halogen	*Reaction conditions*	*Equation*	*Reactivity*
Fluorine	Explodes when mixed.	$F_2 + H_2 \longrightarrow 2HF$	↑
Chlorine	Explodes when mixed in sunlight.	$Cl_2 + H_2 \longrightarrow 2HCl$	
Bromine	Reaction requires heat.	$Br_2 + H_2 \longrightarrow 2HBr$	
Iodine	Even at 400°C, reaction is not complete.	$I_2 + H_2 \rightleftharpoons 2HI$	

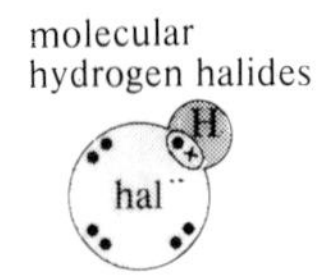

molecular hydrogen halides

In industry, a jet of hydrogen can be made to burn safely in chlorine gas according to the equation in Table 33.3. This is the usual industrial preparation of hydrogen chloride.
In the laboratory, however, you would not use this method to produce hydrogen chloride. Instead, you would react sodium chloride with concentrated sulphuric acid:

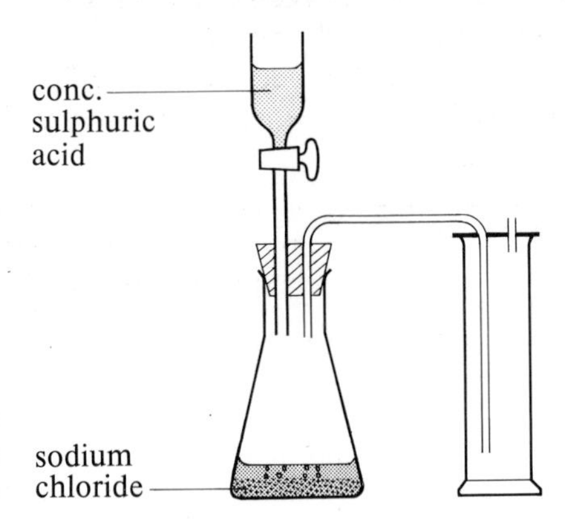

Fig. 2
Making hydrogen chloride in the lab

$$NaCl + H_2SO_4 \longrightarrow NaHSO_4 + HCl\uparrow$$

(NaHSO$_4$: sodium hydrogen sulphate)

Compounds with other non-metals

The reactions of other non-metals with the halogens are often rather complicated.

Carbon, phosphorus, and sulphur burn freely in chlorine gas. The main reactions are:

$C + 2Cl_2 \longrightarrow CCl_4$ tetrachloromethane
$P_4 + 6Cl_2 \longrightarrow 4PCl_3$ phosphorus(III) chloride
$S_8 + 4Cl_2 \longrightarrow 4S_2Cl_2$ sulphur(I) chloride

The products are all unpleasant-smelling liquids with molecular properties.

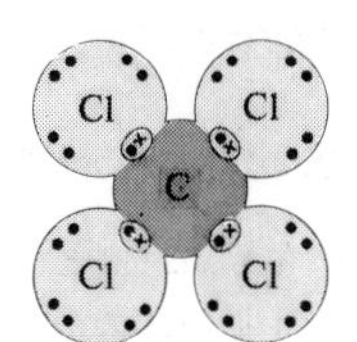

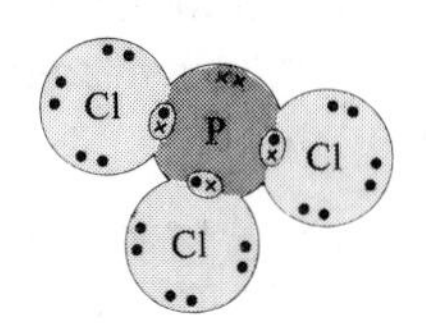

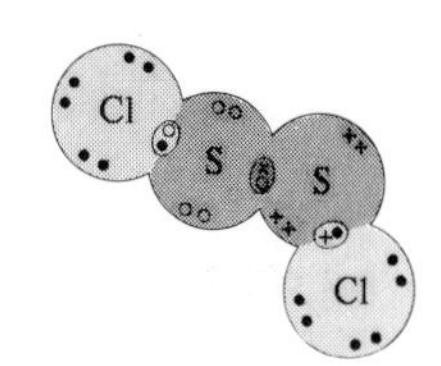

33.4 Solubility and acid/base chemistry

The elements in water

In Table 33.1 you saw that all the halogens are soluble in water. Fluorine is so reactive that it decomposes water:

$$2F_2 + 2H_2O \longrightarrow \underset{\text{hydrofluoric acid}}{4HF} + O_2\uparrow$$

The other halogens react with water also. As you would expect, chlorine is the most reactive halogen after fluorine:

$$Cl_2 + H_2O \rightleftharpoons \underset{\text{hydrochloric acid}}{HCl} + \underset{\text{hydrogen chlorate(I)}}{HOCl}$$

The halogens react with water to give acidic products; we shall look at this reaction more closely later. But it is worth remembering now the effect halogens have on damp litmus paper: first the paper turns red, then it is rapidly bleached. The paper goes red because of the acidic products made by reaction between the halogens and the water in it—hydrochloric acid, for example. It is then bleached because the halogens react with the dyes in litmus.

The hydrogen halides as acids

All the hydrogen halides dissolve extremely well in water. They produce solutions that turn blue litmus red and conduct electricity.

The 'fountain experiment' was described on page 135. It works because hydrogen chloride is extremely soluble in water. It is about 17 000 times more soluble than air at 0°C. A solution of hydrogen chloride in water is called **hydrochloric acid**.

To make a solution of hydrochloric acid in the laboratory, you begin by making the gas, using the method shown in Fig. 2. Then, instead of collecting the gas in a jar, lead it straight into water. However, you cannot just put the end of the delivery tube into water: if you do, the water will rush back into the reaction flask, as in the fountain experiment. A funnel, turned upside-down, prevents this 'suck-back':

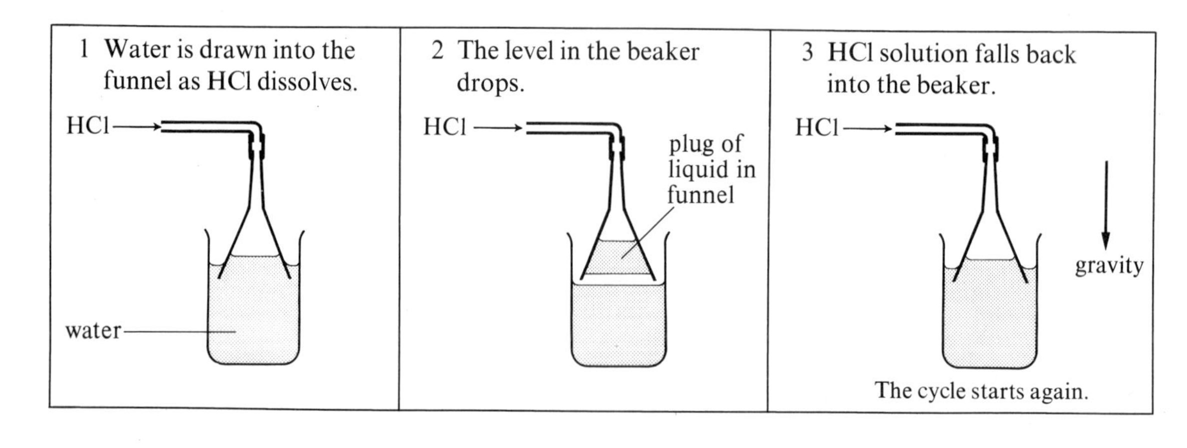
1 Water is drawn into the funnel as HCl dissolves.
HCl
water
2 The level in the beaker drops.
HCl
plug of liquid in funnel
3 HCl solution falls back into the beaker.
HCl
gravity
The cycle starts again.

Hydrochloric acid shows all the properties that we expect of acids—

(i) the acid can be neutralised to make chloride salts:

Acid	metal oxide		salt	water
$2HCl$	$+MgO$	$\longrightarrow$	$MgCl_2$	$+H_2O$
$2H_3O^+_{(aq)}$	$+ MgO_{(s)}$	$\longrightarrow$	$Mg^{2+}_{(aq)}$	$+ 3H_2O_{(l)}$

Acid	metal hydroxide		salt	water
$2HCl$	$+Cu(OH)_2$	$\longrightarrow$	$CuCl_2$	$+2H_2O$
$2H_3O^+_{(aq)}$	$+ Cu(OH)_{2\,(s)}$	$\longrightarrow$	$Cu^{2+}_{(aq)}$	$+ 4H_2O_{(l)}$

Acid	metal carbonate		salt	water	carbon dioxide
$2HCl$	$+CaCO_3$	$\longrightarrow$	$CaCl_2$	$+H_2O$	$+CO_2\uparrow$
$2H_3O^+_{(aq)}$	$+ CaCO_{3\,(s)}$	$\longrightarrow$	$Ca^{2+}_{(aq)}$	$+3H_2O_{(l)}$	$+ CO_{2\,(g)}$

(ii) it also reacts with reactive metals:

$$2HCl + Mg \longrightarrow MgCl_2 + H_2\uparrow$$
$$2H_3O^+_{(aq)} + Mg_{(s)} \longrightarrow Mg^{2+}_{(aq)} + H_{2\,(g)} + 2H_2O_{(l)}$$

Solubility of the metal halides

Most metal halides are soluble in water. Two important exceptions are the halides of silver and lead.

We use the insolubility of silver chloride to test whether an unknown substance contains a chloride. A precipitation reaction happens if silver nitrate solution is added to a metal chloride solution:

Chloride test

1. Dissolve the unknown substance.
2. Add some dilute nitric acid to the solution.
3. Add a few drops of silver nitrate solution.
4. An immediate white precipitate of silver chloride will appear, if the unknown substance is a chloride.

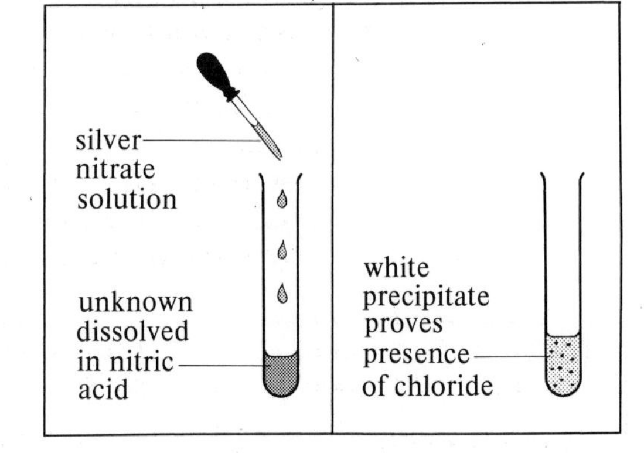

For example:

soluble	+ soluble		soluble	insoluble
$NaCl$	$+ AgNO_3$	$\longrightarrow$	$NaNO_3$	$+AgCl\downarrow$
$Cl^-_{(aq)}$	$Ag^+_{(aq)}$	$\longrightarrow$		$AgCl_{(s)}$

33.5 Redox chemistry

The elements as oxidising agents The pure halogens have the oxidation number zero. But halogen compounds normally contain the halogen with an oxidation number of –I, e.g. NaCl.

$$2Na + Cl_2 \longrightarrow 2NaCl$$

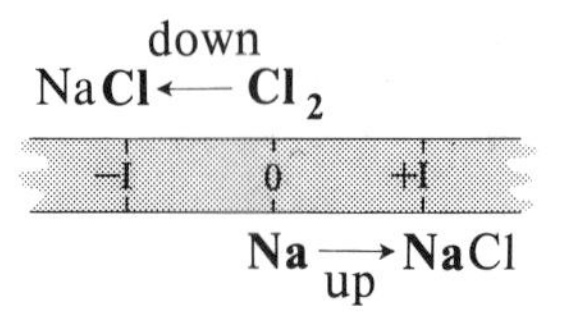

Sodium's oxidation number has gone up during the reaction, so it has been oxidised. Chlorine has acted as an oxidising agent. Chlorine can also oxidise:

other gases—chlorine oxidises hydrogen sulphide gas, for example;
substances in solution—chlorine oxidises iron(II) chloride in solution.

These reactions are shown below.

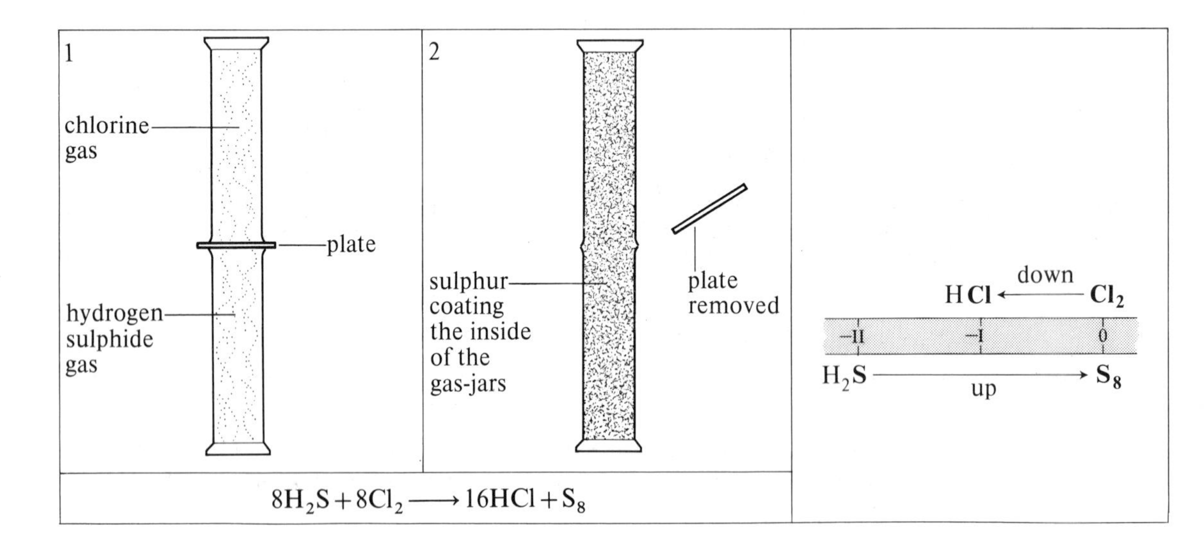

$$8H_2S + 8Cl_2 \longrightarrow 16HCl + S_8$$

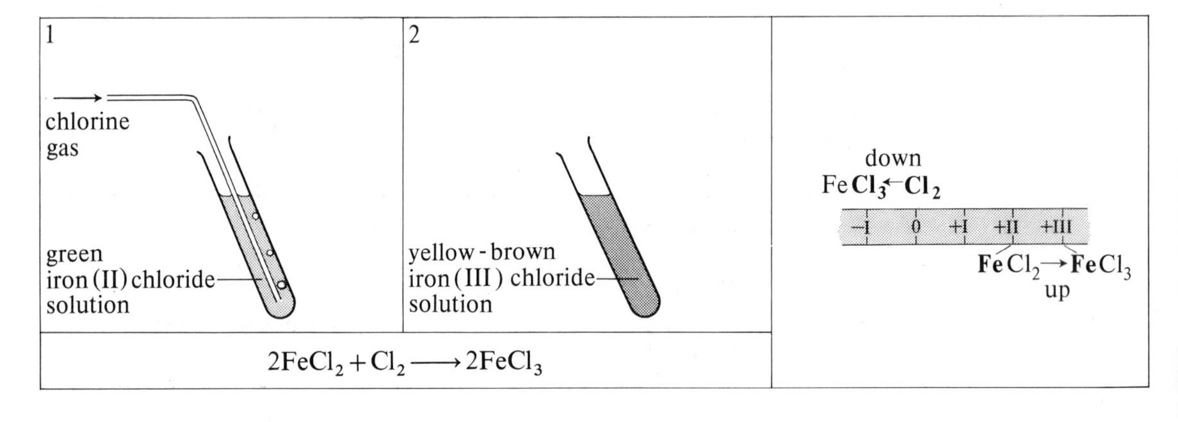

$$2FeCl_2 + Cl_2 \longrightarrow 2FeCl_3$$

Halides as reducing agents

A halide is a compound of a halogen with one other element. The oxidation number of the halogen in a halide is normally $-I$ (for example $\overset{-I}{KI}$ and $\overset{-I}{HBr}$). Sometimes a halide reacts with another substance to produce the pure halogen as a product, oxidation number zero. Therefore the halides have been oxidised, and must be acting as reducing agents. Some examples are:

(i) Hydrogen chloride reduces manganese(IV) oxide, in the laboratory preparation of chlorine.

$$MnO_2 + 4HCl \longrightarrow MnCl_2 + Cl_2 + 2H_2O$$

down
$MnCl_2 \leftarrow MnO_2$
−I 0 +I +II +III +IV
$HCl \rightarrow Cl_2$
up

(ii) Potassium iodide reduces chlorine.

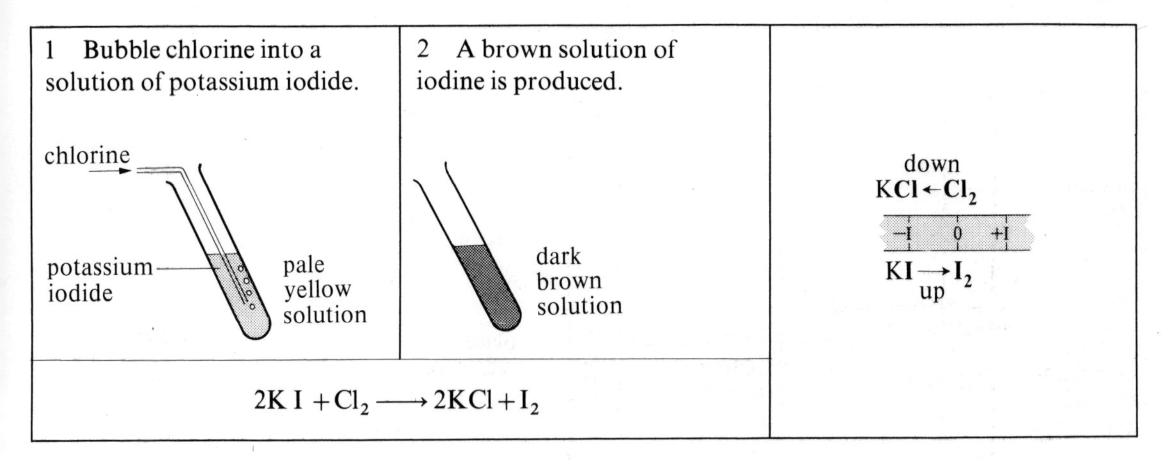

In this last reaction, a more reactive non-metal (chlorine) is displacing a less reactive non-metal (iodine). You can read more about displacement on pages 181 and 182.

Disproportionation

On page 394 we looked at the effect that the pure halogens have on water:

$$Cl_2 + H_2O \rightleftharpoons HCl + HOCl$$

HOCl is an unusual compound of chlorine. It contains chlorine with a *positive* oxidation number:

Formula	*Oxidation number*	*Name*
HOCl	H +I; O −II; Cl +I (total zero)	Hydrogen chlorate(I)

A strange thing has happened to the oxidation number of chlorine during this reaction: it has gone both up and down!

A special word is used to describe this type of reaction: **disproportionation. We say a substance has disproportionated if it contains an element whose oxidation number has gone both up and down during a single reaction.**
Chlorine disproportionates in water to hydrogen chloride and hydrogen chlorate(I). Compare this with the decomposition of hydrogen peroxide on page 376.

33.6 Special features

Uses Halogen compounds have a particularly wide range of uses in everyday life. Here are a few of the more important ones:

(i) **Silver halides in photography**
A freshly-made sample of silver chloride, made in a precipitation reaction, is white. If it is left in bright light, it quickly darkens: the reaction is speeded up by light. This effect is even more obvious with silver bromide. Silver halides—especially silver bromide—form the basis of most photographic film.

(ii) **Fluorides and tooth-decay**
The enamel on our teeth contains a compound that can react with the fluoride ion. If small amounts of fluoride are added to drinking-water, it makes tooth enamel stronger. This means that the teeth resist acid attack better, and so there is less tooth-decay.

(iii) **Organic fluorine compounds as propellants and refrigerants**
Aerosol cans contain 'propellants'. These are easily compressible and unreactive gases, which can force out whatever substance is in the can. There is some concern that these propellants may damage the ozone layer high in the atmosphere. This would let more radiation in, and might lead to an increase in skin cancers.

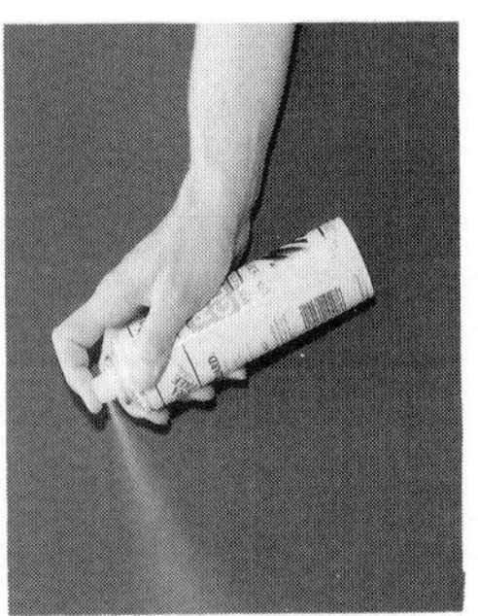

Fridges contain 'refrigerants', which are also gases that are easy to compress. In both cases, the gas most commonly used is CCl_2F_2. Its trade-name is Freon.

(iv) **Non-stick linings for cooking pans**

The non-stick surface put onto cooking pans is a polymer consisting of a carbon chain with fluorine atoms bonded to it. It is called polytetrafluoroethene.

(v) **Dry cleaning liquids**

All the non-polar liquids used in the dry-cleaning trade are organic chlorine compounds. The two most common ones have the formulas CCl_3CH_3 and $CCl_2{:}CCl_2$.

(vi) **Insecticides**

Many insecticides are organic chlorine compounds. The most famous is DDT. This insecticide eliminated the lice on refugees after the second world war, so that there were none of the typhoid epidemics that caused so many deaths after the first war. The typhoid germs were carried by the lice. DDT also cut down the number of deaths from malaria by killing mosquitos. In one year it reduced deaths from malaria by 34%. Unfortunately, DDT is chemically very stable, and so it remains in the environment. Fish, birds, and other animals high in the food-chain build up concentrations of DDT in their fat. This results in higher death rates and lower fertility in the animals.

A replacement to DDT, which has a similar structure, but is water soluble and less stable, has been made. This sort of substance, that does not build up in the environment, is described as being biodegradable.

(vii) **Herbicides and weedkillers**

A range of organic chlorine compounds have been made that imitate the growth hormones in plants. They stimulate the leaf growth of the plant, which then dies because it has not got the root system to support so much leaf. One of the most infamous is the *agent orange*, which was banned in America as being too dangerous, but was used to defoliate the forests of Vietnam.

Others are used as weedkillers in agriculture. They are usually known by the initials of their chemical name, such as 2,4,D and 2,4,5,T.

(viii) **Anaesthetics and drugs**

Two of the oldest anaesthetics are chloroform and chloral hydrate, a pair of chlorine compounds. Modern anaesthetics work more quickly and do not have the side effects of these drugs, but they are mostly chlorine compounds as well.

Many other drugs and germicides used in medicine are halogen

Reaction summary for chlorine

Extraction

In industry: electrolysis of molten or dissolved chlorides.
$2Cl^- \longrightarrow Cl_2\uparrow + 2e^-$ at the anode.
In the laboratory: $4HCl + MnO_2 \longrightarrow MnCl_2 + 2H_2O + Cl_2\uparrow$

Reactions of the element

(i) Oxidising other elements:
$2Na + Cl_2 \longrightarrow 2NaCl$
$2Al + 3Cl_2 \longrightarrow Al_2Cl_6$
$H_2 + Cl_2 \longrightarrow 2HCl$
$C + 2Cl_2 \longrightarrow CCl_4$
$2P + 3Cl_2 \longrightarrow 2PCl_3$ (or PCl_5 depending on supply of chlorine)

(ii) Oxidising compounds:
$2FeCl_2 + Cl_2 \longrightarrow 2FeCl_3$
$2NaBr + Cl_2 \longrightarrow 2NaCl + Br_2$
$2KI + Cl_2 \longrightarrow 2KCl + I_2$
$H_2S + Cl_2 \longrightarrow 2HCl + S$

(iii) Disproportionation: $Cl_2 + H_2O \longrightarrow HCl + HOCl$

The hydride

It is a very soluble acidic gas: $HCl + H_2O \longrightarrow H_3O^+ + Cl^-$
The solution reacts with metals and bases, as is usual for acids.

The chlorides

They are all soluble except AgCl, $PbCl_2$ and Hg_2Cl_2.
They are usually stable to heat.

Comparing halogen reactions

(i) Precipitation of silver halides:
$AgNO_3 + NaCl \longrightarrow NaNO_3 + AgCl\downarrow$ (white)
$AgNO_3 + NaBr \longrightarrow NaNO_3 + AgBr\downarrow$ (white)
$AgNO_3 + NaI \longrightarrow NaNO_3 + AgI\downarrow$ (yellow)

(ii) Precipitation of lead halides:
$Pb(NO_3)_2 + 2NaCl \longrightarrow 2NaNO_3 + PbCl_2\downarrow$(white)
$Pb(NO_3)_2 + 2NaBr \longrightarrow 2NaNO_3 + PbBr_2\downarrow$ (cream)
$Pb(NO_3)_2 + 2NaI \longrightarrow 2NaNO_3 + PbI_2\downarrow$ (yellow)

(iii) Displacement of a halogen by a more reactive halogen:
$Cl_2 + 2KBr \longrightarrow 2KCl + Br_2$
$Cl_2 + 2KI \longrightarrow 2KCl + I_2$
$Br_2 + 2KI \longrightarrow 2KBr + I_2$

Questions

1 (a) Draw a fully labelled diagram of a chlorine atom.
(b) Use the diagram to explain both chlorine's group number and the fact that chlorine ions have a charge of −1.
(c) Draw a diagram to show the electronic structure of a chlorine molecule.

2 Suggest a reason why iodine is a solid at room temperature, while bromine is a liquid, and chlorine and fluorine are gases.

3 Chlorine is produced industrially by electrolysing brine solution.
(a) Draw a diagram of the type of cell used.
(b) Explain how the cell works.
(c) Name the reactants and products.
(d) Write equations for the reactions happening in the cell.

4 Chlorine gas is bubbled into an aqueous solution of potassium iodide.
(a) What would you see?
(b) Write an equation for the reaction.
(c) The reaction is often called either a *displacement reaction* or a *redox reaction*. Explain why both terms can be used.
(d) Why is chlorine more reactive than iodine?

5 (a) Chlorine is *not* made in the laboratory by electrolysing brine. What method is used?
(b) Give an equation for the reaction.
(c) Sketch the apparatus that you would use to carry out the reaction.

6 Potassium bromide is a typical ionic salt.
(a) Why is it called a *salt*? What would you react together in the laboratory to make this salt.
(b) Draw a diagram to show the bonding in this salt.
(c) State three properties that are typical of ionic salts.

7 (a) Give equations for two reactions which produce hydrogen chloride gas.
(b) Which reaction would you use to produce a sample of the gas in the laboratory?
(c) Draw the apparatus you would need to make a *solution* of hydrogen chloride in water. Explain your choice of apparatus.
(d) Would you expect a solution of hydrogen chloride in water to:
 (i) conduct electricity
 (ii) affect litmus paper
 (iii) react with copper
 (iv) react with copper oxide
 (v) react with magnesium
(e) if the hydrogen chloride had been dissolved in a non-polar solvent like toluene instead of water, which of the properties above would be different? Explain.

8 Chlorine combines directly with aluminium.
(a) Draw the apparatus you would use in the laboratory to carry out this combination.
(b) Describe what you would see during the reaction.
(c) Write down the equation for this reaction.

9 Describe what you would observe if the following were each added to a gas jar of chlorine:
(a) some universal indicator
(b) some potassium iodide solution
(c) some iron(II) sulphate solution
(d) a piece of burning magnesium
(e) a burning wax taper

10 What would you observe if the following pairs of solutions were added together? In each case give the equation for the reaction:
(a) sodium chloride and silver nitrate
(b) lead nitrate and potassium chloride
(c) chlorine water and potassium bromide

11 Compounds of the halogens are used in many everyday applications.
Write a short paragraph about their uses in: (a) photography (b) preventing tooth-decay
(c) aerosols

Index

Where one reference in a group is more important than the others, it is set in **bold** type.

Answers to numerical questions

Chapter 6

1 10^{-3} cm
2 99·99%
8 35·5

9 75% $^{63}_{29}Cu$; 25% $^{65}_{29}Cu$
10 (b) 20%; 80%
11 (a) 10·5% of each
(b) 24·3

Chapter 19

7 (b) -239 kJ mol^{-1}
8 (a) 44 g
(b) 220 000 kJ
(c) 352 g
9 (a) +46·5 kJ
(b) −93 kJ

Chapter 20

3 (d) 1·3 min
(e) 1·6 min
(f) 120 cm^3

Chapter 22

3 (a) 108
(b) 24
(c) 1
(d) 28
4 (a) 16
(b) 58·5
(c) 40
(d) 133·5
(e) 63
5 (a) 1:19
(b) 7:16
(c) 12:1
(d) 24:5
6 (a) H, 5%; F, 95%
(b) Al, 90%; H, 10%
(c) Ca, 71·4%; O, 28·6%
(d) Si, 87·5%; H, 12·5%
(e) Ca, 40%; C, 12%; O, 48%
(f) H, 11·1%; O, 88·9%
(g) C, 75%; H, 25%
7 (a) Fe, 77·8%; O, 22·2%
(b) Fe, 70%; O, 30%
(c) Fe, 72·4%; O, 27·6%
(d) Na, 33·3%; N, 20·3%; O, 46·4%
(e) Na, 27·0%; N, 16·5%; O, 56·5%
(f) Fe, 36·8%; S, 21·1%; O, 42·1%
8 (a) Ca, 54·1%; O, 43·2%; H, 2·7%
(b) Cu, 25·5%; S, 12·8%; O, 57·7%; H, 4·0%
(c) N, 35·0%; H, 5·0%; O, 60·0%
10 (a) SiH_4
(b) Cu_2S
(c) CO
(d) CO_2
(e) MgO
(f) KO_2
11 (a) $CaCO_3$
(b) HClO
(c) $HClO_2$
(d) $HClO_3$
(e) $HClO_4$
12 (a) KNO_3
(b) $NaNO_2$
(c) $NaHCO_3$
(d) KOH
(e) $Ca(OH)_2$

Chapter 23

1 (a) 2*L*
(b) C, 0·5 *L*; O, *L*
(c) 0·01 *L*
(d) $\frac{L}{40}$ or 0·025 *L*
(e) 0·1 *L*
(f) 0·005 *L*
(g) 0·001 *L*
(h) 0·0278 *L*
2 (a) 80 g
(b) 12 g
(c) 254 g
(d) 25·6 g
(e) 25 g
3 (a) 0·5 mol
(b) 0·1 mol
(c) 0·1 mol
(d) 0·25 mol
(e) 0·01 mol
4 (a) 0·5 mol
(b) 0·05 mol
(c) 0·01 mol
(d) 0·1 mol
(e) 1 mol
(f) 0·005 mol
5 (a) 44·8 dm^3
(b) 11·2 dm^3
(c) 5·6 dm^3
(d) 67·2 dm^3
6 (a) 27·6 dm^3
(b) 45·9 dm^3
(c) 0·0224 dm^3
(d) 0·291 dm^3
(e) 0·121 dm^3
(f) 3·05 dm^3
(g) 0·0280 dm^3
7 (b) 150 cm^3
(c) 100 cm^3
(d) 540 mm
(e) 2·96 dm^3
8 (b) (i) 0·965 dm^3
(ii) 1·32 dm^3
(iii) 0·912 dm^3
(iv) 1·05 dm^3
9 (a) 206 cm^3
(b) 27·3 dm^3
(c) 8·45 dm^3
(d) 0·377 mol
10 (a) 0·1 mol dm^{-3}
(b) 0·01 mol dm^{-3}
(c) 0·2 mol dm^{-3}
(d) 0·1 mol dm^{-3}
(e) 0·1 mol dm^{-3}
11 (a) 4·25 g
(b) 0·913 g
(c) 2·45 g
(d) 1 g
(e) 3·95 g
(f) 7·35 g

Chapter 24

4 4 g
5 11·2 g
6 2·8 g
7 70 g
8 15·1 g
9 11·2 dm^3
10 42 dm^3
11 56·7 g
12 130 cm^3
13 150 $cm^3 O_2$; 100 $cm^3 CO_2$
14 800 $cm^3 O_2$; 500 $cm^3 CO_2$
15 50 cm^3

Chapter 25

10 (a) 60 dm^3
(b) 70 g